# Geophysical Monograph Series

Including

**IUGG Volumes**
**Maurice Ewing Volumes**
**Mineral Physics Volumes**

# Geophysical Monograph Series

# Atmospheric Science Across the Stratopause

David E. Siskind
Stephen D. Eckermann
Michael E. Summers
*Editors*

American Geophysical Union
Washington, DC

## Published under the aegis of the AGU Books Board

**Library of Congress Cataloging-in-Publication Data**
Atmospheric science across the stratopause / David E. Siskind, Stephen D. Eckermann, Michael
E. Summers, editors.
    p. cm. -- (Geophysical monograph ; 123)
    Includes bibliographical references.
    ISBN 0-87590-981-7
    1. Stratosphere. I. Siskind, D. E. II. Eckermann, Stephen D. III. Summers, Michael E. IV.
Series.

QC881.2 .S8 .A86 2000
551.51'4--dc21

                                                                    00-045093

ISBN 0-87590-981-7
ISSN 0065-8448


Copyright 2000 by the American Geophysical Union
2000 Florida Avenue, N.W.
Washington, DC 20009

# CONTENTS

# CONTENTS

# CONTENTS

# PREFACE

In recent years, two separate geophysical research cultures have developed in the United States to study the atmosphere: one of space scientists, who focus on the mesosphere, thermosphere and ionosphere, and the other of atmospheric scientists, who focus on the troposphere and stratosphere. The boundary between these two research domains roughly coincides with the stratopause (50 km). While the division of the atmosphere into layers and boundaries serves as a useful way to characterize the various chemical and dynamical processes that distinguish these layers, these boundaries are not impermeable. To understand such critical issues as global change, geophysicists must study the atmosphere as an integrated system. The purpose of this monograph is to highlight those studies that consider the coupling of these two regions and thus bring together two scientific specialties (atmospheric science and space science) that are often considered separately.

What interests us about the interface between atmospheric science and space science? The upper stratosphere and mesosphere are regions where extraterrestrial forcings (energetic solar photons and magnetospheric electrons) interact directly with forcings from the lower atmosphere (e.g., anthropogenic pollution) or even the surface (as in the case of volcanic perturbations). Several kinds of couplings are explored in this volume. First, dynamical coupling arises because the meteorology of the upper atmosphere is strongly influenced by the upward propagation of planetary waves, gravity waves, and tides from the lower atmosphere. Chemical coupling results from the upward transport of tropospheric trace constituents (including such pollutants such as the ozone-depleting chlorofluorocarbons and various greenhouse gases), which can change the chemical composition of the middle atmosphere. While most dynamical and chemical coupling is forced from below, scientists now recognize that in the polar regions the forcing can come from above and that large-scale unmixed descent of upper atmospheric air occurs each winter. Energetic coupling arises from direct radiative and electrical energy propagation into the middle atmosphere from both above and below.

This volume presents a hierarchy of papers on these and other topics in three categories. First, a collection of tutorials introduces topics of fundamental importance. Thereafter, longer review articles and shorter research papers give a more focused summary of the state-of-the-art in various aspects of coupling across the stratopause. One of the key motivations in middle atmospheric research is the issue of future trends such as the continued loss and eventual recovery of ozone and possible widespread climatic changes due to increases in greenhouse gases. Such anthropogenic changes must be isolated from natural interannual variability such as the solar cycle or single-event perturbations like a volcanic eruption. In recognition of the importance of this topic, we devote a separate section to several aspects of interannual variability.

The idea for this monograph grew out of an AGU Chapman Conference entitled "Atmospheric Science Across the Stratopause," held in Annapolis, Maryland, in April 1999. The papers presented in this volume are derived from a selected subset of the papers presented at that conference. We owe the high quality of the papers in this monograph to the diligence, expertise and rigorous standards of the reviewers, acknowledged alphabetically below. The original impetus for bridging the domains of these two scientific cultures came from Jack Kaye, of NASA's Office of Earth Science, and Mary Mellot, of NASA's Office of Space Science. Finally, we thank the Naval Research Laboratory in general for its support and specifically, R. Conway, R. Meier, and M. Lindsey for their encouragement and administrative assistance in producing this monograph.

David E. Siskind
Stephen D. Eckermann
Michael E. Summers

*Naval Research Laboratory*

# Reviewers

| | |
|---|---|
| M.J. Alexander | G. Mount |
| D. Allen | G. Nedoluha |
| J. Bacmeister | W. Norton |
| M. Baldwin | A. Parrish |
| D. Broutman | V. Pasko |
| R.T. Clancy | A. Plumb |
| R. Conway | E. Remsberg |
| G. Crowley | C. Randall |
| R. deZafra | R. Roble |
| J. Doyle | J. Rodriguez |
| T. Dunkerton | K. Rosenlof |
| F. Eparvier | D. Rusch |
| T. D. Fairlie | J. Russell III |
| J. Forbes | R. Salawitch |
| L. Froidevaux | B. Sandor |
| R. Garcia | S. Solomon |
| K. Hamilton | A. Smith |
| J. Holton | M. Stevens |
| C. Jackman | S. Strahan |
| R. Lieberman | D. Suszcynsky |
| R. Lindzen | J. Thayer |
| R. Link | S. Thorpe |
| J. Lumpe | R. Vincent |
| C. McLandress | D. Waugh |
| F. Mills | J. Whiteway |
| K. Minschwaner | D. Wuebbles |
| S. Miyahara | Y. Yung |
| M. Mlynczak | J. Ziemke |
| D. Marsh | X. Zhu |

# Introduction

David E. Siskind, Stephen D. Eckermann, and Michael E. Summers[1]

*E. O. Hulburt Center for Space Research, Naval Research Laboratory, Washington DC*

## HISTORICAL CONTEXT

Researchers of the middle and upper atmosphere often refer colloquially to the altitude range from about 40-150 km as "the ignorosphere" (the ignored layer). This label is intended to illustrate the relative lack of data available from this altitude region due to the difficulty of making in-situ measurements. The bottom of the ignorosphere corresponds to the highest altitudes sampled by balloons, whereas the top of the ignorosphere corresponds to the lowest altitudes to which orbiting satellites can regularly descend. In between, the only in-situ data are isolated rocket samplings.

It can be argued, however, that labeling the entire 40-150 km region "the ignorosphere" is an overstatement. Certainly the upper portion (~80-150 km) has been observed via several remote sensing techniques for many decades. These observations include optical measurements of the aurora and airglow between 85-110 km (Chamberlain, 1961), radar measurements of the ionosphere and of the ionization produced by meteor trails and visual observations of both noctilucent clouds from 80-85 km (Humphreys, 1933) and meteor trails from 70-100 km (see Goody [1958], Murgatroyd [1957] and Hamilton [1999] for summaries). From an historical perspective, as we will discuss, the 40-80 km region would appear to be the last frontier of atmospheric science. Since one theme of this monograph is to bridge a scientific and cultural divide between research communities above and below the stratopause, a review of this history can be both interesting and instructive.

[1]Now at Institute of Computational Sciences and Informatics (CSI), Department of Physics and Astronomy, and the Center for Earth Observing and Space Research (CEOSR), George Mason University, Fairfax, VA

Atmospheric Science Across the Stratopause
Geophysical Monograph 123

Middle atmospheric research began with the announcement in 1902 by M. Teisserene de Bort that the atmospheric temperature no longer decreased with increasing altitude above 8-11 km. As discussed by Goody [1958], this was quite a surprise, since the conventional wisdom at the time was that the tropospheric temperature decrease must necessarily continue until the absolute zero of outer space was reached. For this reason de Bort's kite and balloon observations met with a great deal of skepticism, foremost of which was that his thermometers were being heated by accidental exposure to sunlight. It was only after his measurements were repeated at night that their accuracy was accepted.

Initially, this region of the atmosphere was called the "isothermal layer" [Gold, 1909]; however, de Bort later suggested "stratosphere" (stratified sphere) and "troposphere" (turning or mixing sphere). The boundary between the two was called the tropopause by Sir Napier Shaw in 1926 [Labitzke, 1999; note however, Goody [1958] says that Shaw merely popularized the term]. Once the data were accepted, it did not take long for an explanation to follow. Gold [1909] appears to have been the first to emphasize the effects of radiative absorption and to discuss ozone in this context. However, it took many decades before the actual temperature structure of the middle atmosphere was well established.

After the concept of radiative equilibrium for the stratosphere was appreciated, it was felt that the middle atmosphere must be isothermal, presumably because the 10-20 km region for which good data existed is roughly isothermal at mid-latitudes [Lindemann and Dobson, 1922; see Chapman [1951] for a discussion]. By the 1920s this view was challenged as several pieces of evidence suggested that the temperature must reach a maximum and then begin falling again. One clue was that gunfire from the French battlefields during World War I was often heard in London. It was shown that this was likely due to refraction of sound from high altitudes, consistent with the existence of a warm layer above 40 km. A second hint was that the atmospheric density at 100 km inferred from meteor trails was over an order of magnitude higher than what one would expect if the temperature profile remained constant from the tropopause up to the altitude where the trails are observed (70-100 km) [Lindemann and Dobson, 1922]. On the other hand, the auroral spectra of Vegard [1932] suggested temperatures near 220 K. From this one can infer a region of increasing and then decreasing temperatures. In the 1930s, such a temperature profile was accepted for two reasons, one later to shown to be incorrect and one later proven correct. The incorrect reason was that the prevailing theories of atmospheric tidal propagation required a warm layer between 30 and 60 km (Chapman and Lindzen [1970] discuss the problems with this theory). The other (correct) reason was that it is consistent with the possibility of ice clouds at 82 km [Humphreys, 1933; Thomas, 1991].

The question then became "What is the value of the temperature maximum?" The first calculation by Gowan [1928] yielded values over 500 K. Chapman [1930] used photochemical theory to point out that the quantity $O/O_3$ should increase above 40 km and thus Gowan's assumed ozone concentration and calculated temperatures were too high. Chapman's work has long been recognized to be pioneering for many reasons. From the perspective of this monograph, it is significant for being the first study to couple photochemical theory to the middle atmospheric thermal budget. Several papers in this monograph demonstrate the modern application of Chapman's ideas [Brasseur et al., Mlynczak, Roble, all this volume]. A second calculation by Gowan [1947], using a more realistic ozone profile, yielded cooler temperatures at 50 km, around 400-450 K. While these values are now known to be excessive, they were not greatly different from the indirect inferences from meteor observations of about 375 K at 60 km and hence were considered plausible. Even as late as 1954, Goody [1958] was presenting a standard temperature profile which peaked at well over 300 K at 60 km. It was not until the end of the 1950s when sufficient information on the solar flux and the relevant atmospheric cross sections became available that realistic assessments of the thermal budget could be made [Murgatroyd and Goody, 1958]. By this time, enough rocket data of atmospheric pressure existed to show that the middle atmospheric temperature maximum was at or below 300 K [Murgatroyd, 1957].

Figure 1, taken from Goody's book, is a useful summary of the state of knowledge of the middle atmospheric temperature profile at the dawn of the space age. It shows an overly warm temperature maximum, but interestingly a temperature minimum at 75 km that is not too far off (although too low in altitude by about

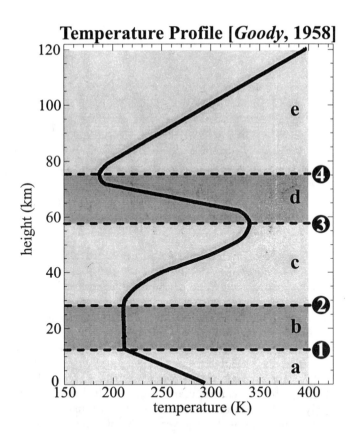

**Temperature Profile [*Goody*, 1958]**

**Figure 1.** Approximate atmospheric temperature profile from *Goody* (1958, Fig. 2). Reprinted with the permission of Cambridge University Press. Several suggested naming conventions for the indicated levels and regions (a-e, 1-4) are summarized in Table 1.

10 km). In light of this relative lack of knowledge about the stratopause region, it is not surprising that the very term "stratopause" did not achieve its modern definition until the late 1950s. Between 1950 and 1959, there were three competing sets of nomenclature for the middle atmosphere, summarized in Table 1, adapted from Table I of Goody's book. The left hand column refers to the indicated levels and layers shown in Figure 1. Note that Chapman [1950] was the first to give the term "mesopause" its modern definition; however, what we now call "stratopause," he called "mesopeak," while Goody did not even name it! Marcel Nicolet [1960] appears to have been the first to suggest that the temperature maximum be called "stratopause."

During the 1960s, one can discern two separate avenues of middle and upper atmospheric research. One branch dealt with the physics and chemistry of the aurora and airglow and was necessarily confined to the atmosphere above 80 km [Chamberlain, 1961]. A second, largely independent community observed and modeled

Table 1. Atmospheric Nomenclature (from Goody, 1958)

| Level in Fig. 1 | Goody (1958) | Flohn and Penndorf (1950) | Chapman (1950) |
|---|---|---|---|
| a | Troposphere | Advection layer | Troposphere |
| 1 | Tropopause | Tropopause Layer | Tropopause |
| b | Lower stratosphere | Isothermal layer | Stratosphere |
| 2 | " " | — | Stratopause |
| c | Upper stratosphere | Warm layer | Mesoincline (*Mesosphere*) |
| 3 | " " | Ozonopause | Mesopeak ( " ) |
| d | " " | Upper mixed layer | Mesodecline ( " ) |
| 4 | Stratopause | Upper Tropopause | Mesopause |
| e | Ionosphere | Ionosphere | Thermosphere/Ionosphere |

the lower portion of the stratosphere from a meteorological perspective. The interest here was in the transport of lower stratospheric ozone and of radioactive debris from atmospheric nuclear tests [see Mahlman and Moxim, 1977]. Thus one might identify a pattern of neglecting the 40-80 km altitude region. While this may be an overgeneralization (certainly some early studies did combine the stratosphere and mesosphere as well as photochemistry and dynamics, e.g. Leovy [1964]), it is interesting to note how much of the first research on ozone depletion in the upper stratosphere was performed by researchers who started out at altitudes well above the upper stratosphere [Stolarski and Johnson 1972; Cicerone, 1974; Stolarski and Cicerone, 1974].

With the launch of the Upper Atmospheric Research Satellite (UARS) in 1991 and the maturation of several ground-based remote-sensing techniques (lidar, radar, microwave radiometry), a wealth of data has now become available for the entire stratosphere and mesosphere. Similarly the development of sophisticated two and three dimensional models has encouraged a broader focus. Researchers now recognize that an integrated view of the atmosphere is necessary if we are to address critical problems such as global change and anthropogenic forcing of atmospheric structure and composition.

However, traces of the historical schism outlined above still remain. At NASA, funding for stratospheric research is administered by the Office of Earth Science while mesospheric research is funded by the Office of Space Science. Just as aeronomers colloquially refer to the "ignorosphere," one often hears (only partially in jest) of a "fundopause" at altitudes typically around the stratopause, which falls in between the main areas of interest of these two NASA Divisions. In recognition of this division of interests and funding, Dr. Jack Kaye of NASA's Office of Earth Science and Dr. Mary Mellot of NASA's Office of Space Science suggested a

workshop to explicitly discuss cross-stratopause issues. Such a workshop, under the auspices of the Naval Research Laboratory, was held in Annapolis in April, 1999. The papers in this volume are derived from a selected subset of the papers presented at that meeting.

## SCOPE OF THE MONOGRAPH

The monograph is divided into five sections. The first consists of four overview tutorial papers. The next four sections expand upon the topics in each of the four tutorials and consist of papers related to: (1) chemistry and energetics; (2) dynamics; (3) long term changes, and; (4) new experiments and models. Each section contains both longer review articles and shorter research papers.

The tutorial section presents four papers: two which cover global numerical modeling and emphasize an integrated view of the stratosphere and mesosphere and two which highlight differences in the controlling dynamics and energetics of the stratosphere and mesosphere. The first paper, by Brasseur et al., provides an overview of model calculations of several possible perturbations to the basic state of the middle atmosphere. These include natural perturbations, such as volcanic eruptions (of which the June 1991 eruption of Mt. Pinatubo is the most recent example) or solar variability, and anthropogenic perturbations such as the emission of radiatively or chemically active gases ($CO_2$, $CH_4$ and the ozone-depleting chlorofluorocarbons). Their main analytic tool is a two dimensional (2D) model which simulates both the stratosphere and mesosphere. The development of such 2D models has been a major catalyst in unifying stratospheric and mesospheric research. The paper by Holton and Alexander discusses the roles of different waves in forcing the global transport circulation of the middle atmosphere. In the stratosphere, large-scale Rossby waves provide the necessary momentum deposition to drive the transport circulation while in the mesosphere, dynamical "pumping" by small-scale gravity waves is predominant. These gravity waves are in turn filtered by the zonal flow in the underlying stratosphere. The paper by Mlynczak discusses the energy budget of the mesosphere where, due to the lower air density, collisional processes no longer dominate the exchange of energy between radiatively active species and their environment. He also points out that exothermic chemical reactions generate more heat in the upper mesosphere than is provided directly by solar radiation. These are fundamental differences between the mesosphere and stratosphere. Finally, Roble discusses a preliminary effort to develop a single three-dimensional global model of the atmosphere from the ground to the exosphere. By coupling a climate model of the troposphere and stratosphere with a general circulation model of the mesosphere, thermosphere and ionosphere, he shows that much of the variability of the upper atmosphere has its origin in the lower atmosphere. With the continued advancement of computing capabilities, his coupled model is likely a forerunner of what will become the standard for atmospheric modeling in the 21st. century.

The section on chemistry and energetics contains four review papers and three research papers. The paper by Heavner et al. focuses on a new topic in middle atmospheric research, that of visible discharges of electricity observed in the middle atmosphere. It is not yet clear how important such discharges are to the global atmosphere; a preliminary estimate by Heavner et al. suggests about 0.1% of the mesospheric heat budget of Mlynczak. The paper by Puertas et al. discusses the transition from a well-mixed lower atmosphere to diffusive separation in the upper atmosphere by focusing on its implications for CO and $CO_2$. They show that diffusive separation of $CO_2$ occurs at lower altitudes than for the other lighter atmospheric constituents, but a persistent problem in modeling the absolute abundances of CO and $CO_2$ remains. The paper by Siskind discusses the descent of upper atmospheric odd nitrogen into the polar winter stratosphere. He shows, in the south polar region in particular, that signatures of upper atmospheric nitric oxide are evident as low as 25-30 km. He argues that this is one of the most clearcut examples to date of downward propagation of upper atmospheric chemical perturbations. The review paper by Summers and Conway and the research paper by Canty and Minschwaner discuss the most poorly understood (yet possibly most important) reactant in the middle atmosphere, OH. Global measurements of this species only became available several years ago. As Summers and Conway discuss, OH measurements can be a useful

indicator of the water vapor abundance. Unfortunately as both papers also show, neither the OH altitude profile nor the variation of the total column abundance are quantitatively understood. Finally, Pesnell et al. look at the effects of particle precipitation and show that the effects on ozone may be less than previously thought.

Four review papers and three contributed research articles deal with the dynamics of the stratosphere-mesosphere system. Fritts and Werne review recent advances in our understanding of how gravity waves break down into turbulence, as revealed by recent high resolution three-dimensional nonlinear numerical model experiments. They stress a transition in the nature of the instabilities with height, from slower dynamically-unstable waves yielding Kelvin-Helmholtz instabilities at lower altitudes, to faster convectively-unstable waves and rapid overturning and mixing at mesospheric altitudes. The full implications for transport and mixing are not clear, however, and await further higher-resolution simulations. Garcia reviews recent advances in modeling the morphology of waves generated by deep convection in the equatorial troposphere, and the role of these waves in driving semiannual oscillations (SAOs) of the equatorial upper stratosphere and mesosphere. He also investigates how these waves may interact with the quasi-biennial oscillation (QBO) of the equatorial lower stratosphere to yield a QBO modulation of the mesospheric SAO. Hagan reviews advances in tidal modeling over the past decade, showing how new modeling tools and data have provided a better understanding of the dominant forcing and dissipation mechanisms for the diurnal tide. Strong interactions between diurnal tides and other processes, such as planetary and gravity waves and mesospheric airglow and inversion layers, have also emerged, which require ongoing modeling efforts to understand fully. Yudin et al. use a tidal model to investigate observed interannual variability of the diurnal tide in the mesosphere at equinoxes. Their results suggest that year-to-year changes in tidal dissipation play a significant role in producing this variability. Russell and Pierce review how chemical tracer measurements by the HALOE instrument on UARS, analyzed with the aid of a global trajectory model, reveal how mesospheric and thermospheric air masses descend into the stratosphere at high southern latitudes during winter. Descent appears most clearly in the data near the center of the vortex, varies interannually, and can penetrate as low as 25 km. The effect of high-latitude dynamics on temperatures is addressed in two observational papers by Duck et al. and Gerrard et al. Duck et al. analyze lidar measurements of temperatures from an Arctic site (80°N) during winter while Gerrard et al. present year-round temperatures from a lidar in Greenland (67°N). Both report considerable variability over yearly, monthly and nightly time scales, most notably during winter, highlighting the importance of dynamical temperature variability on both long and short time scales.

Possibly the most pressing question for atmospheric researchers as a whole is: "How will the atmosphere change in the coming decades?" In recognition of the importance of this question, this monograph devotes a separate section to the topic of long-term changes. Four review papers discuss different aspects of the problem. The paper by Hamilton discusses several aspects of the natural variability of the atmosphere, including free variations, such as the QBO, and forced variations, such as those induced by volcanic perturbations or the solar cycle. Jucks and Salawitch discuss the interaction between natural and anthropogenic variability and how each might affect the evolution of upper atmospheric ozone over the next decade. They show that changes in several atmospheric constituents, such as odd nitrogen ($NO_y$), $H_2O$ or $CH_4$, could either mitigate or enhance the expected slow recovery of upper stratospheric ozone due to decreasing chlorine abundances. Interannual changes in $NO_y$, $H_2O$ and $CH_4$ have already been observed; as yet it is unclear whether they are of natural or anthropogenic origin. This question is addressed in the next two papers, by Randel et al. and Nedoluha et al.. Randel et al.'s paper focuses on satellite data, while Nedoluha et al. focus on ground-based data. Two of the most surprising observations of the last 8 years are the decrease of upper stratospheric $CH_4$ from 1992-1996 and the increase of $H_2O$ in the entire middle atmosphere over the same period. Randel shows that these changes are not representative of decadal-scale trends but are episodic. Both Randel et al. and Nedoluha et al. consider the possibility of changes in transport; however, they do not rule out other influences, such as from solar variability or as-yet-unidentified changes in the temperature of the tropical tropopause.

The final section presents the results of new experiments and new models. Three of the papers, the review by Grossmann and the research papers by Ward et al. and by Tan and Eckermann, discuss data from the Cryogenic Infrared Spectrometers and Telescopes for the Atmosphere (CRISTA) experiment. As reviewed by Grossman, this shuttle-borne satellite instrument made infrared limb measurements of a suite of atmospheric gases at high vertical and horizontal resolution. The resulting data have shown complex wave structures in temperature, ozone and other atmospheric constituents at both large (Ward et al.) and small (Tan and Eckermann) spatial scales. Tan and Eckermann and Ward et al. model these wave-related features. The CRISTA experiment is important because its use of multi-azimuth limb scanning will also be implemented by the next generation High Resolution Dynamics Limb Sounder (HIRDLS) on board the EOS AURA satellite. Another new experimental technique to measure atmospheric OH, far-infrared (terahertz) spectroscopy, is described by Englert et al. Like CRISTA, this experimental approach will also be implemented on the EOS AURA satellite by the EOS/Microwave Limb Sounder (MLS) to provide global OH data from the middle atmosphere. Yee et al. present new ozone measurements from the Mid-Course Space Experiment (MSX). They combine measurements of both the refraction and extinction of starlight to obtain an ozone density profile from 10-100 km. Finally, the paper by Zhu et al. presents results from a new model that looks at the coupling between photochemistry and tidal transport. This coupling is likely to be a major area of research once data from the upcoming NASA/Thermosphere-Ionosphere-Mesosphere-Energetics and Dynamics (TIMED) satellite become available.

*Acknowledgments.* The historical section of this paper benefitted from conversations with Bob Conway and comments from Kevin Hamilton. This work was funded by the Office of Naval Research. Table 1 and Figure 1 were reprinted with permission of Cambridge University Press.

## REFERENCES

Chamberlain, J. W., Physics of the aurora and airglow, Academic Press, 704 pp, 1961.

Chapman, S., On ozone and atomic oxygen in the upper atmosphere, *Phil. Mag. S.7, 10,* 369, 1930.

Chapman, S., Upper atmospheric nomenclature, *J. Atmos. Terr. Phys.,1,* 121, 1950.

Chapman, S. Some phenomena of the upper atmosphere, *Proc. Phys. Soc., B, 64,* 833, 1951.

Cicerone, R. J., Photoelectrons in ionosphere-radar measurements and theoretical computations, *Rev. Geophys., 12,* 259, 1974.

Flohn, H., and R. Penndorf, *Bull. Amer. Meteor. Soc., 31,* 71, 1950.

Gold, E., The isothermal layer of the atmosphere and atmospheric radiation, *Proc. Roy. Soc. Lond., 82,* 43, 1909

Goody, R. M., *The physics of the stratosphere,* Cambridge Univ. Press, (reprinted edition), 1958.

Gowan, E. H., The effect of ozone on the temperature of the upper atmosphere-I, *Proc. Roy. Soc. Lond. A, 120,* 655, 1928.

Gowan, E. H., Ozonosphere temperatures under radiation equilibrium, *Proc. Roy. Soc. Lond. A, 190,* 219, 1947.

Hamilton, K. P., Dynamical coupling of the lower and middle atmosphere: historical background to current research, *J. Atmos. Sol.-Terr. Phys., 61,* 73, 1999.

Humphreys, W. J., Nacreous and noctilucent clouds, *Mon. Weather Rev., 61,* 228, 1933.

Labitzke, K. G. and H. van Loon, *The Stratosphere: Phenomena, History and Relevance*, Springer-Verlag, 1999.

Lindemann, F. A., and G. M. B. Dobson, A theory of meteors and the density and temperature of the outer atmosphere to which it leads, *Proc. Roy. Soc. Lond. A, 102*, 411, 1922.

Murgatroyd, R. J., Winds and temperatures between 20 km and 100 km - a review, *Quart. J. Roy. Meteor. Soc., 83*, 417, 1957.

Murgatroyd, R. J., and R. M. Goody, Sources and sinks of radiative energy from 30 to 90 km, *Quart. J. Roy. Meteor. Soc., 84*, 225, 1958.

Nicolet, M., The properties and constitution of the upper atmosphere, in *Physics of the Upper Atmosphere*, J. Ratcliffe ed., Academic Press, 1960.

Stolarski, R. S., and R. J. Cicerone, Stratospheric chlorine: a possible sink for ozone, *Canad. J. Chem., 52*, 1610, 1974.

Stolarski, R. S., and N. P. Johnson, Photoionization and photoabsorption cross-sections for ionospheric calculations, *J. Atmos. Terr. Phys., 34*, 1691, 1972.

Thomas, G. E., Mesospheric clouds and the physics of the mesopause region, *Rev. Geophys., 29*, 553, 1991.

---

D.E. Siskind and S.D. Eckermann, Naval Research Laboratory, Code 7641, Washington, DC, 20375, siskind@uap2.nrl.navy.mil

M.E. Summers, Department of Physics and Astronomy, George Mason University, Fairfax, VA, 22030

# Natural and Human-Induced Perturbations in the Middle Atmosphere: A Short Tutorial

Guy P. Brasseur, Anne K. Smith, Rashid Khosravi, Theresa Huang, and Stacy Walters

*National Center for Atmospheric Research, Boulder CO*

**Simon Chabrillat and Gaston Kockarts**

*Belgian Institute for Space Aeronomy, Brussels, Belgium*

This tutorial paper presents a new two-dimensional model of the atmosphere (0 to 120 km) that describes the interactions between dynamical, radiative and chemical processes. The model is used to assess the response of the atmosphere (e.g., ozone concentration, temperature) to natural perturbations such as volcanic eruptions or solar variability, and to anthropogenic emissions of radiatively or chemically active gases. For example, it is shown that, above 25 km altitude, the depletion of ozone in response to the emissions of man-made chlorofluorocarbons (1990 conditions) reaches a maximum of 10-20% at 45 km altitude, with the highest values in polar regions. A large ozone reduction (typically 30%) is found in the mesosphere near 80-85 km in response to the methane increase over the 1850-1990 period. The predicted ozone reduction has been accompanied by middle atmosphere cooling. These anthropogenic perturbations are predicted to have comparable magnitudes to natural changes (such as those produced by solar variability).

## INTRODUCTION

Since the beginning of the industrial revolution (around 1850), the energy consumption by the world population has increased by a factor of 80 and the industrial production has grown by 2 orders of magnitude. Over the same period, the population has increased from 1.2 to 6.0 billion. Since 1950, the absolute population growth has been larger than during the entire previous historical period.

Today, approximately 80 percent of the world's primary energy is supplied as fossil fuel, and the release into the atmosphere of combustion products such as carbon dioxide ($CO_2$), carbon monoxide ($CO$), nitrogen oxides ($NO$ and $NO_2$) and sulfur dioxide ($SO_2$) contributes to the degradation of air quality. In addition, changes in land-use associated with the development of intensive agriculture,

and specifically tropical deforestation and wetland destruction, have also perturbed significantly the chemical composition of the atmosphere. Biomass burning in the tropics during the dry season, often in relation to agricultural practices, is a major source of regional and even global air pollution. Finally, the increasing use of industrially-manufactured chemical compounds, such as the chlorofluorocarbons (CFCs), has been an important source of chemical perturbations in the middle atmosphere.

The various impacts of these anthropogenic compounds are now well documented; they include the formation in the troposphere of secondary photochemical pollutants (e.g., ozone and other oxidants) and of aerosols (sulfates, nitrates, etc.) with well-established consequences for human health, visibility and crop productivity. The increasing acidity of precipitation has produced sometimes dramatic effects on aquatic ecosystems. The observed increase in the atmospheric concentration of long-lived greenhouse gases (carbon dioxide, methane, nitrous oxides, chlorofluorocarbons (see Table 1) and of shorter-lived

Atmospheric Science Across the Stratopause
Geophysical Monograph 123

**Table 1.** Concentration Changes in the Atmosphere

|            | Pre-Industrial | 1950  | 1990   | Lifetime (yrs) |
|------------|---------------|-------|--------|----------------|
| $CO_2$ (ppmv)    | 280.00 | 310.00 | 354.00 | 50–200 |
| $CH_4$ (ppmv)    | 0.79   | 1.10   | 1.72   | 8      |
| $N_2O$ (ppbv)    | 288.00 | 295.00 | 310.00 | 120    |
| CFC-11 (pptv) | 0.00   | small  | 280.00 | 50     |
| CFC-12 (pptv) | 0.00   | small  | 484.00 | 100    |

aerosols has perturbed considerably many biogeochemical cycles and remains an important source of global and regional climate change. Finally, the observed depletion of stratospheric ozone and specifically the formation of a springtime ozone hole in Antarctica since the beginning of the 1980's are the consequence of the release in the atmosphere of anthropogenic chlorofluorocarbons. *Randel* [this issue] discusses interannual changes in stratospheric constituents.

A comprehensive study of the impacts of human activities on the global environment requires a detailed description of physical, chemical and biological processes in the atmosphere, ocean and continental biosphere. Comprehensive Earth system models are being developed to address these issues. Considerable efforts have been made in the last two decades to assess the impact of human-induced perturbations using in most cases zonally-averaged two-dimensional (2-D) models. For example, the response of the stratosphere to the emissions of CFCs has been assessed in *WMO* [1999] using the 2-D models of *Weisenstein et al.* [1998], *Bekki and Pyle* [1994], *Randeniya et al.* [1997], *Jackman et al.* [1996], *Kinnison et al.* [1994], *Grooss et al.* [1998], *Pitari et al.* [1993], *Smyshlyaev and Yudin* [1995], *Velders* [1995], etc. Three-dimensional studies of ozone depletion have also been performed based on the chemical transport models of *Granier and Brasseur* [1991], *Pitari et al.* [1992], *Mahfouf et al.* [1993], *Austin and Butchart* [1994], *Lefèvre et al.* [1994], *Steil et al.* [1998], *Shindell et al.* [1998], and others. Other model studies have focused on the impact of high speed civil transport [see e.g., *IPCC*, 1999] and the current fleet of subsonic aircraft [see e.g., *Friedl et al.*, 1997; *Brasseur et al.*, 1998]. The impact of natural perturbations has also been considered: model simulations of the effects of solar variations have been reported by *Brasseur and Simon* [1981], *Huang and Brasseur* [1993], *Brasseur* [1993], *Fleming et al.* [1995], *Jackman et al.* [1996]. The effects of charged-particle precipitation have been considered by *Jackman et al.* [1996] and *Callis et al.* [1997].

In this paper, we examine how anthropogenic emissions of chemical compounds at the surface affect the chemical composition (and specifically the concentration of ozone) and the thermal structure of the middle atmosphere (stratosphere and mesosphere). We focus on the effect of increasing concentrations of carbon dioxide, chloro-

fluorocarbons and methane on ozone and temperature, but do not consider in the present study the atmospheric impacts of other human perturbations such as the injection of $NO_x$ by future commercial supersonic aircraft or by large nuclear explosions. Potential effects of increasing levels of nitrous oxide are also not discussed. We compare the magnitude of several anthropogenic effects with natural perturbations and specifically with the chemical and thermal effects of volcanic eruptions and solar variability. This assessment will be performed using a coupled two-dimensional model of the atmosphere called Simulations of Chemistry, Radiation and Transport of Environmentally Important Trace Species (SOCRATES) that represents many of the couplings between chemical, radiative and dynamical processes in the middle atmosphere. Modeling of the coupled atmospheric region is further discussed by *Roble* and by *Zhu* [this issue].

## BRIEF DESCRIPTION OF THE SOCRATES MODEL AND SELECTED MODEL RESULTS

The model used in the present study is two-dimensional (latitude, altitude) and extends from pole to pole and from the surface to approximately 120 km altitude. The model has undergone substantial update and expansion from that described in *Brasseur et al.* [1990]. The spatial resolution is 5 degrees in latitude and 1 km in altitude. The dynamical equations are expressed in the transformed Eulerian mean framework with the height coordinate expressed by the pressure altitude. The global circulation is calculated above 2 km altitude, where boundary conditions are specified based on the downward control principle of *Haynes et al.* [1991]. The zonal mean momentum forcing is provided by the dissipation of wavenumber 1 and 2 planetary waves, gravity waves, tides, and in the troposphere by mesoscale systems. The interaction between planetary waves and the mean flow as well as the quasi-horizontal mixing associated with wave dissipation ($K_{yy}$ eddy diffusion coefficient) are derived by a simple Rossby wave model. Gravity wave and tidal wave breaking (momentum forcing and associated $K_{zz}$ vertical eddy diffusion coefficient) are parameterized according to the scheme of *Lindzen* [1981]. The momentum forcing associated with weather systems in the troposphere is specified according to climatological values. Large scale advective transport of chemical compounds in the middle atmosphere and in the troposphere (above 2 km) is treated using the semi-Lagrangian formalism of *Smolarkiewicz and Rasch* [1991]. Molecular diffusion of heat and chemical species accounts for vertical exchanges in the lower thermosphere. A simple parameterization of convection in the tropics and in frontal systems [*Langner et al.*, 1990] accounts for vertical transport of chemical compounds in the troposphere. Horizontal ($K_{yy}$) and vertical ($K_{zz}$) eddy diffusion coefficients are specified in the troposphere to account for rapid exchanges of chemical compounds in this part of the atmosphere.

Diabatic heating by absorption of solar radiation by ozone and molecular oxygen is treated using the two-stream method of *Toon et al.* [1989], and accounts for aerosol and cloud radiative effects. Chemical heating associated with the exothermicity of chemical reactions affecting oxygen and hydrogen systems is taken into account. Cooling by infrared emission below 60 km is calculated by a modified version of Malkmus band model [*Briegleb*, 1992]. Cooling effects of water vapor, carbon dioxide, ozone, methane, nitrous oxide, and CFC-11 and -12 are accounted for. Above 60 km, the formalism of *Fomichev et al.* [1993] is used to account for the non-LTE effects of $CO_2$, $H_2O$, and ozone. Latent heat release associated with cloud condensation in the troposphere is specified from climatological studies.

The chemical scheme accounts for a total of 53 compounds including those belonging to the oxygen, hydrogen, nitrogen, chlorine, and bromine families with 183 chemical and photochemical reactions, including 7 heterogeneous conversion mechanisms operating on sulfate aerosol and polar stratospheric cloud particles. In addition to its stratospheric production by the nitrous oxide ($N_2O$) oxidation, nitric oxide (NO) is produced by ionic reactions in the thermosphere, by cosmic ray precipitation in the lower stratosphere and by lightning activity in the troposphere. The model solves the chemical equations using an implicit algorithm with a timestep of approximately 3 hours, accounting for diurnal variations in the solar flux and hence in the concentration of chemical species. Lower boundary conditions are expressed either as a specified concentration (based on observations) or an emission flux, or (for certain species) as a surface deposition velocity at the surface. Conditions at the upper boundary (120 km) are generally expressed by a zero flux, except for species like atomic oxygen, atomic hydrogen and nitric oxide, for which a concentration is prescribed.

To illustrate some of the results provided by the SOCRATES model, we present in Figure 1 the calculated zonally averaged values of several dynamical parameters including temperature, net diabatic heating rate, zonal wind component, meridional mass stream function (transformed Eulerian mean), and momentum forcing due to dissipation of planetary and gravity waves. Figures 2a–d show the mean meridional distribution of several chemical species: water vapor, ozone, hydroxyl radical, and $ClO_x$ (chlorine atom and chlorine monoxide). The water vapor shown in Figure 2a is characterized by a relatively constant mixing ratio between 20 and 70 km and a sharp decrease around 70-90 km due to rapid photolysis by the solar Lyman $\alpha$ line. The mixing ratio maximum found near 55 km altitude is due to the stratospheric oxidation of methane. The corresponding concentrations derived by the model (about 6 ppmv) are in good agreement with the values provided by the Halogen Occultation Experiment (HALOE) observations onboard the Upper Atmosphere Research Satellite (UARS) [see *Harries et al.*, 1996]. The calculated

OH number density (24 hour diurnal average) is shown in Figure 2b. The maxima predicted by the model at 45 km and 75 km altitude are consistent with the observations of *Jucks et al.* [1998] and *Summers et al.* [1997]. The reactive chlorine (Cl + ClO) mixing ratio shown in Figure 2c reaches a maximum of 400-800 pptv at 40 km altitude, consistent with, for example, the observations of *Michelsen et al.* [1996]. The calculated ozone mixing ratio (see Figure 2d) reaches a maximum of 10 ppmv at 30 km in the tropics in good agreement with space observations. Values of typically 0.5-1.0 ppmv are calculated between 60 and 70 km and are consistent with HALOE observations [*Russell et al.*, 1993; *Brühl et al.*, 1996]. A more detailed discussion of these results and a validation of the model are presented in *Brasseur et al.* (G. Brasseur, manuscript in preparation, 2000). It is important to emphasize, however, the limitations associated with two-dimensional models. Since the planetary wave structure in the longitudinal direction and the diurnal variation in photochemical quantities cannot be explicitly represented (as in three-dimensional models), these effects are accounted for through approximate formulations or parameterizations. The model is not meant to represent short-term daily variations, but rather simulates the atmospheric evolution on monthly time scales.

## NATURAL PERTURBATIONS

The chemical composition and thermal structure of the middle atmosphere can be significantly perturbed by natural phenomena: large volcanic eruptions affect mostly the lower stratosphere while periodic changes in short-wave solar irradiance have a pronounced impact in the thermosphere and mesosphere. We briefly review the response of the atmosphere to these perturbations.

*Volcanic Eruptions*

The occasional injection into the atmosphere of sulfur produced by massive volcanic eruptions (such as the eruptions of El Chichón in 1982 and of Mt. Pinatubo in 1991) has important—but transient—effects on the Earth climate and on the chemical composition of the stratosphere. As sulfur dioxide released by the active volcanoes is converted into sulfate particles (in about 1 month), an enhanced fraction of the incoming solar radiation is scattered back to space, while an increased amount of long-wave terrestrial radiation emitted by the surface and the troposphere is absorbed in the lower stratosphere. After the eruption of Mt. Pinatubo, global tropospheric temperatures were reduced [*Houghton et al.*, 1996; *Hansen et al.*, 1996], while the temperature in the layer above the tropopause (15–25 km) was enhanced by 1–3 K [*Angell*, 1993]. For eruptions taking place at low latitudes, such changes in the radiative heating result in enhanced upward motions in the tropics (stronger Hadley

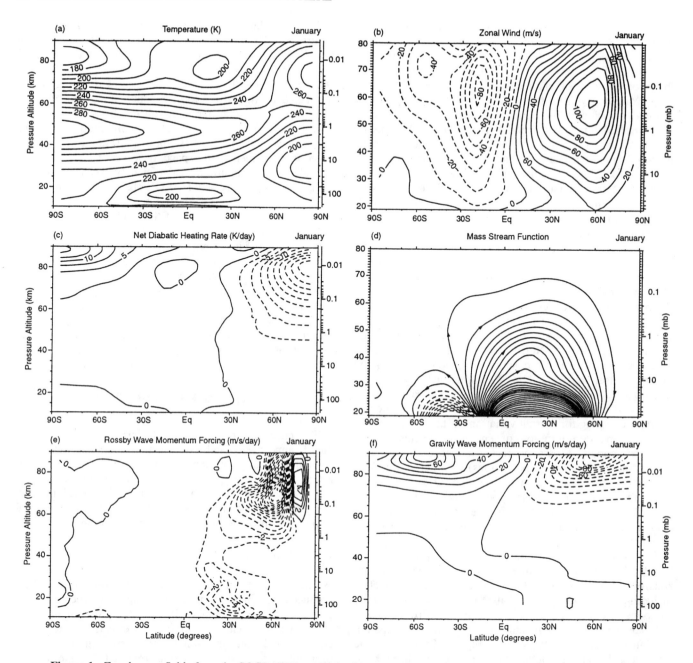

**Figure 1.** Zonal mean fields from the SOCRATES model for January. a) temperature (K); b) zonal wind (m/s), c) net diabatic heating rate (K/day); d) transformed Eulerian mean mass stream function (showing the direction of the flow); and e) Rossby wave momentum deposition (m s⁻¹/day); and f) gravity wave momentum deposition (m s⁻¹/tday).

circulation) and consequently a decrease in the ozone column over the equator [*Tie et al.*, 1994b] during the first months following the eruption. The change in the photolysis rates as the solar light is scattered by the volcanic cloud also contributes to a reduction in the tropical ozone column.

The enhanced aerosol loading resulting from the eruption also provides additional sites for heterogeneous chemical reactions to occur. Such processes tend to reduce the concentration of gas phase nitrogen oxides in favor of nitric acid, and to activate chlorine from relatively inert HCl and ClONO$_2$ reservoirs into reactive ClO. At the same time, the concentration of the hydroxyl radical is enhanced [*Tie et al.*, 1994b]. Figure 3 shows the time evolution of the sulfate surface area density during 3 years following the eruption of Mt. Pinatubo, as calculated by the SOCRATES model as

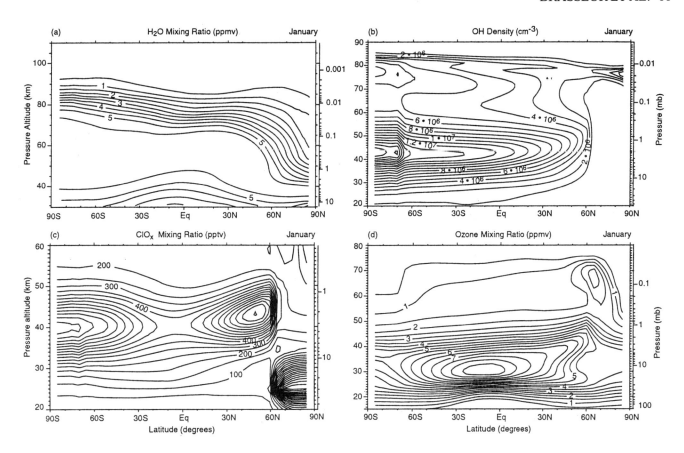

**Figure 2.** Zonal mean mixing ratios of trace chemical species from the SOCRATES model for January.  a) $H_2O$; b) OH;
c) $ClO_x$; and d) $O_3$.  The maximum mixing ratio of $ClO_x$ in the Arctic region at 25 km altitude is 1.2 ppbv.

a function of height at 60°N.  In the simple microphysical model used in this study and described in detail by *Tie et al.* [1994a], the aerosol particles are assumed to be spherical liquid droplets with a size distribution represented by 25 discrete bins covering a range of particle radii from 0.01 μm to 2.56 μm (with a particle volume doubling from one bin to the next).  The aerosol surface area densities reach 15 microns/cm² in the lower stratosphere after about 6 months. Significant perturbations to the chemical fields appear after a few months and persist for two years.  Chemical perturbations are cyclic and have maximum effect in the late winter to spring season.  This is even more apparent in Figure 4, which shows a time series of the column ozone change in the model and compares it with observations.

*Thermospheric Variability*

Middle atmosphere concentrations of several chemical compounds including atomic oxygen, atomic hydrogen and nitric oxide, which have relatively long lifetimes in the thermosphere, have a dependence on transport downward from the thermosphere.  Hence they may be subject to fluctuations associated for example with wave propagation

and dissipation (Rossby waves, gravity waves, tides, etc., see *Holton and Alexander*; *Fritts and Werne*; *Hagan*; and *Yudin*, this issue).  In addition, enhancement in solar activity or episodic solar disturbances may increase the production rate and the concentration of these species [see e.g., *Pesnell*, this issue].  One important issue is to assess how deep in the atmosphere these perturbations propagate.

To address this question, we consider the case of thermospheric nitric oxide.  Above 100 km altitude, this compound is produced primarily by the photodissociation of molecular nitrogen ($N_2$) and by dissociative ionization (energetic particles) of $N_2$ and $O_2$.  During solar disturbances, the mean concentration of NO may be increased by a factor 5-10.  The perturbation NO can be advected or diffused downward [see *Siskind*, this issue], but its penetration is strongly limited by the photolytic destruction of NO which provides an efficient loss mechanism for nitric oxide in the mesosphere.  The downward transport of odd nitrogen is, however, possible during winter in the polar night regions where no photolysis takes place [see also *Russell and Pierce*, this issue].

Figure 5 shows the response of NO in the middle atmosphere when the concentration of NO at 120 km

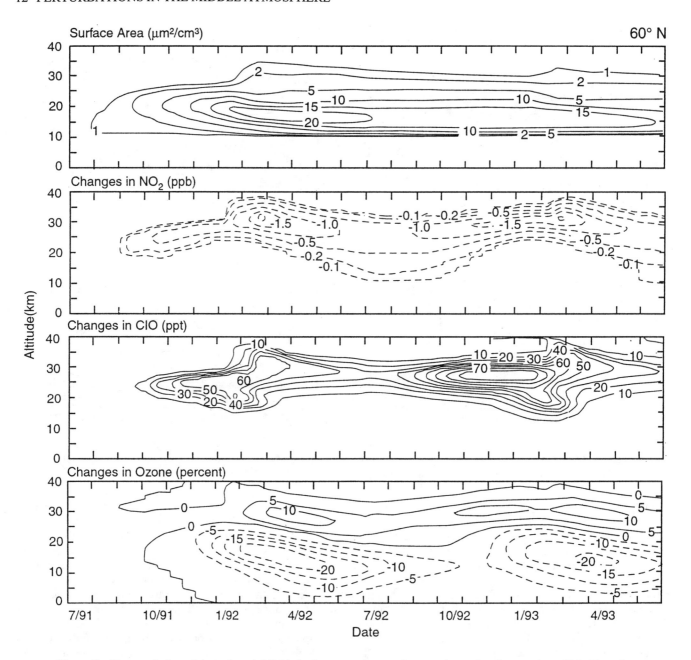

**Figure 3.** Time evolution of several model fields in the troposphere and stratosphere after the eruption of Mt. Pinatubo. a) aerosol surface area (microns/cm²); b) perturbation in NO₂ (ppbv); c) perturbation in ClO (pptv); and d) perturbation in O₃ (percent). The perturbation fields are the difference between model runs with Pinatubo effects and a control case.

altitude (upper boundary of the model) is uniformly multiplied by a factor of 10. This result is obtained after a 5 year model integration (half the period of the solar cycle) and should therefore be regarded primarily as a sensitivity test rather than a detailed simulation of the real evolution of NO. It is interesting to note that the thermospheric perturbation propagates down to the winter stratopause. As the loss for NOₓ in this region of the atmosphere is very low, this perturbation tends to remain present for several

months and, when the sun returns in spring and summer, to produce a substantial depletion of high latitude ozone near 50 km altitude.

*Solar Variability*

Changes in the intensity of short-wave solar radiation (ultraviolet and X-rays) affect the photochemical production and loss rates of several chemical species in the

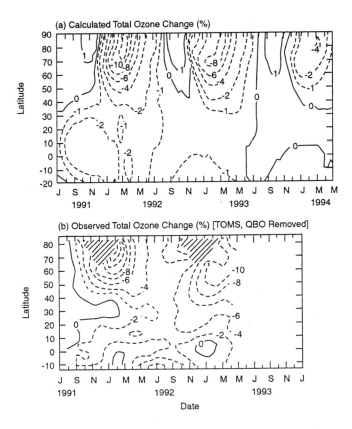

(a) Calculated Total Ozone Change (%)

(b) Observed Total Ozone Change (%) [TOMS, QBO Removed]

Date

**Figure 4.** Time and latitude cross-section of the calculated (above) and observed (below) percentage change in the total column amount of ozone in the Northern Hemisphere following the eruption of Mt. Pinatubo.

upper and middle atmosphere. In the case of ozone, for example, the production rate by $O_2$ photolysis (wavelength shorter than 242 nm) is significantly enhanced from the thermosphere to the middle stratosphere when solar activity increases. At the same time, the increased photolysis of water vapor (at Lyman alpha and in the spectral region of the $O_2$ Schumann-Runge bands) produces additional OH and $HO_2$ radicals [see also *Summers*, this issue], enhancing the destruction rate of ozone in the mesosphere. Since solar absorption by ozone is a major heat source in the atmosphere, a response of the temperature to solar variability should be expected.

Changes in the solar irradiance are associated for example with the 11-year solar cycle. The uneven distribution of sunspots on the Sun produces a 27-day variation in the solar flux, (apparent rotation period of the Sun). Figure 6 shows the 11-year solar variability adopted in the model, with values of typically 70% at Lyman alpha, 15% at 160 nm, 8% at 200 nm, 3-4% at 250 nm and less than 1% beyond 300 nm.

The effects of solar variability on middle atmosphere parameters are schematically represented in Figure 7. First, one has to consider the impact of solar variability on

photolysis frequencies for molecules that absorb radiation at wavelengths less than 300 nm. This is, for example, the case for molecular oxygen whose photodissociation leads to the formation of atomic oxygen and subsequently of ozone. The production of O and $O_3$ in the thermosphere,

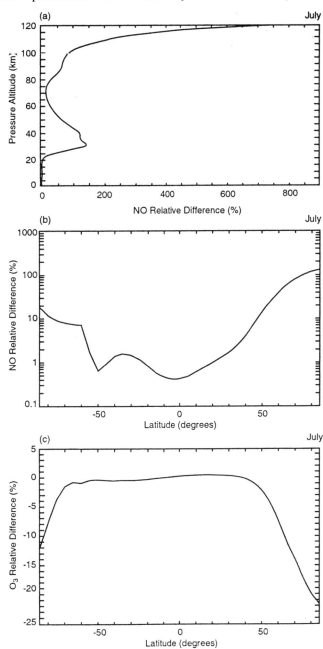

**Figure 5.** Perturbation of the NO density in the middle atmosphere resulting from a tenfold increase of thermospheric NO at 120 km. The response of NO (in percent) is shown in July as a function of altitude at 80N (a) and as a function of latitude at 40 km (b). The response of ozone (percent) is also shown as a function of latitude at 40 km in July (c).

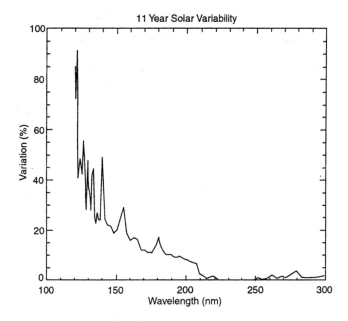

**Figure 6.** Variability in percent of the short wavelength solar radiation at the top of the atmosphere associated with the 11-year solar cycle.

mesosphere and stratosphere is therefore enhanced during high solar activity periods. Enhanced solar activity also accelerates the photolytic destruction of water vapor above 70 km altitude by radiation in the Lyman-alpha solar line. Water photolysis is a source of H, OH and $HO_2$, which contribute substantially to the destruction of mesospheric O and $O_3$. The ozone destruction in the mesosphere is therefore also enhanced during high solar activity.

Figures 8 presents a calculation of the response of O, $O_3$, $H_2O$, and OH to an increase of solar activity from its minimum to its maximum value [*Huang and Brasseur*, 1993]. As chemical constituents in the mesosphere and upper stratosphere respond to solar variability more rapidly than the 11 year period, we compare the results provided by two model integrations at quasi-stationary state (5 year model runs): one for solar maximum conditions and the second one for solar minimum conditions. As expected, the model shows that ozone and atomic oxygen concentrations increase in the thermosphere and in the upper stratosphere due to enhanced photolysis of $O_2$. In the mesosphere, near 70 km altitude, the decrease in the concentrations is associated with the enhancement by more than 25% in the OH concentration. Note also that from solar minimum to solar maximum conditions, the water vapor density is reduced by 60% and more in the vicinity of and above the mesopause.

A second potential consequence of enhanced solar activity is a change in the diabatic heating rate (solar and chemical heating). This change is expected to be produced by the variation in the intensity of the incoming solar

energy and in the concentration of absorbing gases (specifically ozone). The resulting changes in the temperature, as calculated by the SOCRATES model, are typically 1-1.5 degrees at the stratopause, 6-10 degrees at the mesopause, and more than 20 degrees at 115 km altitude (Figure 9). The values calculated near 40-50 km are somewhat smaller than the values deduced from observations by *Hood et al.* [1993] and *Keating et al.* [1994].

It is important to note that the ozone and temperature responses to solar activity are strongly coupled. Not only do changes in ozone affect the temperature through changes in diabatic heating, but in addition, ozone concentrations are directly dependent on temperature. This last effect results from the strong temperature dependence of chemical rate constants for reactions such as $O + O_3$ and changes in the air density associated with a modified thermal structure of the atmosphere (hydrostatic adjustment).

Finally, the third mechanism by which solar activity affects the middle atmosphere is provided by the enhanced photochemical and ionic production of nitric oxide in the thermosphere and its downward transport in the winter polar region (Figure 5).

## ANTHROPOGENIC PERTURBATIONS

We now assess the potential changes in the chemical composition and temperature of the middle atmosphere in response to human-induced perturbations. We examine specifically the response of the stratosphere and

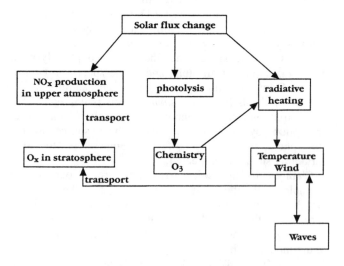

**Figure 7.** Schematic representation of the impact of solar flux changes on the composition and structure of the middle atmosphere.

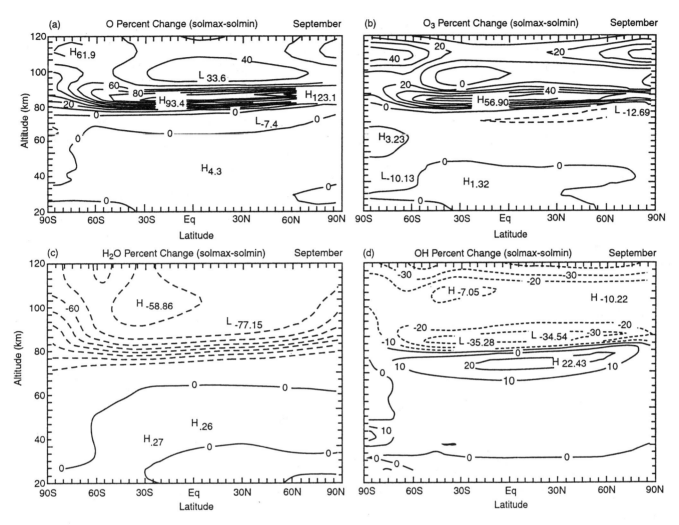

**Figure 8.** The calculated change in percent of O, O₃, H₂O, and OH during September in response to an increase of solar activity from its minimum to its maximum value.

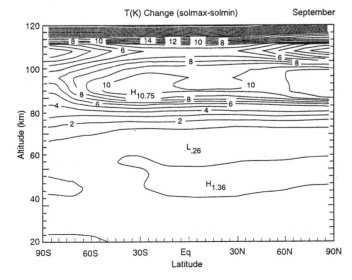

mesosphere to increasing concentrations of carbon dioxide ($CO_2$), chloro-fluorocarbons (CFCs), and methane ($CH_4$).

*Carbon Dioxide*

Approximately 6 Gtons of $CO_2$ are released every year in the atmosphere as a result of fossil fuel burning. Although some of this anthropogenic carbon is absorbed by the ocean and the continental biosphere, approximately half of this input remains in the atmosphere. As $CO_2$ is transported to the upper layers of the atmosphere, it enhances the amount of terrestrial energy that is radiated to space and hence contributes to middle and upper atmosphere cooling. To

**Figure 9.** The calculated change in temperature (K) during September in response to an increase of solar activity from its minimum to its maximum value.

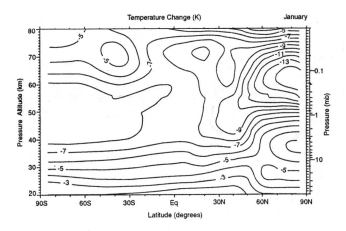

**Figure 10.** The calculated change in temperature (K) during January in response to a doubling of atmospheric $CO_2$.

assess the sensitivity of the temperature in the vicinity of the stratopause to changes in atmospheric $CO_2$, we consider the steady-state response of the atmosphere to a doubling in the concentration of $CO_2$. Figure 10 shows the calculated response in the temperature in the middle atmosphere: a cooling of about 8 K is predicted at 50 km during summer. During winter, the temperature is reduced by up to 14 K at 60 km in the polar region.

The cooling of the stratopause leads to enhanced ozone concentration in the upper stratosphere and lower mesosphere. The model estimate of the relative ozone change is shown in Figure 11. Relative changes at 45 km are 14% near the equator and up to 18-19% at high latitudes. At 60 km, the ozone increase is of the order of 6%. The dominant cause of the predicted ozone increase is a decrease in the reaction rate of several of the reactions that contribute to ozone loss.

## Chlorofluorocarbons

The presence in the atmosphere of industrially manufactured halocarbons (and specifically chlorofluorocarbons (CFCs)) provides a source of reactive chlorine in the middle atmosphere. In the upper stratosphere, the catalytic destruction process of ozone is provided by the following reactions:

$$Cl + O_3 \rightarrow ClO + O_2$$
$$ClO + O \rightarrow Cl + O_2$$

$$\text{Net:} \quad O_3 + O \rightarrow 2O_2$$

Several model calculations [e.g., *Jackman et al.*, 1996; *Considine et al.*, 1998] have been performed to assess the ozone response to increasing concentrations of CFCs. The

long-term changes derived by these models are in reasonable agreement with ozone trends derived from space observations [see e.g., *WMO*, 1999]. Figure 12 shows the changes in the ozone concentration and temperature calculated by SOCRATES and resulting from the present level of anthropogenic chlorine in the atmosphere, compared with preindustrial chlorine amounts. The ozone reduction at 40-45 km altitude is of the order of 15% in the tropics and reaches maxima of 22 and 26% at the summer and winter poles, respectively. The spatial distribution of the upper stratospheric ozone depletion is in qualitative agreement with the ozone trends presented by *Randel and Wu* [1999] and derived from observations by the Stratospheric Aerosol and Gas Experiments (SAGE I and II) for the 1979–1997 period. The cooling of the stratopause derived by the model is of the order of 3–4K. Future changes in upper stratospheric ozone are further discussed by *Jucks and Salawitch* [this issue].

## Methane

Although methane is a greenhouse gas, which tends to cool the stratosphere, the atmospheric response to increasing methane includes elements not present with the response to increasing $CO_2$. Methane oxidation leads to higher water and OH concentrations in the upper stratosphere and mesosphere and hence to less ozone at these altitudes. Figure 13 shows the change in the concentration of these 3 compounds when the concentration of $CH_4$ is increased from its pre-industrial to its present value (the concentration of all other compounds is assumed to remain equal to pre-industrial values). Note the large increase (more than 40%) in the OH concentration and the corresponding ozone decrease (more than 30%) at 82 km at high latitudes during summer. The reaction of methane

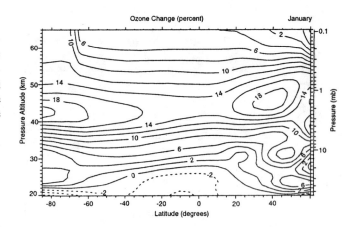

**Figure 11.** The calculated change in ozone (percent) during January in response to a doubling of atmospheric $CO_2$.

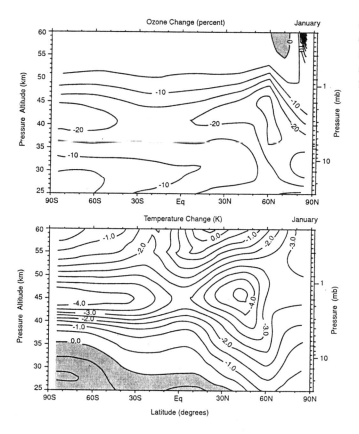

**Figure 12.** The calculated change in ozone (percent) and temperature (K) during January in response to the increase in atmospheric chlorine from preindustrial times to the present.

with Cl converts active chlorine to HCl. The net effect is a reduction of ozone in the mesosphere where the $HO_x$ reaction cycles are dominant, but can be an increase of ozone in the stratosphere where chlorine destruction is dominant. Figure 14 shows the response of the middle atmosphere ClO and ozone to the increase in methane from pre-industrial to present values for a contemporary level of chlorine. As expected, ClO abundances are reduced by as much as 15–30% at 40–50 km by the enhanced concentrations of methane. As a result, the concentration of ozone is enhanced below 45 km altitude. Above this height, the effect of chlorine on ozone becomes limited and the calculated ozone reduction results from the increased OH and $HO_2$ concentration in response to higher methane concentrations.

## CONCLUSIONS

Human activities (fossil fuel consumption, biomass burning, land-use changes, industrial processes) have produced very substantial changes in the chemical composition of the atmosphere on the global scale. Some of the perturbed gases penetrate into the middle atmosphere, where the changes affect the dynamical structure of the atmosphere and the composition of a number of trace gases. Some of the trends have not been

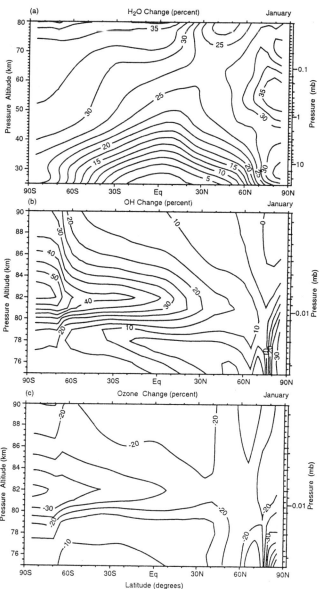

**Figure 13.** The calculated percentage change in water vapor, OH and ozone concentrations during January in response to the increase in atmospheric methane from preindustrial times to the present. The concentrations of all other compounds ($CO_2$, $N_2O$, CFCs) are assumed to be representative of pre-industrial conditions.

# 18  PERTURBATIONS IN THE MIDDLE ATMOSPHERE

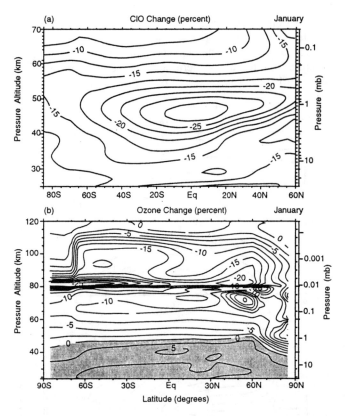

**Figure 14.** The calculated percentage change in ClO and ozone concentrations when methane is enhanced from its pre-industrial to its present value, and for a level of CFCs that is representative of present conditions.

well observed at this point. However, numerous models, including the two-dimensional SOCRATES model described in this paper, predict that significant changes may have already occurred due to the perturbations in $CO_2$, chlorine and methane. Figures 15a–b show that ozone and temperature could have changed significantly between the pre-industrial era and present time. Ozone depletion maximizes near 45 km in response to CFC's emissions (15–20% near the pole and 10–15% in the tropics and at mid-latitude). Large ozone depletion are also predicted in the mesosphere (about 30%) near 80–85 km primarily in response to methane increase. Temperature decreases are 6–10 K decrease in the vicinity of the stratopause and 2–6 K in the mesosphere.

These anthropogenic changes need to be compared to natural variations such as those produced by the 11-year solar cycle and major volcanic eruptions. These are predicted to have comparable magnitudes to some of the anthropogenic changes and can be easier to identify from observations because of the cyclic (in the case of solar variability) or catastrophic (in the case of the Mt. Pinatubo eruption) nature of the perturbation. Successes in the

simulation of the atmospheric response to these natural perturbations lends credence to the model predictions of the response to the gradual increase in anthropogenic trace gases.

A number of the examples presented for temperature or composition change suggest that the impact may be magnified with increasing altitude. *Thomas* [1996] has suggested that, because of this magnification, the impact of global composition changes might first be identified unambiguously in the mesosphere. As seen from the cases presented here, no part of the middle atmosphere remains unchanged when any of these natural or anthropogenic external changes are investigated. The analysis of atmospheric changes requires a global approach; the vertical propagation of the atmospheric response to variability is not well understood but could be a key factor.

*Acknowledgments.* This work was supported in part by the National Aeronautics and Space Administration Office of Space Science Grant 1998-067. The National Center for Atmospheric Research is operated by the University Corporation for Atmospheric Research under the sponsorship of the National Science Foundation.

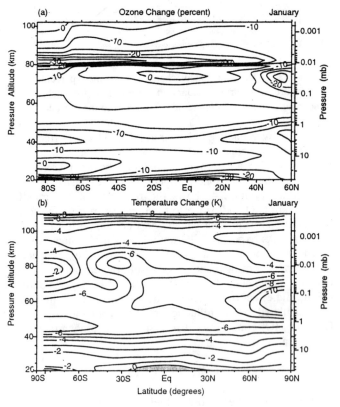

**Figure 15.** The calculated change in the ozone concentration (percent) and in temperature (K) from preindustrial time to the present.

# REFERENCES

Angell, J. K., Comparisons of stratospheric warming following Agung, El Chichón and Pinatubo volcanic eruptions, *Geophys. Res. Lett., 20,* 715–718, 1993.

Austin, J., and N. Butchart, The influence of climate change and the timing of stratospheric warmings on Arctic ozone depletion, *J. Geophys. Res., 99,* 1127-1145, 1994.

Bekke, S., and J. A. Pyle, A two-dimensional modeling study of the volcanic eruption of Mount Pinatubo, *J. Geophys. Res., 99,* 18,861-18,869, 1994.

Brasseur, G., The response of the middle atmosphere to long-term and short-term solar variability: A two-dimensional model, *J. Geophys. Res., 98,* 23,079-23,090, 1993.

Brasseur, G., and P. C. Simon, Stratospheric chemical and thermal response to long-term variability in solar UV irradiance, *J. Geophys. Res., 86,* 7343-7362, 1981.

Brasseur, G., et al., An interactive chemical dynamical radiative two-dimensional model of the middle atmosphere, *J. Geophys. Res., 95,* 5639-5655, 1990.

Brasseur, G. P., et al., European Scientific Assessment of the Atmospheric Effects of Aircraft Emissions, *Atmos. Environ., 32,* 2329-2422, 1998.

Briegleb, B. P., Longwave band model for thermal radiation in climate studies, *J. Geophys. Res., 97,* 11475–11485, 1992.

Brühl, C., et al., Halogen Occultation Experiment ozone channel validation, *J. Geophys. Res., 101,* 10,217-10,240, 1996.

Callis, L. B., et al., On the origin of midlatitude ozone changes: Data analysis and simulations for 1979-1993, *J. Geophys. Res., 102,* 1215-1228, 1997.

Considine, D. B., et al., Interhemispheric asymmetry in the 1 mbar $O_3$ trends: An analysis using an interactive zonal mean model and UARS data, *J. Geophys. Res., 103,* 1607-1618, 1998.

Fleming, E. L., et al., The middle atmospheric response to short and long term solar UV variations: Analysis of observations and 2D model results, *J. Atmos. Terr. Phys., 57,* 333-365, 1995.

Fomichev, V. I., A. A. Kutepov, R. A. Akmaev, and G. M. Shved, Parameterization of the 15 μm $CO_2$ band cooling in the middle atmosphere (15–115 km), *J. Atmos. Terr. Phys., 55,* 7–18, 1993.

Friedl, R. R. ,et al., *Atmospheric Effects of Subsonic Aircraft: Interim Assessment Report of the Advanced Subsonic technology Program,* NASA Ref. Publ. 1400, National Aeronautics and Space Administration, Washington, D. C., 1997.

Fritts, D. C., and J. Werne, Turbulent dynamics and mixing due to gravity waves in the lower and middle atmosphere, this issue, 2000.

Granier, C., and G. Brasseur, Ozone and other trace gases in the Arctic and Antarctic regions: Three-dimensional model simulations, *J. Geophys. Res., 96,* 2995-3011, 1991.

Grooss, J.-U., C. Brühl, and T. Peter, Impact of aircraft emissions on tropospheric and stratospheric ozone: Part I, Chemistry and 2-D model results, *Atmos. Environ., 32,* 3173-3184, 1998.

Hagan, M. E., Modeling atmospheric tidal propagation across the stratopause, this issue, 2000.

Hansen, J., et al., A Pinatubo climate modeling investigation, in *Global Environmental Change,* NATO ASI Ser., I., G. Fiocco, D. Fua, and G. Visconti, Eds., pp. 233–272, Springer-Verlag, New York, 1996.

Harries, J. E., et al., Validation of measurements of water vapor from the Halogen Occultation Experiment (HALOE), *J. Geophys. Res., 101,* 10,205-10,216, 1996.

Haynes, P. H., et al., On the "downward control" of extratropical diabatic circulations by eddy-induced mean zonal forces, *J. Atmos. Sci., 48,* 651-678, 1991.

Holton, J. R., and M. J. Alexander, The role of waves in the transport circulation of the middle atmosphere, this issue, 2000.

Hood, L. L., J. L. Tirikowic, and J. P. McCormack, Quasi-decadal variability of the stratosphere: Influence of long-term solar ultraviolet variations, *J. Atmos. Sci., 50,* 3941–3958, 1993.

Houghton, J. T., et al., *Climate Change 1995,* IPCC Report, Cambridge University Pres, 1996.

Huang, T. Y. W., and G. P. Brasseur, Effect of long-term solar variability in a two-dimensional interactive model of the middle atmosphere, *J. Geophys. Res., 98,* 20,413-20,427, 1993.

IPCC (Intergovernmental Panel on Climate Change), *Aviation and the Global Atmosphere: A Special Report of IPCC Working Groups I and III,* edited by J. T. Houghton, D. Lister, J. Penner, M. McFarland, D. Griggs, and D. Dokken, Cambridge University Press, Cambridge, U.K., 1999.

Jackman, C. H., et al., Past, present and future modeled ozone trends with comparisons to observed trends, *J. Geophys. Res., 101,* 28,753-28,767, 1996.

Jucks, K. W., et al., Observations of OH, $HO_2$, $H_2O$, and $O_3$ in the upper stratosphere: Implications for $HO_x$ photochemistry, *Geophys. Res. Lett., 25,* 3935-3938, 1998.

Jucks, K. W., and R. J. Salawitch, Future changes in upper stratospheric ozone, this issue, 2000.

Keating, G. M., L. S. Chiou, and N. C. Hsou, Coupling between middle atmosphere trend estimates and solar effects in ozone vertical structure, *Adv. Space Res., 14,* 2001–2009, 1994.

Kinnison, D. E., et al., The chemical and radiative effects of the Mt. Pinatubo eruption, *J. Geophys. Res., 99,* 25,705-25,731, 1994.

Langner, J., H. Rodhe, and M. Olofsson, Parameterization of subgrid scale vertical tracer transport in a global two-dimensional model of the troposphere, *J. Geophys. Res., 95,* 13691–13706, 1990.

Lefèvre, F., et al., Chemistry of the 1991-92 stratospheric winter: Three-dimensional model simulations, *J. Geophys. Res., 99,* 8183-8195, 1994.

Lindzen, R. S., Turbulence and stress owing to gravity wave and tidal breakdown, *J. Geophys. Res., 86,* 9707–9714, 1981.

Mahfouf, J. F., et al., Response of the Météo-France climate model to changes in $CO_2$ and sea surface temperature, *Clim. Dyn., 9,* 345-362, 1993.

Michelsen, H. A., et al., Stratospheric chlorine partitioning: Constraints from shuttle-borne measurements of [HCl], [$ClNO_3$], and [ClO], *Geophys. Res. Lett., 23,* 2361-2364, 1996.

Pesnell, W. D., et al., Energetic electrons and their effects on stratospheric and mesospheric ozone in May 1992 and beyond, this issue, 2000.

Pitari, G., S. Palermi, G. Visconti, and R. G. Prinn, Ozone response to a $CO_2$ doubling: Results from a stratospheric circulation model with heterogeneous chemistry, *J. Geophys. Res., 97,* 5953-5962, 1992.

Pitari, G., V. Rizi, L. Ricciardulli, and G. Visconti, High-speed civil transport impact: The role of sulfate, nitric acid trihydrate, and ice aerosols studies with a two-dimensional model including aerosol physics, *J. Geophys. Res., 98,* 23,141-23,164, 1993.

Randel, W. J., and F. Wu, A stratospheric ozone trends data set for global modeling studies, *Geophys. Res. Lett., 26,* 3089–3092, 1999.

Randel, W. J., et al., Interannual changes in stratospheric constituents and global circulation derived from satellite data, this issue, 2000.

Randeniya, L. K., et al., Heterogeneous $BrONO_2$ hydrolysis: Effect on $NO_2$ columns and ozone at high latitudes in summer, *J. Geophys. Res., 102,* 23,543-23,557, 1997.

Roble, R. G., On the feasibility of developing a global atmospheric model extending from the ground to the exosphere, this issue, 2000.

Russell, J. M. III, et al., The Halogen Occultation Experiment, *J. Geophys. Res., 98*, 10,777-10,797, 1993.

Russell, J. M. III, and R. B. Pierce, Observations of polar descent and coupling in the thermosphere, mesosphere and stratosphere provided by HALOE, this issue, 2000.

Shindell, D. T., D. Rind, and P. Lonergan, Increased polar stratospheric ozone losses and delayed eventual recovery due to increasing greenhouse gas concentrations, *Nature, 392*, 589-592, 1998.

Siskind, D. E., On the coupling between middle and upper atmospheric odd nitrogen, this issue, 2000.

Smolarkiewicz, P. K., and P. J. Rasch, Monotone advection on the sphere: An Eulerian versus semi-Lagrangian approach, *J. Atmos. Sci., 48*, 793-810, 1991.

Smyshlyaev, S. P., and V. A. Yudin, Numerical simulation of the aviation release impact on the ozone layer, *Izv. Atmos. Oceanic Phys., 31*, 116-125, 1995.

Steil, B., et al., Development of a chemistry module for GCMs: First results of a multi-annual integration, *Ann. Geophys., 16*, 205-228, 1998.

Summers, M. E., et al., Implications of satellite OH observations for middle atmospheric $H_2O$ and ozone, *Science, 277*, 1967-1970, 1997.

Summers, M. E., and R. R. Conway, Insights into middle atmospheric hydrogen chemistry from analysis of MAHRSI OH observations, this issue, 2000.

Thomas, G. E., Global change in the mesosphere-lower thermosphere: Has it already arrived? *J. Atmos. Terr. Phys., 58*, 1629–1656, 1996.

Tie, X. X., X. Lin, and G. Brasseur, Two-dimensional coupled dynamical/ chemical/microphysical simulation of the global distribution of El Chichón volcanic aerosols, *J. Geophys. Res., 99*, 16,779-16,792, 1994a.

Tie, X. X., G. P. Brasseur, B. Briegleb, and C. Granier, Two-dimensional simulation of Pinatubo aerosol and its effect on stratospheric ozone, *J. Geophys. Res., 99*, 20545–20562, 1994b.

Toon, O. B., C. P. McKay, T. P. Ackerman, and K. Santhanam, Rapid calculation of radiative heating rates and photo-dissociation rates in inhomogeneous multiple scattering atmosphere, *J. Geophys. Res., 94*, 16287–16301, 1989.

Velders, G. J. M., *Scenario Study of the Effects of CFC, HCFC, and HFC Emissions on Stratospheric Ozone,* Report No. 722201 006, National Institute of Public Health and the Environment, The Netherlands, 1995.

Weisenstein, D. K., et al., The effects of sulfur emissions from HSCT aircraft: A 2-D model intercomparison, *J. Geophys. Res., 103*, 1527-1547, 1998.

WMO: Scientific Assessment of Ozone Depletion: 1998, World Meteor-ological Organization, Global Ozone Research and Monitoring Project, Report No. 44, Geneva, 1999.

Yudin, V. A., et al., Interannual variability of the diurnal tide in the low-latitude mesosphere and lower thermosphere during equinoxes: Mechanistic model interpretation of the 1992-96 HRDI measurements, this issue, 2000.

Zhu, X., et al., Coupled models of photochemistry and dynamics in the mesosphere and lower thermosphere, this issue, 2000.

Guy P. Brasseur, Theresa Huang, Rashid Khosravi, Anne K. Smith, and Stacy Walters, National Center for Atmospheric Research, P. O. Box 3000, Boulder, Colorado 80307

Simon Chabrillat, and Gaston Kockarts, Belgian Institute for Space Aeronomy, 3 avenue Circulaire, 1180 Brussels, Belgium

# The Role of Waves in the Transport Circulation of the Middle Atmosphere

James R. Holton

*University of Washington, Seattle, WA*

M. Joan Alexander

*Colorado Research Associates, Boulder, CO*

Zonal forces generated by breaking planetary and gravity waves are crucial to understanding large-scale dynamics and transport in the middle atmosphere. These forces, sometimes called "wave-drag", excite a pumping action that drives a meridional transport circulation. In the stratosphere the primary wave drag is exerted by planetary waves in the winter hemisphere. The transport circulation is characterized by rising motion at low latitudes, poleward drift, and sinking in high latitudes. In the mesosphere, on the other hand, the primary wave drag is exerted by breaking gravity waves. The mesospheric transport circulation is pole-to-pole; there is rising motion in the summer hemisphere, cross equatorial drift, and sinking in the winter hemisphere. The stratospheric and mesospheric transport circulations are consistent with the observed temperature structure, and can account for much of the observed behavior of long-lived trace constituents in the middle atmosphere.

## 1. INTRODUCTION

Viewed globally, the average circulation of the atmosphere features mean winds that are primarily zonal; that is, they tend to blow roughly parallel to latitude circles. In the extratropical troposphere the zonal winds are eastward directed, with characteristic speeds in the range of 10-40 m s$^{-1}$. In the middle atmosphere the direction reverses seasonally; in winter the winds are eastward and in summer westward. Maximum speeds are of order 60-80 m s$^{-1}$ in the upper stratosphere and lower mesosphere. Advection by this east-west flow causes trace constituents with isolated sources (volcanic emissions, for example) to become fairly

evenly distributed in longitude around the globe within a few weeks. Thus, the lowest order role of global transport is to homogenize tracer distributions in the longitudinal direction. For this reason meridional and vertical variations of long-lived trace constituents tend to be much larger than zonal variations. It is thus useful to zonally average constituent distributions and to focus on the variations in the meridional plane (i. e., in height-latitude cross-sections). Zonal averaging reduces the dynamics from three dimensions to two dimensions (height and latitude). It is conceptually equivalent to transport in a hypothetical zonally symmetric circulation, provided that the symmetric circulation model incorporates driving of the zonally symmetric flow by forces induced by the asymmetric component of the flow (i. e., by atmospheric eddies, or waves). Such wave-induced forces are crucial for understanding the general circulation and transport in the real atmosphere.

Atmospheric Science Across the Stratopause
Geophysical Monograph 123

Wave-induced forces are not only responsible for driving the transport circulation in the meridional plane, but also have a profound influence on the observed temperature distribution. In the lower stratosphere there are temperature minima at the equator and at both the summer and winter poles, and there are temperature maxima in midlatitudes. In the middle and upper stratosphere there is a monotonic increase of temperature from the winter pole to the summer pole, qualitatively consistent with the latitudinal distribution of solar radiation. In the mesosphere, on the other hand, temperature decreases monotonically from the winter pole to the summer pole, a situation that cannot possibly be explained by the distribution of solar radiation [*Leovy*, 1964].

In the absence of atmospheric motions each vertical column of the atmosphere would establish an equilibrium between solar heating and infrared cooling. Careful calculations indicate that in such a radiative equilibrium state the pole-to-equator temperature difference would be much greater than observed in the stratosphere. In high latitudes in the winter hemisphere the observed temperature near the stratopause is about 100 K warmer than the radiative equilibrium temperature. On the other hand, near the summer mesopause observed temperatures are about 100 K colder than radiative equilibrium.

The extremely low radiative equilibrium temperature in the polar winter stratosphere reflects the absence of solar radiation in the polar night. To attain radiative equilibrium the polar stratosphere would have to cool until there was a balance between absorption of the weak upwelling radiation from the polar troposphere and emission of infrared radiation to space. In the case of the summer mesosphere radiative equilibrium would require sufficient warming so that infrared emission could balance the strong solar radiative absorption. In both situations dynamical processes involving adiabatic heating and cooling through vertical motion must be invoked to explain the dramatic differences between the radiative equilibrium and the observed temperature distribution.

## 2. WAVES AND WAVE-INDUCED FORCES IN THE MIDDLE ATMOSPHERE

As noted above, the global-scale circulation of the middle atmosphere is strongly dependent on the net zonal force exerted by longitudinally asymmetric motions. The dominant zonally asymmetric motions contributing to the zonal forcing in the middle atmosphere are not random turbulent fluctuations, but rather consist of various types of atmospheric waves. Wave activity may propagate horizontally and vertically over large distances from wave sources

to regions where transience, nonlinear wavebreaking, or radiative or turbulent dissipation causes momentum transfer to the mean flow. Waves can thus provide a strong non-local influence on the zonal momentum. In order to better understand the nature of wave interactions with the mean flow, and of wave transport processes, it is necessary to first discuss the nature of the major types of wave motions in the atmosphere.

Wave motions in the atmosphere result from a balance between inertia and restoring forces acting on fluid parcels displaced from their equilibrium latitudes or altitudes. For the waves of importance to the global-scale circulation of the middle atmosphere the primary restoring forces are provided by the fluid buoyancy (gravity waves), and by the isentropic gradient of potential vorticity (Rossby waves). Potential vorticity is an important conserved property for adiabatic frictionless flows in rotating fluids. It is somewhat analogous to spin angular momentum in solid body mechanics. Normally potential vorticity increases monotonically from the South Pole to the North Pole. The meridional gradient of potential vorticity tends to resist meridional displacement of fluid parcels.

### 2.1. Rossby Waves

The wave type that is of most importance for large-scale meteorological processes in the troposphere and the stratosphere is the *Rossby wave*, or *planetary wave*. In its most general sense the Rossby wave is a potential vorticity conserving motion that owes its existence to the basic state meridional gradient of potential vorticity. The mechanism of the Rossby wave can be illustrated by considering the simple case of a homogeneous, incompressible fluid of uniform depth. For such a fluid (referred to as a *barotropic fluid*) the horizontal velocity is nondivergent, potential vorticity conservation is equivalent to conservation of absolute vorticity, where absolute vorticity is the sum of the relative vorticity owing to the rotation of the fluid and the planetary vorticity owing to the rotation of the earth [e. g., see *Holton*, 1992].

Consider a closed chain of fluid parcels initially at rest in such a barotropic fluid and aligned along a circle of constant latitude. Now the absolute vorticity $\eta$ *is* given by $\eta = \zeta + f$, where $\zeta$ is the relative vorticity and $f = 2\Omega \sin\phi$ is the Coriolis parameter, which is equal to the local vertical component of planetary vorticity. Here, $\Omega$ is the angular speed of rotation of the earth and $\phi$ is latitude. Assuming that $\zeta = 0$ at time $t_0$, suppose that the chain of fluid parcels undergoes a sinusoidal meridional displacement as shown in Figure 1. Let $\delta y$ be the meridional displacement of a fluid parcel from the original latitude at $t_1$. Then at $t_1$

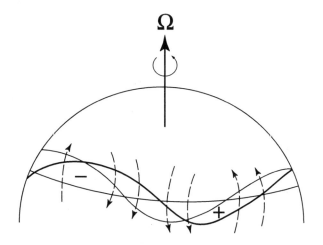

**Figure 1.** Perturbation vorticity field and induced velocity field (dashed arrows) for a sinusoidal displacement of a chain of fluid parcels from its mean latitude. $\Omega$ denotes the angular velocity of the earth.  Heavy wavy line shows original perturbed position, light line shows the westward displacement caused by advection of the pattern by the induced meridional velocity field.

conserva-tion of absolute vorticity implies that at each point along the chain of parcels

$$(\zeta + f)_{t_1} = f_{t_0}$$

or

$$\zeta_{t_1} = f_{t_0} - f_{t_1} = -\beta \delta y \tag{1}$$

where $\beta \equiv df/dy$ is the planetary vorticity gradient at the original latitude.

From (1) it is evident that if the chain of parcels is subject to a meridional displacement that varies sinusoidally in longitude, conservation of absolute vorticity requires that the resulting perturbation vorticity be positive (i.e., cyclonic) for a southward displacement and negative (anticyclonic) for a northward displacement.

This perturbation vorticity field will induce a meridional velocity field, which advects the chain of fluid parcels southward west of the vorticity maximum and northward east of the vorticity minimum, as indicated in Figure 1. Thus, the fluid parcels oscillate back and forth about their equili-brium latitude, and the pattern of poleward and equatorward parcel  displacement propagates to the west. The phase speed of the  westward propagation can be calculated for this situa-tion if we assume that the parcel displacement field depends only on $x$ and $t$ and that the mean zonal flow vanishes:

$$\delta y = a \sin k(x - ct)$$

where $a$ is the displacement amplitude, $k$ is the zonal wavenumber, and $c$ is the phase speed. But by definition of the meridional velocity,

$$v = \frac{D}{Dt}(\delta y) = -kca \cos k(x - ct)$$

The disturbance vorticity can then be related to the parcel displacement field as

$$\zeta = \partial v/\partial x = k^2 ca \sin k(x - ct)$$

Thus, substitution into (1) yields $c = -\beta/k^2$, which confirms that the phase propagation is westward, and is highly dependent on the horizontal scale of the meridional parcel displacement field.

This westward propagating vorticity field constitutes a Rossby wave. The meridional gradient of absolute vorticity resists meridional displacements, and provides the restoring force mechanism for Rossby waves. Rossby waves differ from other wave types such as gravity waves because the zonal phase propagation of gravity waves can be either eastward or westward, while the phase propagation of Rossby waves can only be westward relative to the mean flow.

For studies of the middle atmosphere it is vertically propagating waves that are of most interest. Vertically propagating Rossby waves are forced modes generated in the troposphere by flow over continental scale topography, by continent-ocean heating contrasts, and by nonlinear interactions among transient tropospheric wave disturbances. For this case linear wave theory based on conservation of potential vorticity [e. g., *Andrews et al.*, 1987] shows that for waves stationary relative to the ground, the vertical wavenumber $m$ satisfies the relation

$$m^2 = \frac{N^2}{f^2}\left[\frac{\beta}{\bar{u}} - \left(k^2 + l^2\right)\right] - \frac{1}{4H^2} \tag{2}$$

where $l$ is the meridional wavenumber $N^2$ is the buoyancy frequency squared, and $H$ the density scale height.

For known $\bar{u}$, $k$, and $l$, (2) gives the vertical structure of the waves.  Thus, for topographically forced solutions $m^2 > 0$ only if

$$0 < \bar{u} < u_c \text{ where } u_c = \beta \left[k^2 + l^2 + \frac{f^2}{\left(4N^2 H^2\right)}\right]^{-1}. \tag{3}$$

$u_c$ is the critical zonal wind value for which $m = 0$. For $m^2 > 0$, waves can propagate vertically, while for $m^2 \leq 0$ the waves are trapped. Thus, for stationary waves, vertical propagation exists only in the presence of mean westerly winds that are less than the critical value, $u_c$. This result is referred to as the *Charney-Drazin criterion*. For example, for a wave with meridional scale 12,000 km, and planetary wave number 1 [one wavelength around a latitude circle], this theory gives $u_c \approx 60$ m s$^{-1}$. More accurate calculations with spherical geometry and realistic winds give $u_c \approx 100$ m s$^{-1}$. It is clear from (3) that $u_c$ decreases rapidly as $k$ and $l$ increase, i.e., the critical value is determined largely by horizontal wave scales. In practice only zonal wavenumbers 1 and 2 propagate significantly into the extratropical stratosphere, and this happens only in the winter hemisphere where $\bar{u} > 0$. Therefore, the Charney-Drazin criterion provides an approximate explanation for the absence of stationary planetary waves in the summer stratosphere and the dominance of waves of zonal wavenumbers 1 and 2 in the winter stratosphere.

Linear wave theory has proved to provide an excellent model for understanding the generation and propagation of atmospheric waves in a wide variety of contexts. Linear theory can also describe certain wave, mean-flow interactions associated with wave transience (i. e., local growth or decay of wave amplitude), and wave damping (i. e., thermal damping due to radiation or mechanical damping due to small scale turbulence). The extratropical wave-induced zonal forcing in the middle atmosphere is, however, primarily a result of *nonlinear* wavebreaking. Thus, to better understand the role of waves in the global circulation in the atmosphere it is necessary to consider the process of wave-breaking.

Linear waves are characterized by the back and forth undulation of material contours. For adiabatic, inviscid conditions such undulations are reversible. For example, if the planetary wave shown schematically in Figure 2 is stationary, then the streamlines ($S_1$, $S_2$, $S_3$) coincide with material contours, and undulate without undergoing net displacement in the meridional or vertical directions over a wave period. Under such conditions waves can play no direct part in driving the global circulation. (They may, however, still contribute to net changes in trace constituent distributions if chemical sources or sinks vary along the undulating material contours, as might occur, for example, if the undulations move parcels in and out of the polar night.)

Rossby waves may have periods comparable to the radiative timescale, but since Rossby waves generally have very large vertical scales and weak temperature perturba-

tions, radiative damping of Rossby waves is very inefficient. For Rossby waves, the primary process leading to a net forcing of the global circulation is nonlinear wave breaking.

The term *wave breaking* for Rossby waves simply refers to a rapid, irreversible deformation of material contours. Since potential vorticity is approximately conserved in Rossby waves, isolines of potential vorticity on isentropic surfaces approximate material contours, and wavebreaking can best be illustrated by considering the field of potential vorticity. Wavebreaking may occur when the disturbance fields reach amplitudes for which nonlinear effects can no longer be neglected in the dynamical equations. For example, if nonlinear terms are included, the equation for conservation of quasi-geostrophic potential vorticity becomes

$$\left(\frac{\partial}{\partial t} + \bar{u}\frac{\partial}{\partial x}\right)q' + v'\frac{\partial \bar{q}}{\partial y} = -u'\frac{\partial q'}{\partial x} - v'\frac{\partial q'}{\partial y} \qquad (4)$$

where the wave potential vorticity is related to the velocity stream-function by

$$q' = \nabla^2 \psi' + \frac{f_0^2}{\rho_0}\frac{\partial}{\partial z}\left(\frac{\rho_0}{N^2}\frac{\partial \psi'}{\partial z}\right)$$

and the meridional gradient of the basic state potential vorticity is related to the planetary vorticity gradient and the mean zonal wind by

$$\frac{\partial \bar{q}}{\partial y} = \beta - \frac{\partial^2 \bar{u}}{\partial y^2} - \frac{f_0^2}{\rho_0}\frac{\partial}{\partial z}\left(\frac{\rho_0}{N^2}\frac{\partial \bar{u}}{\partial z}\right)$$

Now for neutral (constant amplitude) waves propagating relative to the ground at zonal phase speed $c$, the variation of phase in time and space is given by $\phi = k(x - ct)$, where $k$ is the zonal wavenumber, and it is readily verified that

$$\frac{\partial}{\partial t} = -c\frac{\partial}{\partial x}$$

so that in the linearized version of (4) there is a balance between advection of the disturbance potential vorticity $q'$ by the Doppler shifted mean wind and the advection of mean potential vorticity by the disturbance meridional wind:

$$(\bar{u} - c)\frac{\partial q'}{\partial x} = -v'\frac{\partial \bar{q}}{\partial y} \qquad (5)$$

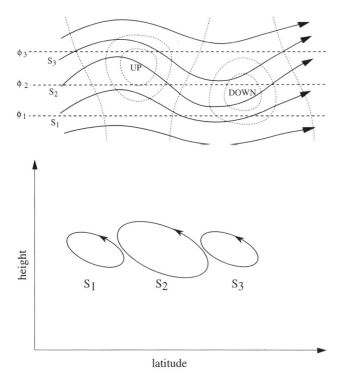

**Figure 2.** Parcel motions for an adiabatic vertically propagating stationary planetary wave in a westerly zonal flow. In upper panel solid lines labeled $S_1$, $S_2$, $S_3$ are parcel trajectories (coincident with streamlines for a staionary wave), heavy dashed lines labeled $\phi_1$, $\phi_2$, $\phi_3$ are latitude circles, light dashed lines are contours of vertical velocity field. Lower panel shows projection of parcel oscillations on the meridional plane.

The validity of the linear approximation can thus be assessed by comparing the sizes of the two terms on the right side of (4) with either term in (5). Linearity holds provided that

$$|\bar{u} - c| >> |u'| \qquad (6a)$$

and

$$\partial\bar{q}/\partial y >> |\partial q'/\partial y| \qquad (6b)$$

Basically these criteria require that the angle with respect to the zonal direction through which material contours fluctuate must remain small. That is, the slope of the material contours in the $x, y$ plane must be small.

In an atmosphere with constant mean zonal wind vertically propagating Rossby waves have amplitudes increasing exponentially in height. Thus, at some altitude the disturbance amplitude will become sufficiently large so that wavebreaking must occur. In the real atmosphere, however,

the mean zonal flow varies in both latitude and in height, and this variation is crucial for understanding the distribution and mean-flow forcing provided by Rossby wave breaking. The simplest example of Rossby wave breaking occurs in the presence of a *critical surface* along which the Doppler shifted phase speed of the wave vanishes $(c - \bar{u} = 0)$.

In that case (6a) can not be satisfied even for small amplitude waves. For understanding of wave behavior near critical lines it is helpful to generalize the Rossby wave analysis above to a situation in which the zonal wind depends on latitude and height, $\bar{u} = \bar{u}(y,z)$. The $x$ and $t$ dependence can then be separated by seeking solutions for the velocity stream function $\psi'$ of the form

$$\psi' = e^{z/2H} \text{Re}\left[\Psi(y,z)e^{ik(x-ct)}\right]. \qquad (7)$$

Substitution into (4) yields an expression for $\Psi(y, z)$:

$$\frac{\partial^2 \Psi}{\partial y^2} + \frac{f_0^2}{N^2}\frac{\partial^2 \Psi}{\partial z^2} + n_k^2 \Psi = 0 \qquad (8)$$

where the small vertical variation of $N^2$ is neglected, and

$$n_k^2(y,z) = (\bar{u} - c)^{-1}\partial\bar{q}/\partial y - k^2 - f_0^2/\left(4HN^2\right) \qquad (9)$$

If (9) is transformed to the coordinates $(\tilde{y},\tilde{z}) \equiv (y, Nz/f_0)$ it has the same form as the equation governing the two-dimensional propagation of light waves in a medium whose refractive index is $n_k$. The propagation of wave activity in that case can be shown to be along rays that behave somewhat like light rays. Thus, wave activity will tend to propagate toward regions of large positive $n_k^2$, and avoid regions of negative $n_k^2$. For stationary Rossby waves $(c = 0)$ of low zonal wavenumber, $n_k^2$ is positive in a region with westerly winds that are not too strong and increases to infinity along a critical surface where the mean flow vanishes. Thus, the index of refraction for wave activity is positive in the winter hemisphere, but increases rapidly toward the equatorial zero wind line. As a result Rossby wave activity tends to propagate upward and equatorward, and wavebreaking occurs in the vicinity of the equatorial critical line. Careful analysis of this situation reveals that meridional propagation can occur in the region of positive zonal flow (providing an equatorward flux of wave activity), but the waves must decay with distance from the critical line in the negative zonal flow region. Thus, there must be a substantial convergence of the flux of wave

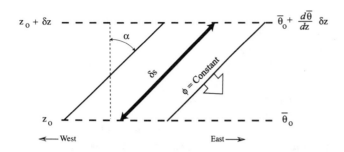

**Figure 3.** Parcel oscillation path (heavy arrow labeled $\delta s$) for eastward and upward propagating gravity waves with phase lines tilted at an angle $\alpha$ to the vertical.

activity in the vicinity of the critical line. The critical line is surrounded by a critical layer in which com-plex nonlinear processes dominate, and mean-flow forcing occurs.

The Rossby wave-breaking process described above is associated with a westward force on the stratospheric mean flow. The associated mixing is quasi-isentropic, mixing air parcels horizontally over large distance scales in the mid-latitude "surf zone" *McIntyre and Palmer* [1984], but giving rise to negligible vertical mixing.

### 2.2. Gravity Waves

Internal gravity waves are vertically propagating waves associated with the buoyancy restoring force in stably stratified fluids. Parcel oscillations in these waves are parallel to the phase lines, and hence perpendicular to the direction of phase propagation as indicated in Figure 3. Thus, for example, the phase surface corresponding to the maximum negative temperature perturbation occurs one quarter cycle after the maximum upward motion, hence, coinciding with maximum upward displacement (and adiabatic cooling) of fluid parcels. The physical mechanism for internal gravity wave propagation can be understood by considering a fluid parcel that is displaced a distance $s$ along a line tilted at an angle to the vertical as shown in Figure 3. The parcel will undergo a vertical displacement $\delta z = \delta s \cos \alpha$. For such a parcel the buoyancy force per unit mass, which is directed in the vertical, is just $-N^2 \delta z$, where $N = (g d\theta/dz)^{1/2}$ is the buoyancy frequency, and $\theta$ is the background potential temperature. The component of the buoyancy force perpendicular to the parcel path is balanced exactly by the pressure gradient force, which is itself perpendicular to the parcel path. The parcel acceleration is thus given by the component of the buoyancy force parallel to the parcel path:

$$-N^2(\delta s \cos \alpha)\cos \alpha = -(N \cos \alpha)^2 \delta s$$

The momentum equation for the parcel oscillation is then

$$\frac{d^2(\delta s)}{dt^2} = -(N \cos \alpha)^2 \delta s$$

which has the general solution $\delta s = \exp[\pm i(N \cos \alpha)t]$. Thus, the parcel executes a simple harmonic oscillation at the frequency $v = N \cos \alpha$. This frequency depends only on the static stability (measured by the buoyancy frequency $N$) and the angle of the phase lines to the vertical.

Detailed analysis reveals that the dispersion relationship for linear gravity waves (neglecting the Coriolis force) is:

$$(v - \bar{u}k)^2 = \hat{v}^2 = N^2 k^2 \left( m^2 + k^2 + \frac{1}{4H^2} \right)^{-1} \quad (10)$$

where $\hat{v} = v - \bar{u}k$ is referred to as the Doppler shifted frequency or *intrinsic frequency*; $k$ is the horizontal wavenumber; and $\bar{u}$ is the component of the wind in the direction of wave propagation here referred to as the $x$ direction. For many observed gravity waves $|m| \gg 1/(2H)$ so that the last term in the denominator can be neglected. However, waves near the mesopause observed for decades in images of infrared and visible nightglow emissions [*Moreels and Herse*, 1977; *Taylor and Hapgood*, 1988] have short horizontal and long vertical wavelengths by nature of the observation technique. Analysis of these images has shown that some of the observed waves are at or near the point of total internal reflection where $m \to 0$ [*Isler et al.*, 1997; *Swenson et al.*, 2000] and the momentum and energy fluxes also go to zero. For these waves, the dispersion relation (10) must be used. For most other observed gravity waves, $|m| \gg 1/(2H)$ so that

$$\hat{v} \approx \pm Nk/\left(m^2 + k^2\right)^{1/2} \approx \pm N \cos \alpha \quad (11)$$

where $\alpha$ is again the angle of phase lines to the vertical. By convention the wavenumber $k$ is always positive. Thus, if the frequency $v$ is positive the horizontal phase speed $c = v/k$ is also positive. For waves in which phase lines slope upward in the positive $x$ direction in the $(x, z)$ plane (as in Figure 3) the vertical wavenumber $m$ must then be negative. This can be easily verified by noting that along lines of constant phase

$$\phi = kx + mz - vt = \text{Constant} \quad (12)$$

so that for $t$ constant and $k$ positive, $m$ must be negative if $\phi$ is to be constant for $x$ and $z$ both increasing.

As is well known, wave energy propagates with the group velocity $\mathbf{c}_g \equiv (c_{gx}, c_{gz})$, which for the dispersion relation (11) can be expressed as

$$\left(c_{gx}, c_{gz}\right) = \left(\frac{\partial v}{\partial k}, \frac{\partial v}{\partial m}\right) = (\bar{u}, 0) \pm \frac{Nm}{\left(m^2 + k^2\right)^{3/2}}(m, -k) \quad (13)$$

For the eastward propagating wave of Figure 3 the negative root applies and $m$ is negative so that the group velocity relative to the mean flow $\bar{u}$ is eastward and upward. Since $|c_{gz}/(c_{gx} - \bar{u})| = |k/m|$ the group velocity vector relative to the mean flow is parallel to lines of constant phase. Thus, energy propagates eastward and upward relative to the mean flow as phase propagates eastward and downward—phase and energy propagate in opposite directions in the vertical!

Gravity waves with intrinsic periods greater than a few hours are influenced by the Coriolis effect, which causes parcel trajectories to be elliptical rather than linear. When the Coriolis force is included in the linearized equations, the dispersion relationship (11) becomes approximately

$$\hat{v}^2 = f^2 + N^2\left(k^2 + l^2\right)\big/m^2 \quad (14)$$

where $k$ is the zonal wavenumber and $l$ is the meridional wavenumber. For vertical propagation ($m^2 > 0$) the intrinsic frequency must satisfy the inequality $|f| < |\hat{v}| < N$, so that rotation imposes a lower limit on the frequency of vertically propagating gravity waves, while static stability imposes an upper limit.

If the horizontal axes are chosen so that $l = 0$ (i. e., the $x$ axis is aligned perpendicular to phase lines), the ratio of vertical to horizontal group velocity components can be expressed as

$$\left|c_{gz}\big/\left(c_{gx} - \bar{u}\right)\right| = |k/m| = \left(\hat{v}^2 - f^2\right)^{1/2}\big/N$$

Thus, for a given intrinsic frequency the energy of an inertia-gravity wave propagates more closely to the horizontal than is the case for a pure gravity wave. It can also be shown that the ratio of the meridional to zonal velocity perturbations is

$$\frac{\tilde{v}}{\tilde{u}} = -i\left(\frac{f}{\hat{v}}\right)$$

so that parcel oscillations trace out an ellipse whose ratio of minor to major axes approaches unity as intrinsic frequency approaches the Coriolis frequency.

The wave-induced zonal forcing in the mesosphere has generally been ascribed to dissipation by gravity waves with horizontal wavelengths in the range of order ~10 – 1,000 km, and such waves also appear to be important in the stratosphere. Observations show these waves can be generated in the troposphere by flow over topography and by convection and weather fronts. Theoretical studies also suggest shear instability and geostrophic adjustment processes are other likely mechanisms through which gravity waves are generated. For linear, adiabatic, inviscid gravity waves, the momentum flux is constant with height; but the wave amplitude grows exponentially with height. Even for waves with very small amplitudes near their source levels in the troposphere, this exponential growth can lead to very large amplitudes in the mesosphere as suggested in Figure 4.

As in the case of Rossby waves, gravity waves remain linear provided that the slopes of material surfaces perturbed by the waves remain small. For adiabatic waves, isentropic surfaces are material surfaces, and the level of gravity wave breaking can be estimated by defining the breaking criterion as

$$\frac{\partial \theta}{\partial z} = \frac{\partial \bar{\theta}}{\partial z} + \frac{\partial \theta'}{\partial z} = 0 \quad (15)$$

where $\bar{\theta}$ and $\theta'$ designate the background and disturbance potential temperature fields, respectively. At the level where (15) holds isentropic surfaces become vertical, and with any further increase in amplitude the wave will become convectively unstable as shown schematically in Figure 5.

The altitude at which a vertically propagating gravity wave breaks can be estimated approximately from the linear wave solution. According to the hydrostatic relationship

$$\partial\Phi/\partial z = RTH^{-1} = R\left(\theta e^{-\kappa z/H}\right)H^{-1} \quad (16)$$

Separating the geopotential field into mean and disturbance components, $\Phi = \bar{\Phi} + \Phi'$ we obtain from (15) and (16)

$$\frac{\partial \theta}{\partial z} = \frac{\partial \bar{\theta}}{\partial z} + \frac{\partial \theta'}{\partial z} \approx \frac{H}{R}e^{\kappa z/H}\left[N^2 + \Phi'_{zz}\right] \quad (17)$$

where $N^2$ is again the buoyancy frequency squared. Finally, substituting the linear wave solution

$$\Phi' = \Phi_0 \exp\left[(z - z_0)/(2H) + i(kx + mz - vt)\right]$$

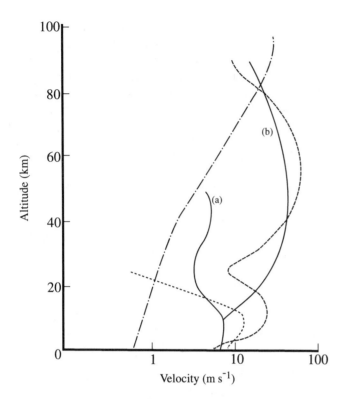

**Figure 4.** Schematic vertical profiles of wind amplitudes corresponding to various types of atmospheric motions in midlatitudes. Solid line: planetary waves (a) summer, (b) winter; dashed: zonal mean; dotted: synoptic scale; dotted-dashed: gravity waves. (Courtesy of Professor T. Matsuno.)

into (17) and again using the approximation $|m| >> (2H)^{-1}$ we find that breaking is initiated (that is, $\partial\theta/\partial z = 0$) when

$$N^2 = m^2 \Phi_0 e^{(z_b - z_0)/2H} \qquad (18)$$

so that the breaking level is given by $z_b \approx 2H \ln|N^2 m^{-2} \Phi_0^{-1}|$ where $\Phi_0$ is the wave amplitude at $z_0$. Under these same assumptions, the breaking level criterion can also be written $u' = |c - \bar{u}|$, where $u'$ is the zonal wind perturbation associated with the wave, [*Fritts*, 1984] analogous to the condition (6a). Inspection of the breaking level criteria reveals that for larger $m$ (i. e., smaller vertical scale) or larger amplitude waves, the breaking level is lower in the atmosphere.

When mean wind variations are included in the theory, the expressions for the breaking altitude, and for the momentum flux convergence at the breaking level depend on the Doppler shifted phase speed. Since $m \approx \pm N(c - \bar{u})^{-1}$ even very small amplitude waves will break as they approach their critical level where the vertical wavenumber becomes infinite. Fritts and Werne (2000) show that the mechanism of wave instability will vary with height.

Convective instability at higher intrinsic frequency (and longer vertical wavelength) will be more likely in the upper mesosphere while dynamic instability at low intrinsic frequency (and shorter vertical wavelength) should dominate in the lower stratosphere. They show that this follows from the observed vertical variations of dominant wave frequencies and energy dissipation rate time scales. These variations with height also imply changing effects on the background atmosphere with height.

In the lower stratosphere, gravity wave amplitudes will be relatively small because they are closer in vertical distance to their sources in the troposphere. For small amplitude waves, instability will tend to occur at larger $m$ (smaller vertical wavelength) and smaller intrinsic frequency $\hat{v}$ and phase speed $(c - \bar{u})$. *Fritts and Werne* [2000] show that the mixing associated with wave breaking in the lower stratosphere at large $m$ (small $c - \bar{u}$) will only mix relatively thin layers of the atmosphere.

Gravity wave breaking also forces accelerations in the background flow. The wave driven force "drags" the flow toward the phase speed $c$ of the breaking wave. For low amplitude waves propagating vertically through background winds that increase from relatively low speeds in the lower stratosphere to larger speeds above, the waves that become unstable will be those propagating downstream that are approaching their critical levels as $c - \bar{u}$ becomes

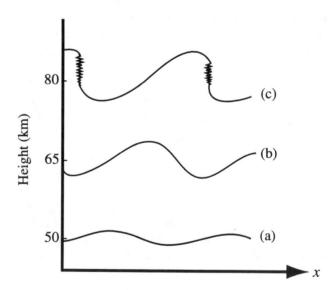

**Figure 5.** Schematic diagram illustrating the breaking of vertically propagating internal gravity waves in the mesosphere. The curves labeled (a), (b), and (c) denote material surfaces (approximately equal to potential temperature surfaces). At the levels of (a) and (b) linear wave theory applies. At the level of (c) nonlinear wavebreaking is occurring, accompanied by turbulence and mixing.

small. The wave dissipation for such waves will, therefore, accelerate the wind toward the larger phase speed, and the sign of the wave driven force will have the same sign as the background wind [*Alexander and Rosenlof*, 1996]. These tendencies would not be true for much larger amplitude waves such as the stationary waves that can occur in the upper troposphere and lower stratosphere over mountain ranges.

In the mesosphere, gravity wave amplitudes will have grown relatively large and instability can occur at smaller $m$ (larger vertical wavelength). The wave breakdown via convective instability can mix parcels of air vertically over a deeper layer through which composition may initially vary substantially. Smaller $m$ also goes with larger intrinsic phase speed $|c - \bar{u}|$. Wave breaking can even occur where $c$ and $\bar{u}$ have opposite sign. This wave breaking at large $|c - \bar{u}|$ is the mechanism believed responsible for the reversal in sign of the middle atmosphere jets near the mesopause. The waves again drag the winds toward the phase speed, but in the upper mesosphere this is a decelerating force. It has therefore been fairly successfully described with a Rayleigh friction parameterization in some models, whereas the stratospheric gravity wave accelerating force cannot be described with Rayleigh friction.

The above discussion makes clear that the breaking levels for gravity waves can depend dramatically on the seasonally varying zonal wind distribution. A filtering of waves by the stratospheric winds will also occur. Thus, westward propagating waves are readily transmitted through the westerly winds of the winter stratosphere while eastward propagating waves encounter critical levels, unless they have zonal phase speeds greater than the maximum wind speed. During the summer, on the other hand, westward propagating waves of low to moderate phase speeds cannot propagate through the stratospheric easterlies, but eastward propagating waves can readily reach the mesosphere. Hence, in the mesosphere, an eastward drag force is exerted in summer and a westward drag force is exerted in the winter. These account for the reversal of the mean zonal wind above the stratospheric winter westerly and summer easterly maxima and the meridional parcel drift from summer to winter pole described in the following sections.

The nature of the gravity waves that drive the transport circulation in the middle atmosphere can now be summarized as follows. Waves propagating in the zonal direction will be those most important in driving the middle atmosphere transport circulation. Mountain waves have been observed with large amplitude in the lower stratosphere, but waves generated by flow over topography will have zonal phase speeds near zero. Such waves are likely to break in the upper troposphere or lower stratosphere where the mean winds reach a local minimum in winter. In summertime even small amplitude mountain waves cannot propagate far into the stratosphere before encountering critical levels. Higher phase speed waves likely dominate in the mesosphere and in the summer stratosphere. Model studies show high phase speed waves can be generated in convective storms [*Alexander and Holton*, 1997]. Some observations in the middle atmosphere have related the occurrence of high phase speed waves to weather systems [*Taylor and Hapgood*, 1988] and deep convective clouds [*Dewan et al*, 1998; *McLandress et al*, 2000]. High phase speed waves likely also play an important role in driving the equatorial quasi-biennial and semi-annual oscillations in the winds. The details of the wave spectrum generated by the different gravity wave sources and their variability are important factors in modeling gravity wave effects in the middle atmosphere. Current global models make crude assumptions about these details. The results of future research in this area will be important to global models that include such gravity wave effects.

## 3. TRANSPORT CIRCULATION OF THE STRATOSPHERE

Synoptic-scale motions decay rapidly with height above the tropopause, so that, as suggested in Figure 4, throughout most of the stratosphere the primary motion systems are planetary in scale. Above the lowest layers of the stratosphere, the global-scale circulation is dominated by a mean flow that is eastward in the winter hemisphere, and westward in the summer hemisphere. During the equinoxes the flow tends to be eastward, but weaker than during the winter.

As a consequence of the rapid rotation of the earth and the strong static stability of the atmosphere, the pressure field is nearly in geostrophic balance with the wind field and in hydrostatic balance with the temperature field. Thus, the zonal flow is proportional to the latitudinal pressure gradient and the temperature is proportional to the vertical pressure gradient. Hence, the vertical shear of the zonal wind is proportional to the latitudinal temperature gradient:

$$f \frac{\partial u}{\partial z} \propto -\frac{\partial T}{\partial \phi}$$

where $f (\equiv 2\Omega \sin \phi)$ is the Coriolis parameter (with $\Omega$ the angular velocity of the earth), $T$ is the temperature, $u$ is the zonal wind, $z$ is altitude, and $\phi$ is latitude. In the winter stratosphere the observed latitudinal temperature gradient is much weaker than the radiative equilibrium gradient, the zonal winds in this region of the atmosphere tend to be much weaker than those that would be in thermal wind balance with the radiative equilibrium temperature distribution. This departure of the mean zonal flow from its radiatively determined state strongly suggests that some significant decelerating force must influence the mean zonal flow distribution in the winter stratosphere. Why such a deceleration must cause temperatures to depart from radiative equilibrium and why this departure is important for transport are major themes of middle atmosphere dynamics.

In the summer hemisphere the flow in the stratosphere is nearly parallel to latitude circles. Large-scale wave motions have very small amplitudes; a slow ageostrophic (cross-isobaric) zonally symmetric meridional drift is the main cause of meridional transport in the stratosphere in this season. In the winter (especially in the Northern Hemisphere) the flow is disturbed by planetary-scale waves of zonal wavenumbers 1 and 2. Thus, the flow in the winter hemisphere departs significantly from zonal symmetry. The mean flow distribution for the Northern Hemisphere features a cyclonic polar vortex that is distorted by quasi-stationary planetary waves that form an anticyclonic disturbance called the Aleutian high.

At times the planetary waves in the Northern winter stratosphere may amplify dramatically over a short span of time, and produce rapid meridional transport, which leads to rapid deceleration of the mean zonal flow, and an accompanying *sudden stratospheric warming* in the polar region. In a major warming the temperature at the 10 hPa level (about 31 km) may increase by as much as 60°C in less than a week. Such major warmings occur only in the Northern Hemisphere, and only once every other year or so. Warmings of a less intense nature occur throughout the winter in both hemispheres, and lead to episodic pulses in the meridional transport. In the Southern Hemisphere the polar vortex region appears to be isolated from such events throughout the winter so that there is very little transport of heat or tracers into the vortex from lower latitudes. The resulting dynamical and chemical isolation of the region is important for occurrence of the Antarctic springtime ozone hole.

The zonal flow deceleration associated with sudden warming of the polar stratosphere represents a temporary enhancement of the normal westward wave-drag force in the winter stratosphere. Breaking of planetary (Rossby) waves primarily produces this westward zonal force. These waves produce a westward force that is hemispheric in scale. Owing to the rapid rotation of the earth, the response to a westward force is to produce a poleward drift. The

Coriolis force associated with the poleward drift approximately balances the wave-drag force, and maintains the zonal winds weaker than they would be in the absence of wavebreaking. Further, by mass continuity, the meridional drift leads to upward motion in low latitudes and downward motion in high latitudes. Thus, the extratropical wave-drag force acts as a sort of "pump"; it pulls air upward in the tropics, and pushes air downward at high latitudes. Through adiabatic cooling accompanying the rising motion, the low latitudes are maintained at temperatures below radiative equilibrium; while through adiabatic warming accompanying the sinking motion, the high latitudes are maintained at temperatures above radiative equilibrium in the winter hemisphere.

Planetary waves excited by orographic and thermal forcing in the troposphere cannot propagate into the westward wind regime of the summer stratosphere. In the summer stratosphere wave disturbances are much weaker than in the winter stratosphere. There is evidence, however, [*Alexander and Rosenlof*, 1996] of a weak westward accelerating force exerted by gravity wavebreaking, which drives a weak poleward transport circulation in the summer stratosphere.

The meridional and vertical circulation produced by wavebreaking (see Figure 6) advects trace constituents meridionally and vertically. It is this circulation that is here referred to as the *transport circulation* (it is often referred to in the literature as the Brewer-Dobson circulation in honor of the work of *Brewer* [1949] and *Dobson* [1956]. It should be stressed that the transport circulation arises in response to the zonal forcing by wavebreaking in the stratosphere. Radiation does not force the circulation. Rather, radiation acts like a "spring" that resists the dynamical driving by the wave forcing. Because the upward motion in the tropics is accompanied by radiative heating, and the downward circulation at high latitudes is accompanied by radiative cooling, this meridional circulation is sometimes referred to as the *diabatic circulation*. Since, however, this terminology suggests radiative forcing, it is best avoided.

An implication of the above conceptual model of the transport circulation is that for sufficiently long timescales the mass and constituent transport across isentropes in the lower stratosphere is largely controlled by wave-induced zonal forcing in the overlying middle atmosphere. Thus, the upwelling in the tropical stratosphere is not caused by local diabatic heating; rather, it is nonlocally forced by the extratropical pump [*Haynes, et al.*, 1991; *Holton, et al.*, 1995]. Through adiabatic cooling the upwelling maintains temperatures below radiative equilibrium in the tropical middle atmosphere, and thus produces net radiative heating.

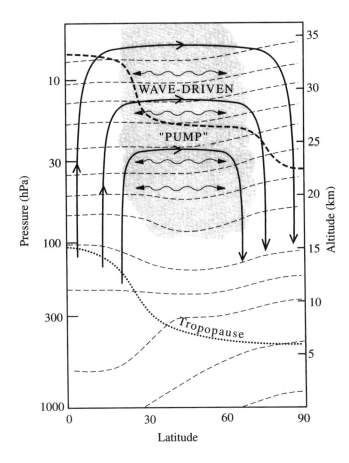

**Figure 6.** Schematic latitude-height cross section showing the wave-driven transport circulation (solid lines with arrows) of the winter stratosphere. Shading indicates region of planetary wave-breaking called the "surf zone". Light dashed lines are isentropes; dotted line indicates the tropopause. Wavy double-headed arrows indicate meridional mixing by wavebreaking. Heavy dashed line shows a constant mixing ratio surface for a long-lived tracer.

This heating must, however, be regarded as a consequence rather than a cause of the nonlocally driven upward motion. (This paradigm is subject to some caveats in the equatorial region [*Plumb and Eluszkiewicz*, 1999]).

The observed annual cycle of temperature in the tropical lower stratosphere provides strong support for this paradigm of upwelling driven nonlocally by extratropical wave-induced forces. Owing to adiabatic cooling, the lowest temperatures should occur at the time of strongest upwelling, and highest temperatures at the time of weakest upwelling. In fact, tropical lower stratospheric temperatures are observed to vary annually, with lowest temperatures over the entire tropics during Northern Hemisphere winter, and highest during Northern Hemisphere summer [*Yulaeva, et al.*, 1994]. This annual cycle is consistent with the annual

cycle in the extratropical wave-induced stratospheric forcing, which is stronger in the Northern winter than in the Southern winter owing to the strong Rossby-wave activity of the Northern Hemisphere winter.

## 4. TRANSPORT CIRCULATION OF THE MESOSPHERE

In the mesosphere breaking of small-scale vertically propagating gravity waves produces the zonal wave-drag force. Unlike Rossby waves, which as described above produce a one-signed (westward directed) forcing, gravity waves can produce either eastward or westward zonal forcing. Gravity waves that propagate eastward relative to the zonal flow produce eastward forcing when they break, while waves that propagate westward relative to the zonal flow produce westward forcing. Gravity waves break in the stratosphere as they approach their critical levels where the phase speed relative to the mean flow (i. e., the Doppler shifted phase speed) approaches zero. Thus, zonal winds in the stratosphere act as filters for upward propagating gravity waves. Only waves with sufficiently large Doppler shifted phase speed can survive to propagate into the mesosphere. Hence, owing to the seasonal reversal in zonal winds in the stratosphere, gravity wave breaking in the mesosphere is dominated by eastward propagating waves in the summer and westward propagating waves in the winter. The resulting forcing tends to decelerate the mean zonal flow toward zero in both the summer and winter hemispheres. For solstice conditions, the mean zonal flow is approximately in steady state, so that time derivatives in the governing equations are negligible and the zonal momentum and thermodynamic balances can be approximated as:

$$\begin{bmatrix} \text{Coriolis Force} \\ \text{(meridional drift)} \end{bmatrix} \approx \begin{bmatrix} \text{Zonal forcing} \\ \text{(wavebreaking)} \end{bmatrix}$$

$$\begin{bmatrix} \text{Adiabatic heating} \\ \text{(vertical motion)} \end{bmatrix} \approx \begin{bmatrix} \text{Radiative cooling} \\ \text{(solar + infrared)} \end{bmatrix}$$

From these relations it is clear that if gravity wave drag, and hence the zonal wave-drag force vanishes in the mesosphere, the meridional transport circulation must vanish and the temperature will be in radiative equilibrium. Thus, the existence of a meridional circulation again depends on wave-induced zonal forcing.

A balance between the zonal wave-drag force and the Coriolis force caused by the meridional drift maintains the observed mean zonal wind distribution. As illustrated in Figure 7, the meridional drift is from the summer to the

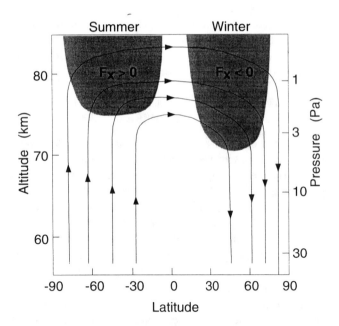

**Figure 7.** Schematic latitude-height cross section showing the solstice season pole-to-pole wave-driven transport circulation in the mesosphere (solid lines with arrows). Shading shows regions of gravity wave breaking with zonal forcing direction indicated by $F_x > 0$ for summer and $F_x < 0$ for winter.

winter hemisphere. This is a consequence of the strong eastward forces exerted by gravity wave-breaking in the summer hemisphere and westward forces in the winter hemisphere. These are balanced by a westward Coriolis acceleration in the summer hemisphere and an eastward Coriolis acceleration in the winter hemisphere .

In summary, the presence of a gravity wave-induced zonal force in both the summer and winter mesosphere exerts a drag on the mean zonal flow, causing a meridional drift down the gradient of the zonal mean geopotential field, from the summer pole to the winter pole. This meridional motion, through mass continuity, gives rise to vertical motion. Through adiabatic heating in winter and cooling in summer this field of vertical motion drives temperatures above their radiatively-determined values in winter and below in summer. Thus the gravity wave-induced force acts as a dynamical pump in the mesosphere to drive fluid upward in the summer hemisphere across the equator and downward in the winter hemisphere. The strength of this pumping is confirmed by noting that near the mesopause (~80 km) the observed temperature in July at the winter pole is ~90 K warmer than radiative equilibrium and at the summer pole is about 60 K colder than radiative equilibrium. It is only because of this dynamically induced cooling of the summer mesopause that noctilucent clouds can form near the mesopause at high latitudes in

summer. At the equinoxes, the circulation of the solstice season mesosphere is replaced by a much weaker two-cell circulation with tropical rising and high latitude sinking associated with weak westward wave drag in both hemispheres.

## 5. IMPLICATIONS FOR TRACE CONSTITUENT TRANSPORT

Transport processes are conveniently divided between those processes that involve mean motions of the atmosphere, or *advection,* and those processes that may be characterized as turbulent, or diffusive in nature. In the case of point sources, such as volcanic eruptions, the distinction is quite clear; advection moves the center of mass of the plume along the direction of the average wind, while turbulent diffusion disperses the plume in the plane orthogonal to the average wind. On a global scale, however, the distinction between advective and diffusive processes is not always clear.

Here it is assumed that the zonally averaged global-scale mass circulation (referred to above as the transport circulation) represents the advective motion, while breaking Rossby and gravity waves provide the turbulent mixing. It was argued above that in the middle atmosphere the global-scale transport circulation is itself driven by wave-induced zonal forces associated with Rossby waves and gravity waves. Thus, the advective and diffusive transport can not be treated as independent processes. Understanding of transport involves both wave and mean-flow transport effects.

In dynamical studies it is usual to characterize a chemical constituent by the volume mixing ratio (or mole fraction), defined as $\chi \equiv n_T / n_A$, where $n_T$ and $n_A$ designate the number densities for the trace constituent and air, respectively. The mixing ratio is conserved following the motion in the absence of sources and sinks, and hence satisfies the simple tracer continuity equation

$$\frac{D\chi}{Dt} = S$$

where $S$ designates the sum of all chemical sources and sinks, and the time derivative is taken following the motion. Letting the meridional and vertical components of the transport circulation be designated as $(\bar{v}^*, \bar{w}^*)$ the zonally averaged tracer continuity equation can be expressed as [e. g., *Andrews et al.,* 1987]

$$\frac{\partial \overline{\chi}}{\partial t} = -\bar{v}^* \frac{\partial \overline{\chi}}{\partial y} - \bar{w}^* \frac{\partial \overline{\chi}}{\partial z} + \bar{S} + \frac{1}{\rho_0} \mathbf{\nabla} \cdot \mathbf{M}$$

Here, overbars represent zonal averages, y is the northward coordinate, $\rho_0$ is air density and **M** represents the meridional and vertical eddy fluxes of tracer. Thus, the first and second terms on the right side represent meridional and vertical advection by the transport circulation, respectively. The third term represents the net zonally averaged chemical sources and sinks, and the fourth term represents the diffusive effects of eddies plus any advective effects not represented by the transport circulation.

In order to appreciate the role of the wave-induced global circulation in determining the distribution of long-lived tracers in the middle atmosphere, it is useful to consider a hypothetical atmosphere in which there are no wave motions, and hence no wave-induced zonal force. In that case the middle atmosphere would relax to radiative equilibrium, the transport circulation would vanish, and the distribution of the tracer would be determined at each altitude by a balance between slow upward diffusion and photochemical destruction. Thus, tracer mixing ratio surfaces would, in an annual mean, tend to be close to horizontal. This is to be contrasted to observed distributions which are characterized by mixing ratio surfaces that are bowed upward in the tropics and slope downward toward both poles.

The global-scale circulation provides a good qualitative and quantitative model for the observed distribution of long-lived trace constituents such as methane and nitrous oxide. These constituents, as shown by the heavy dashed line in the schematic of Figure 6, have elevated mixing ratio surfaces in the tropical upwelling region and downward displaced mixing ratio surfaces in the high latitude subsidence region. These features are especially notable in the winter hemisphere, where planetary wave breaking is strong. In the surf zone at midlatitudes wave breaking not only generates a westward zonal force driving the mean meridional mass circulation, but also causes meridional transport and mixing along the isentropes thereby flattening the surfaces of constant mixing ratio as shown in Figure 6.

The above transport paradigm is clearly supported by long-lived tracer observations from the Upper Atmosphere Research Satellite (see review by *Dessler et al.*[1998]) . For example, methane shows surfaces of constant mixing ratio that are displaced upward in the tropics and slope downward toward the poles in the extratropics consistent with advection by the cross-isentropic global-scale circulation. In middle latitudes there are regions in which tracer mixing ratio surfaces are nearly horizontal, reflecting the horizontal homogenizing role of meridional dispersion by planetary wave breaking in the surf zone. The surf zone is bounded at both low and high latitudes by strong meridional tracer gradients. The existence of such gradients is evidence that

there is only weak mixing into and out of the tropics and into and out of the polar winter vortex. Thus these locations are sometimes referred to as "transport barriers" (e. g., [*Polvani, et al.,* 1995]). The strong winds, and strong wind shears that occur along the transport barriers at the subtropical and polar edges of the surf zone act to suppress wave breaking, and hence to minimize mixing and sustain the strong gradients at those locations. On the other hand, rapid meridional dispersion occurs in the "surf-zone" of the winter hemisphere extratropics where planetary wave activity is strong [*McIntyre and Palmer*, 1984].

A fundamental aspect of the conceptual model of Figure 6 is the occurrence of an upwelling region in the tropics. This upwelling is driven nonlocally by extratropical wave breaking, but is isolated from meridional mixing with air from the extratropics by the subtropical transport barrier [*Plumb*, 1996]. It was pointed out earlier that the upward transport in the tropics must undergo an annual cycle, with stronger tropical upwelling in northern winter than in northern summer. As a consequence the tropical tropopause temperature varies annually. The coldest temperatures occur during northern winter when the upward mass flux is at its maximum, and warmest temperatures occur in northern summer when the upward mass flux is at its minimum. Because the saturation mixing ratio for water is strongly temperature dependent, and air passing through the tropical tropopause tends to be "freeze dried" to its saturation mixing ratio at the temperature of the tropopause, the mixing ratio for water vapor entering the tropical stratosphere also undergoes a strong annual cycle. As air is pulled upward across the tropical tropopause by the wave-induced global circulation the saturation mixing ratio at the tropopause is imprinted on the rising air in a manner similar to the imprint of a signal on a magnetic tape as the tape passes the recording head. The imprinted signal (water vapor mixing ratio) is then pulled upward into the stratosphere by the nonlocally driven upwelling. In the absence of mixing the seasonal cycle should be preserved as the air is slowly pulled upward into the middle atmosphere. Satellite data [*Mote, et al.,* 1996] show that patterns of alternating low and high water vapor mixing ratio move upward with height at a rate of about a kilometer per month, and that this tape-recorder signal is recognizable at least to the 10 hPa level.

The fact that the tape-recorder signal remains distinct for more than a year, as the air is pulled upward, is strong evidence that the mixing in of air from higher latitudes must be rather slow and that vertical mixing in the tropical stratosphere is very small. Recent work suggests that there is significant mixing between the tropics and higher latitudes over the course of a year between the tropopause and about 20 km altitude, but that in-mixing is weak above that level until about the 30 km level [*Mote, et al.,* 1998]. But, owing to the decrease of density with altitude, and the tendency of Rossby waves to be ducted toward the high refractive index produced by weak tropical winds, planetary wave intensity in the subtropics increases in the middle and upper stratosphere. Wave breaking then becomes common, and increased transport and mixing occurs across the sub-tropics. Thus, the tape recorder signal is gradually erased as air is drawn upward above the 30 km level. This mixing of air across the subtropical transport barrier occurs primarily when wave breaking associated with large amplitude Rossby waves irreversibly pulls filaments of tropical air into the extratropics, and filaments of extratropical air into the tropics. Similar processes of mixing caused by fila-mentation occur at the edge of the winter polar vortex.

The influence of the extratropical pumping by planetary wave drag extends well into the mesosphere in the winter hemisphere. The transport circulation lofts air to the meso-sphere in the tropics, where it drifts poleward and returns to the stratosphere primarily in the polar regions [*Bacmeister, et al.,* 1995]. Planetary wave mixing across the boundary of the polar vortex in the upper stratosphere and lower meso-sphere tends to modify tracer distributions in the polar winter region. In the following summer these modified dis-tributions are advected upward into the mesosphere by the gravity wave induced mesospheric transport circulation shown schematically in Figure 7.

Whereas planetary wavebreaking produces meridional mixing, gravity wavebreaking produces vertical mixing [*Lindzen,* 1981]. The efficiency of vertical tracer transport by gravity wave generated turbulent mixing, compared to advection by the transport circulation remains uncertain [*McIntyre,* 1989]. However, it is clear that vertical diffusion associated with gravity wave breaking must be important for transporting constituents downward from the thermo-sphere into the mesosphere. For long-lived tracers, such as water vapor, meridional and vertical advective transport by the global transport circulation appears to dominate over vertical diffusion throughout most of the mesosphere [*Holton and Schoeberl,* 1988]. A detailed understanding of the relative roles of the wave-driven global transport circu-lation and small-scale turbulent mixing in the mesosphere awaits better global observations of both dynamical pro-cesses and trace constituent behavior.

*Acknowledgements.* This work was supported by the NASA Atmospheric Chemistry and Analysis Program, NASA Grant NAG-1-2193.

## REFERENCES

Alexander, M. J. and J. R. Holton, A model study of zonal forcing in the equatorial stratosphere by convectively induced gravity waves, *J. Atmos. Sci.,* 54, 408-419, 1997.

Alexander, M. J. and K. H. Rosenlof, Nonstationary gravity wave forcing of the stratospheric zonal mean wind, *J. Geophys. Res.,* 101, 23,465-23,474, 1996.

Andrews, D. G., J. R. Holton and C. B. Leovy, *Middle Atmospheric Dynamics,* Academic Press, Orlando, 489, pp., 1987.

Bacmeister, J. T., M. R. Schoeberl, M. E. Summers, J. R. Rosenfield and X. Zhu, Descent of long lived trace gases in the winter polar vortex, *J. Geophys. Res. 100,* 11669-11684, 1995.

Brewer, A. M., Evidence for a world circulation provided by the measurements of helium and water vapor distribution in the stratosphere, *Quart. J. Roy. Meteor. Soc.,* 75, 351-363, 1949.

Dessler, A. E., M. D. Burrage, J.-U. Grooss, J. R. Holton, J. L. Lean, S. T. Massie, M. R. Schoeberl, A. R. Douglass and C. H. Jackman, Selected Science highlights from the first 5 years of the Upper Atmosphere Research Satellite (UARS) program, *Revs. Geophys.,* 36, 183-210, 1998.

Dewan, E. M., et al., MSX satellite observations of thunderstorm-generated gravity waves in mid-wave infrared images of the upper stratosphere, *Geophys. Res. Lett.,* 25, 939-942, 1988.

Dobson, G. M. B., Origin and distribution of polyatomic molecules in the atmosphere, *Proc. Roy. Soc. London, A236,* 187-193, 1956.

Fritts, D. C., Gravity wave saturation in the middle atmosphere: A review of theory and observations, *Rev. Geophys.,* 22, 275-308, 1984.

Fritts, D. C., and J. A. Werne, Turbulence dynamics and mixing due to gravity waves in the lower and middle atmosphere, (this volume), 2000.

Haynes, P. H., C. J. Marks, M. E. McIntyre, T. G. Shepherd and K. P. Shine, On the "downward control" of extratropical diabatic circulations by eddy-induced mean zonal forces, *J. Atmos. Sci.,* 48, 651-678, 1991.

Holton, J. R., An *Introduction to Dynamic Meteorology,* Academic Press, Orlando, 511 pp., 1992.

Holton, J. R., P. H. Haynes, M. E. McIntyre, A. R. Douglass, R. B. Rood and L. Pfister, Stratosphere-troposphere exchange, *Reviews of Geophysics,* 33, 403-439, 1995.

Holton, J. R. and M. R. Schoeberl, The role of gravity wave generated advection and diffusion in the transport of tracers in the mesosphere, *J. Geophys. Res.,* 93, 11075-11082, 1988.

Isler, J.R., M.J. Taylor, and D.C. Fritts, Observational evidence of wave ducting and evanescence in the mesosphere, *J. Geophys. Res.,* 102, 26,301-26,313, 1997.

Leovy, C. B., Simple models of thermally driven mesospheric circulation, *J. Atmos. Sci.*, *21*, 327-341, 1964.,

Lindzen, R. S., Turbulence and stress owing to gravity wave and tidal breakdown, *J. Geophys. Res.*, *86*, 9707-9714, 1981.

McIntyre, M. E., On dynamics and transport near the polar mesopause in summer, *J. Geophys. Res.*, *94*, 14617-14628, 1989.

McIntyre, M. E. and T. N. Palmer, The 'surf zone' in the stratosphere, *J. Atmos. Terr. Phys.*, *46*, 825-849, 1984.

McLandress, C., M.J. Alexander, and D.L. Wu, Microwave Limb Sounder observations of gravity waves in the stratosphere: A climatology and interpretation, *J. Geophys. Res.*, (in press), 2000.

Moreels, G., and M. Herse, Photographic evidence of waves around the 85 km level, *Planet. Space Sci.*, 25, 265-273, 1977.

Mote, P. W., T. J. Dunkerton, M. E. McIntyre, E. A. Ray, P. H. Haynes and J. M. I. Russell, Vertical velocity, vertical diffusion, and dilution by midlatitude air in the tropical lower stratosphere, *J. Geophys. Res.*, *103*, 8651-8666, 1998.

Mote, P. W., K. H. Rosenlof, M. E. McIntyre, E. S. Carr, J. C. Gille, J. R. Holton, K. H. Kinnersley, H. C. Pumphrey, J. M. I. Russell, and J. W. Waters, An atmospheric tape-recorder: the imprint of tropical tropopause temperatures on stratospheric water vapor, *J. Geophys. Res. 101*, 3989-4006, 1996.

Plumb, R. A., A "tropical pipe" model of stratospheric transport, *J. Geophys. Res. 101*, 3957-3972, 1996.

Plumb, R. A. and J. Eluszkiewicz, The Brewer-Dobson circulation: dynamics of the tropical upwelling, *J. Atmos. Sci.*, *56*, 868-890, 1999.

Polvani, L. M., D. W. Waugh and R. A. Plumb, On the subtropical edge of the stratospheric surf zone, *J. Atmos. Sci.*, *52*, 1288-1309, 1995.

Swenson, G.R., M.J. Alexander, and R. Haque, Dispersion imposed limits on atmospheric gravity waves in the mesosphere: Observations from OH Airglow, *Geophys. Res. Lett.*, *27*, 875-878, 2000.

Taylor, M. J. and M. A. Hapgood, Identification of a thunderstorm as a source of short period gravity waves in the upper atmospheric nightglow emissions, *Planet. Space Sci.*, 36, 975-985, 1988.

Yulaeva, E., J. R. Holton and J. M. Wallace, On the cause of the annual cycle in the tropical lower stratospheric temperature, *J. Atmos. Sci., 51*, 169-174, 1994.

James R. Holton, Department of Atmospheric Sciences, Box 351640, University of Washington, Seattle, WA 98195

Joan Alexander, Colorado Research Associates, 3380 Mitchell Lane, Boulder, Colorado

# A Contemporary Assessment of the Mesospheric Energy Budget

Martin G. Mlynczak

*Atmospheric Sciences Research, NASA Langley Research Center, Mail Stop 420, Hampton, VA 23681-2199*

The energy budget of the terrestrial mesosphere is a frontier of research in the atmospheric sciences. A fundamental difference between the mesospheric energy budget and that in lower atmospheric regions is that the primary mesospheric heating and cooling mechanisms involve processes that are far removed from local thermodynamic equilibrium (LTE). In the past decade there has been a great advance in the knowledge of many of these processes. For example, we currently believe that exothermic chemical reactions generate more heat in the upper mesosphere than is provided directly by solar radiation. We also believe that radiative emission and cooling by carbon dioxide ($CO_2$) is governed primarily by collisional energy transfer from atomic oxygen, and that the $CO_2$ molecule is a variable species in the upper mesosphere. Airglow is thought to significantly reduce the heating efficiencies of both solar radiation and exothermic reactions. Dynamical processes including gravity waves and tides are also clearly important. To date there is not an extant data set from which the mesospheric energy budget can be confidently derived on a global basis, and thus we know very little concerning the relative importance of the various sources and sinks of energy from observation. In this paper we will review the current status of knowledge of the energy balance in the mesosphere and will contrast it with the rather well-known energy balance in the stratosphere. Contemporary calculations are given of solar energy deposition and heating, infrared radiative cooling, exothermic chemical reactions, and the global mean radiative balance, as derived from the Solar-Mesosphere Explorer data set.

## 1. INTRODUCTION

A fundamental scientific problem is to understand the balance and flow of energy through the atmosphere and Earth system. The energy budget governs the thermal structure, the dynamics, and the chemical composition of the atmosphere. It is interesting to note that the first quantitative space-based measurements of the Earth system were of the planet's energy balance made by the low-resolution sensors designed by V. E. Suomi at the University of Wisconsin in the late 1950's and early 1960's [*Suomi and Shen*, 1962]. These sensors provided the first detailed look at the balance of incoming solar energy and outgoing infrared energy and permitted the first quantitative evaluation of the planetary albedo. The results of these pioneering experiments confirmed that the Earth at low latitudes receives more solar energy than is radiated back to space and that more energy at high latitudes is radiated back to space than is received through solar insolation [*VonderHaar and Suomi*, 1969]. Thus, for equilibrium to be maintained in the planetary climate, equator-to-pole energy transport is required and this is the driver of the Earth's weather.

Atmospheric Science Across the Stratopause
Geophysical Monograph 123

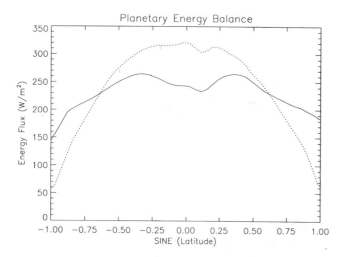

**Figure 1a**. Zonal mean absorbed solar radiation (dashed line) and outgoing infrared radiation (solid line) for the entire planet, as measured by the Earth Radiation Budget Experiment, for 1986. The data are displayed against the sine of latitude so that the abscissa is proportional to area.

It is also possible to examine the balance of energy in the terrestrial middle atmosphere. From measurements of the thermal structure, the chemical composition, and models of the infrared cooling processes, it is possible to compute the gain and loss of energy in the stratosphere and mesosphere. Shown if Figure 1 is just such a computation. Figure 1a represents the annual energy balance of the entire planet (absorbed solar radiation and emitted infrared radiation) as measured for the year 1986 by the Earth Radiation Budget Experiment [*Barkstrom*, 1984]. Figure 1b represents the balance of absorbed solar radiation and emitted infrared radiation for the stratosphere at equinox [*Mlynczak et al.*, 1999b]. Figure 1c depicts the balance between absorbed solar energy, energy deposited by exothermic chemical reactions, and emitted infrared radiation in the mesosphere, also at equinox. In all three panels the vertically integrated rate (W/m$^2$) of solar heating and infrared cooling is shown against the sine of the latitude such that the abscissa is proportional to area. These figures illustrate that the basic principle of an excess of absorbed solar radiation over infrared loss at low latitudes with an excess of infrared emission over solar insolation at high latitudes holds not only for the entire planet but also for the individual regions of the middle atmosphere. We can conclude from this that the need for equator-to-pole energy transport is a requirement for the entire atmosphere below 90 km. Figure 1 also illustrates the scale of energy driving the various atmospheric regions. On a planetary scale we are concerned with energies on the order of several hundred W/m$^2$, while in the stratosphere we are concerned with 10-15 W/m$^2$, and in the mesosphere with 4-5 mW/m$^2$. The nearly five order of magnitude

difference in the energy budgets of the troposphere and mesosphere reflects the corresponding differences in air density between these regions.

To say that the energy budget of the mesosphere remains a frontier in the atmospheric sciences does not imply that it has not been studied in depth. There is a long history of modeling studies and observations of mesospheric energetics extending back nearly 45 years. In fact, the basic processes that determine the mesospheric energy balance were generally known over 40 years ago. *Murgatroyd and Goody* [1958, hereafter MG] made the first computation of the mesospheric energy balance and considered absorption of solar ultraviolet radiation and infrared emission by carbon dioxide and ozone. MG included non-local thermodynamic equilibrium (non-LTE) effects in their computation of radiative cooling by CO$_2$. MG also recognized the importance of chemical potential energy and exothermic chemical reactions as sources of heat and purposely limited the altitude range of their calculation to below 90 km to avoid errors in the computed heating rates. *Kellog* [1961] postulated that the warm winter mesopause was at least partially maintained by energy release by exothermic reactions involving atomic oxygen transported from the thermosphere. *Hines* [1965] discussed the role viscous dissipation of internal gravity waves in the energy balance. *Shved* [1972] discussed the role of airglow in reducing the amount of energy available for heat in solar energy deposition and exothermic reactions.

Knowledge of the basic processes mentioned above has been refined over the years, and with the advent of large-scale computing facilities, numerical models of the

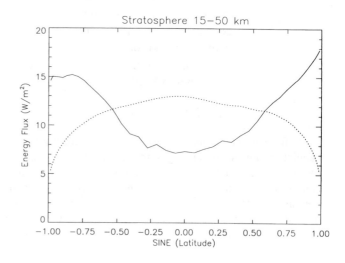

**Figure 1b**. Zonal mean absorbed solar radiation (dashed line) and outgoing infrared radiation (solid line) for the stratosphere, 15 to 50 km, as derived from the Upper Atmosphere Research Satellite, at equinox. The data are displayed against the sine of latitude so that the abscissa is proportional to area.

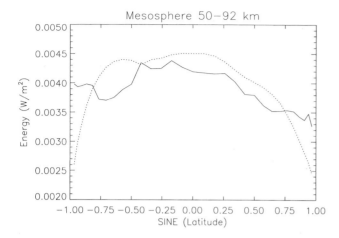

**Figure 1c.** Zonal mean absorbed solar radiation (dashed line) and outgoing infrared radiation (solid line) for the mesosphere, 50 to 92 km, as derived from the Solar-Mesosphere Explorer satellite, at equinox. The data are displayed against the sine of latitude so that the abscissa is proportional to area.

stratosphere, mesosphere, and lower thermosphere have been developed. Among the most detailed models to date is the three-dimensional Thermosphere-Ionosphere-Mesosphere-Electrodynamics General Circulation Model (TIME-GCM) which now extends from 30 to 500 km [*Dickinson et al.*, 1984; *Roble et al.*, 1987a; *Roble*, 1995; *Roble*, 2000, *this volume*]. Numerous two-dimensional models have been developed [e.g., *Garcia et al.*, 1992; *Garcia and Solomon*, 1994; *Apruseze et al*, 1984, 1982]. Such models include radiative, chemical, and dynamical processes in great detail. These models have been used to demonstrate the importance of the variability of solar energy [*Garcia et al.*, 1984], the role of Joule heating [*Roble* et al., 1987b], the role of exothermic chemical reactions [*Garcia and Solomon*, 1983], the importance of non-LTE infrared radiative transfer [*Apruzese et al.*, 1984], and the effects of gravity wave breaking [e.g., *Schoeberl et al.*, 1983].

In essence, the major radiative, chemical, and dynamical processes that govern the energy balance in the mesosphere are thought to be known. However, we still do not have a quantitative, global understanding of the magnitude and relative importance of these processes and their overall role in determining the structure and variability of the mesosphere.

Our goal in this paper is to provide a contemporary assessment of the status of knowledge of the mesospheric energy budget. A comprehensive study of the stratospheric energy balance has recently been completed [*Mertens et al.*, 1999; *Mlynczak et al.*, 1999b] using global observations of the stratosphere from instruments aboard the Upper Atmosphere Research Satellite (UARS). The planetary scale energy balance has been observed from space over

the past 15 years by the Earth Radiation Budget Experiment [*Barkstrom*, 1984] and is currently being observed by the Clouds and the Earth's Radiant Energy System (CERES) experiment [*Wielicki et al.*, 1996] as part of NASA's Earth Observing System (EOS). The mesospheric energy balance will be observed through space-based measurements to be made as a fundamental goal of NASA's Thermosphere-Ionosphere-Mesosphere Energetics and Dynamics (TIMED) mission beginning in late 2000 or early 2001. We intend to review the present knowledge and recent progress made in understanding of the mesospheric energy balance as a preview to global observations from which we hope to close the mesospheric energy budget.

We begin by reviewing the basic principles of conservation of energy as it relates to the mesosphere, followed by in-depth review of the physical processes including discussion of significant recent advances and important historical contributions. Calculations of these processes are then given followed by a summary of our shortcomings and plans to resolve these through observations, modeling, and laboratory research.

## 2. CONSERVATION OF ENERGY IN THE MESOSPHERE

In order to evaluate the energy balance in the mesosphere we must consider input of solar and terrestrial radiation, input of energy associated with dynamical phenomena, output of energy in the form of infrared and airglow emission, storage of energy in latent chemical form, conversion of energy, and energy transport. These processes are not new or different, especially the radiative phenomena, and are considered in almost any study of atmospheric energetics. However, the mesosphere is fundamentally different from the lower atmosphere in that, above about 60 km, collisional processes no longer dominate energy exchange between radiatively active molecules and their environment. This fact results in two important consequences. First, solar energy is not completely thermalized locally and immediately, and second, radiative cooling is not solely dependent on the thermal structure. Stated differently, non-local thermodynamic equilibrium (non-LTE) processes must be considered in the computation of the mesospheric energy balance for both solar heating and infrared cooling. Shown in Figure 2 is a diagram indicating the major flows of energy in the mesosphere. Solar energy (primarily electromagnetic radiation, but particle inputs cannot be totally neglected in the very upper mesosphere) is absorbed by $O_2$, $O_3$, and $CO_2$. The wavelengths that are absorbed range from the ultraviolet Lyman $\alpha$ at 121.5 nm absorbed by $O_2$ to the mid-infrared at 4.3 $\mu$m absorbed by $CO_2$.

## Energy Conservation in the Mesosphere

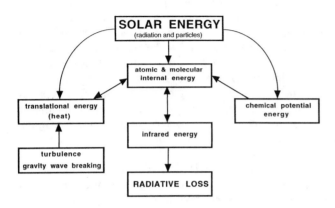

**Figure 2**. The basic flow of energy in the mesosphere.

Listed in Table 1 are the primary absorption features relevant to solar radiative energy inputs in the mesosphere.

As indicated in Figure 2, absorbed solar energy may be initially apportioned into three pools: translational energy (heat), internal energy of electronically or vibrationally excited photolysis product species, and chemical potential energy. The latter is defined as the solar energy used to break the chemical bonds and thus dissociate the absorbing species. Listed in Table 2 are the fractions of absorbed energy apportioned to each of the three pools immediately upon photolysis of ozone and molecular oxygen. For the main absorption bands of $O_3$ and $O_2$ in the mesosphere most energy initially shows up as chemical potential energy or internal energy and not as heat. Consequently, the heating rate primarily depends on the disposition of the energy in the chemical and internal energy pools. Also indicated in Figure 2 are heating processes associated with dynamical phenomena such as turbulent dissipation of breaking gravity waves. These phenomena are thought to provide substantial energy input to the mesosphere [e.g., *Lubken*, 1997], at least at some locations and times.

To demonstrate the significance of the energy in the chemical energy pool, we show in Figure 3 a plot of the ratio of the chemical potential energy to the mean thermal energy $(3/2 \ kT)$ as a function of altitude. At 80 km this ratio is about 1% and it rapidly increases with altitude such that above 85 km it exceeds 1.0, meaning that there is more energy in chemical potential form than thermal form. The data in Figure 3 were calculated using the Garcia-Solomon 2-D model [*Garcia et al.*, 1992; *Garcia and Solomon*, 1994]. The chemical potential energy, primarily carried by atomic oxygen, may not be realized until quite sometime after the original photon deposition due to the long chemical lifetime of O above about 75 km, allowing for significant transport of energy to occur through atomic oxygen transport. An evaluation of the energy balance in the mesosphere must account explicitly for the production, transport, and conversion of chemical energy.

The energy in the internal energy pool has essentially two fates. It may be physically quenched to heat through collisions or it may be radiated in the form of airglow emission, primarily from $O_2(^1\Delta)$, $O_2(^1\Sigma)$, and also from $CO_2$ at 4.3 μm. This emission significantly reduces the amount of energy available for heat thus giving a heating efficiency substantially less than 1.0. In addition, exothermic reactions and recombinations may form excited product species (such as $OH(\upsilon)$) which may radiate some of the original chemical potential energy before it is physically quenched to heat, thus reducing the chemical heating efficiency. Detailed models of production and loss of excited product species are required to accurately assess the heating due to solar and chemical processes. A complete description of the deposition and conversion of solar (and chemical) energy to heat and radiation is given by *Mlynczak and Solomon* [1993].

We have mentioned the potential for significant reduction in solar and chemical heating rates due to airglow emission from excited photolysis or reaction/recombination product species. It is important to emphasize that these airglow emissions, while removing large amounts of energy from the atmosphere, do not result in atmospheric cooling. The airglow energy is never in the translational energy pool. True radiative cooling occurs only when there is a conversion of translational energy to internal energy of a radiatively active species followed by spontaneous emission of radiation and loss of energy from the volume of atmosphere under consideration. Often the process of radiative cooling involves the consideration of complex radiative exchange between atmospheric layers, as indicated by the vertical double arrow in Figure 2. Radiative cooling in the mesosphere is driven primarily by non-LTE processes in the carbon dioxide molecule. Cooling by ozone (9.6 μm) and water vapor (6.3 μm

**Table 1.** Primary absorbers, features, and wavelength ranges solar radiation absorption in the mesosphere.

| Absorber | Feature | Wavelength Range |
|---|---|---|
| $O_3$ | Hartley band | 203 - 305 nm |
|  | Huggins band | 305 - 397 nm |
|  | Chappuis band | 397 - 850 nm |
| $O_2$ | Schumann-Runge continuum | 130 - 175 nm |
|  | Schumann-Runge bands | 175 - 200 nm |
|  | Herzberg continuum | 200 - 240 nm |
|  | Lyman α | 121.5 nm |
|  | Atmospheric band | 762 nm |
| $CO_2$ | Mid-infrared bands | 2.0, 2.7, and 4.3 μm |

**Table 2**. Initial disposition of energy absorbed by ozone and molecular oxygen.

|                    | Heat | Chemical Potential | Internal |
|--------------------|------|--------------------|----------|
| Ozone              |      |                    |          |
| Hartley band       | 13%  | 24%                | 63%      |
| Huggins band       | 76%  | 24%                | 0%       |
| Chappuis band      | 76%  | 24%                | 0%       |
|                    |      |                    |          |
| Oxygen             |      |                    |          |
| Sch.-Runge contin. | 1%   | 72%                | 27%      |
| Sch.-Runge band    | 24%  | 76%                | 0%       |
| Lyman α            | 30%  | 50%                | 20%      |

vibration-rotation bands and the far-infrared rotational bands) is important in the lower mesosphere. In the lower thermosphere the cooling eventually transitions from $CO_2$-dominated to that dominated by the NO molecule in the vibration-rotation bands at 5.3 µm. Note the fine structure lines of atomic oxygen in the far-infrared are also important in lower thermospheric cooling. With the exception of the far-infrared $H_2O$ and possibly the atomic oxygen lines, non-LTE occurs and must be considered in evaluating the cooling rates.

The degree of departure from LTE in the mesosphere is obtained by examining the vibrational or electronic temperatures of the upper states of the significant emissions. Shown for illustration in Figure 4 are the electronic temperatures of the $O_2(^1\Delta)$ state which radiates at 1.27 µm and of the $O_2(^1\Sigma)$ state which radiates at 762 nm. These states are produced subsequent to ozone photolysis and are central to determining the efficiency of solar heating. Over this altitude range the kinetic temperature is usually between 320 and 150 K. In contrast, the electronic temperatures of these two molecular oxygen energy levels are well over 1000 K, which is consistent with the prodigious emission of airglow by these species in the daytime.

The occurrence of non-LTE places an additional requirement on the information necessary to solve the radiative transfer problems in order to attain accurate values of the infrared cooling, the solar heating, the chemical heating, or the airglow emission intensity. Specifically, the rates of exchange of atomic and molecular internal and translational energy must be well-known. These rates, which are computed analogously to bimolecular chemical reaction rates, are the product of the exchange rate "constant" and the density of the two colliding species. In many (if not most) cases, the rate "constants" are poorly known at mesospheric temperatures, while accurate knowledge of the rates (generally better than 10-20% for most) is required to accurately solve the non-LTE problems, and hence, accurately determine the energy balance of the mesosphere. Recent work has defined the

outstanding problems in non-LTE energy transfer kinetic parameters and we will discuss these later in this paper.

As mentioned above, the energy budget of the mesosphere is not well known in contrast to the energy balance of the stratosphere. This is primarily on account of the paucity of global observations from which to properly derive the mesospheric energy budget. In Table 3 we give a comparison of the major properties of the energy budgets of the stratosphere and mesosphere, which together comprise the middle atmosphere. As indicated by the double question marks in the table, there are some fundamental properties of the mesospheric energy balance that we just don't know. Primarily, we do not know, from observation, if the mesosphere is in global mean radiative equilibrium on timescales longer than a month. *Mlynczak et al.* [1999b] demonstrated that, to within the accuracy of the UARS observations, the stratosphere is very close to global mean radiative equilibrium on a monthly basis. Closely related to the global mean energy balance is the role of turbulent dissipation in the mesosphere. If this process is a significant, continuous global source of energy, then global mean radiative equilibrium cannot be expected to hold. It is primarily this question to which the upcoming TIMED mission can make the greatest contribution to advancing our knowledge of the middle atmosphere.

## 3. CURRENT ASSESSMENT OF THE MESOSPHERIC HEAT BALANCE

In this section we will review the status of knowledge of the major energy budget terms, including discussion of some of the major advances made over time. We emphasize that the intent here is not to provide an exhaustive, historical review of the literature, but rather to

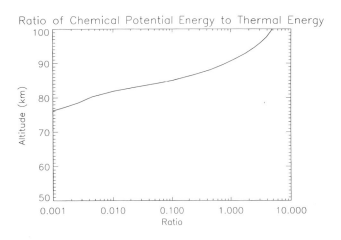

**Figure 3**. Ratio of chemical potential energy to mean thermal energy (3/2 kT), as a function of altitude, as derived from the Garcia-Solomon 2-D model.

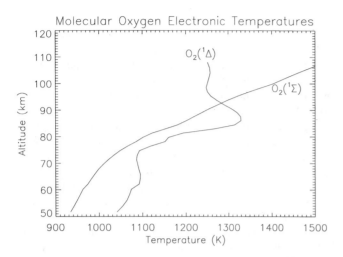

**Figure 4**. Electronic temperatures (kelvin) of the $O_2(^1\Delta \rightarrow {}^3\Sigma)$ transition at 1.27 μm and the $O_2(^1\Sigma \rightarrow {}^3\Sigma)$ transition at 762 nm.

give sufficient reference so that the progress and current status is properly represented. We will also show contemporary calculations of many of the processes which are based on observations made by the Solar-Mesosphere Explorer (SME) experiment that observed temperatures and airglow emissions in the mesosphere in the 1980s.

The first attempt to compute both solar heating and infrared cooling rates in the stratosphere and mesosphere for the purpose of diagnosing the radiative balance and the meridional circulation was in the pioneering work of *Murgatroyd and Goody* [1958]. Since that time there have been numerous studies in addition to significant advances in the knowledge of the distribution and abundance of atmospheric radiatively active species, radiative transfer modeling, spectroscopy, energy transfer kinetics, and basic physical processes.

Perhaps the most prescient contributions to the study of the mesospheric energy budget were made by P. Crutzen thirty years ago in two little-known references. First, in a discussion contained in *Houghton*, [1970] of a previous paper by Houghton [*Houghton*, 1969], Crutzen showed that

atomic oxygen may be very important in facilitating infrared emission by water vapor and in particular, infrared radiative cooling by carbon dioxide in the upper mesosphere. Second, *Crutzen* [1971] demonstrated that exothermic reactions involving both odd-oxygen and odd-hydrogen species, including the reaction of atomic hydrogen and ozone, play a substantial role in the heat budget of the mesosphere. These ideas have turned out to be fundamental. Crutzen's suggestion regarding atomic oxygen was prompted by laboratory measurements of energy transfer between atomic oxygen and molecular oxygen reported by *Kiefer and Lutz* [1967] and of reactions involving $CO_2$ and $O(^1D)$ reported by *Zipf* [1969]. In essence, the rate constant for translation-to-vibration energy transfer is very large for collisions between atomic oxygen and species such as $O_2$, $CO_2$, and NO. Crutzen recognized this and the impact it could have on the upper atmosphere radiation balance, demonstrating that the intensity of infrared radiation emitted by species such as $CO_2$ and $H_2O$ depends directly on the distribution of and collisions with a highly reactive and variable chemical species, atomic oxygen. It is often said that chemistry, dynamics, and radiation are coupled, but Crutzen's work gives new meaning to the term "coupling" as minor chemical species are shown to directly influence and alter radiation through energy transfer upon collisions.

*3.1 Solar Heating Rates*

The computation of solar heating rates requires (in principle) knowledge of the solar irradiance, the absorber amount, and the absorption cross section for each absorber as a function of wavelength. As indicated in Table 1 the primary absorbers in the mesosphere are ozone, molecular oxygen, and carbon dioxide. Typically absorption of ultraviolet radiation by ozone and molecular oxygen has always been considered. However, absorption in the near- and mid-infrared by $O_2$ (762 nm) and $CO_2$ (2.0, 2.7, and 4.3 μm) cannot be neglected. Shown in Figure 5 is a

**Table 3**. Comparison of stratospheric and mesospheric heat budgets.

| Process | Stratosphere | Mesosphere |
|---|---|---|
| Infrared cooling | $CO_2$, $O_3$, $H_2O$, *LTE* | $CO_2$, $O_3$, $H_2O$, *non-LTE* |
| Solar heating | $O_2$, $O_3$, $CO_2$, *LTE* | $O_2$, $O_3$, $CO_2$, *non-LTE* |
| Airglow losses | negligible | substantial |
| Heating efficiency | 1.0 | < 1.0 |
| Particle inputs | negligible | at high latitudes |
| Radiation balance | Global mean radiative equilibrium | ?? |
| Chemical heating | Immediate and local | Non-immediate, non-local |
| Turbulence/Dynamics | Negligible in global mean | ?? |
| $CO_2$ abundance | Constant mixing ratio in altitude | Variable mixing ratio in altitude |

computation of the solar heating due to absorption by $O_2$ and $CO_2$ in the infrared compared with heating due to absorption by ozone in the Hartley band. In the middle mesosphere the $CO_2$ and $O_3$ heating are comparable and the $O_2$ near-infrared heating accounts for an additional 10% over the $CO_2$ and $O_3$ heating. In all three cases full non-LTE calculations are required to account for substantial airglow loss. In the case of $CO_2$, the radiative coupling between atmospheric layers must also be taken into account. The infrared solar heating by $O_2$ and $CO_2$ is reviewed in detail by *Mlynczak and Marshall* [1996] and by *Lopez-Puertas et al.* [1990], respectively.

To compute the rate of heating due to the absorption of solar ultraviolet radiation the following general expression is used to compute the rate of energy deposition $\partial Q/\partial t$

$$\frac{\partial Q}{\partial t} = JN(h\nu - D) \quad (1)$$

where J is the photolysis rate (photons per second), N is the absorber density, $h\nu$ is the energy of the absorbed photon, and D is the energy required to dissociate the molecule. (Note that in the case of absorption by $CO_2$ in the mid-infrared D is zero.) The rate of heating $\partial T/\partial t$ at a given altitude or pressure is derived from the first law of thermodynamics from the expression

$$\varepsilon \frac{\partial Q}{\partial t} = \rho C_p \frac{\partial T}{\partial t} \quad (2)$$

where $\varepsilon$ is the heating efficiency at the specified altitude or pressure, $\rho$ is the density, and $C_p$ is the heat capacity at constant pressure. *Mlynczak and Solomon* [1993] give a simple parameterization of the efficiencies of solar heating as functions of pressure. The efficiency is less than 1.0 due to the loss of energy by airglow generated directly and indirectly by excited photolysis product species. It should be noted that the role of airglow in reducing the heating efficiency and the solar heating rates was also considered by *Apruzese et al.* [1984].

It has recently been shown by *Mlynczak* [1999] that the oxygen airglows provide a direct measure of solar energy deposition and heating independent of knowledge of the solar irradiance, the absorber amount, and the absorber cross sections. To illustrate this principle, we consider the $O_2(^1\Delta)$ airglow at 1.27 μm and assume (for illustration only) that it is solely generated as a consequence of ozone photolysis. With these assumptions, it is straightforward to show that

$$O_2(^1\Delta) = \frac{JO_3}{A + kO_2} \quad (3)$$

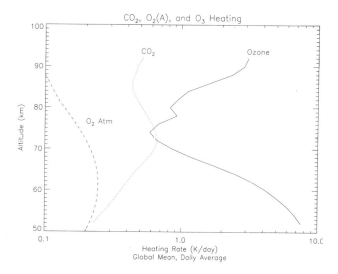

**Figure 5**. Global mean, daily average heating rates due to absorption of solar ultraviolet radiation by ozone, absorption of infrared radiation by carbon dioxide, and absorption of near-infrared radiation by molecular oxygen in the atmospheric band.

where J is the ozone photolysis rate, $O_3$ is the ozone density, A is the Einstein coefficient for spontaneous emission by $O_2(^1\Delta)$, k is the rate of physical quenching of $O_2(^1\Delta)$ by $O_2$, and $O_2$ is the molecular oxygen density. Combining (1), (2), and (3) we obtain

$$\frac{\partial T}{\partial t} = \frac{\varepsilon}{\rho C_p}(h\nu - D)(1 + \frac{kO_2}{A})V_\Delta \quad (4)$$

In (4) $V_\Delta$ is the volume emission rate of the $O_2(^1\Delta)$ airglow emission. From (4) we see that the heating rate (kelvins per day) may be derived directly from measurements of the oxygen airglow independent of knowledge of the solar irradiance, the ozone density, and the ozone absorption cross sections. This is strictly true only when the ozone abundance and the $O_2(^1\Delta)$ are jointly in photochemical steady state. All that is required is an accurate measure of the airglow emission and the atmospheric density. The reason for this is that the airglow is directly proportional to the product of the ozone photolysis rate and the ozone density $(JO_3)$ and this product contains the basic information on energy deposition. The complete theory including all sources of $O_2(^1\Delta)$ is developed in *Mlynczak* [1999].

*Mlynczak et al.* [2000a] have applied this technique to measurements made by the Near Infrared Spectrometer (NIRS) instrument which flew on the Solar-Mesosphere Explorer (SME) instrument in the 1980's. They show that energy deposition rates in the Hartley band of ozone derived directly from the airglow are in good agreement (generally better than 20% in the mesosphere) with those calculated directly by the Garcia-Solomon 2-D

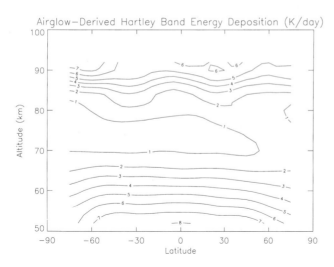

**Figure 6**. Zonal mean, daily average heating rates (kelvin per day) due to absorption of solar radiation by ozone in the Hartley band.

photochemical model [*Garcia et al.*, 1992; *Garcia and Solomon*, 1994]. Only at mid- to high latitudes in the winter hemisphere is the upper mesospheric energy deposition derived to be consistently and significantly larger than predicted by the model at almost all altitudes. Agreement between rates of energy deposition (when calculated in K/day on a pressure surface) is identical to agreement between ozone volume mixing ratios. Previous studies [e.g., *Solomon et al.*, 1983a] have shown that photochemical models tend to significantly underestimate the ozone abundance in the mesosphere and this has led to the concept of the "model deficit" of ozone. The results presented by *Mlynczak et al.* [2000a] suggest that the model deficit, at least in the comparisons shown, is not as significant as once thought. Apparently changes in model photochemistry, SME data calibrations, and airglow kinetic model parameters are responsible for achieving the current agreement. The odd-hydrogen chemistry of the mesosphere, thought to be central to determining the ozone abundance, is reviewed by *Summers and Conway* [2000, *this volume*]. Because ozone is a prime player in the mesospheric energy budget through its absorption of solar ultraviolet radiation and its participation in exothermic chemical reactions, both as a reactant and a product, accurate knowledge of the ozone abundance is fundamental to solving the energy balance problem.

Shown in Figure 6 is a zonal contour of the daily average solar heating rate (K/day) due to absorption in the Hartley band of ozone for March 1983. The heating rates are derived from the SME airglow independent of knowledge of the ozone abundance, the ozone absorption cross sections, and the solar irradiance. The heating rates exhibit the expected features, namely large heating rates

just above the stratopause where the ozone abundance is large, small heating rates in the mid-mesosphere where the ozone abundance experiences a local minimum, and large heating in the upper mesosphere corresponding with the secondary ozone maximum.

*3.2 Chemical Heating Rates*

As shown in Table 2, large pools of chemical potential energy are created upon photolysis of $O_2$ or $O_3$. In the case of the Schumann-Runge continuum, only 1% of the available photon energy is immediately made available for heat, while 72% shows up initially as chemical potential energy. Thus it should be no surprise that chemical reactions should play an important role in the energetics of the mesosphere. Long-range transport and subsequent recombination of atomic oxygen has long been known to occur in the upper mesosphere and lower thermosphere. *Kellog* [1961] suggested that the warm winter mesopause is maintained at least partially by energy release associated with recombination of atomic oxygen transported from the sunlit hemisphere.

The role of heating by exothermic chemical reactions involving odd-hydrogen species was apparently first mentioned by *Crutzen* [1971] and then again by *Brasseur and Offerman* [1986]. *Mlynczak and Solomon* [1991, 1993] carried out detailed studies of the role of mesospheric heating due to exothermic chemical reactions. They found that seven chemical reactions, listed in Table 4, are significant sources of heat. In particular, the reaction of atomic hydrogen and ozone may be the single largest source of heat between 83 and 95 km altitude. Furthermore, the heating due to exothermic chemical reactions is competitive with, and between about 70 and 95 km, exceeds the heating due directly to solar radiation, as computed from the Garcia-Solomon 2-D model and is shown in Figure 7.

The daily average chemical heating rate (in kelvin per day) at equinox due to the 7 reactions listed in Table 4 is shown in Figure 8. These rates are computed from the abundance of atomic hydrogen and atomic oxygen derived

**Table 4**. Exothermic reactions important in the mesospheric energy budget.

| Reaction | Exothermicity (kcal/mole) |
|---|---|
| $H + O_3 \rightarrow OH + O_2$ | -76.90 |
| $H + O_2 + M \rightarrow HO_2 + M$ | -49.10 |
| $O + HO_2 \rightarrow OH + O_2$ | -53.27 |
| $O + OH \rightarrow H + O_2$ | -16.77 |
| $O + O + M \rightarrow O_2 + M$ | -119.40 |
| $O + O_2 + M \rightarrow O_3 + M$ | -25.47 |
| $O + O_3 \rightarrow O_2 + O_2$ | -93.65 |

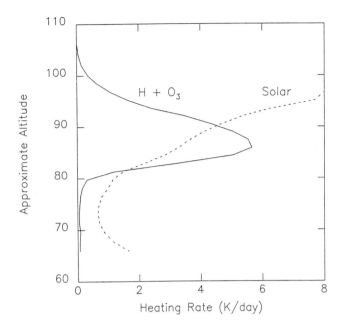

**Figure 7**. Comparison of daily average heating rates due to the absorption of solar ultraviolet radiation and to the reaction of atomic hydrogen and ozone, as computed from the Garcia-Solomon 2-D model, after *Mlynczak and Solomon* [1993].

from the SME airglow data [*Thomas*, 1990]. We assume photochemical steady state at noon and midnight in order to derive the heating due to all seven reactions. This assumption permits computation of heating from all reactions using only the H and O data. Over the 80 to 92 km region covered by the SME data the daily average heating ranges from 4 K/day to 16 K/day, which is in reasonable agreement with estimates from model computations (e.g., *Mlynczak and Solomon*, [1993]). The total solar plus chemical heating rates (daily average) for the mesosphere between 50 and 92 km at equinox are shown in Figure 9. The sharp gradient in heating just below 80 km is likely due to the merging of two separate data sets, one of solar heating, and one of chemical heating, which do not have the same vertical resolution.

### 3.3 Radiative Cooling Rates

Radiative cooling in the mesosphere is accomplished primarily through emission by carbon dioxide at 15 μm. Between about 75 km and 110 km emission by $CO_2$ is the only significant cooling mechanism. In the lower mesosphere cooling by ozone and by water vapor is important in addition to cooling by $CO_2$. The modern-era computations of radiative cooling by carbon dioxide began with the doctoral theses of *Drayson* [1966] and *Kuhn* [1966]. Both Drayson and Kuhn evaluated non-LTE cooling by carbon dioxide. Drayson in particular was

among the first to apply line-by-line computations in infrared radiative cooling calculations, a practice that is standard today.

As mentioned above, Crutzen [*Houghton*, 1970] suggested that atomic oxygen would be important in facilitating infrared radiative cooling. The key to understanding just how large this effect could be depends on the rate constant for vibrational relaxation by atomic oxygen,

$$CO_2(010) + O \rightarrow CO_2(000) + O + (667 \text{ cm}^{-1}) \quad (5)$$

This rate is very difficult to measure in the laboratory owing to the difficulty of simultaneously producing vibrationally excited $CO_2$ in the presence of atomic oxygen and then determining the decay rates, especially at low temperatures (~160 K) typical of the mesosphere. Crutzen, in his comment on Houghton's paper, showed that the calculated cooling rate at 100 km altitude could vary by a factor of 4 depending on the value of the O-$CO_2$ quenching rate. The upper limit for this rate assumed by Crutzen as 6 x $10^{-8}$ atm sec or 6.2 x $10^{-13}$ cm$^3$s$^{-1}$, which is about a factor of 4 lower than the middle of the contemporary range of this rate. The temperature dependence of this rate is unknown. We wish to emphasize that while the rate of quenching of $CO_2$ by O is often focused upon (because it is the quantity that is measurable in the laboratory), it is the collisional excitation of $CO_2$ by O that provides the internal energy that is radiated, thereby cooling the atmosphere. The collisional excitation rate is determined from the collisional quenching rate by applying the principle of detailed balance.

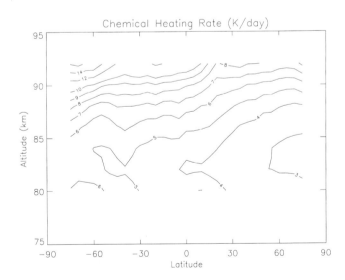

**Figure 8**. Zonal mean, daily average rates of heating due to the seven exothermic chemical reactions listed in Table 4, as derived from SME data.

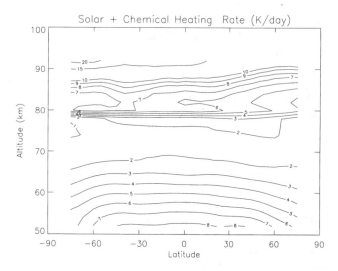

**Figure 9.** Zonal mean, daily average rates of heating due to the seven exothermic chemical reactions and to the absorption of solar radiation.

The next major advance in the understanding of the non-LTE nature of $CO_2$ emission came from the paper by *Sharma and Wintersteiner* [1990]. They analyzed measurements of infrared radiation emitted by $CO_2$ made during a sounding rocket flight. Sharma and Wintersteiner concluded that the quenching rate of $CO_2$ vibrations by atomic oxygen must be much larger than previously assumed in order to explain the observed radiance measurements. They suggested a rate constant of $6 \times 10^{-12}$ $cm^3s^{-1}$ at 300 K, which is more than an order of magnitude larger than previous estimates. *Shved et al.* [1991] measured a rate of $1.4 \times 10^{-12}$ $cm^3s^{-1}$ in the laboratory at 300 K, thus confirming a "large" value for this rate. We note that this large value was particularly relevant to planetary science in that *Keating and Bougher* [1992] used these results to reduce much of the uncertainty in the heat budget of the Venus thermosphere.

The results of *Sharma and Wintersteiner* were extended by *Rodgers et al.* [1992] and *Lopez-Puertas et al.*, [1992b] who analyzed data from the Atmospheric Trace Molecule Spectroscopy (ATMOS) experiment that flew on the space shuttle in the 1980's. ATMOS is a solar occultation experiment capable of measuring the populations of the ground and first excited states of the $CO_2$. *Rodgers et al.* used these measurements to directly calculate the vibrational temperatures of the fundamental and first hot bands of the carbon dioxide bending ($\nu_2$) mode. They showed the vibrational temperatures of these two states, to within the uncertainty of the measurements, to be essentially equal to the kinetic temperature below 95 km. Based on the detailed models of *Lopez-Puertas et al.* [1986a, 1986b, 1992a], the only plausible explanation for this "near-LTE" behavior could be a large rate of

collisional quenching of $CO_2$ vibrations by atomic oxygen. It is now thought that the value of this rate is between 1.5 and $6 \times 10^{-12}$ $cm^3s^{-1}$. A more complete review of measurements and inferences of this critical rate is given in *Mlynczak et al.* [2000b].

In order to compute the rate of $CO_2$ cooling in the mesosphere the $CO_2$ abundance must also be known. $CO_2$ has a long lifetime against photolysis in the mesosphere, and, based on photochemical considerations, the standard expectation is that the $CO_2$ concentration should be nearly well-mixed below the turbopause ($\sim 105$ km). Above the turbopause diffusive separation occurs based on the molecular mass of each species and the mixing ratio is anticipated to deviate from the well-mixed value. This standard picture has been called into question by the analyses reported by *Lopez-Puertas et al.* [1989] based on observations of non-LTE $CO_2$ emission at 4.3 μm made by the the Stratospheric and Mesospheric Sounder (SAMS) instrument on the Nimbus VII satellite and by the Improved Stratospheric and Mesospheric Sounder (ISAMS) instrument on the UARS satellite. These analyses show that the $CO_2$ mixing ratio in the upper mesosphere is not well-mixed and if fact begins to deviate from well-mixed as low as 80 km. At present there is nothing to suggest that the effect is an artifact due to some misunderstood process or mechanism in the non-LTE model of $CO_2$ vibrations. Thus, the reasons for the observed, steep fall-off in the volume mixing ratio of $CO_2$ are not well-understood. The status of our understanding of mesospheric $CO_2$ is given by *Lopez-Puertas et al.* [2000, *this volume*].

Shown in Figure 10 is a computation of the radiative cooling rate in kelvin per day due to emission by carbon

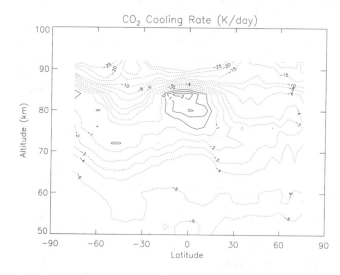

**Figure 10.** Rates of radiative cooling, in kelvin per day, due to infrared emission by carbon dioxide, as calculated from the SME data.

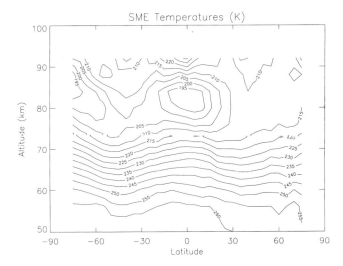

**Figure 11.** Zonal mean temperatures (at 3 p.m. local time) as measured by the SME experiment.

dioxide. The computations employ the SME temperature climatology [*Clancy et al.*, 1994] at equinox and the $CO_2$ mixing ratios inferred from the SAMS and ISAMS measurements. The non-LTE model of *Lopez-Puertas et al.* [1986a, 1986b, 1992a] was used to carry out the radiative transfer and cooling rate computations. From this figure we see that the effect of $CO_2$ emission is to cool the mesosphere. Of interest however is the region of net heating near the equator near 80 km. This feature is likely not a permanent feature of the atmosphere otherwise, when solar heating is accounted for, would result in a region in which radiation is constantly acting to increase its temperature.

In order to explain the net heating feature at 80 km in the radiative cooling we must first recall that the SME satellite was in a sun-synchronous orbit. That implies that the observations of were made at the same local time each day. In the specific case of SME, the observations were made at about 3 p.m. local time. Because of tidal processes, there are large diurnal variations in temperature at some latitudes and altitudes in the mesosphere. Based on the Global Scale Wave Model [*Hagan et al.*, 1995] the effects of tides at 83 km near the equator at 3 p.m. local time produce a region which is cooler than the atmosphere above or below. This is shown directly in Figure 11 that shows the SME temperatures from which the cooling rates were computed. There is a region of locally cooler temperatures near the equator at 83 km. The effect of infrared radiative heating of this region is quite simply explained by the convergence of net radiative flux from the warmer regions above and below. This analysis demonstrates the importance of tidal and temporal effects on the energetics. When analyzing TIMED data, for example, it will be imperative to compute the energy

balance as a function of local time before any spatial averaging is undertaken. We note that *Mlynczak et al.* [1999b] showed large errors in radiative cooling start to occur in the lower mesosphere if spatial averaging of temperatures and constituents is done before temporal averaging of the energetics.

*3.4 Net Radiative and Chemical Energy Balance*

The SME data, owing to their sun-synchronous sampling, will not allow us to get a definitive picture of the net radiation balance in the mesosphere. However, it is still instructive to examine the net radiation balance from the computations shown above. Shown in Figure 12 is the net heating due to deposition of solar ultraviolet radiation, exothermic chemical reactions, and infrared radiative cooling at equinox. As expected, the results show a large area of net heating between 80 and 88 km which extends from 75 S to about 30 N, with a maximum of about 8 kelvin per day. The atmosphere above 88 km over this latitude range essentially shows a net cooling everywhere, exceeding 25 kelvin per day at 90 km at the equator. The net heating in the lower mesosphere is rather weak by comparison, between –2 kelvin per day and +2 kelvin per day.

Shown in Figure 13 is the global mean energy balance derived from the above data. The solid curve is the sum of the solar and chemical globally averaged heating rates and the dot-dash curve is the global mean infrared cooling rate. The global mean net heating is shown by the dashed curve. This figure shows an excess of infrared cooling over solar heating between 55 and 75 km at this local time, followed by an excess of solar and chemical heating over infrared

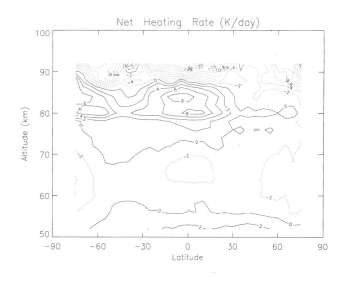

**Figure 12.** Net radiative and chemical heating in the mesosphere as measured by the SME experiment.

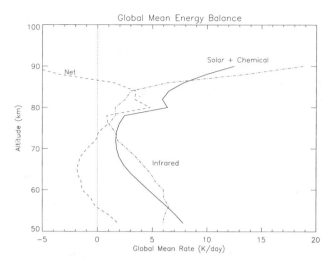

**Figure 13**. Global mean solar and chemical heating, infrared radiative cooling, and net heating as measured by the SME experiment.

cooling between 75 and 86 km. Infrared cooling exceeds the solar heating above 86 km.

Definitive analysis of the net radiation balance is not possible with this data due to the lack of temporal sampling. However, we could speculate that the excess of infrared cooling over solar heating for this month might be an indicator of another energy source, such as turbulent dissipation, operating in the mesosphere. If this were so, the mesosphere might be expected to radiatively cool at a rate that is larger than the solar and chemical heating rate. In addition, the large cooling in the upper mesosphere near 90 km may be indicative of the atomic oxygen enhanced cooling by carbon dioxide. Confirmation of either of these speculations must await more accurate data from the TIMED mission.

We conclude this section with a discussion of further applications of the heating and cooling data and mention a special need. It is typical to use the net radiative heating such as is shown in Figure 12 to compute the diabatic meridional circulation [e.g., *Dunkerton*, 1978]. *Mlynczak et al.* [1999b] and references therein have done this for the stratosphere. However, application of this technique requires implicitly that the global mean net radiative balance be zero, i.e., global mean radiative equilibrium must hold. If not, as in Figure 13, large ad-hoc mass adjustments are required to achieve a psuedo-equilibrium such that the continuity and momentum equations can be solved. *Mlynczak et al.* [1999b] did not face this problem in the stratosphere with the UARS data as the net heating on a monthly-average basis was very close to zero. We must determine the state of the mesosphere which we expect to encounter (e.g., radiative/chemical/dynamical equilibrium) to place a reasonable physical constraint on

the net heating in order to derive physically meaningful meridional circulations from the energy balance. If the mesosphere is consistently observed, on monthly to seasonal timescales, to have an excess of infrared cooling then such observations may provide information on the magnitude of other heat sources. Over a sufficiently long period of time the global mean heating on a pressure surface must tend to zero otherwise the pressure surface would continually be moving up or down, which is unphysical. Techniques must be developed to derive the meridional circulation from the heat balance in the absence of global mean radiative equilibrium.

## 4. OTHER SOURCES OF ENERGY AND DYNAMICAL INFLUENCES

We have considered in detail the solar, chemical, and radiative sources and sinks of energy in the previous sections. The purpose of this section is to highlight some other energy sources that are more somewhat more difficult to quantify or that perhaps are less important on a global scale. We will specifically mention Joule heating, possible heating associated with sprites, and effects of wave breaking.

In Table 3 we mentioned particle inputs as possible sources of energy in the mesosphere. In particular, Joule heating associated with solar proton events can produce heating rates in the mesosphere rivaling that associated with the absorption of solar ultraviolet radiation by ozone [*Roble et al.*, 1987b]. A solar proton event can also lead to catalytic destruction of odd-oxygen in the mesosphere [*Solomon et al.*, 1983b], reducing the amount of ozone and thus the amount of solar ultraviolet heating. Events of this magnitude, while producing significant alterations of the mesospheric energy balance on a local scale, do not occur with great frequency.

Other more common events that may deposit significant amounts of energy in the mesosphere include sprites. *Heavner et al.* [2000, *this volume*] estimates that as much as 1 gigajoule of energy per sprite can enter in the mesosphere, at a frequency of about 1 per second, meaning about 1 gigawatt per sprite. As sprites occur above tropospheric thunderstorms, the amount of energy they can deposit in the global mesosphere could be quite large. Further research is needed to quantify the role of energy deposition in the mesosphere associated with sprites.

Measurements and theoretical modeling efforts have also suggested that dynamical processes can lead to the heating or cooling of the mesosphere [e.g., *Lubken*, 1997; *Gardner et al.*, 1998; *Liu et al.*, 2000]. For example, the existence of turbulence in the mesosphere is likely evidence of dissipation of some wave feature, and likely leads to local heating. In addition, the observed thermal inversion layers in the middle and upper mesosphere appear to develop on timescales that require heating rates much larger than can be attributed solely to radiative or chemical effects as we currently understand them. *Liu et al.* [2000] showed that heating rates as large as 10 kelvin per hour (compare with the daily average solar and chemical heating rates shown earlier) could occur due to breaking gravity waves.

To define the magnitude of dynamical processes as sources or sinks of energy in the mesosphere, on a global basis, is a very difficult task. Unlike the more conventional solar, chemical, and infrared sources and sinks of energy, dynamical processes have no measurable radiative signal associated with them, and hence are impossible to measure with conventional remote sensing techniques. It is true that the effects of dynamics may be manifest, for example, in the structure of the observed airglow. However, airglow variations are simply reflecting the changes in composition and temperature due to the dynamics, and not heating or cooling associated with a dynamical phenomenon.

In the context of a space-based remote sensing mission such as TIMED, the only way to assess dynamical sources and sinks is to accurately determine the radiative and chemical sources and sinks that can be measured directly simultaneous with the temperature profile. Combined with wind measurements, the net effect of dynamical processes can be inferred as a residual. This approach, which is effectively what will be used to analyze the TIMED data, will not yield any information on a specific dynamical process, but should be sufficient to determine the extent to which dynamical phenomena are significant sources of energy and thereby influence the large scale thermal structure of the mesosphere.

## 5. LABORATORY MEASUREMENTS AND THE MESOSPHERIC HEAT BUDGET

It would be mistaken to infer from this paper the idea that the mesospheric energy balance will be solved with the current suite of radiative/chemical/dynamical models and observations from the TIMED mission. At the very heart of our ability to close the energy budget is knowledge of the rates of collisional energy transfer, nascent distribution fractions, rates of chemical reaction, and Einstein coefficients for spontaneous emission and absorption used to model the non-LTE energy level populations of $O_2$, $O_3$, $CO_2$, $H_2O$, OH, NO, and O. These various parameters are typically measured in the laboratory or computed by various quantum mechanical techniques. We have seen already the sensitivity of mesospheric cooling to the rate of energy transfer from vibration to translation in collisions between atomic oxygen and carbon dioxide. To understand the mesospheric energy budget, thermal structure, and chemical composition, an enormous number of rates are

required to be known accurately. The radiation fields and our ability to predict and understand them depend on this knowledge. Furthermore, as is the case with the TIMED mission, observations from space of various radiation fields are being used to retrieve the temperature, the density, the chemical abundances, and the energy balance. For each observation to be made by the TIMED mission, the radiation field is in non-LTE. Accurate interpretation of these radiation measurements can be made only if the energy transfer rate constants mentioned above are accurately known. However, many of the rates are only poorly known. For example, the rate of quenching of molecular oxygen vibrations in collisions with atomic oxygen has values in the literature which differ by a factor of 1600.

Given the existing uncertainty in energy transfer rates, a series of papers has appeared in the literature over the last several years [*Mlynczak et al.*, 2000b, 1999a, 1998; *Mlynczak and Zhou*, 1998; *Mlynczak and Olander*, 1995; *McDade*, 1998]. The express purpose of these papers is to define the accuracy to which the kinetic and spectroscopic parameters associated with non-LTE mechanisms must be known in order to derive accurately temperature, water vapor, ozone, and terms of the energy budget from satellite radiance measurements. Many of the required measurements, such as the $O-CO_2$ quenching rate at mesospheric temperatures, are driving the state-of-the-art in laboratory and theoretical techniques.

To illustrate the magnitude of the challenge in determining the required rates we consider the triatomic ozone molecule with over 260 bound vibrational levels below the dissociation limit. Ozone is chemically excited up to the dissociation limit upon its formation by the recombination of O and $O_2$. There are over 34,000 possible collisional quenching rates to be known for all possible vibrational relaxation pathways from the highest lying bound state to the ground vibrational state. Clearly the rates for ozone cannot all be measured in the laboratory and therefore require a combination of theoretical modeling and measurement.

NASA and the U. S. National Science Foundation are presently funding theoretical analysis of physical quenching of ozone, and laboratory measurements of physical quenching and energy transfer involving molecular oxygen, nitric oxide, and the hydroxyl radical. In addition, there is an effort at the University of Canterbury in New Zealand to measure the temperature dependence of the physical quenching of $CO_2$ by O at mesospheric temperatures [L. Phillips, U. Canterbury, private communication]. To a large degree, these efforts have been spurred on by the needs of the TIMED mission referenced above. We should soon have an array of key kinetic rates for use in studying the mesospheric energy balance from state-of-the-art measurements.

## 6. CLOSING THE ENERGY BUDGET

Closing the mesospheric energy budget is one of the primary goals of NASA's upcoming TIMED mission. By "closing" we mean that the terms of the energy budget are known globally through observation, and that photochemical models of the mesosphere reproduce the observed behavior with fidelity and a minimum of "tuning." We have every expectation that the heat budget will be closed at the end of the TIMED mission, at least on monthly to seasonal timescales. The TIMED mission is designed to measure the thermal structure and key radiative and chemical sources and sinks of energy. In addition, vector winds will also be measured. The TIMED mission samples a region known to contain large diurnal variations in temperature and chemistry with only one satellite in a high inclination orbit. Thus temporal sampling is clearly an issue although the TIMED satellite will precess through local time over a period of 2 months. Proper calculation and averaging of the components of the energy budget as functions of time is essential. It must be understood that much of the role of dynamics may end up being inferred from TIMED for this reason.

## 7. SUMMARY

There has been tremendous progress in the theoretical understanding of the myriad of heating and cooling processes that govern the energy budget of the mesosphere. To summarize the current knowledge we state the following, given in no particular order of preference:

1.  It is believed that heating due to exothermic chemical reactions rivals and at times, exceeds that associated directly with the deposition of solar ultraviolet radiation. This is result is readily apparent based on the disposition of absorbed solar energy as indicated in Table 2.
2.  Airglow emission substantially reduces the amount of energy available for heat in both solar energy deposition and chemical energy deposition.
3.  The fundamental and first hot bands of the bending ($v_2$) mode of carbon dioxide are nearly in LTE below 100 km altitude. They are maintained near LTE by the efficient conversion of translational energy to internal energy upon collisions between atomic oxygen and carbon dioxide.
4.  Collisions involving atomic oxygen are important in determining the strength of infrared emission by water vapor (via interaction with $O_2(v)$) and nitric oxide. This along with (3) above demonstrates a direct coupling of chemistry and radiation.
5.  Turbulent dissipation and other dynamical phenomena may substantially alter the mesospheric energy balance on a global scale.
6.  Full understanding of the mesospheric energy balance will require accurate knowledge of numerous energy transfer kinetic parameters. These have been defined and now being addressed through laboratory measurement or theoretical computation.
7.  Carbon dioxide, the primary infrared radiator responsible for mesospheric cooling, is a variable species in the mesosphere. This effect is perhaps caused by some dynamical phenomenon as the photochemical lifetime of $CO_2$ is too long to account for such variations.
8.  The molecular oxygen and hydroxyl airglow emissions, despite being far removed from LTE, are direct measures of energy deposition due to absorption of solar ultraviolet radiation and to release of energy by exothermic chemical reactions.
9.  The ozone abundance in the mesosphere, long thought to be underestimated by photochemical models, is apparently in good agreement with the Garcia-Solomon 2-D model at this point.
10. The TIMED mission, with its planned measurement of temperatures, winds, and energy sources and sinks, should provide the requisite information to close the global mesosphere energy budget on monthly to seasonal timescales.

With this knowledge we are well-prepared to analyze the measurements to be provided by the TIMED mission. However, it is our feeling that despite our best efforts to date to understand the mesosphere, Nature will find more ways to surprise and delight us to achieve an even better understanding. It is this continuing challenge to which we look forward.

*Acknowledgments.* I would like to thank NASA Langley Research Center for support through its Thompson Fellowship for a sabbatical year during which much of the work and thought contained in this paper was undertaken. I am especially grateful to S. Solomon and R. Roble for many opportunities and insights gained over the years. I would like to acknowledge J. Barker, R. Drayson, R. Garcia, M. Hagan, W. Kuhn, M. Lopez-Puertas, J. Meriwether, J. Russell, D. Siskind, and M. Summers for assistance over the years. M. Lopez-Puertas kindly provided the infrared cooling rates and solar heating rates associated with $CO_2$.

## REFERENCES

Apruzese, J. P., M. R. Schoeberl, and D. F. Strobel, Parameterization of IR cooling in a middle atmosphere dynamics model 1. Effects on the zonally averaged circulation, *J. Geophys. Res.*, 87, 8951-8966, 1982.

Apruzese, J. F., M. R. Schoeberl, and D. F. Strobel, Parameterization of IR cooling in a middle atmosphere dynamics model. 1. Effects on the zonally averaged circulation, *J. Geophys. Res.*, 87, 8951-8966, 1984.

Barkstrom, B. R., The earth radiation budget experiment (ERBE), *Bull. Amer. Met. Soc.*, 65, 1170-1185, 1984.

Brasseur, G., and D. Offerman, Recombination of atomic oxygen near the mesopause: Interpretation of rocket data, *J. Geophys. Res.*, 91, 10,818-10,824, 1986.

Clancy, R. T., D. W. Rusch, and M. T. Callan, Temperature minima in the average thermal structure of the middle mesosphere (70-80 km) from analysis of 40- to 92-km SME global temperature profiles, *J. Geophys. Res.*, 99, 19,001-19,020, 1994.

Crutzen, P. J., Energy conversions and mean vertical motions in the high latitude summer mesosphere and lower thermosphere, in *Mesospheric Models and Related Experiments,* edited by G. Fiocco, pp 78-88, D. Reidel, Norwell, Mass., 1971.

Dickinson, R. E., E. C. Ridley, and R. G. Roble, Thermospheric general circulation with coupled dynamics and composition, *J. Atmos. Sci.*, 41, 205-219, 1984.

Drayson, S. R., Calculation of longwave radiative transfer in planetary atmospheres, Ph.D. thesis, Univ. of Michigan, 110 pp., 1967.

Dunkerton, T., On the mean meridional mass motions of the stratosphere and mesosphere, *J. Atmos. Sci.*, 35, 2325-2333, 1978.

Garcia, R. R., F. Stordal, S. Solomon, and J. T. Kiehl, A new numerical model of the middle atmosphere. 1. Dynamics and transport of tropospheric gases, *J. Geophys. Res.*, 97, 3559-3585, 1992.

Garcia, R. R., S. Solomon, R. G. Roble, and D. W. Rusch, A numerical study of the response of the middle atmosphere to the 11 year solar cycle, *Planet. Space Sci.*, 411-423, 1984.

Garcia, R. R., and S. Solomon, A new numerical model of the middle atmosphere. 2. Ozone and related species, *J. Geophys. Res.*, 99, 12,937-12,951, 1994.

Garcia, R. R., and S. Solomon, A numerical model of the zonally averaged dynamical and chemical structure of the middle atmosphere, *J. Geophys. Res.*, 88, 1379-1400, 1983.

Gardner, C. S., and W. Yang, Measurements of the dynamical cooling rate associated with the vertical transport of heat by dissipating gravity waves in the mesopause region at the Starfire Optical Range, NM, *J. Geophys. Res.*, 103, 16,909-16,927, 1998.

Hagan, M. E., J. M. Forbes, and F. Vial, On modeling migrating solar tides, *Geophys. Res. Lett.*, 22, 893-896, 1995.

Heavner, M. et al., Sprites, jets, and elves: Visible energy transport across the stratopause, *this volume*, 2000.

Hines, C. O., Dynamical heating of the upper atmosphere, *J. Geophys. Res.*, 70, 177-183, 1965.

Houghton, J. T., Absorption and emission by carbon-dioxide in the mesosphere, *Q. J. R. Meteorol. Soc.*, 96, 767-770, 1970.

Houghton, J. T., Absorption and emission by carbon-dioxide in the mesosphere, *Q. J. R. Meteorol. Soc.*, 85, 1-20, 1969.

Keating, G. M., and S. Bougher, Isolation of major Venus thermospheric cooling mechanism and implications for Earth and Mars, *J. Geophys. Res.*, 97, 4189, 1992.

Kellog, W. W., Chemical heating above the mesopause in winter, *J. Meteorol.*, 18, 373-381, 1961.

Kiefer, J. H., and Lutz, R. W., The effect of oxygen atoms on the vibrational relaxation of oxygen, 11th Symposium on Combustion, The Combustion Institute, Pittsburgh, PA, 67-76, 1967.

Kuhn, W. R., Infrared radiative transfer in the upper stratosphere and mesosphere, Scientific report, Dept. Astro-Geophysics, Univ. of Colorado, 159 pp., 1966.

Liu, H., M. E. Hagan, and R. G. Roble, Local mean state changes due to gravity wave breaking modulated by diurnal tide, *J. Geophys. Res.*, in press, 2000.

Lopez-Puertas, M., M. A. Lopez-Valverde, R. R. Garcia, and R. G. Roble, A review of $CO_2$ and $CO$ abundances in the mesosphere and thermosphere, *this volume*, 2000.

Lopez-Puertas, M., M. A. Lopez-Valverde, C. P. Rinsland, and M. R. Gunson, Analysis of the upper atmosphere $CO_2(v_2)$ vibrational temperatures retrieved from ATMOS/Spacelab 3 observations, *J. Geophys. Res.*, 97, 20,469-20,478, 1992.

Lopez-Puertas, M., M. A. Lopez-Valverde, and F. W. Taylor, Vibrational temperatures and radiative cooling of the $CO_2$ 15 μm bands in the middle atmosphere, *Quart. J. Roy. Meteor. Soc.*, 118, 499-532, 1992.

Lopez-Puertas, M., M. A. Lopez-Valverde, and F. W. Taylor, Studies of solar heating by $CO_2$ in the upper atmosphere using a non-LTE model and satellite data, *J. Atmos. Sci.*, 47, 809-822, 1990.

Lopez-Puertas, M., and F. W. Taylor, Carbon dioxide 4.3 μm emission in the Earth's atmosphere: A comparison between Nimbus 7 SAMS measurements and non-LTE radiative transfer calculations, *J. Geophys. Res.*, 13,045-13,068, 1989.

Lopez-Puertas, M., R. Rodrigo, A. Molina, and F. W. Taylor, A non-LTE radiative transfer model for infrared bands in the middle atmosphere. I. Theoretical basis and its application to $CO_2$ 15 μm bands, *J. Atmos. Terr. Phys.*, 48, 729-748, 1986.

Lopez-Puertas, M., R. Rodrigo, J. J. Lopez-Moreno, and F. W. Taylor, A non-LTE radiative transfer model for infrared bands in the middle atmosphere. II. $CO_2$ (2.7 and 4.3 μm) and water vapor (6.3 μm) bands and $N_2(1)$ and $O_2(1)$ vibrational levels, *J. Atmos. Terr. Phys.*, 48, 749-764, 1986.

Lubken, F. J., Seasonal variation of turbulent energy dissipation rates at high latitudes as determined by in-situ measurements of neutral density fluctuations, *J. Geophys. Res.*, 102, 13,441-13,456, 1997.

McDade, I.C., Laboratory measurements required for upper atmosphere remote sensing of atomic oxygen, *Adv. Space. Res.*, 21, 1998.

Mertens, C. J., M. G. Mlynczak, R. R. Garcia, and R. W. Portmann, A detailed evaluation of the stratospheric heat budget. I. Radiation transfer, *J. Geophys. Res.*, 104, 6021-6038, 1999.

Mlynczak, M. G., and C. J. Mertens, M. Lopez-Puertas, P.P. Wintersteiner, R. H. Picard, and J. Winick, Kinetic requirements for the measurement of temperature at 15 μm under non-LTE conditions, *Geophys. Res. Lett.*, revised, 2000b.

Mlynczak, M. G., R. Roble, R. Garcia, and M. Hagan, Solar energy deposition rates in the mesosphere derived from airglow observations: Implications for the ozone deficit problem, *J. Geophys. Res.*, revised, 2000a.

Mlynczak, M. G., A new perspective on the molecular oxygen and hydroxyl airglow emissions, *J. Geophys. Res.*, 104, 27,535-27,543, 1999.

Mlynczak, M. G., Mertens, C. J., R. R. Garcia, and R. W.

Portmann, A detailed evaluation of the stratospheric heat budget. II. Global radiation balance and diabatic circulations, *J. Geophys. Res.*, 104, 6039-6066, 1999b.

Mlynczak, M.G., D. K. Zhou, M. Lopez-Puertas, G. Zaragoza, and J. M. Russell III, Kinetic requirements for the measurement of water vapor at 6.8 μm under non-LTE conditions, *Geophys. Res. Lett.*, 26, 63-66, 1999a.

Mlynczak, M. G., and D. K. Zhou, Kinetic and spectroscopic requirements for the measurement of mesospheric ozone at 9.6 μm under non-LTE conditions, *Geophys. Res. Lett.*, 25, 639-642, 1998.

Mlynczak, M. G., D. K. Zhou, S. M. Adler-Golden, Kinetic and spectroscopic requirements for the inference of chemical heating rates and atomic hydrogen densities from OH Meinel band measurements, *Geophys. Res. Lett.*, 25, 647-650, 1998.

Mlynczak, M. G., and B. T. Marshall, A reexamination of the role of solar heating in the $O_2$ atmospheric and infrared atmospheric bands, *Geophys. Res. Lett.*, 23, 657-660, 1996.

Mlynczak, M. G., and D. S. Olander, On the utility of the molecular oxygen dayglow emissions as proxies for middle atmospheric ozone, *Geophys. Res. Lett.*, 22, 1,377-1,380, 1995.

Mlynczak, M. G., and S. Solomon, A detailed evaluation of the heating efficiency in the middle atmosphere, *J. Geophys. Res.*, 98, 10,517-10,541, 1993.

Mlynczak, M. G., and S. Solomon, Middle atmosphere heating by exothermic chemical reactions involving odd-hydrogen species, *Geophys. Res. Lett.*, 18, 37-40, 1991.

Murgatroyd, R. J., and R. M. Goody, Sources and sinks of radiative energy from 30 to 90 km, *Q. J. R. Meteorol. Soc.*, 87, 225-234, 1958.

Roble, R.G., On the feasibility of developing a global atmospheric model extending from the ground to the exosphere, *this volume* 2000.

Roble, R. G., Energetics of the Mesosphere and Thermosphere, The Upper Mesosphere and Lower Thermosphere: A review of experiment and theory, Geophysical Monograph No. 87, 1-21, American Geophysical Union, 1995.

Roble, R. G., *et al.*, Joule heating in the mesosphere and thermosphere during the July 13, 1982 solar proton event, *J. Geophys. Res.*, 92, 6083-6090, 1987b.

Roble, R. G., E. C. Ridley, and R. E. Dickinson, On the global mean structure of the thermosphere, *J. Geophys. Res.*, 8745-8758, 92, 1987a.

Rodgers, C. D., F. W. Taylor, A. H. Muggeridge, M. Lopez-Puertas, and M. A. Lopez-Valverde, Local thermodynamic equilibrium of carbon dioxide in the upper atmosphere, *Geophys. Res. Lett.*, 19, 589-592, 1992.

Schoeberl, M.R., D. F. Strobel, and J.P. Apruzese, A numerical model of gravity wave breaking and stress in the mesosphere, *J. Geophys. Res.*, 88, 5249-5259, 1983.

Sharma, R. D., and P. P. Wintersteiner, Role of carbon dioxide in cooling planetary atmospheres, *Geophys. Res. Lett.*, 17, 2201-2204, 1990.

Shved, G. M., L. E. Khvorostovskaya, I.Yu. Potekhin, A.I. Dem'yanikov, A. A. Kutepov, and V. I. Fomichev, Measurement of the quenching rate constant of $CO_2(0110)$-O collisions and its significance for the thermal regime and radiation in the lower thermosphere, *Atmos. Ocean. Phys.*, 27, 295, 1991.

Shved, G. M., Role of airglow in the cooling of the atmosphere near the mesopause, *Geomagn. Aeron.*, 3, 500-501, 1972.

Solomon, S., G. C. Reid, D. W. Rusch, and R. J. Thomas, Mesospheric ozone depletion during the solar proton even of July 13, 1982, *Geophys. Res. Lett.*, 10, 257-261, 1983b.

Solomon, S., D. W. Rusch, R. J. Thomas, R. S. Eckman, Comparison of mesospheric ozone abundances measured by the Solar Mesosphere Explorer and model calculations, *Geophys. Res. Lett.*, 10, 249-252, 1983a.

Summers, M. and R. Conway, Insights into middle atmosphere odd-hydrogen chemistry from analysis of MAHRSI observations of hydroxyl (OH), *this volume*, 2000.

Suomi, V. E., and W. C. Shen, Horizontal variation of infrared cooling and the generation of eddy available potential energy, *J. Atmos. Sci.*, 20, 62-65, 1962.

Thomas, R. J., Atomic hydrogen and atomic oxygen density in the mesopause region: Global and seasonal variations deduced from Solar Mesosphere Explorer near-infrared observations, *J. Geophys. Res.*, 95, 16,457-16,476, 1990.

Vonder Haar, T. H., and V. E. Suomi, Measurements of the Earth's radiation budget from satellites during a five-year period. Part I. Extended time and space means, *J. Atmos. Sci.*, 28, 305-314, 1969.

Wielicki, B. A., B. R. Barkstrom, E. F. Harrison, R. B. Lee III, G. L. Smith, and J. E. Cooper, Clouds and the Earth's radiant energy system (CERES): An Earth Observing System experiment, *Bull. Amer. Met. Soc.*, 77, 853-868, 1996.

Zipf, E. C., The collisional deactivation of metastable atoms and molecules in the upper atmosphere, *Can. J. Chem.*, 47, 1863-1870, 1969.

Martin G. Mlynczak, Atmospheric Science Research, NASA Langley Research Center, Mail Stop 420, Hampton, VA 23681-2199. m.g.mlynczak@larc.nasa.gov

# On the Feasibility of Developing a Global Atmospheric Model Extending From the Ground to the Exosphere

R. G. Roble

*High Altitude Observatory, National Center for Atmospheric Research, Boulder, Colorado*

It is well known that solar EUV and UV forcing and auroral heat and momentum sources have a significant effect on thermospheric and ionospheric structure and dynamics. Yet the observed variability in these regions appears to be more than can be accounted for by considering only these processes. It is also known that the upward propagating diurnal and semi-diurnal tides affect the thermosphere and ionosphere and there are also studies that show that large scale planetary waves, such as the Quasi-two day wave, and 5 and 15 day waves produce signals as high as the ionospheric F-region. Furthermore, planetary wave and gravity wave breaking in the mesosphere has been shown to affect the stratosphere through "downward control." Most models of the middle atmosphere have their upper boundary in the 70–150 km region whereas most models of the thermosphere/ionosphere have their lower boundary near 80–95 km. This region of the atmosphere is highly dynamic making the specifications of thermal, compositional, dynamical and electrodynamical boundary conditions difficult. There is an important need for the development of a model of the entire atmosphere that can be used to investigate couplings between atmospheric regions and solar-terrestrial interactions. A crude attempt has been made to develop such a model by coupling the NCAR Community Climate Model (CCM3) and the Thermosphere-Ionosphere-Mesosphere-Electrodynamics General Circulation Model (TIME-GCM). This coupled model has been run for two years and the findings from this simulation indicate that important couplings between atmospheric regions occur. These results suggest that a model of the entire atmosphere is feasible and it will be needed to address many of the important problems that will arise in the new millenium.

## 1. INTRODUCTION

The Earth's atmosphere is continuous between the ground and exosphere yet for certain practical reasons the modeling of the atmosphere is usually divided into distinct regions, such as the coupled thermosphere/ionosphere or the coupled lower and middle atmosphere. Lower and upper boundaries of these models are usually in the vicinity of 70–130 km altitude, a highly variable region of the atmosphere that has been studied by ground-based and rocket instruments locally but has been relatively unexplored on a global basis. The Upper Atmosphere Research Satellite (UARS), however, is now providing measurements that are defining the basic global temperature, compositional, and dynamic structure of the upper meso-

Atmospheric Science Across the Stratopause
Geophysical Monograph 123

sphere and lower thermosphere for the first time. Furthermore, the measurements are showing considerable coupling between atmospheric regions such as the mesosphere/lower thermosphere and the stratospheric QBO, semi-annual variations, planetary wave filtering of gravity waves and many others.

These measurements provide a challenge for theoretical and numerical models to characterize this critical region of the atmosphere that links the highly variable solar and auroral controlled region of the upper thermosphere with the highly variable dynamically controlled lower atmosphere. It is a difficult region to impose boundary conditions for either the upper or lower atmosphere models. Numerical models are required that simulate the properties of both the thermosphere/ionosphere and the lower/middle atmosphere regions self-consistently and allow couplings between various atmospheric regions.

The National Center for Atmospheric Research (NCAR) Thermosphere-Ionosphere-Mesosphere-Electrodynamics General Circulation Model (TIME-GCM) that extends between 30 and 500 km is an attempt to bridge this traditional gap between upper and lower atmosphere models. It incorporates all of the features developed in previous general circulation models of the upper atmosphere, the TGCM [*Dickinson et al.*, 1981, 1984], the TIGCM [*Roble et al.*, 1988], and the TIE-GCM [*Richmond et al.*, 1992] and couples it to a mesospheric and upper stratospheric extension in the TIME-GCM [*Roble and Ridley*, 1994]. The T refers to thermosphere, I ionosphere, M mesosphere, and E electrodynamics in each of the GCM characterizations described above. This model, however, does not extend into the troposphere and stratosphere and thus must represent lower atmospheric variability by specifying boundary conditions at 10 mb. Currently the model only includes a zonally averaged seasonal variation and Hough Mode tidal forcings for the diurnal (1,1) tide and the semidiurnal (2-2 through 2-6) tidal components.

The general characteristics of some of the lower and upper atmosphere models will first be described. Some results from the TIME-GCM simulation of the coupling of the mesosphere, thermosphere, and ionosphere will then be presented. Finally, some results from a crude attempt to couple a model of the lower and upper atmosphere will be described. The overall progress of these efforts suggest that it is feasible and timely to consider developing a model of the entire atmosphere from the ground-to-exosphere to study various important scientific problems and the nature of the couplings between atmospheric regions.

## 2. MIDDLE ATMOSPHERE MODELS

There are two types of 3-D Middle Atmosphere GCM's that have been used to study the dynamics and structure of the middle atmosphere. The first are mechanistic models of the middle atmosphere with a lower boundary near the tropopause and an upper boundary in the 70–150 km range. These models usually cover a more limited altitude range and are forced by specified boundary conditions. Some of these models and their simulations have been described by [*Akmaev et al.*, 1996; *Geller et al.*, 1997; *Roble and Ridley*, 1994; *Smith and Brasseur*, 1991; *Butchart et al.*, 1982; *Berger and von Zahn*, 1999]. The other type are general circulation models that extend between the ground and also the 70–150 km altitude region. These models are self-consistent and phenomena throughout the atmosphere are generated internally. They do, however, require parameterizations for sub-grid scale phenomena. Some of these models are described by [*Boville*, 1995; *Hamilton*, 1995, 1996; *Hamilton et al.*, 1995; *Hunt*, 1991; *McLandress*, 1997; *Miyahara and Forbes*, 1994; *Rind et al.*, 1988a, b; *Manzini and McFarlane*, 1998]. Most of the above models solve only for the dynamics and temperature structure. Composition is usually specified or included in off-line models. Progress is being made toward coupling of dynamics and composition in these models but a fully self-consistent coupling has not yet been made. These models include the dynamics of the highly variable lower atmosphere but not the processes operating in the upper atmosphere nor the solar and auroral variability that begin to affect the upper layers. Most of these models include a rigid lid and sponge layers at the upper boundary. This influences the solution near the upper boundary and extends somewhat lower into the middle atmosphere [*Shepherd et al.*, 1996; *Lawrence*, 1997]. In addition, gravity wave momentum deposition in the upper atmosphere can influence the dynamics at lower altitudes through "downward control" [*Haynes et al.*, 1991; *Garcia and Boville*, 1994].

There is clearly a need to increase the height of the upper boundary of these models well into the thermosphere and include solar and auroral processes as well as other physical and chemical processes such as ion drag, molecular diffusion, molecular thermal conditivity, electrodynamic interactions with the ionosphere and others.

## 3. UPPER ATMOSPHERE MODELS

Several different upper atmosphere 3-D models have been developed over the years to study processes in the thermosphere and ionosphere and especially their dynamic response to solar and auroral variability [*Mayr and Volland*, 1966; *Fuller-Rowell and Rees*, 1980; *Dickinson et al.*, 1981; *Mikkelsen and Larsen*, 1993; *Namgaladze et al.*, 1990]. These models have been used for thermospheric research and in the analysis of satellite and ground-based data for many years. In this brief paper, I will present results from one of these models the NCAR TIME-GCM to illustrate the feasibility of developing an atmospheric model that extends from the ground-to-exosphere.

## 4. TIME-GCM

The NCAR Thermosphere-Ionosphere-Mesosphere-Electrodynamics General Circulation Model (TIME-GCM) is the latest in a series of three-dimensional time-dependent models that have been developed over the past two decades to simulate the circulation, temperature, and compositional structure of the upper atmosphere and ionosphere. It combines all of the previous features of the TGCM [*Dickinson et al.*, 1981, 1984], TIGCM [*Roble et al.*, 1988], and TIE-GCM [*Richmond et al.*, 1992] that were described previously and the model has been extended downward to 30 km altitude including aeronomical processes appropriate for the mesosphere and upper stratosphere, as described by *Roble and Ridley* [1994], *Roble et al.* [1987] and *Roble* [1995]. The essential differences between the model that was described in those previous papers and the model used for these simulations include the following:

1. The chemical reaction rates for the aeronomic scheme described by *Roble* [1995] have been updated to be consistent with the JPL-97 compilation [*DeMore et al.*, 1997].

2. The simplified methane oxidation scheme described by *Roble* [1995] has been modified to be consistent with the chemistry described by *LeTexier et al.* [1988].

3. The background diffusion that was used in the model for the simulations described by *Roble and Ridley* [1994] has been reduced by two orders of magnitude, consistent with the findings of *Akmaev et al.* [1996] in their simulation of the diurnal tide. The background diffusion and Rayleigh friction are now very small throughout the model ($\sim 10^3$ cm$^2$ s$^{-1}$; $1.0 \times 10^{-9}$ s$^{-1}$ respectively).

4. A new gravity wave parameterization, similar to the one used in the NCAR Community Climate Model (CCM3) [*Kiehl et al.*, 1997] is now being used in the TIME-GCM. It is a modified *Lindzen* [1981] scheme

that replaces the *Fritts and Lu* [1993] parameterization used previously in the model [*Roble and Ridley*, 1994]. We also include the calculated gravity wave flux through the 30 mb surface determined by *McFarlane* [1987] to represent orographic forcings primarily in the Northern Hemisphere during winter.

5. The CO$_2$ infrared cooling parameterization has been updated to include the model of *Fomichev et al.* [1998] to account for a variable CO$_2$ mixing ratio that is important for non-LTE processes in the upper mesosphere and lower thermosphere. All calculations assume an O-CO$_2$ vibrational relaxation rate of $3 \times 10^{-12}$cm$^{-3}$s$^{-1}$ that seems to work reasonably well on all of the terrestrial planetary thermospheres [*Bougher et al.*, 1998].

6. The tidal forcing at the lower boundary near 30 km that represents diurnal and semi-diurnal components excited in the troposphere have been specified using results from the Global Scale Wave Model (GSWM) [*Hagan*, 1996]. The amplitudes and phases of the propagating diurnal (1,1) and semi-diurnal tide (2-2 through 2-6) were obtained from the GSWM at 30 km altitude for various seasons and used as lower boundary amplitudes and phases in the TIME-GCM.

7. Solar Ionization rates are calculated using the EUVAC solar flux model and absorption cross-sections from *Richards et al.* [1994]. Solar photodissociation rates for the mesosphere and upper stratosphere are determined using the parameterizations given in *Brasseur and Solomon* [1986] and *Zhao and Turco* [1997].

8. In addition to the diurnal and semi-diurnal tidal forcing at the lower boundary the zonally averaged geopotential height and temperature at 10 mb from the empirical model of *Flemming et al.* [1988] were used to specify the latitudinal distribution of these quatities for the month under consideration. The zonal and meridional winds at the lower boundary were calculated using the zonally averaged geopotential heights. The above are improvements made to the original model described by *Roble and Ridley* [1994] to better represent various physical and chemical processes and obtain better agreement with climatology and UARS data.

## 5. TIME-GCM CALCULATED STRUCTURE FOR EQUINOX AND DECEMBER SOLSTICE CONDITIONS

The TIME-GCM was used to simulate a perpetual March equinox and December solstice for solar cycle medium conditions appropriate to the time of UARS measurements during the March/April and December/

January 1992/1993 periods as described by *McLandress et al.* [1996]. Although perpetual March equinox and December solstice simulations may not be realistic because of changing seasons, they are consistent approximations when compared with the binned UARS data that span a two month period. This time period was needed to retrieve the diurnal cycle because of the slow UARS orbital precession as discussed by *McLandress et al.* [1996].

For the model simulations a daily and 3 month time averaged solar F10.7 flux value of 150 was used to specify the solar spectral irradiance in the solar EUV and UV flux model embedded within the TIME-GCM [*Roble*, 1995]. A steady auroral forcing with ionospheric convection specified by a 45 KV cross-polar cap potential drop and auroral particle precipitation hemispheric power input of 6 GW was used in the auroral model described by *Roble and Ridley* [1987]. We also assume steady climatological seasonal and tidal forcings with no superimposed planetary wave activity at the lower boundary.

The gravity wave forcing at the lower boundary is similar to that used by *Garcia and Solomon* [1985] having a latitudinal distribution of sin |2L| where L is latitude. Five waves were used in each of the cardinal directions with a 20 m/s offset in the eastward direction, consistent with the findings of *Medvedev et al.* [1998]. Since these simulations were done without considering the influence of planetary waves, gravity wave drag is the primary drag on the neutral winds in the upper stratosphere and lower mesosphere, but ion drag becomes increasingly important in the thermosphere. For the equinox simulation, the amplitude of the latitudinal distributions (sin |2L|) is the same in both hemispheres.

With these boundary conditions and specified inputs for the parameterizations all other physical and chemical processes are calculated self-consistently. For example, heating rates are calculated using the calculated ozone distribution and the amount of solar energy absorbed at each grid point and time step. Similarly IR cooling is determined using the calculated $CO_2$ and O distributions as well as model calculated temperatures. Other quantities are also self-consistently determined.

The model starts from arbitrary initial conditions and is run until a steady state diurnally reproducible solution is achieved. Since our main interest is in the UARS data above 60 km, a steady state pattern emerges after about 20 days model simulation time. These results are then used for comparisons with UARS data, empirical models and other measurements made by ground-based observatories for similar geophysical conditions.

The UARS observed and TIME-GCM calculated zonal and meridional winds at 12 LT for the equinox pe-

riod are shown in Figure 1. Although the model calculates winds from 30–500 km only the 60–200 km region is shown to be consistent with the UARS measurements. The observed and calculated meridional winds clearly show a strong diurnal tide with a vertical wavelength of about 25 km and with maximum winds near 20 degrees latitude. The diurnal tidal amplitude increases with altitude reaching around 70 m/s near 100 km before being dissipated by molecular diffusion and ion drag. Above about 100 km the thermospheric meridional winds are driven by the absorption of solar EUV and UV radiation and at high latitudes aurora ion convection and auroral particle heating. Thus a gradual transition occurs between the lower atmosphere forced mesosphere and solar and auroral forced thermosphere and ionosphere near 100 km.

The observed and calculated zonal winds display a similar transition. In the mesosphere the dominant feature influencing the wind is the strong semi-annual wind near 80 km in the equatorial region. In the model this semi-annual wind is forced by superimposing a momentum source with a Gaussian distribution in height and latitude with a magnitude of -6 m/s/day. For this perpetual equinox simulation the model does not generate a semi-annual variation and to match the UARS observations of a strong westward zonal jet near 80 km a momentum source must be artificially imposed.

The strong westward zonal winds in the equatorial region filter the gravity waves allowing the eastward propagating waves to penetrate the strong westward jet and deposit their eastward momentum near 110 km. This eastward momentum deposition forces an eastward equatorial jet at that altitude. The dissipating diurnal tide in the lower thermosphere has been shown by *Forbes et al.* [1993] to force a westward wind. The eastward winds at the equator near 100 km require a momentum source from gravity waves and/or fast Kelvin waves as discussed by *Lieberman and Hays* [1994] and *Lieberman and Riggin* [1997]. Above 110 km the thermosphere displays a solar driven diurnal variation with strong auroral influences at high latitude.

The observed and calculated zonal and meridional winds for December solstice conditions are shown in Figure 2. The zonal winds are westward in the summer hemisphere and eastward in the winter hemisphere with reversals above the regions of maximum mid-latitude winds in the mesosphere. The reversals extend into the lower thermosphere. The TIME-GCM simulation requires a factor of four larger gravity wave forcing in the winter hemisphere compared with the summer hemisphere. The summer amplitude is a factor of two smaller than for equinox whereas it is a factor of two larger than equinox in the winter hemisphere. Part of this drag is

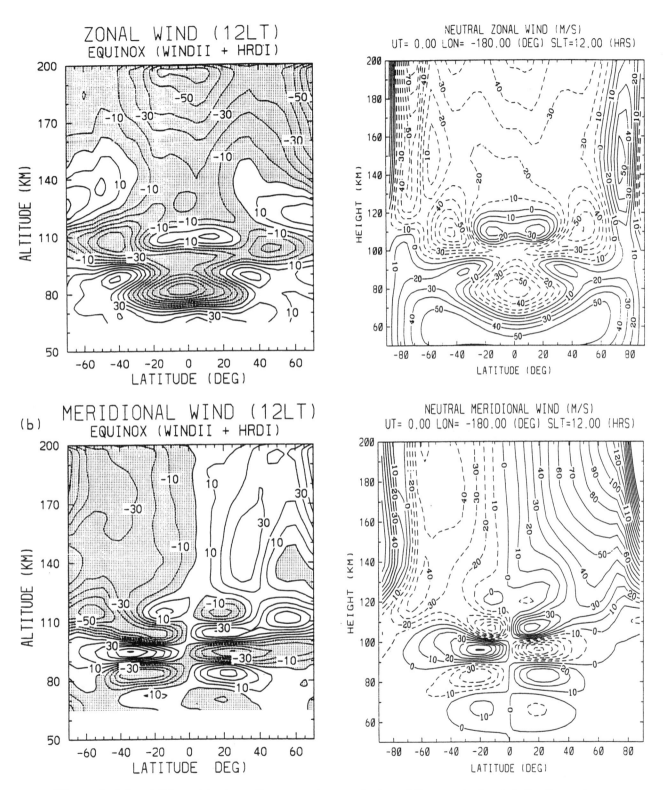

**Figure 1.** (a) and (b) combined zonally averaged mesospheric and thermospheric winds for the equinox (February 12 to May 3, 1993 at 1200 h local time for (a) zonal and (b) meridional wind components [*McLandress et al.*, 1996]. (c) and (d) TIME-GCM calculated zonal and meridional winds for equinox with similar geophysical conditions.

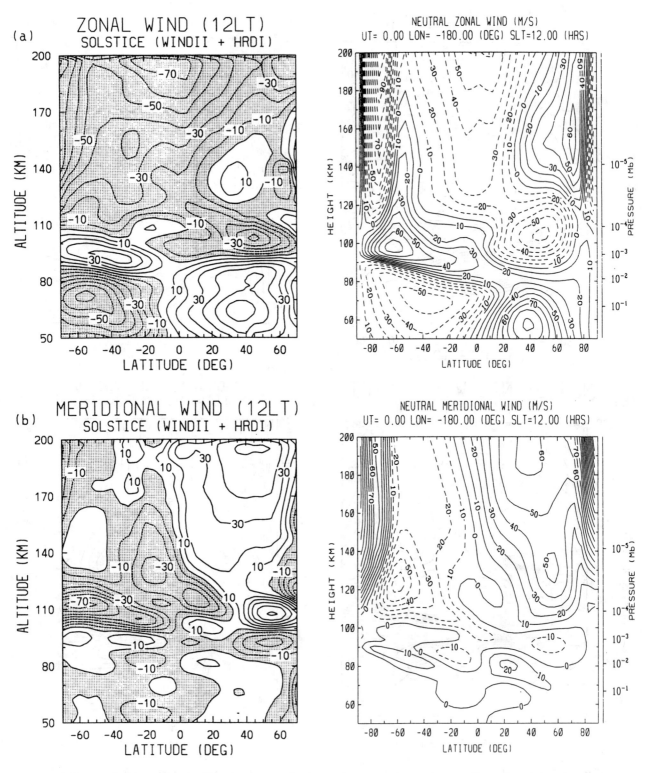

**Figure 2.** Same caption as Figure 1 except for December solstice conditions. The solstice data are for November 19, 1993 to February 4, 1993 [*McLandress et al.*, 1996].

contributed by the mountains forcing a gravity wave flux in the Northern Hemisphere winter. The model does not generate any large scale planetary waves to slow the mean winds and therefore gravity waves provide the main drag acting on the atmosphere for these simulations. Above 110 km, the zonal winds are mainly diurnal, driven by strong EUV and UV forcing and at high latitudes by auroral processes.

The solstice meridional winds have a much smaller tidal component compared with equinox conditions. The zonal average winds (not shown here) are mainly from the summer-to-winter hemisphere in the upper thermosphere and in the mesosphere. In the lower thermosphere between 100 and 120 km, however, the winds in the summer polar region are from the equator toward the summer pole in response to the strong gravity wave forcing of the wind reversal in the summer hemisphere. There is also a relatively strong meridional jet of about 10–20 m/s flowing equatorward from the summer mesopause region near 80 km altitude.

A sample of some of the other fields calculated by the TIME-GCM for December solstice conditions is shown in Figure 3. The calculated neutral gas temperature shows a warm summer stratosphere and a cold summer mesosphere. The temperature contours of the cold summer mesopause region tilt downward to near 85 km at mid-latitudes. In the winter hemisphere the mesopause is near 100 km. The calculated water vapor mixing ratio is shown in Figure 3b. Maximum water vapor mixing ratios occur in the summer mesosphere and minimum values occur in the descending polar winter hemisphere. The calculated $CO_2$ mixing ratio for solstice conditions is shown in Figure 3c. In general the mixing ratio of $CO_2$ remains constant to about 90 km in the summer hemisphere and decreases in the polar vortex of the winter hemisphere. Air with low mixing ratio $CO_2$ is transported downward in the winter polar region.

The calculated diurnal variation of the ozone mixing ratio in the equatorial region is shown in Figure 3d. Below 70 km there is a small diurnal variation in the ozone mixing ratio with maximum values near sunset and minimum values near dawn. Above 70 km there is a very large diurnal variation with maximum values at night illustrating a strong ozone photodissociation into atomic oxygen during daylight and a rapid recombination back to ozone at night.

There are many more atmospheric fields calculated by the TIME-GCM, including about 30 other atmospheric and ionospheric constituents that are not shown in this brief paper. The constituents are: [$O_2$, $N_2$, O, $O_3$, $N(^2D)$, $N(^4S)$, NO, $NO_2$, $H_2O$, $H_2$, $CH_4$, H, $HO_2$, OH, $CO_2$, CO, He, Ar, Na, $O(^1D)$, $O_2(^1\Sigma_g)$, $O_2(^1\Delta_g)$, $H_2O_2$, $O^+(^2P)$, $O^+(^2D)$, $O^+(^4S)$, $H^+$, $NO^+$, $O_2^+$, $N_2^+$, $N^+$, and Ne]. In general, the model does appear to provide a reasonable simulation of both middle and upper atmosphere properties and it is able to simulate the complex coupling processes that exist between the mesosphere and thermosphere. This demonstrates the feasibility of constructing a model that connects the middle and upper atmospheres.

## 6. ANNUAL VARIATION

With the equinox and solstice simulation parameters established and constrained by the UARS data, the TIME-GCM was then run to simulate the seasonal variation during the year. For this year's simulation the solar and auroral forcing were held constant at the values discussed above, therefore, the calculated variation of temperature, winds, and composition are due entirely to the seasonal variation of solar forcing. The derived parameters, such as the gravity wave forcings from the above UARS studies, were allowed to vary smoothly (sinusoidaly) between the equinox and solstice conditions that were described earlier. At the lower boundary the tidal forcing from the lower atmosphere was artificially held fixed during the year and only the zonally averaged latitudinal gradients of geopotential and temperature were allowed to vary in accordance with the variations specified by the Flemming et al. [1988] climatological model at 10 mb. The artificial forcing of the semiannual variation in the mesosphere was also allowed to vary smoothly between equinox and solstice conditions. It was anticipated that the model would not generate a semiannual variation because of missing physics of the type described by Mengel et al. [1995] and Garcia and Sassi [1999]. Model histories were recorded daily at 00 UT and hourly histories for certain fields were obtained every 10 days throughout the year.

The calculated yearly variation of temperature at 50, 85, and 120 km altitude are shown in Figure 4. In the stratosphere, the maximum temperatures occur over the summer pole and minimum values in the winter. At 85 km the opposite occurs giving a cold summer and warm winter mesopause. Then at 120 km the temperature distribution is similar to the stratosphere with a warm summer and cold winter polar region. The variation of the zonal winds follow a similar patter with eastward winds in winter and westward winds in summer for the stratosphere and lower thermosphere and the opposite for the mesopause region. The calculations also show an abrupt transition in the dynamics in the vicinity of equinox particularly in the upper mesosphere. This abrupt transition is clearly seen in the yearly variation of the atomic oxygen distribution shown in Figure 4. For the equinox transition going from winter-to-

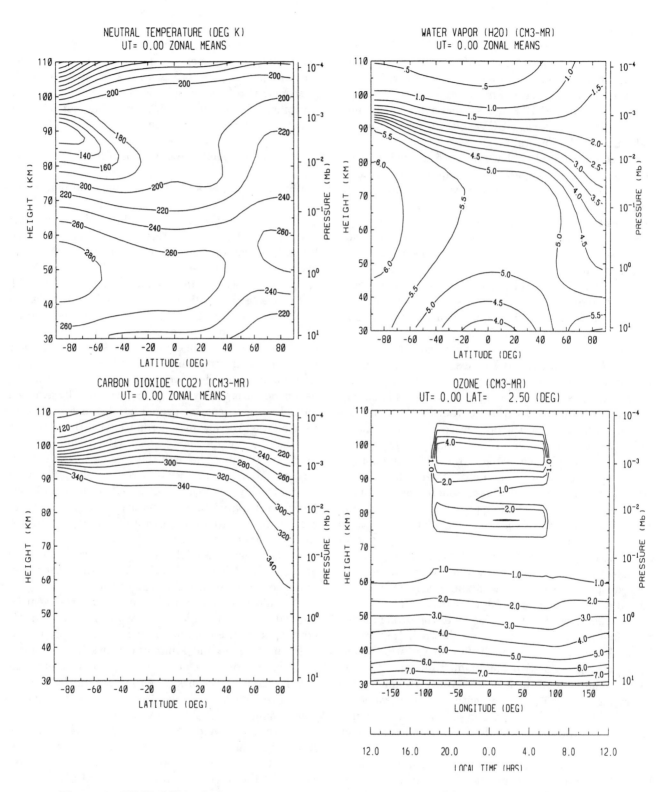

**Figure 3.** TIME-GCM calculated zonally average (a) neutral gas temperature, (K), (b) $H_2O$ mixing ratio, (c) $CO_2$ mixing ratio and (d) ozone mixing ratio around the 2.5 N latitude circle. The calculations are for the December solstice simulation.

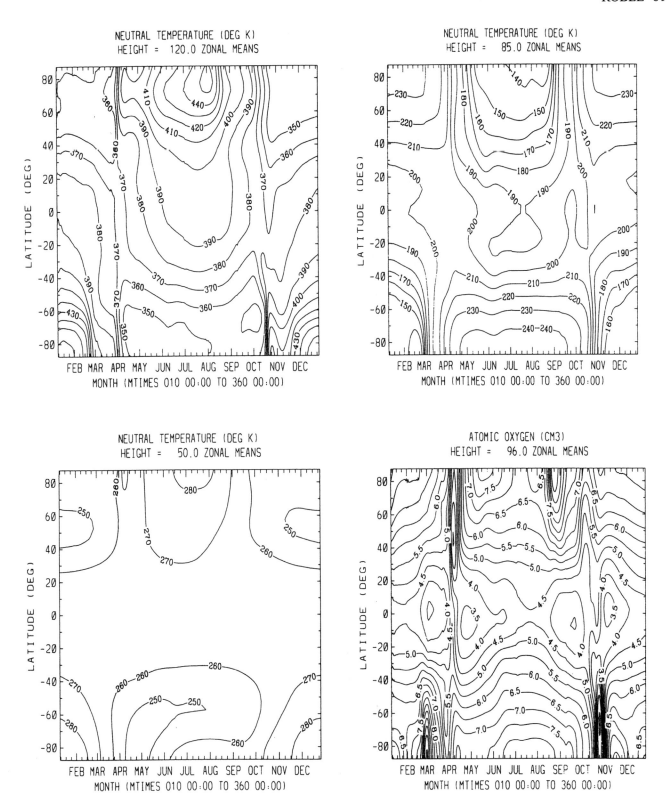

**Figure 4.** TIME-GCM variation of zonal average neutral gas temperature, (K) over a year at (a) 120 km, (b) 90 km, (c) 50 km altitude, and (d) TIME-GCM calculated zonally averaged atomic oxygen number density (cm$^{-3}$) at 96 km over a year.

summer there is an abrupt decrease in atomic oxygen at high- to mid-latitudes occurring close to the time that the zonal winds reverse their winter-to-summer direction. For the summer-to-winter transition there is first a large enhancement in the oxygen densities and then a much smaller decrease compared to the other seasonal variation. The winter-to-summer transition is similar to the atomic oxygen greenline variation observed by *Shepherd et al.* [1999] from both UARS satellite and ground-based data. The atomic oxygen at this altitude has a minimum in the equatorial region throughout the year. In analyzing the model output both the abrupt temperature and atomic oxygen changes near equinox are related to changes in the residual circulation caused by variations in gravity wave forcing in the upper mesosphere/lower thermosphere. Prior to March equinox in the northern hemisphere the gravity wave forcing in the MLT is westward and the residual circulation is poleward and downward. Around equinox, the gravity wave forcing in this region becomes eastward because of filtering of the mesospheric winds and the residual circulation becomes equatorward and upward. This decreases the atomic oxygen in the polar region.

The calculated seasonal variation with the TIME-GCM indicates that a model of the coupled upper stratosphere, mesosphere, thermosphere, and ionosphere can reasonably simulate the complex coupling processes between different atmospheric regions further demonstrating the feasibility for constructing a model of the entire atmosphere.

## 7. FLUX-COUPLED TIME-GCM/CCM3

To obtain some insight of how the variability of the lower atmosphere affects the upper atmosphere, the TIME-GCM has been flux-coupled to the NCAR Community Climate Model (CCM3). The CCM3 has been described by *Kiehl et al.* [1998]. It is a spectral model with a horizontal T42 spectral resolution (approximately 2.8 by 2.8 degree transform grid) and has 18 levels in the vertical extending between the ground and 2.9 mb. The model time step is 20 minutes and includes a diurnal cycle, where radiative fluxes are calculated every hour.

The TIME-GCM, on the other hand, is a finite difference grid point model with 4th order horizontal differencing on a 5 by 5 degree latitude/longitude grid. It has 45 levels in the vertical extending from 10 mb to above 500 km with a vertical resolution of 2 grid points per scale height. The model time step is 5 minutes and for these runs the lower boundary has been raised to 2.9 mb, the upper boundary pressure of CCM3. A diur-

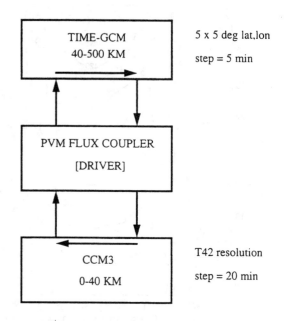

TIME-GCM / CCM3 (0-500 KM)

Fields exchanged: T, U, V, H, W, $H_2O$, $CH_4$

**Figure 5.** Schematic of the flux-coupled TIME-GCM and CCM3 and the PVM Message Passing Driver.

nal cycle for all chemical species and physical processes operates self-consistently within the model.

To couple the two models a message passing flux-coupler is used to synchronize the model time steps and provide the interpolation of quantities in both time and space that are passed between the two models as shown in Figure 5. Thus information at the CCM3 upper boundary is transferred to the lower boundary of TIME-GCM and vice versa. The actual physical quantities that are used in the transfer are given in the figure. Since the time constants are much longer in the CCM3, the combined models are started from a 10 year run of CCM3 in a stand alone simulation. Except for certain long-lived chemical species in the TIME-GCM, the temperature and dynamics of the middle and upper atmosphere adjust to an imposed lower boundary forcing in about 20 days of simulation time. Therefore, the coupled models are allowed to adjust for several months before histories are recorded for analysis. The primary motivation for this initial investigation is to determine how variability generated in the lower atmosphere propagates into the upper atmosphere and ionosphere. The replacement of the rigid lid upper boundary of CCM3 with the TIME-GCM does affect the upper stratosphere layers within CCM3 somewhat, but generally does not propagate too deeply into the stratosphere. I consider the flux-coupling of two dissimilar models at an interface in the free atmosphere only an exploratory exercise to obtain some idea on how a self-consistent model of the entire atmosphere would behave. It is basically a feasibility study to determine just how processes in the lower atmosphere affect the upper atmosphere. From previous studies of the upper atmosphere, using the TIME-GCM only it is clear that solar and auroral variability alone cannot represent the variability observed by ground-based and satellite instruments especially the day-to-day variability in the upper mesosphere and lower thermosphere. Another motivation for the development of a GCM of the entire atmosphere is to examine how deep solar-terrestrial effects propagate into the Earth's atmosphere.

The flux-coupled TIME-GCM/CCM3 model was run for a total simulation time of two years. The results show that the lower atmosphere introduces significant variability into the upper atmosphere by upward propagation, dissipation and reflection and interference of large scale waves and the filtering of gravity waves by dynamical structures in the lower atmosphere. A snapshot of the zonal winds along the 42.5 N latitude circle for day 311 in the year 2 simulation is shown in Figure 6. In the troposphere, there are considerable zonal wind fluctuations up to the lower stratosphere. The mesospheric jet is about 90 m/s near 55 km with wave activity propagating upward through the mesosphere and into the lower thermosphere. Strong wave activity is especially apparent between 100 and 200 km and above 200 km the diurnal variation caused by intense solar EUV heating is seen but with considerable variability. This figure clearly indicates that thermosphere and ionosphere dynamics is influenced by disturbances propagating upward from the lower atmosphere with wave activity in the zonal wind field across the stratopause.

The calculated temperature, zonal wind, atomic oxygen and nitric oxide number densities are shown in Figure 7 for the 2.5 N latitude circle on day 84 of the second year simulation. It is seen that there exists a considerable amount of variability that maximizes in the 80–200 km altitude region and that all atmospheric variables are affected. The source of this variability is related to weather systems and other disturbances generated in the troposphere. Large-scale Rossby waves are evanescent above the troposphere and even (stationary) planetary Rossby waves can propagate to higher altitudes only in winter. What does propagate vertically in any season are (1) small-scale gravity waves (forced by processes connected with weather systems, flow over

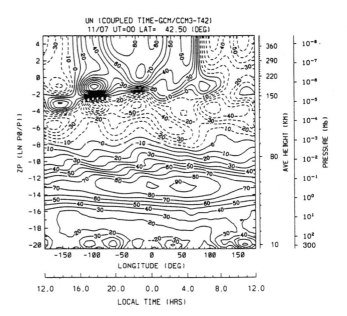

**Figure 6.** The calculated zonal winds along the 42.5 N latitude circle at 00 UT on day 312 of the flux-coupled TIME-GCM/CCM3 run.

mountains, etc.) and (2) in the tropics, equatorially-trapped Kelvin and inertia-gravity waves [*Garcia and Sassi*, 1999] propagate upward into the upper mesosphere and lower thermosphere and produce variability on horizontal scale-sizes of thousands of kilometers. These structures are similar to those observed by the WINDII and HRDI instruments onboard the UARS satellite [e.g., *Shepherd et al.*, 1999].

The greatest variability occurs in the upper mesosphere and lower thermosphere. In the upper thermosphere this variability is damped by molecular diffusion, thermal conductivity and ion drag and the strong solar EUV and UV and auroral forcings overwhelm the effects from the lower atmosphere. Studies by *Mendillo et al.* [1998] indicate that at F-region altitudes during geomagnetic quiet conditions, the variation in the F-region electron density is on the order of 10–30% and is caused by disturbances propagating upward from the lower atmosphere.

Other results from the flux-coupled model run show that the seasonal variation in the thermosphere-ionosphere system is influenced by the seasonal variations in the lower atmosphere that occurs on a different time scale. For example, at the stratopause the winter zonal winds in the southern hemisphere are faster and persist longer than in the northern hemisphere and this influences the mesosphere and thermosphere transition times. The diurnal tide calculated by the flux-coupled model is similar to the tide observed by HRDI and

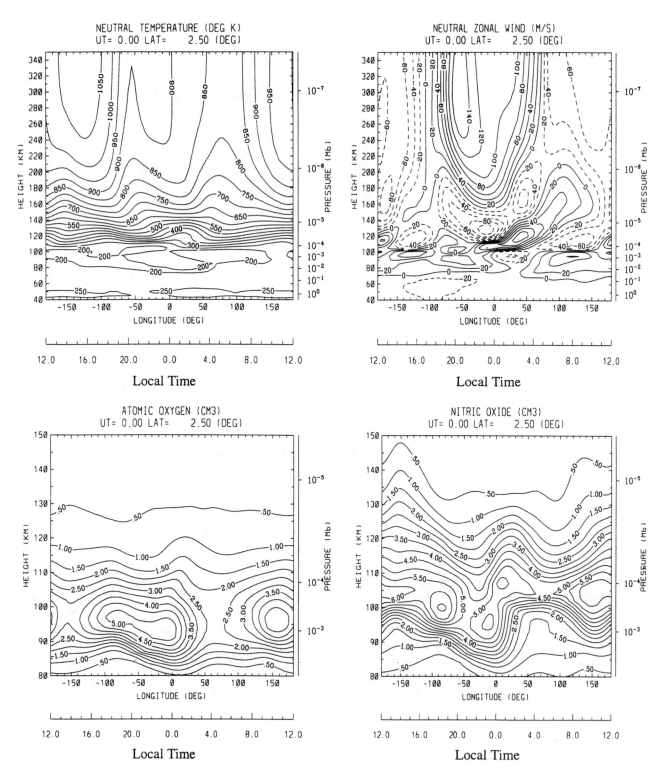

**Figure 7.** Calculated (a) neutral gas temperature (K), (b) zonal wind, m/s, (c) atomic oxygen number density ($\times 10^{-11}$) and (d) nitric oxide number density ($\times 10^{-7}$) cm$^{-3}$, around the 2.5 N latitude circle for day 84 of the year 2 simulation with the flux-coupled TIME-GCM/CCM3.

WINDII with maximum amplitudes occurring in the spring equinox period and minimum tidal amplitudes at solstice. The flux-coupled tidal amplitudes, however, are about a factor of 2 smaller than the observed values.

## 8. GROUND-TO-EXOSPHERE GCM

The general success of the models in simulating the dynamics, temperature and compositional structure of various atmospheric regions and the demonstration of the importance of coupling between atmospheric regions indicate that it is desirable to construct a model of the entire atmosphere, for example, from the ground/ocean interface to the exosphere. The success of the various component models in simulating their respective regions also indicate that it is feasible to construct such a global model. The importance of such a model is apparent when one wants to study such problems as:

- Physical and Chemical Interactions – A comprehensive global model will be useful for studying phenomena that involve couplings between vertical layers, such as the QBO, semiannual oscillation in the stratosphere and mesosphere, downward transport of aurora generated NO to the stratosphere, variability in the dynamo generated electric fields, propagation and interaction of gravity waves with tides, planetary waves, QBO, SAO, etc. and the chemical transport of various long-lived species and others.

- Climate Variability and Climate Change – The impact of stratospheric variability on climate and the role of the stratosphere in climate change are currently open questions. The dynamics of the stratosphere are dominated by the interaction of dynamical forcing by waves propagating upward from the troposphere and radiative forcing by solar heating due to ozone. Much of the vertically propagating gravity wave activity goes directly into the mesosphere and affects the stratosphere through "downward control." Research on the climate effects of the stratosphere requires a model which extends from the surface through the mesosphere, including interactive ozone chemistry.

- Climate Response to Solar Variability – Several studies have suggested that much of the climate variability over the last several centuries can be explained by variations in solar irradiance. Satellite observations of solar irradiance over the last 20 years show that most of the variation is in the UV radiation (wavelengths shorter than 200 nm).

Radiation at these wavelengths is almost entirely absorbed above the middle stratosphere suggesting that impacts on surface climate must come from changes in middle atmosphere chemistry and dynamics.

- Space Weather – Variations in thermospheric and ionospheric properties are partly driven by solar and auroral variations and partly by forcing from the lower atmosphere. Understanding these effects requires a model that extends upward from the surface with realistic variability from the lower atmosphere interacting with variability from solar and auroral forcings.

- Satellite Observations – There are many satellite projects that have a need for a comprehensive model that predicts dynamics, temperature, compositional and airglow structure. These include the UARS, TIMED, EOS, COSMIC, and DOE satellites as well as various ground-based programs such as CEDAR, GEM, and RISE.

It thus appears feasible and timely to construct a model of the entire atmosphere to study these and other important problems including couplings across the stratopause.

## REFERENCES

Akmaev, R. A, J. M. Forbes, and M. E. Hagen, Simulation of tides with a spectral mesosphere/lower thermosphere model, *Geophys. Res. Lett.*, *23*, 2173-2176, 1996.

Berger, U., and U. von Zahn, The two-level structure of the mesopause: A model study, *J. Geophys. Res.*, *104*, 22,083-22,093, 1999.

Boville, B. A., Middle atmosphere version of CCM2 (MACCM2): Annual cycle and interannual variability, *J. Geophys. Res.*, *100*, 9017-9039, 1995.

Brasseur, G., and S. Solomon, *Aeronomy of the Middle Atmosphere*, 452 pp., D. Reidel Publishing Co., second edition, Dordrecht, Holland, 1986.

Butchart, N., S. A. Clough, T. N. Palmer, and P. J. Trevelyan, Simulations of an observed stratospheric warming with quasigeostrophic refractive index as a model diagnostic, *Quart. J. R. Met. Soc.*, *108*, 475-502, 1982.

DeMore, W. B., et al., Chemical kinetics and photochemical data for use in stratospheric modeling, Evaluation Number 10, *JPL-Publication 94-1*, NASA-JPL, Pasadena, CA, 1994.

Dickinson, R. E., E. C. Ridley, and R. G. Roble, A three-dimensional general circulation model of the thermosphere, *J. Geophys. Res.*, *86*, 1499-1512, 1981.

Dickinson, R. E., E. C. Ridley, and R. G. Roble, Thermospheric general circulation with coupled dynamics and composition, *J. Atmos. Sci.*, *41*, 205-219, 1984.

Flemming, E. L., S. Chandra, M. R. Schoeberll, and J. J.

Barnett, Monthly mean global climatology of temperature, wind, geopotential height, and pressure for 0–120 km, *NASA Tech. Memo TM-100697*, 85 pp., Goddard Space Flight Center, Greenbelt, MD, 1988.

Fomichev, V. I., and G. M. Shved, Parameterization of the radiative flux divergences in the 9.6μm $O_3$ band, *J. Atmos. Terr. Phys.*, *47*, 1037-1049, 1985.

Fomichev, V. I., J.-P. Blanchet, and D. S. Turner, Matrix parameterization of the 15μm $CO_2$ band cooling in the middle and upper atmosphere for variable $CO_2$ concentration, *J. Geophys. Res.*, *103*, 11,505-11,528, 1998.

Forbes, J. M., R. G. Roble, and C. G. Fesen, Acceleration, heating and compositional mixing of the thermosphere due to upward-propagating tides, *J. Geophys. Res.*, *98*, 311-321, 1993.

Fritts, D. C., and W. Lu, Spectral estimates of gravity wave energy and momentum fluxes, II, Parameterization of wave forcing and variability, *J. Atmos Sci.*, *50*, 3695-3713, 1993.

Fuller-Rowell, T. J., and D. Rees, A three-dimensional time-dependent global model of the thermosphere, *J. Atmos. Sci.*, *37*, 2545-2567, 1980.

Garcia, R. R., and B. A. Boville, "Downward Control" of the mean meridional circulation and temperature distribution of the polar winter stratosphere, *J. Atmos. Sci.*, *51*, 2238-2245, 1994.

Garcia, R. R., and F. Sassi, Modulation of the mesospheric semiannual oscillation by the quasibiennial oscillation, *Earth Planets Space*, *51*, 563-569, 1999.

Garcia, R. R., and S. Solomon, The effect of breaking gravity waves on the dynamics and chemical composition of the mesosphere and lower thermosphere, *J. Geophys. Res.*, *90*, 3850-3868, 1985.

Geller, M. A., V. A. Yudin, B. V. Khattatov, and M. E. Hagan, Modeling the diurnal tide with dissipation derived from UARS/HRDI measurements, *Ann. Geophysicae, 15*, 1198-1204, 1997.

Hagan, M. E., Comparative effects of migrating solar sources on tidal signatures in the middle and upper atmosphere, *J. Geophys. Res.*, *101*, 21,213-21,222, 1996.

Hamilton, K., Aspects of mesospheric simulation in a comprehensive general circulation model, in *The Upper Mesosphere and Lower Thermosphere: A Review of Experiment and Theory, Geophys. Mono., 87*, (R. M. Johnson and T. L. Killeen, eds.), 255-264, 1995.

Hamilton, K., Comprehensive meteorological modelling of the middle atmosphere: A tutorial review, *J. Atmos. Terr. Phys.*, *58*, 1591-1627, 1996.

Hamilton, K., R. J. Wilson, J. D. Mahlman, and L. J. Umscheid, Climatology of the SKYHI troposphere-stratosphere-mesosphere general circulation model, *J. Atmos. Sci.*, *52*, 5-43, 1995.

Haynes, P., C. Marks, M. McIntyre, T. Shepherd, and K. Shine, On the "Downward Control" of extratropical diabatic circulations by eddy-induced mean zonal forces, *J. Atmos. Sci.*, *48*, 651-678, 1991.

Hunt, B. G., A simulation of the gravity wave characteristics and interactions in a diurnally varying model atmosphere, *J. Meteor. Soc. Jpn.*, *68*, 145-161, 1991.

Kiehl, J. T., J. J. Hack, G. B. Bonan, B. A. Boville, D. L. Williamson, and P. J. Rasch, The National Center for Atmospheric Research Community Climate Model: CCM3, *J. Climate, 11*, 1131-1149, 1998.

Lawrence, B. N., Some aspects of the sensitivity of stratospheric climate simulation to model lid height, *J. Geophys. Res.*, *102*, 23,805-23811, 1997.

LeTexier, H., S. Solomon, and R. R. Garcia, The role of molecular hydrogen and methane oxidation in the water vapor budget, *Q. J. R. Meteorol. Soc.*, *114*, 281-295, 1988.

Lieberman, R. S., and P. B. Hays, An estimate of the momentum deposition in the lower thermosphere by the observed diurnal tide, *J. Atmos. Sci.*, *51*, 3094-3115, 1994.

Lieberman, R. S., and D. Riggin, High resolution Doppler imager observations of Kelvin waves in the equatorial mesosphere and lower thermosphere, *J. Geophys. Res, 102*, 26,117-26,130, 1997.

Lindzen, R. S., Turbulence and stress owing to gravity wave and tidal breakdown, *J. Geophys. Res.*, *86*, 9707-9714, 1981.

Manzini, E., and N. A. McFarlane, The effect of varying the source spectrum of a gravity wave parameterization in a middle atmosphere general circulation model, *J. Geophys. Res.*, *103*, 31,523-31,539, 1998.

McFarlane, N. A., The effect of orographically excited gravity wave drag on the general circulation of the lower stratosphere and troposphere, *J. Atmos. Sci.*, *44*, 1775-1800, 1987.

McLandress, C. M., Seasonal variability of the diurnal tide: Results from the Canadian Middle Atmosphere General Circulation Model, *J. Geophys. Res.*, *102*, 29,747-29,764, 1997.

McLandress, C., G. G. Shepherd, B. H. Solheim, M. D. Burrage, P. B. Hays, and W. R. Skinner, Combined mesosphere/thermosphere winds using WINDII and HRDI data from the Upper Atmosphere Research Satellite, *J. Geophys. Res.*, *101*, 10441-10453, 1996.

Medvedev, A. S., G. P. Klaassen, and S. R. Beagley, On the role of an anisotropic gravity wave spectrum in maintaining the circulation of the middle atmosphere, *Geophys. Res. Lett.*, *25*, 509-512, 1998.

Mendillo, M., H. Rishbeth, R. G. Roble, E. Damboise, and J. Wroten, Ionospheric variability originating from tropospheric and stratospheric sources, paper presented at the Spring AGU Meeting, Baltimore, MD, May 26-29, 1998.

Mengel, J. G., H. G. Mayr, K. L. Chan, C. O. Hines, C. A. Reddy, N. F. Arnold, and H. S. Porter, Equatorial oscillations in the middle atmosphere generated by small scale gravity waves, *Geophys. Res. Lett.*, *22*, 3027-3030, 1995.

Mikkelsen, I. S., and M. F. Larsen, Comparisons of spectral thermospheric general circulation model simulations and E and F region chemical release wind observations, *J. Geophys. Res.*, *98*, 3693-3709, 1993.

Miyahara, S., and J. M. Forbes, Interactions between diurnal tides and gravity waves in the lower thermosphere, *J. Atmos. Terr. Phys.*, *56*, 1356-1373, 1994.

Namgaladze, A. A., Yu. N. Koren'kov, V. V. Klimenko, I. V. Karpov, F. S. Bessarb, V. A. Surotkin, T. A. Glushcenko, and N. M. Naumova, A global numerical model of the thermosphere, ionosphere, and protonosphere, *Geomagnetism and Aeronomy, 30*, 515-521, 1990.

Richards, P. G., J. A. Fennelly, and D. G. Torr, EUVAC:

A solar EUV flux model for aeronomical calculations, *J. Geophys. Res., 99*, 8981-8992, 1994.

Richmond, A. D., E. C. Ridley, and R. G. Roble, A thermosphere/ionosphere general circulation model with coupled electrodynamics, *Geophys. Res. Lett., 19*, 601-604, 1992.

Rind, D., R. Suozzo, N. K. Balachandran, A. Lacis, and G. Russell, The GISS global climate – middle atmosphere model, I, Model structure and climatology, *J. Atmos. Sci., 45*, 329-370, 1988a.

Rind, D., R. Suozzo, and N. K. Balachandran, The GISS global climate – Middle atmosphere model, II, Model variability due to interactions between planetary waves, the mean circulation and gravity wave drag, *J. Atmos. Sci., 45*, 371-386, 1988b.

Roble, R. G., Energetics of the mesosphere and thermosphere, in *The Upper Mesosphere and Lower Thermosphere: A Review of Experiment and Theory, Geophys. Mono., 87*, (R. M. Johnson and T. L. Killeen, eds.), 1-21, 1995.

Roble, R. G., and E. C. Ridley, An auroral model for the NCAR thermosphere general circulation model (TGCM), *Annales. Geophysicae, 5A*, (6), 369-382, 1987.

Roble, R. G., and E. C. Ridley, A thermosphere-ionosphere-mesosphere-electrodynamics general circulation model (TIME-GCM): Equinox solar cycle minimum simulations (30–500 km), *Geophys. Res. Lett., 21*, 417-420, 1994.

Roble, R. G., E. C. Ridley, and R. E. Dickinson, On the global mean structure of the thermosphere, *J. Geophys. Res., 92*, 8745-8758, 1987.

Roble, R. G., E. C. Ridley, A. D. Richmond, and R. E. Dickinson, A coupled thermosphere/ionosphere general circulation model, *Geophys. Res. Lett., 15*, 1325-1328, 1988.

Shepherd, G. G., J. Stegman, P. Epsy, C. McLandress, G. Thuillier, and R. H. Wiens, Springtime transition in lower thermospheric atomic oxygen, *J. Geophys. Res., 104*, 213-223, 1999.

Smith, A. K., and G. P. Brasseur, Numerical simulation of the seasonal variation of mesospheric water vapor, *J. Geophys. Res., 96*, 7553-7563, 1991.

Zhao, X., and R. P. Turco, Photodissociation parameterization for stratospheric photochemical modeling, *J. Geophys. Res., 102*, 9447-9459, 1997.

R. G. Roble, High Altitude Observatory, National Center for Atmospheric Research, 3450 Mitchell Lane, Boulder, CO 80301. (e-mail: roble@ucar.edu)

# Sprites, Blue Jets, and Elves: Optical Evidence of Energy Transport Across the Stratopause

Matthew J. Heavner, [1] Davis D. Sentman, Dana R. Moudry, and Eugene M. Wescott

*Geophysical Institute, University of Alaska Fairbanks*

Carl L. Siefring and Jeff S. Morrill

*Naval Research Laboratory*

Eric J. Bucsela

*Raytheon ITSS*

Sprites, blue jets, blue starters, and elves are recently documented optical evidence of previously unknown forms of upward electrical energy transport across the stratopause. These energetic processes have not been incorporated into most models or descriptions of middle- and upper-atmospheric dynamics, in part because the details of the processes themselves are still poorly understood. The earliest (1995) ground based red spectral observations of neutral molecular nitrogen emissions from sprites indicate a low energy phenomena compared to the neutral and ionized emissions observed in lightning or aurora. However, recent sprite observations of ionized molecular nitrogen emissions indicate the presence of higher energy processes. In 1998, the EXL98 aircraft campaign characterized the blue emissions of sprites, blue jets, and elves. Aircraft measurements include filtered images, at 427.8 nm ($N_2^+$(1NG)) and 340.7 nm ($N_2$(2PG)), as well as NUV/blue spectral observations between 320-460 nm, while ground based time-resolved photometric measurements were also made in this wavelength range. We discuss the filtered and spectral NUV observations in conjunction with earlier red (640-920 nm) spectral and filtered photometer observations. The identification of ionized nitrogen emissions indicates processes with electron energies of at least 18.6 eV to produce these emissions, assuming excitation is directly from the $N_2$ ground state. This paper provides middle- and upper-atmospheric scientists an introduction to these recently discovered phenomena and the

---

[1]Now at Los Alamos National Labs.

Atmospheric Science Across the Stratopause
Geophysical Monograph 123
Copyright by the American Geophysical Union

current best estimates of the energetic contributions of these phenomena to the atmosphere above the tropopause.

## 1. INTRODUCTION

A serendipitous video observation of what was then called a "cloud-to-stratosphere discharge" (Franz et al., 1990) provided the first recorded image of optical emissions in the middle- and upper-atmosphere associated with thunderstorms, and launched a rapidly evolving field of both observational (Sentman, 1998; Rodger, 1999) and theoretical efforts (reviewed in Rowland, 1998). Since then a number of phenomena have been discovered, which are now called sprites, blue jets and blue starters, and elves. The common sources of energy driving these processes in the stratosphere and mesosphere are tropospheric thunderstorms and lightning. The electric field generated by these storms heats ambient electrons which result in electron impact excitation of the ambient atmosphere leading to the observed optical emissions. Thus, these emissions are evidence of rapid transmissions of electromagnetic energy from the troposphere, across the stratopause, to altitudes as high as the lower thermosphere/ionosphere. In this paper, we discuss optical and temporal characteristics of sprites, blue jets and starters, and elves. Our primary focus will be on the most recent observations to provide insight into the energetics of these events. Modeling efforts will also be discussed with regard to energy implications. Although there are numerous issues remaining to be resolved prior to including sprites in our global understanding of the middle- and upper-atmosphere, this paper provides atmospheric scientists information necessary to consider possible mesospheric effects of these newly discovered phenomenon.

The optical energy (between 395-700 nm) radiated by a sprite has been estimated to be 12-60 kJ per event (Sentman et al., 1995b). The radiated optical energy represents only a small fraction of the total energy deposited in the middle- and upper-atmosphere. This energy was calculated by convolving spectral observations with the camera response. Based on these values of radiated energy and a number of assumptions about the kinetic processes which affect the emissions in this wavelength range, estimates of the total energy deposition into $O_2$ and $N_2$ (vibrational and electronic states), range from $\sim$250 MJ to $\sim$1 GJ per event (see section 3.1 for details of this optical based energy estimate). For comparison, recent calculations of the available energy in a sprite, made using electrostatic field magnitude considerations, yields 1-10 MJ. One of the major goals of current research is to reduce the broad range of these estimates by way of more accurate observation.

A second measure of the energetics is the determination of the distribution of energy required to produce the observed emissions, that is, what is the energy associated with the processes occurring in the middle atmosphere which cause the optical emissions? We present evidence of ionized emissions from $N_2^+$ which have an energy threshold of 18.6 eV assuming excitation is occurring directly from the ground electronic state of $N_2$.

Molecular nitrogen ($N_2$) is the major constituent ($\sim$80%) of the atmosphere in the region (20-90 km, or $\sim$10 pressure scale heights) where blue jets, blue starters, sprites, and elves occur. $N_2$ is a spectroscopically rich molecule with many excited electronic states which produce significant emission. These provide a good diagnostic for studying the energetic processes occurring in sprites, blue jets, elves, in a similar manner as has been done with airglow and aurora. The primary optical emission of sprites is from the molecular nitrogen first positive group ($N_2(1PG)$) (Mende et al., 1995; Hampton et al., 1996) although we will discuss other $N_2$ and $N_2^+$ emissions which have been recently observed.

The optical emissions observed in sprites, blue jets, and elves are caused by collisions of energetic (heated) electrons with neutrals. These emissions are similar to those observed in the aurora (Vallance Jones, 1974) and are discussed in detail in the following sections. A fundamental difference between auroral and sprite observations involves the altitude where these emissions are produced. At lower altitudes, where sprites are produced, quenching and other collisional processes play an important role in determining the details of the observed emissions (Morrill and Benesch, 1996). The energies of the various $N_2$ electronic states are used in conjunction with the observed emissions to determine the energetics of processes associated with sprites, blue jets/blue starters, and elves. State energies are often characterized in terms of monoenergetic electrons as can be measured in the laboratory while physical processes in nature generally have a more complex energy distribution. Modeling and analysis efforts typically describe an electron distribution similar to a Boltzmann distribution with characteristic energy of $\sim$1 eV but modified to also have an additional 'high' energy tail (a Druvysteyn distribution has been suggested by Green et al. (1996)).

## 2. PHENOMENOLOGY

Since initial documentation in 1989, three types of phenomena occurring above thunderstorms have been

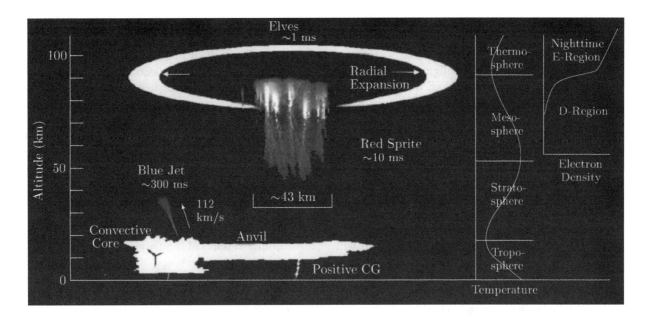

**Figure 1.** The three types of transient optical events above thunderstorms recorded in the past decade are summarized. Sprites are brief emissions occurring between approximately 40-95 km, associated with large positive cloud-to-ground lightning discharges. Blue jets are cones of light which propagate upward from the top of the electrically active convection core of thunderstorms at an average velocity of 112 km/s to a terminal altitude of ~37 km. Elves occur over both positive and negative CG's at the base of the ionosphere and expand radially to horizontal diameters as large as ~500 km and vertical widths of several km.

documented: sprites, blue jets, and elves, as illustrated in Figure 1. Also shown in this figure are typical vertical temperature and electron density profiles which provide context for the three types of phenomena. Two characteristics separating the phenomena into these three groups are duration and altitude. Sprites span the altitude between ~40-95 km and last ~10-100 ms. Elves, while not currently triangulated for accurate altitude determination, are estimated to occur between 75-95 km altitudes, lasting less than 1 ms. Blue jets appear from cloud tops (~20 km) and propagate upwards in the shape of an expanding cone to altitudes of ~40 km, over a period of ~200-300 ms. Blue starters appear to propagate upward from storm tops, and are similar to jets. However, at altitudes of ~25 km or less, the starters extinguish rather than propagating to 40 km as with a blue jet.

Lightning discharges are classified into four types, based on direction of propagation (cloud-to-ground or ground-to-cloud) and the net charge removed (positive or negative) (Berger, 1978). Negative cloud-to-ground discharges remove negative charge from the thunderstorm to ground and constitute approximately 90% of lightning activity. Positive cloud-to-ground discharges constitute almost all the remainder of lightning activity

(positive and negative ground-to-cloud discharges are rare, occurring from tall human structures or mountains). Positive cloud-to-ground discharges often occur away from the most electrically active convective cores of thunderstorms, and generally are during the later stages of large thunderstorms. Sprites are observed in association with large positive cloud-to-ground (CG) lightning rather than over the more electrically active convective core (Boccippio *et al.*, 1995).

Great progress has been made in the theoretical understanding of sprites, elves, blue jets and blue starters since the first video recordings. These phenomena can be classified by the source of the electric fields that generate them. Experimentally sprites and elves are clearly associated with lightning discharges, which can be detected by instruments on the ground (e.g., the National Lightning Detection Network - NLDN). Jets and starters are the least understood of the phenomena, don't appear to be associated with detectable lightning, and may be associated directly with breakdown near charge centers in the storm (similar to normal lightning). In a thunderstorm, electrical charges are accumulated in a number of charge centers by a slow process, typically minutes or 10's of minutes (MacGorman and Rust, 1998). A lightning discharge is a rapid rearrangement of these charges and consequently intense electric fields are generated. These fields have two primary sources, one electromagnetic and one electrostatic.

The current change ($di/dt$) in a lightning stroke radiates a large electromagnetic pulse (EMP). The EMP fields propagate upward, with amplitude decreasing inversely with distance $1/r$ (like the far field of an antenna) until reaching the bottom of the ionosphere. These fields are not large enough to cause breakdown in the atmosphere at low altitudes, but the atmospheric density and, thus, the field required for breakdown decreases exponentially with altitude (the scale height in the middle atmosphere is ~7 km). The theoretical studies discussed below (originating with work by Inan *et al.* (1991)) show that breakdown from the EMP field will occur in the region between the bottom of the ionosphere at about 95 km and about 75 km, consistent with some of the larger current changes associated with lightning. The EMP fields are associated with the generation of elves. The 100 $\mu$s duration of elves is close to the 10-100 $\mu$s duration of the high current portions of typical lightning strokes. Elves also have a large horizontal extent (100's of km) matching expectations from the $1/r$ drop in field strength.

Sprites are associated with the Quasi-Electrostatic (QE) fields of lightning, as first suggested by Wilson (1925). Their sources are the electrostatic fields from the charges in thundercloud. A lightning stroke rapidly removes charge from the thunderstorm, resulting in a rapid change in the charge configuration of the storm giving rise to an electrostatic field. These 'near-fields' fall-off as $r^{-3}$ from the source (the dipole made up of cloud charge and ground image) and have an amplitude proportional to the charge moment ($Q \times L$). If the upper atmosphere were not conducting, only a change in the electrostatic field level that is delayed by the speed of light propagation time would be observed. However, in the stratified conducting atmosphere the behavior is more complex. Free charge carriers in the atmosphere respond to the new electrostatic field configuration and 'shield-out' the field. This process can be accomplished on approximately the local relaxation time scale ($\sigma/\epsilon_0$) which is 10's of ms at 80 km and a few seconds at 40 km altitude. Because of the very different time scales for cloud charging (minutes), electrical relaxation (milliseconds to seconds) and lightning discharge (microseconds to milliseconds) the QE fields in the upper atmosphere have a rapid onset followed by a decay to a lower level. The field from the original cloud charge configuration does not significantly penetrate to high altitudes because a shielding charge configuration has developed

during the slow thunderstorm charge buildup. A lightning stroke changes the electrostatic field of the thunderstorm. The initial amplitude of the QE field is essentially the same as the electrostatic field change expected in a non-conducting atmosphere since the atmosphere does not have time to relax. The typical lifetime of sprites (10's of milliseconds) corresponds well with the relaxation time at higher altitudes 60-90 km. Positive cloud-to-ground lightning flashes tend to neutralize greater charge from higher altitudes than other types of lighting, partially explaining the strong correlation between sprites and positive flashes.

## 2.1. Sprites

Anecdotal reports of flashes above thunderstorms from ground based observers and airplane pilots date back to the late 19th century (Vaughan and Vonnegut, 1989, and references therein). Early considerations of thunderstorm effects on the middle atmosphere included breakdown ionization (Wilson, 1925). The first recorded observation of non-lightning optical emissions associated with thunderstorm activity was made on July 6, 1989 with an intensified low light level camera being calibrated for a sounding rocket campaign (Franz *et al.*, 1990; Winckler, 1996). This initial evidence led to several large campaigns to characterize sprites, many of these were centered around the Yucca Ridge Field Station in Colorado (Lyons, 1995, 1996) and aircraft flights over the central plains of the United States (Sentman *et al.*, 1995b; Wescott *et al.*, 1995).

The brightest features of sprites (the 'body') occur at ~70 km altitude. Often bright 'branches' diverge upward from the 'body' of the sprite. Further, there is usually also 'hair' topping the sprite, with no presently resolvable structure, and 'tendrils' extend to regions below the 'body' of the sprite. The upper and lower extremes in altitude have been triangulated to extend above 95 km and below 40 km (Sentman *et al.*, 1995b; Wescott *et al.*, 1998b). The horizontal extent of a sprite event can be as large as 50 km with a great deal of both vertical and horizontal structure in sprites. Recently, a telescope/imaging system recorded filament-like structures only several tens of meters wide (the resolution of the imaging system) in the 'body' of sprites (Gerken *et al.*, 2000). In general, as the resolving capabilities of sprite observing instrumentation has increased, further structure has been found indicating sprites are highly structured on a fine scale rather than simply a diffuse glow.

Intensified imaging systems used for sprite, jet, and elves observations have primarily been standard television video rate systems (providing 17 ms resolution from interlaced fields). One of the first optical observations of sprites was made in 1993 using photometers (with 100 μs resolution) in a comparative study of optical and Very Low Frequency (VLF) electric field observations (Winckler *et al.*, 1996). In 1995, observations of sprites and elves using a 4 channel 15 μs resolution photometer were made by Fukunishi *et al.* (1996). The photometers were occasionally run with a red-pass filter (passing light with wavelength greater 650 nm). One channel in this study was operated with a wide field of view, while three channels were positioned in a vertical array. This study determined that sprites generally begin at ~70 km and then propagate both upward and downward. Sprites were also observed to start with predominantly blue emissions (lasting less than 100 μs) followed by a much longer red emission from sprites (Takahashi *et al.*, 1995). Photometers filtered for specific $N_2$ groups operated at 1.3 ms time resolution were used to make measurements in 1995-1998 (Armstrong *et al.*, 1998) and are discussed further in the energetics section. In 1996, a blue filtered imager made observations in conjunction with a 50 μs resolution blue filtered photometer (Suszcynsky *et al.*, 1998). The photometer observations and high speed imager observations (with 1 ms resolution). Stanley *et al.* (1999) show that sprites develop and evolve on sub-millisecond time scales. All the spectral observations discussed later in this paper are made using video imaging systems (17 ms resolution) and therefore average over much of the temporal evolution of sprites.

## 2.2. Blue Jets and Blue Starters

On July 1, 1994 University of Alaska scientists aboard aircraft were observing a thunderstorm over Arkansas as part of a three week sprite campaign. Because of the storm/aircraft geometry (the cameras were viewing away from the sunset), the observations began about one hour earlier than most other nights. Fifty-nine examples of luminous columns of light propagating up from the storm tops were recorded during a 22 minute period. At the time of observation, these new phenomena were obviously different from sprites in two ways: (1) the propagation speed was such that the upward motion was easily apparent at video rates and (2) the emissions were blue, as observed by an intensified color camera. Due to the upward motion and the color of the phenomena, they were given the name blue jets. Triangulation analysis of 34 blue jets observed simultaneously by both aircraft found an upper altitude of 37.2±5.3 km and an upward velocity of 112 ± 24 km/s (Wescott *et*

*al.*, 1995). The blue jets do not occur in association with either positive or negative cloud-to-ground lightning as reported by the National Lightning Detection Network. The blue jets do occur over regions of storms actively producing negative cloud-to-ground discharges and heavy hail activity, but the jets are not temporally coincident with a single discharge.

A similar phenomena, blue starters, propagate up from the top of the thunderstorm (Wescott *et al.*, 1996). However, starters terminate at an average altitude of ~25.5 km and have a velocity range of 27-153 km/s. Intensified color camera observations record starters as the same color as blue jets. Thirty starters were observed in the same region of the storm as the blue jets, during the same 22 minute interval. On several other nights starters were observed over the active convective core of storms (but no blue jets were recorded). Blue starters were also observed during the EXL98 mission (described in the Energetics section). The starters were observed in blue filtered images and also in the near infrared. The implications of these observations will be discussed below.

## 2.3. Elves

A third phenomena observed above thunderstorms are elves (Emissions of Light and VLF perturbations due to EMP Sources (Fukunishi *et al.*, 1995)), very brief (less than 1 ms) red emissions at the bottom of the ionosphere (at the upper altitudes of sprites). Optical "airglow enhancements" associated with lightning were first reported based on shuttle observations (Boeck *et al.*, 1992) and possibly from sounding rockets (Li *et al.*, 1991). The phenomena were named elves after further documentation and photometer observations determined the brief (~1 ms) lifetime of elves (Fukunishi *et al.*, 1996). Strong Joule heating of the base of the ionosphere by tropospheric lightning EMP was predicted earlier (Inan *et al.*, 1991), and the possibility of optical emissions were first investigated by Taranenko *et al.* (1993b) stimulated by the observations of lightning associated airglow enhancements by Boeck *et al.* (1992). The model for elves is the interaction between the (spherically expanding) EMP from a cloud-to-ground lightning discharge and the (planar) lower ionosphere (Inan *et al.*, 1996b). The radially expanding disc (intersection of the expanding sphere and plane) was measured using a "fly's-eye" array of seventeen photometers operating at 40 kHz (Inan *et al.*, 1997).

Elves occur above thunderstorms, both in conjunction with sprites and alone. Just prior to publication of this manuscript, a distinction between elves and 'sprite halos' has been made. Elves are very brief (less than 1 ms) emissions which generally cannot be observed with imaging systems operating at standard video rates of 30 fields per second. "Sprite halos" are morphologically similar to elves, but last between 3-6 ms, occur at lower altitudes, and are smaller than elves horizontally. Halos observed in the intensified color camera are red, and spectral observations of halos identify the main emissions to be $N_2$(1PG) (Heavner, 2000). Filtered observations of $N_2$(1PG) emission confirm the identification of halos emissions (Mike Taylor, personal communications). To date, triangulation analysis of elves has not been performed. Single station analysis using National Lightning Detection Network lightning location estimates the altitude of elves between 75-95 km, the typical vertical thickness is several km, and the horizontal extent as much as 530 km (Dial and Taylor, 1999). Elves are observed above both positive and negative CG lightning (Barrington-Leigh and Inan, 1999).

## 3. ENERGETICS

Color camera observations of sprites and blue jets from 1994 provided the first information about the energy processes producing middle atmospheric optical emissions (Sentman et al., 1995b; Wescott et al., 1995). The optical signature from sprites appeared primarily in the red channel of the color camera with some blue emissions from the lower portions of sprites. Blue jets appear solely in the blue channel. Spectrographic observations of sprites in 1995 identified the brightest red emissions as the molecular nitrogen first positive group ($N_2$(1PG)) (Mende et al., 1995; Hampton et al., 1996). Ground based blue filtered photometer observations and blue and red filtered observations were made in the summers of 1996, 1997, and 1998 to understand the blue emissions of sprites (Armstrong et al., 1998; Suszcynsky et al., 1998). Most observations of optical emissions above thunderstorms are made at low slant angles (10° or less), so severe Rayleigh scattering is present and atmospheric transmission in the blue is poor (c.f. Section 2 of Morrill et al. (1998)). The EXL98 (Energetics of Upper Atmospheric Excitation by Lightning, 1998) campaign was designed specifically to study microphysical energy processes of sprites, blue jets, and elves. EXL98 (Sentman et al., 1998) was centered on a series of aircraft flights during July 1998 with several intensified cameras. These cameras covered wavelengths from the near-UV (320 nm) to the near-IR (>1500 nm) and provided a unique opportunity to study the energetics of these phenomena. One reason for the use of aircraft in 1998 was to get above the dense atmosphere to facilitate blue observations.

### 3.1. Energetics of Sprites–Observations

Ground based red spectral observations of sprites were made during the northern hemisphere summers of 1995 and 1996. The first measured optical spectrum of a sprite was obtained June 21, 1995 from the Mt. Evans observatory in Colorado (Hampton et al., 1996). The 1995 red spectral observations characterized sprite emissions between 550-850 nm, documenting that the primary optical emissions from sprites is $N_2$(1PG) (Mende et al., 1995; Hampton et al., 1996), a molecular nitrogen group with red emissions brightest between 700-1000 nm. In 1996, spectral observations increased the red range of observations to 1000 nm and attempted to improve the characterization of blue emissions. The observed $N_2$(1PG) emission are due to molecular nitrogen electronic transitions $N_2(B^3\Pi_g \rightarrow A^3\Sigma_u^+)$, with a quenching altitude of ~53 km (see Table 1).

One specific event, from July 24, 1996 at 03:58:23.975 required the inclusion of $N_2^+$(Meinel) emission to generate a reasonable synthetic spectral fit to the observed spectrum (Morrill et al., 1998; Bucsela et al., 2000). The observed spectrum and synthetic spectral fit appear in upper panel of Figure 2. The resulting relative vibrational distribution for the $N_2(B^3\Pi_g)$ and $N_2^+(A^2\Pi_u)$ states are presented in the lower panel of Figure 2. This observation is interesting since the horizontal slit of the spectrograph was measuring sprite emissions from an altitude of 57 km, well below the quenching altitude of $N_2^+$(Meinel) (85-90 km). As discussed by Morrill et al. (1998) and Bucsela et al. (2000) the presence of $N_2^+$(Meinel) emission may be due, in part, to energy transfer processes beyond simple quenching. The possible presence of $N_2^+$(Meinel) emission in the spectra of Mende et al. (1995) and Hampton et al. (1996) was also discussed by Green et al. (1996). The $N_2^+$(Meinel) emission requires at least 16.5 eV to excite the lowest vibration level of the upper state, indicative of higher energetic processes than $N_2$(1PG) emission (requiring only 7.5 eV to excite). Additionally, this event had the strongest signature of blue emissions (as observed using the blue filtered imager, see Figure 2 of Suszcynsky et al. (1998)). Blue emissions, whether $N_2$(2PG) or $N_2^+$(1NG), are indicative of a higher energy process than the red $N_2$(1PG) emissions.

Several molecular nitrogen electronic transitions are summarized in Table 1. The upper and lower electronic states are listed in this table of the emissions. The quenching altitude is the altitude at which 50% of the population of upper electronic state is collisionally deactivated before it undergoes radiative decay (Vallance Jones, 1974). The threshold energy in the table is the energy between the lowest vibrational level of the up-

Table 1. Several neutral and ionized $N_2$ emissions observed in sprites, blue jets, and elves, as well as the aurora. The first observations of sprites showed only $N_2(1PG)$ emission, which requires the lowest threshold energy of the $N_2$ states that emit optically. The threshold energies reported are based on filling the lowest vibration energy level of the electronic state. This table is based on Table 4.7 of Vallance Jones [1974].

| Name | Upper State | Lower State | Lifetime | Quench Alt. | Energy |
|------|-------------|-------------|----------|-------------|--------|
| $N_2(1PG)$ | $N_2(B^3\Pi_g)$ | $N_2(A^3\Sigma_u^+)$ | 6 $\mu$s | 53 km | 7.50 eV |
| $N_2(2PG)$ | $N_2(C^3\Pi_u)$ | $N_2(B^3\Pi_g)$ | 50 ns | 30 km | 11.18 eV |
| $N_2^+(1NG)$ | $N_2^+(B^2\Sigma_u^+)$ | $N_2^+(X^2\Sigma_g^+)$ | 70 ns | 48 km | 18.56 eV |
| $N_2^+(M)$ | $N_2^+(A^2\Pi_u)$ | $N_2^+(X^2\Sigma_g^+)$ | 14 $\mu$s | 85-90 km | 16.54 eV |
| $N_2(VK)$ | $N_2(A^3\Sigma_u^+)$ | $N_2(X^1\Sigma_g^+)$ | 2 s | 145 km | 6.31 eV |

per state and the lowest vibrational level of the ground state of $N_2$, although the peak in the electron impact excitation cross-section occurs at slightly higher energy (Vallance Jones, 1974).

A series of recent photometric studies have examined time-resolved blue/NUV emissions. In 1995, Armstrong et al. (1998) used a filter centered at 431.7 nm with a FWHM (full width, half maximum) of 10.6 nm. In 1996, a second filter centered at 399.2 nm with a FWHM of 9.6 nm was used in conjunction with the 431.7 nm filter. Both of these filters include both $N_2^+(1NG)$ and $N_2(2PG)$ emissions in their bandpass, making definitive observations regarding ionized emissions from sprites difficult. However, the ratio of the two filtered photometers is able to discriminate between lightning, sprites, and elves. The 399.2/431.7 ratio for lightning is approximately unity, in agreement with the expected continuum radiation. For sprites, the 399.2/431.7 ratio is $\sim$2 while elves alone (based on a lower number of observations) give a measured 399.2/431.7 ratio of $\sim$3. The temporal evolution of the ratio of the two filters suggests an initial process with an electron temperature equivalent to $\sim$10 eV (for less than 1 ms) followed by a longer lasting $\sim$1 eV process. Similar observations were made by Suszcynsky et al. (1998) using a 20 nm wide filter centered on 425 nm with a photometer to observe the 427.8 nm $N_2^+(1NG)(0,1)$ emission. However, the filter bandpass also included the lower energy $N_2(2PG)(1,5)$ emission at 426.8 nm as well as several other less intense $N_2(2PG)$ emissions which may have contributed to a portion of the observed signal. A blue filtered camera (response centered at 410 nm with a passband between 350-475 nm) was used in conjunction with the photometer. The important point to note here is that both of these studies indicate the presence of $N_2^+$ emissions during the initial portion of the sprite.

The EXL98 campaign used an aircraft to get above the most dense portion of the atmosphere, in order to characterize the blue emissions of sprites. This involved both broad and narrow band video observation as well as NUV/blue spectral observations. The brightest $N_2^+(1NG)$ emission is at 391.4 nm (1NG, (0,0) band), but instrumental response, atmospheric transmission, and the close 389.4 nm and 394.3 nm $N_2(2PG)$ emissions made the second brightest $N_2^+(1NG)$ emission at 427.8 nm (1NG, (0,1) band) the target for filtered imaging during EXL98. Knowledge of possible contamination by neutral $N_2$ emission (2PG) through the filter at 426.8 nm is critical in both the filter selection and data analysis. This is especially important for observations at the edge of the image due to the shift of the filter response toward the blue with increased angle.

The first NUV/blue spectral observations of sprites were recorded during the mission (Heavner et al., 2000). The primary blue $N_2$ emissions identified in the observed spectra is the $N_2(2PG)$ neutral emission from the $N_2(C^3\Pi_u{\rightarrow}B^3\Pi_g)$ transition. $N_2^+(1NG)$ emission from the $N_2^+(B^2\Sigma_u^+{\rightarrow}X^2\Sigma_g^+)$ transition appears but is very weak in these spectra. An example of the blue spectral observations from a sprite observed at July 28, 1998 06:41:01.278 is presented in Figure 3. The upper panel is from a panchromatic imager with a 13.7° x 10.0° field of view. The black box indicates the NUV spectrograph field of view defined by the spectrograph entrance slit. The lower panel is the observations from the NUV spectrograph. The observed spectrum is plotted with a solid line, and a synthetic fit to the spectrum including both $N_2(2PG)$ and $N_2^+(1NG)$ is overplotted (see Bucsela and Sharp (1997) for a discussion of the fitting technique). The higher energy component ($N_2^+(1NG)$) of the synthetic fit is plotted as a separate curve. The observed blue spectrum is from the "body" of the sprite, while

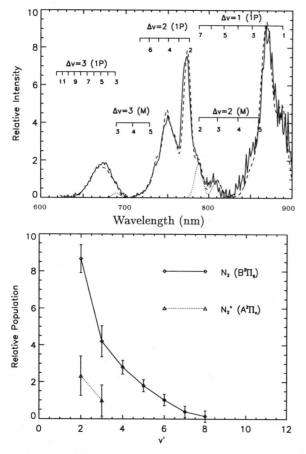

**Figure 2.** Sprite spectrum of the 1996/07/24 03:58:24 event with a least-squares fit consisting of both N$_2$(1PG) and N$_2^+$(Meinel) synthetic spectra. The higher energy N$_2^+$(Meinel) component is shown separately as a dotted line. The vibration distributions determined in the fit are shown in the lower panel. This plot is reproduced from figure 2 of [?].

both color and filtered images of sprites show the majority of blue light is from the lower portions (Sentman *et al.*, 1995b). There is no significant signal indicated in the 427.8 nm image of the sprite body, in agreement with the observed spectrum. The sprite occurred at a great circle distance of 329 km from the EXL98 aircraft (flying at ~14 km altitude).

In contrast to the lack of any blue ionized nitrogen signature in the previous example, a second EXL98 observation, from July 24, 1998 04:57:43 is presented in Figure 4. The upper panel of Figure 4 is an image of the sprite from the wide field-of-view (FOV) camera, with the white square indicating the FOV of the narrow field camera. The lower right panel is the cropped image from the 427.8 nm filtered and the lower left panel is from an identical camera which was not fil-

tered. The 427.8 nm image has been processed with histogram equalization, a method of stretching the dynamic range of the image. The observations of the sprite in the 427.8 nm filtered imager indicates a higher energy process occurring in the tendrils (18.6 eV electrons are required to excite the lowest N$_2^+$(1NG) state from the N$_2$ ground state). The post flight calibration of the EXL98 filtered cameras is currently underway. Once complete these results will provide improved estimates of electron energies occurring in the various portions of sprites.

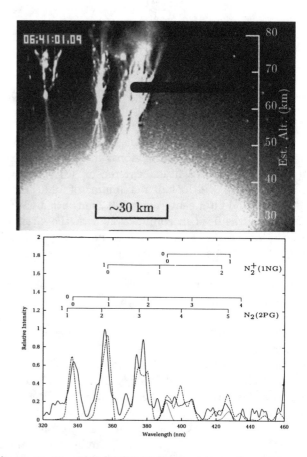

**Figure 3.** The 1998/07/28 06:41:01 sprite as imaged by an unfiltered camera on the EXL98 aircraft is presented in the upper panel. The black box indicates the field of view of the NUV/Blue spectrograph. The solid line in the lower panel is the observed spectrum. A synthetic fit of the spectrum including both N$_2$(2PG) and N$_2^+$(1NG) emissions is shown as a dashed line, while just the N$_2^+$(1NG) contribution to the fit is indicated as a dotted line. This blue spectral observation of a sprite has no ionized emissions above the noise levels of the observation. A 427.8 nm filtered imager recorded no emissions from the sprite, in agreement with the spectral observations. The wavelength of the band head of emissions and the upper and lower vibration levels of the N$_2^+$(1NG) and N$_2$(2PG) transitions are indicated.

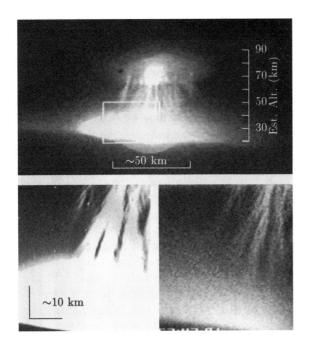

**Figure 4.** The 1998/07/22 04:57:43 sprite, at a great circle distance of ~200 km, as seen by three cameras on the EXL98 aircraft is presented. The upper panel is the unfiltered wide field of view (~40°) camera, illustrating the entire sprite event with a white box indicating the approximate field of view of the other cameras. The lower left image is the cropped image from the unfiltered narrow field of view camera and the lower right image is from the 427.8 nm filtered imager. Significant $N_2^+$(1NG) emission is observed at 427.8 nm indicating energetic processes occurring in the tendrils of sprites. Unfortunately, on July 22, the NUV camera (which recorded the blue spectrum in Figure 3) was not operating in spectrographic mode.

Preliminary estimates of sprite energies can be made from the observed optical emissions. Starting with the 12-60 kJ per sprite reported by Sentman et al. (1995b) the energy per sprite is calculated as outline in the folowing discussion. Stellar calibrations were used to determine absolute photon flux measurements of sprites in the UAF 1994 measurements, yielding total optical energies of ~50 kJ per sprite. The 1994 observations were based on color camera measurements with response between 395-700 nm. The observed emissions are primarily from the $N_2$(1PG). Based on the vibrational distributions determined by the spectral fitting and the camera response convolved with the $N_2$(1PG) spectrum, only ~5% of the total $N_2$(1PG) emission is detected by the color camera. Matching the observed photon flux in the image of the sprite to the observed spectrum yields a photon spectrum of the sprite. The color camera photon calibration assumed all emissions

centered at the peak response of the red channel, or 650 nm.

This energy flux is used to calculate the total energy deposited in molecular nitrogen and oxygen as follows. Geometrical integration of the stellar calibrations yields a total flux $1.81 \times 10^{23}$ photons per sprite. Because only 5% of the photons emitted from radiating $N_2(B^3\Pi_g)$ molecules are within the color camera bandpass, the above flux implies there are $3.62 \times 10^{24}$ radiating $N_2(B^3\Pi_g)$ molecules. Assuming that half of the $B^3\Pi_g$ state $N_2$ molecules which are excited via electron impact are quenched, the total $N_2(B^3\Pi_g)$ population is $7.24 \times 10^{24}$. For 1 eV excitation energy, the ratio of all $N_2$ electronically excited states to $B^3\Pi_g$ electronic state excitation is 3.45 (Slinker and Ali, 1982). This gives $2.5 \times 10^{25}$ total $N_2$ electronically excited molecules. Assuming that the upper states all cascade down to the $N_2(A^3\Sigma_u^+)$ state, each molecule has energy ~6.5 eV following radiative and collisional cascade (Morrill and Benesch, 1996). Therefore the total energy stored in electronic excitation is ~26 MJ.

Energy is also distributed amongst vibrationally excited states within the electronic states, so we calculate the vibrational energy similarly. From above, there are $3.62 \times 10^{24}$ $N_2(B^3\Pi_g)$ molecules. The ratio of $N_2$ vibrationally excited states to the $N_2(B^3\Pi_g)$ state is 2674 for a 1eV electron energy distribution (Slinker and Ali, 1982), so the total number of vibrationally excited $N_2$ states is $1.94 \times 10^{28}$. Assuming the average vibrational energy is 0.3 eV (corresponding to $N_2(X^1\Sigma_g^+)(v=1)$), there is ~930 MJ of energy in the vibrationally excited $N_2$ states. Combining the vibrational and electronic energy calculated above, we find ~950 MJ deposited in molecular nitrogen by a sprite whose optical energy is 50 kJ. Assuming similar excitation in $O_2$, scale by the relative densities of $N_2$ and $O_2$ for a total energy deposition of 1.2 GJ per sprite in the middle- and upper-atmosphere. Considerations of molecular nitrogen energy transfer processes may decrease the above calculation of energy deposition by a factor of four (200-300MJ).

### 3.2. Energetics of Sprites–Models

Three types of models have been proposed to describe sprites using the quasi-electrostatic heating and runaway electron mechanisms (see Rowland (1998) for a review of models). These models take a tropospheric lightning discharge as the initial energy source and propose mechanisms for the propagation of this energy to the electron impact excitation of middle- and upper-atmospheric molecular nitrogen, resulting in the observed optical emissions. The quasi-electrostatic heating model and the EMP-induced breakdown model describe the heating of ambient thermal electrons through electric fields while the runaway electron model proposes the energization of neutrals directly by high energy electrons. In these models, heated/accelerated electrons collisionally transfer energy to the neutrals to produce optical emissions, ionization and possible long-term (hour time scale) heating of the upper atmospheric neutrals (Picard *et al.*, 1997).

The quasi-electrostatic heating model (Pasko *et al.*, 1997) postulates electric fields which are a result of the sudden reconfiguration of thunderstorm charge due to the removal of charge by a lightning discharge. A large positive cloud to ground flash removes positive charge from the upper charge center (*e.g.* 300 C removed from 10 km altitude) which produces a large change in the electric field (greater than 100 V/m at 70 km altitude). The field persists for approximately the local relaxation time (milliseconds at 85 km altitude, but seconds at 45 km altitude). The strength of the field is determined by charge moment ($Q \times L$) of the lightning flash. In order to test the model, experimental determination of charge moment from ELF/VLF (10 Hz - 30 kHz) measurements has been developed (Cummer *et al.*, 1998). Similar to the quasi-electrostatic model is the EMP induced breakdown model which includes the addition of an upward propagating EMP associated with a large lightning stroke that can produce breakdown at altitudes above 60 km (Taranenko *et al.*, 1993a, 1993b; Inan *et al.*, 1996a; Fernsler and Rowland, 1996; Rowland *et al.*, 1995, 1996; Milikh *et al.*, 1995).

The runaway electron model postulates a high energy seed electron (possibly a cosmic ray secondary) accelerated by the electric field above a thunderstorm (Roussel-Dupré *et al.*, 1998; Taranenko and Roussel-Dupré, 1996; Lehtinen *et al.*, 1996). If the seed electron has enough kinetic energy, a collision between the seed electron and a low energy (ambient) electron can result in two relativistic electrons. This process is predicted to produce a beam of avalanching runaway electrons. Optical emissions produced from this mechanism would have a much higher characteristic energy than the quasi-electrostatic heating model predicts.

### 3.3. Energetics of Blue Jets/Starters

Blue jets and starters are upward propagating cones of blue emissions. At the altitudes of blue jets (below $\sim$40 km), the $N_2$(1PG) upper state ($N_2$($B^3\Pi_g$)) is strongly quenched (*e.g.* the upper state is collisionally deactivated more rapidly than decay by spontaneous emission), so no red $N_2$(1PG) emission is observed in blue jets. The blue light is affected strongly by Rayleigh scattering at long path lengths, while the close proximity to the storm required for short path lengths ($\sim$150 km) make blue jet/starter observations a geometrical problem: the blue jets and starters occur over the active cores of storms, so large storm systems may have intervening clouds blocking viewing of blue jets/starters. For these reasons, the dearth of ground based observations of blue jets may not be truly indicative of their frequency of occurrence.

The blue emissions in jets are most likely either from $N_2$(2PG) or $N_2^+$(1NG) emissions. Analysis of color camera observations of blue jets found emissions from the lower energy $N_2$(2PG) alone could not sufficiently account for the observations (Wescott *et al.*, 1998b), so $N_2^+$(1NG) emission were suggested. No blue jets or starters were observed with the NUV/Blue spectrograph on EXL98, but several starters were clearly recorded by the 427.8 nm imager. Blue starters are similar to blue jets, propagating from cloud tops, but terminate abruptly after less than 10 km of upward travel. A blue starter observed July 17, 1998 during the EXL98 campaign is presented in Figure 5. The left panel is a view of the 340.7 nm (a neutral $N_2$(2PG) emission) filtered observation, the center panel is the 427.8 nm filtered image, and the right panel is the unfiltered narrow imager. Although the starter is bright in the 427.8 nm filtered camera, it extends over a much smaller region than the unfiltered or 340.7nm images. This confirms the suggestion of ionized emissions in starters (and likely blue jets as well) although the ionization appears to be confined to the central region. Blue jets possibly have the signature of higher total energy processes than either sprites or elves because of the ionized emissions, and blue jets last 10-100 times longer than other phenomena. Wescott *et al.* (1998a) estimate that the average blue jet or starter transfers about $10^9$ J of energy to the stratosphere.

In addition to blue emission, jets have been observed in the near-IR during EXL98. These emissions are most likely some combinations of $N_2$(1PG) and $N_2^+$(Meinel). Considering the quenching heights in Table 1 $N_2$(1PG) and $N_2^+$(Meinel) emission should be strongly quenched. However, energy transfer processes may enhance the population of the lower vibrational level of the $N_2$(1PG) and $N_2^+$(Meinel) upper electronic states. Another possibility would be emission from atomic species (O and N) as is observed in lightning. The existence of atomic emissions would indicate a much more energetic process than is thought to occur in sprites and elves. This issue will best be resolved by spectroscopic observations of blue jets and starters.

**Figure 5.** Observations of the July 17, 1998 starter from the EXL98 aircraft. The left image is from the 340.7 nm filtered imager. The 427.8 nm filtered image is the center image and the right image is from the unfiltered imager. The 340.7 nm and 427.8 nm images have been histogram equalized to stretch the dynamic range of the images. The background of the 340.7nm and 427.8nm images have been surpressed so that only the emmission associated with the starter can be seen. These three images have been scaled to a common field-of-view. The emission at 427.8 nm from $N_2^+$(1NG) indicates that electron energies of at least 18.6 eV are present in the processes causing blue starters. This emission appears to extend over a smaller region and the neutral (340.7nm) emission.

Two models based on streamer-type phenomena have been proposed to explain blue jets (Pasko *et al.*, 1996; Sukhorukov *et al.*, 1996; Sukhorukov and Stubbe, 1998; Yukhimuk *et al.*, 1998). Based on modeling efforts, a single jet has been postulated to cause local density perturbations of nitric oxide (10%) and ozone (0.5%) at 30 km altitude (Mishin, 1997). As mentioned above, if the processes leading to blue jets and sprites are energetic enough to create atomic species of N and O, the implications for other chemical effects is significantly increased.

### 3.4. Energetics of Elves

Elves have been described by models as heating from the EMP from tropospheric lightning (Inan *et al.*, 1997; Pasko *et al.*, 1998). The optical emissions of elves have been identified as originating from $N_2$(1PG) (Heavner, 2000). The short duration of elves suggests that individual elves are less important energetically than sprites or jets. However, elves (optical emissions) may represent only the high energy component of the effect of lightning discharges transferring energy to the middle atmosphere. Lower energy lightning discharges may not cause optical emissions, but the majority of tropospheric lightning discharges can still cause neutral heating in the middle and upper atmosphere but do not stimulate optical emissions. Before the documentation of elves, the neutral heating of the upper atmosphere by lightning was proposed as a mechanism for creating a long-term (hour time scale) infrared glow above thunderstorms (Inan *et al.*, 1991; Picard *et al.*, 1997). The mechanism described the vibrational excitation of ground state $N_2$ by electron impact followed

by energy transfer via $N_2(X^1\Sigma_g^+)(v>0)\rightarrow CO_2$. Because elves are observed to occur above both positive and negative CG's they may occur more frequently than sprites.

## 4. GLOBAL FREQUENCY

The majority of sprite observations have been made during the northern hemisphere summers and with large thunderstorms over the midwestern high plains of North America. Few observations from a platform which has a global view have been made – a total of 19 sprites have been identified in the video observations from the U.S. space shuttle (Boeck *et al.*, 1998). The distinct signature of a sprite has not been reported in any satellite observations.

In February-March 1995, an aircraft campaign to explore the global distribution of sprites was based in Lima, Peru (Sentman *et al.*, 1995a; Heavner *et al.*, 1995). Although Amazonia is one of the most active lightning regions of the world only ~20 sprites were observed during a two week period. Similar aircraft campaigns over the central United States observed hundreds of events. The observed low activity level is likely due in part to logistical and observational challenges of the particular observations: a border war between Peru and Ecuador was being fought during the campaign, and a system such as the U.S. National Lightning Detection network was not available to identify regions of strong positive lightning activity. The NLDN data greatly improves the identification of likely sprite producing regions of storms. Effects of the nearly horizontal magnetic field of the Earth near the equator is another possible factor for the apparent low frequency of sprite observations in low latitude regions. Theorists postulating a runaway electron beam associated with sprites and blue jets generally agree that a horizontally aligned Earth's magnetic field will deacrease the formation of the runaway electron beams (Lehtinen *et al.*, 1997; Gurevich *et al.*, 1996; Taranenko and Roussel-Dupré, 1997). The proposed QE heating models do not generally address any dependence on the orientation of the local Earth's magnetic field.

Research has also been underway in Australia during southern hemisphere summers since 1997 (Hardman *et al.*, 1998) and recently observations of sprites occurring above Japanese winter thunderstorms have been made (Fukunishi *et al.*, 1999). Intensified cameras aboard the space shuttle have observed at least one sprite above thunderstorms in Africa, as well as several above South America (Boeck *et al.*, 1998).

In order to approach the questions of global rates and distributions of sprites, elves, and blue jets, more observations are required. The identification of a global synoptic detection method of these events would enhance such measurements. Attempting to estimate the global occurrence of sprites is an elusive problem. The latest estimate of the global cloud to ground lightning flash rate of between $10s^{-1}$ and $14s^{-1}$ (Mackerras et al., 1998). Positive flashes are less than 10% of the total lightning flashes, and not every positive flash produces an optically detectable sprite. Therefore a preliminary upper limit for the global occurrence of sprites is estimated at 1 per second. However, this is based on the assumption that the global sprite distribution is similar to global lightning distribution. While current observations are biased geographically, lightning and sprites do not appear to necessarily have the same spatial distribution.

Based on the estimates of one sprite per second globally and one GJ energy per sprite, the global energy deposition in the mesosphere is approximately $2~\mu W/m^2$. Comparing this with Figure 1(c) of Mlynczak (2000), sprites may be .1% of the mesospheric energy budget. The more significant effects are likely to be on smaller spatial scales. Often an active thunderstorm produces several hundred sprites in a few hours, which leads to more significant local heating.

Blue jets are rarely recorded by ground based observers and even aircraft campaigns record many more sprites than blue jets/starters. However, the dearth of jet observations may be partially explained by Rayleigh scattering and observational difficulties described earlier. The long duration of blue jets and the observation of strong $N_2^+(1NG)$ emissions indicate that blue jets may be an important energetic process in the stratosphere. Elves are associated with both positive and negative cloud-to-ground lightning, so the global elves occurrence rate is probably higher than the global sprite rate.

## 5. SUMMARY

In the past decade, several types of optical emissions occurring above thunderstorms have been identified. The emissions span the middle- and upper-atmosphere between thunderstorm tops and ~95 km. Spectroscopic and filtered observations of sprites and blue jets have been presented and discussed. The observed nitrogen emissions indicate electrons with energies of at least 18.6 eV are required to describe some of the observed emissions. A 1 eV Boltzmann electron distribution (modified with a high energy tail component) matches the observations and is physically realistic. Based on observations of the total optical energy emitted by a sprite, we estimate the total energy deposited into ac-

tive molecular nitrogen (both vibrational and electronic state energy) to range from 250 MJ to 1 GJ. Recent observations from EXL98 under current analysis will help clarify and confirm these values. The preliminary estimate on global occurrence rates of sprites, blue jets, and elves, is on the order of 1 per second.

*Acknowledgments.* Dan Osborne, Jim Desroschers, Laura Peticolas, Veronika Besser, and Don Hampton were instrumental to data collection and campaign operations. We also thank the High-frequency Active Auroral Research Program for loan of the Near-IR instruments used in the aircraft missions. Aeroair, Inc., and particularly Jeff Tobolsky, made all UAF aircraft missions fly. MJH acknowledges Dirk Lummerzheim for fruitful discussion. The GI-UAF group was supported by NASA Grant No. NAG5-5125. The GI-UAF, NRL, and Raytheon groups were supported by NASA Grant No. NAG5-5019. JSM was partially supported by the Edison Memorial graduate training program at NRL, EJB was supported as an ASEE postdoc, and CLS was supported by ONR 6.1 funds.

## REFERENCES

Armstrong, R. A., et al., Photometric measurements in the SPRITES '95 & '96 campaigns of nitrogen second positive (399.8 nm) and first negative (427.8 nm) emissions, *J. Atmos. Solar Terr. Physics*, 60(7-9), 787–799, 1998.

Barrington-Leigh, C. P. and U. S. Inan, Elves triggered by positive and negative lightning discharges, *Geophys. Res. Lett.*, 26(6), 683–686, 1999.

Berger, K., Blitzstrom-parameter von aufwärtsblitzen, *Bull. Schweiz. Elektrotech. Ver.*, 69, 353–360, 1978.

Boccippio, D. J., et al., Sprites, Extreme-Low-Frequency transients, and positive ground strokes, *Science*, 269, 1088–1091, 1995.

Boeck, W. L., et al., Lightning induced brightening in the airglow layer, *Geophys. Res. Lett.*, 19, 99–102, 1992.

Boeck, W. L., et al., The role of the space shuttle videotapes in the discovery of sprites, jets and elves, *J. Atmos. Solar Terr. Physics*, 60(7-9), 669–677, 1998.

Bucsela, E., et al., $N_2(B^3\Pi_g)$ and $N_2^+(A^2\Pi_u)$ vibrational distributions observed in sprites, *Geophys. Res. Lett.*, 2000, Submitted.

Bucsela, E. J. and W. E. Sharp, NI 8680 and 8629 Å multiplets in the dayglow, *J. Geophys. Res.*, 102, 2457–2466, 1997.

Cummer, S. A., et al., ELF radiation produced by electrical currents in sprites, *Geophys. Res. Lett.*, 25(8), 1281–1284, 1998.

Dial, R. and M. Taylor, Investigation of lightning induced elves over the great plains, CEDAR 1999 Workshop.

Fernsler, R. and H. Rowland, Models of lightning-produced sprites and elves, *J. Geophys. Res.*, 101, 29653, 1996.

Franz, R. D., R. J. Nemzek, and J. R. Winckler, Television images of a large upward electrical discharge above a thunderstorm system, *Science*, 249, 48–51, 1990.

Fukunishi, H., et al., Lower ionosphere flashes induced by lightning discharges, *EOS Supplement*, 76(46), F114, 1995.

Fukunishi, H., et al., Elves: Lightning-induced transient luminous events in the lower ionosphere, *Geophys. Res. Lett.*, *23*, 2157–2160, 1996.

Fukunishi, H., *et al.*, Occurrences of sprites and elves above the Sea of Japan near Hokuriku in winter, *EOS Supplement*, *80*(46), F217, 1999.

Gerken, E. A., U. S. Inan, C. P. Barrington-Leigh, Telescopic imaging of sprites, *Geophys. Res. Lett.*, in press, 2000.

Green, B. D., *et al.*, Molecular excitation in sprites, *Geophys. Res. Lett.*, *23*(23), 2161–2164, 1996.

Gurevich, A. V., *et al.*, Runaway electrons in the atmosphere in the presence of a magnetic field, *Radio Sci.*, *31*, 1541, 1996.

Hampton, D. L., *et al.*, Optical spectral characteristics of sprites, *Geophys. Res. Lett.*, *23*(1), 89–92, 1996.

Hardman, S. F., *et al.*, Sprites in Australia's Northern Territory, *EOS Supplement*, *79*(45), F135, 1998.

Heavner, M. J., *Optical Spectroscopic Observations of Sprites, Blue Jets, and Elves: Inferred Microphysical Processes and Their Macrophysical Implications*, Ph.D. thesis, University of Alaska, Fairbanks, 2000.

Heavner, M. J., *et al.*, Sprites over Central and South America, *EOS Supplement*, *76*(46), F115, 1995.

Heavner, M. J., *et al.*, NUV/Blue spectral observations of sprites in the 320-460 nm region: $N_2$ (2PG) emissions, *Geophys. Res. Lett.*, 2000, Submitted.

Inan, U., *et al.*, Rapid lateral expansion of optical luminosity in lightning-induced ionospheric flashes referred to as 'elves', *Geophys. Res. Lett.*, *24*(5), 583–586, 1997.

Inan, U. S., T. F. Bell, and J. V. Rodrigues, Heating and ionization of the lower ionosphere by lightning, *J. Geophys. Res.*, *18*, 705–708, 1991.

Inan, U. S., *et al.*, On the association of terrestrial gamma-ray bursts with lightning and implications for sprites, *Geophys. Res. Lett.*, *23*, 1017–1020, 1996a.

Inan, U. S., W. A. Sampson, and Y. N. Taranenko, Space-time structure of optical flashes and ionization changes produced by lightning-EMP, *Geophys. Res. Lett.*, *23*(2), 133–136, 1996b.

Lehtinen, N. G., *et al.*, $\gamma$-ray emission produced by a relativistic beam of runaway electrons accelerated by quasi-electrostatic thundercloud fields, *Geophys. Res. Lett.*, *23*(19), 2645–2648, 1996.

Lehtinen, N. G., *et al.*, A two-dimensional model of runaway electron beams driven by quasi-electrostatic thundercloud fields, *Geophys. Res. Lett.*, *24*(21), 2639–2642, 1997.

Li, Y. Q., *et al.*, Anomalous optical events detected by rocket-borne sensor in the WIPP campaign, *J. Geophys. Res.*, *96*(A2), 1315–1326, 1991.

Lyons, W. A., Low-light video observations of frequent luminous structures in the stratosphere above thunderstorms, *Monthly Weather Review*, *122*, 1940, 1995.

Lyons, W. A., Sprite observations above the U.S. High Plains in relation to their parent thunderstorm systems, *J. Geophys. Res.*, *101*, 29641–29652, 1996.

MacGorman, D. R. and W. D. Rust, *The Electrical Nature of Storms*. Oxford University Press, New York, NY, 1998.

Mackerras, D., *et al.*, Global lightning: Total, cloud and ground flash estimate, *J. Geophys. Res.*, *103*(D16), 19791–19809, 1998.

Mende, S. B., R. L. Rairden, and G. R. Swenson, Sprite spectra: $N_2$1PG band identification, *Geophys. Res. Lett.*, *22*, 2633–2636, 1995.

Milikh, G. M., K. Papadopoulos, and C. L. Chang, On the physics of high altitude lightning, *Geophys. Res. Lett.*, *22*, 85–88, 1995.

Mishin, E., Ozone layer perturbation by a single blue jet, *Geophys. Res. Lett.*, *24*(15), 1919–1922, 1997.

Mlynczak, M. G., A contemporary assessment of the mesospheric energy budget, this volume, 2000.

Morrill, J. S., and W. M. Benesch, Auroral $N_2$ emissions and the effect of collisional processes on $N_2$ triplet state vibrational populations, *J. Geophys. Res.*, *101*, 261–274, 1996.

Morrill, J. S., *et al.*, Time resolved $N_2$ triplet state vibrational populations and emissions associated with red sprites, *J. Atmos. Solar Terr. Physics*, *60*, 811–829, 1998.

Pasko, V. P., U. S. Inan, and T. F. Bell, Blue jets produced by quasi-electrostatic pre-discharge thundercloud fields, *Geophys. Res. Lett.*, *23*(3), 301–304, 1996.

Pasko, V. P., *et al.*, Sprites produced by quasi-electrostatic heating and ionization in the lower ionosphere, *J. Geophys. Res.*, *102*(A3), 4529–4561, 1997.

Pasko, V. P., U. S. Inan, and T. F. Bell, Ionospheric effects due to electrostatic thundercloud fields, *J. Atmos. Solar Terr. Physics*, *60*(7-9), 863–870, 1998.

Picard, R. H., *et al.*, Infrared glow above thunderstorms?, *Geophys. Res. Lett.*, *24*(21), 2635–2638, 1997.

Rodger, C. J., Red sprites, upward lightning and VLF perturbations, *Rev. Geophysics*, *37*(3), 317–336, 1999.

Roussel-Dupré, R., *et al.*, Simulations of high-altitude discharges initiated by runaway breakdown, *J. Atmos. Solar Terr. Physics*, *60*(7-9), 917–940, 1998.

Rowland, H., *et al.*, Lightning drive EMP in the upper atmosphere, *Geophys. Res. Lett.*, *22*, 361–364, 1995.

Rowland, H., R. Fernsler, and P. Bernhardt, Breakdown of the neutral atmosphere in the D-region due to lightning driven electromagnetic pulses, *J. Geophys. Res.*, *101*, 7935–7945, 1996.

Rowland, H. L., Theories and simulations of elves, sprites, and blue jets, *J. Atmos. Solar Terr. Physics*, *60*(7-9), 831–844, 1998.

Sentman, D., Electrical excitation of the middle and upper atmosphere by lightning: Special issue of the Journal of Atmospheric and Solar-Terrestrial Physics, *J. Atmos. Solar Terr. Physics*, *60*(7-9), 667, 1998.

Sentman, D. D., *et al.*, The Peru95 sprites campaign: Overview, *EOS Supplement*, *76*(17), S66, 1995a.

Sentman, D. D., *et al.*, Preliminary results from the Sprites94 aircraft campaign: 1. Red Sprites, *Geophys. Res. Lett.*, *22*, 1205–1208, 1995b.

Sentman, D. D., *et al.*, The EXL98 sprites campaign, *EOS Supplement*, *79*(45), F164, 1998.

Slinker, S. and A. W. Ali, Electron excitation and ionization rate coefficients for $N_2$, $O_2$, NO, N, O, Technical report, Naval Research Labs, 1982.

Stanley, M., *et al.*, High speed video of initial sprite development, *Geophys. Res. Lett.*, *26*(20), 3201–3204, 1999.

Sukhorukov, A. I. and P. Stubbe, Problems of blue jet theories, *J. Atmos. Solar Terr. Physics*, *60*(7-9), 725–732, 1998.

Sukhorukov, A. I., E. A. Rudenchik, and P. Stubbe, Simulation of the strong lightning pulse penetration into the lower ionosphere, *Geophys. Res. Lett.*, *23*, 2911, 1996.

Suszcynsky, D. M., *et al.*, Blue-light imagery and photometry of sprites, *J. Atmos. Solar Terr. Physics*, *60*(7-9), 801–809, 1998.

Takahashi, Y., *et al.*, Spatial and temporal relationship between lower ionospheric flashes and sprites, *EOS Supplement*, *76*(106), 1995.

Taranenko, Y. and R. Roussel-Dupré, Reply, *Geophys. Res. Lett.*, *24*(21), 2645–2646, 1997.

Taranenko, Y. N. and R. Roussel-Dupré, High-altitude discharges and gamma-ray flashes: A manifestation of runaway air breakdown, *Geophys. Res. Lett.*, *23*(5), 571–574, 1996.

Taranenko, Y. N., U. S. Inan, and T. F. Bell, Interaction with the lower ionosphere of electromagetnic pulses from lightning: heating attachment, and ionization, *Geophys. Res. Lett.*, *20*, 1539–1542, 1993a.

Taranenko, Y. N., U. S. Inan, and T. F. Bell, The interaction with the lower ionosphere of electromagnetic pulses from lightning: excitation of optical emissions, *Geophys. Res. Lett.*, *20*, 2675–2678, 1993b.

Vallance Jones, A., *Aurora*. D. Reidel Publishing Co., 1974.

Vaughan, Jr., O. H. and B. Vonnegut, Recent observations of lightning discharges from the top of a thundercloud into the clear air above, *J. Geophys. Res.*, *94*, 13179–13182, 1989.

Wescott, E. M., *et al.*, Preliminary results from the Sprites94 aircraft campaign: 2. Blue Jets, *Geophys. Res. Lett.*, *22*, 1209–1212, 1995.

Wescott, E. M., *et al.*, Blue starters: Brief upward discharges from an intense Arkansas thunderstorm, *Geophys. Res. Lett.*, *23*(16), 2153–2156, 1996.

Wescott, E. M., *et al.*, Blue Jets: their relationship to lightning and very large hailfall, and their physical mechanisms for their production, *J. Atmos. Solar Terr. Physics*, *60*(7-9), 713–724, 1998a.

Wescott, E. M., *et al.*, Observations of 'Columniform' sprites, *J. Atmos. Solar Terr. Physics*, *60*(7-9), 733–740, 1998b.

Wilson, C. T. R., The electric field of a thunderstorm and some of its effects, *Proc. Roy. Soc. London*, *37*, 32D, 1925.

Winckler, J. R., Further observations of cloud-ionosphere electrical discharges above thunderstorms, *J. Geophys. Res.*, *100*, 14335, 1995.

Winckler, J. R., *et al.*, New high-resolution ground-based studies of sprites, *J. Geophys. Res.*, *101*, 6997–7004, 1996.

Yukhimuk, V., *et al.*, Optical characteristics of blue jets produced by runaway air breakdown, simulation results, *Geophys. Res. Lett.*, *25*(17), 3289–3292, 1998.

M.J.Heavner, D.D. Sentman, D.R. Moudry, E.M.Wescott, Geophysical Institute, University of Alaska Fairbanks 903 Koyukuk Drive, Fairbanks AK 99775-7320 (e-mail: heavner@lanl.gov,[dsentman,drm,rocket]@gi.alaska.edu).

C.L.Siefring, J.S.Morrill, Naval Research Laboratory, Washington DC 20375 (e-mail: [siefring@ccs, morrill@shogun].nrl.navy.mil).

E.J. Bucsela, Raytheon ITSS, 4400 Forbes Blvd., Lanham MD (email: eric_j_bucsela@raytheon.com).

# A Review of $CO_2$ and CO Abundances in the Middle Atmosphere

Manuel López-Puertas and Miguel Á. López-Valverde

*Instituto de Astrofísica de Andalucía (CSIC), Granada, Spain*

Rolando R. Garcia and Raymond G. Roble

*National Center for Atmospheric Research, Boulder, Colorado*

A review is presented of $CO_2$ and CO in the middle atmosphere. Knowledge of their abundances is important for understanding the thermal budget and transport processes of the middle atmosphere, and for its remote sounding. For both CO and $CO_2$, several techniques have been used to measure their abundance. Significant improvements have occurred over the last two decades with high quality satellite data now available from the Atmospheric Trace Molecule Spectroscopy (ATMOS) and Improved Stratospheric and Mesospheric Sounder (ISAMS) experiments. The $CO_2$ observations are well explained by models that account for diffusive separation, but a small overestimate above 95 km still remains. TIME-GCM calculations suggest a large depletion in the polar winter, down to near the stratopause which is as yet unconfirmed but which could be significant for remote sensing. The CO observations confirm the basic features of the seasonal and latitudinal variability predicted by dynamical models, particularly the enhancements in polar winter. However, the models underpredict the mesospheric CO abundance at mid-latitudes by a factor of 1.5–3 and fail to reproduce the strong latitudinal and vertical gradients in the polar regions. A reduction in the OH abundance from standard chemical models may partially solve this discrepancy. Finally, a comparison of simultaneously measured $CO_2$ and CO in the upper mesosphere/lower thermosphere with models suggests an eddy diffusion coefficient about 2–3 times smaller than those currently used.

## 1. INTRODUCTION

It is well known that the abundance of carbon dioxide is increasing with time in the lower atmosphere at a rate of ~1.5–2 ppmv year$^{-1}$ in recent years due to fossil fuel combustion and cement production [see, e.g., *Hansen et al.*, 1998; *Keeling and Whorf*, 1998; *Brasseur et al.*, 1999]. This increase propagates upwards to the stratosphere and mesosphere with a time lag which depends on altitude and is about 5-6 years between the troposphere and stratosphere [*Bischof et al.*, 1985]. $CO_2$ is well mixed with a nearly constant volume mixing ratio (vmr) up to the upper mesosphere/lower thermosphere, where it begins to decrease due to the molecular diffusion ($CO_2$ is heavier than the mean molecular weight of the atmosphere), and to destruction by UV solar radiation.

Atmospheric Science Across the Stratopause
Geophysical Monograph 123

CO$_2$ is very important in the stratosphere and in the lower and middle mesosphere. Its infrared emission at 15 $\mu$m is the major radiative cooling mechanism in this region [see, e.g., *Mlynczak*, 2000, this issue], and it has been widely used to remotely sound the temperature and composition of these regions. Despite its importance, it has scarcely been measured at those altitudes. Only in situ aircraft and balloon measurements have been taken, with the major goal of monitoring the upward propagation of the CO$_2$ increase in the troposphere [see, e.g., *Bischof et al.*, 1985]. The CO$_2$ vmr has been traditionally considered to have no significant spatial (latitudinal or seasonal) variability in the middle atmosphere. We present in section 4 model results which question that understanding and discuss the role of the meridional advection on its global distribution.

In the upper mesosphere/lower thermosphere, the CO$_2$ 15 $\mu$m emission also represents the major energy loss of this region (between 90 and 130 km). In addition, satellite instruments, scheduled to be launched within the next two years, are sensitive enough to measure CO$_2$ infrared emissions with good signal to noise ratio in this region, e.g., SABER on the TIMED mission [*Mlynczak and Russell*, 1995] and MIPAS on Envisat [*Endemann and Fischer*, 1993]. The purpose of those measurements is not only to understand the radiative cooling but also to derive the temperature structure. At these altitudes, CO$_2$ 15 $\mu$m emissions are not in local thermodynamic equilibrium (LTE), and this must be taken into account in the temperature retrieval scheme. We now have the computational resources necessary for carrying out such retrievals, and the necessary non-LTE knowledge is expected to be acquired in the near future with new laboratory experiments to determine the CO$_2$(0,1,0)–O($^3$P) collisional rate coefficient. However, the CO$_2$ vmr must also be known. This emphasizes the importance of determining its distribution in the upper atmosphere.

Another aspect of importance is the possible impact of the continuous increase of CO$_2$ on the structure of the upper atmosphere. Because of its large radiative cooling, the increase of CO$_2$ (which is expected to warm the troposphere and cool the stratosphere) should also give rise to an even larger cooling of the upper atmosphere, hence decreasing its temperature and producing a contraction of these upper regions [see, e.g., *Roble and Dickinson*, 1989; *Roble*, 1993]. So far it is unclear if the upper mesosphere/lower thermosphere has become colder in the last few decades [see, e.g., *Danilov*, 1998]. In addition, it remains to be demonstrated that, if such cooling occurs, it is caused by the CO$_2$ increase.

CO$_2$ abundance has been measured more frequently in the upper mesosphere/lower thermosphere, mainly because of its important role in the energy budget, and for understanding its departure from diffusive equilibrium. Early rocket measurements in this region indicated that it was well mixed up to about 100 km and, for this reason, many model calculations in the past have neglected diffusive separation. As a result, the CO$_2$ vmr above about 90 km has been overpredicted compared to recent observations [see, e.g., *Rodrigo et al.*, 1986, and references therein; *Taylor*, 1988; *López-Puertas and Taylor*, 1989]. Although the coefficient of molecular diffusion is well known, the altitude dependence of eddy mixing, parameterized by the eddy diffusion coefficient, and its seasonal/latitude variation, are incompletely understood. We present in section 2 a review of the CO$_2$ measurements in the upper mesosphere/lower thermosphere, and compare them with 2D and 3D model calculations in section 4. We discuss in that section the relative roles of the eddy and molecular diffusion coefficients, how the latter contributes to the rapid fall-off of CO$_2$ vmr above about 85–90 km, and how it could explain the spread in the observed CO$_2$ vmr's in this region.

Carbon monoxide abundances are well documented in the troposphere, both as regards their spatial distribution [see, e.g., the August 1994 special issue of *JGR*], and their interannual trend [*Novelli et al.*, 1998; *Clerbaux et al.*, 1998]. We are interested here in its distribution and variability in the middle atmosphere, which is much less known and understood, specially in a global sense. Theoretical interest on CO as a dynamical tracer in the middle atmosphere goes back to *Hays and Olivero* [1971]. It has also been suggested that its strong tropospheric variability may be affecting the chemistry of the stratosphere [*Brasseur and Solomon*, 1986]. In addition to the study of the CO production terms and transport mechanisms with a number of theoretical models of the middle atmosphere [*Solomon et al.*, 1985; *Fleming et al.*, 1995], CO concentrations have been investigated using quite diverse measurement techniques, both in-situ (balloon and rocket) and remote (from the ground and from satellites), all of which have tried to overcome the difficulties of detecting this constituent's small concentration and relatively weak emission rates above the tropopause.

Current theoretical models show that CO is produced in the stratosphere and lower mesosphere as an end product of methane oxidation and that it only has one important chemical sink, its reaction with OH. In the mesosphere and lower thermosphere the major source is CO$_2$ photolysis. At these high altitudes chemical losses of CO are very minor. Due to this thermospheric source and subsequent downward transport into the middle atmosphere, the CO vmr generally increases with height throughout the middle atmosphere. Chemical time scales increase with altitude steadily throughout the mesosphere, and vertical and meridional transport processes should affect the CO abundances at those altitudes [*Solomon et al.*, 1985; *Fleming et al.*, 1995; *Allen et al.*, 1999]. These studies suggest that carbon monoxide may be a good tracer of mesospheric dynamics, especially at high latitudes and during winter, when OH abundances are smaller, but those regions were highly unexplored until the flight of the ISAMS instrument on board UARS [*López-Valverde et al.*, 1993, 1996].

Carbon dioxide and carbon monoxide are the major carbon compounds of the upper atmosphere and their interconversion is useful for understanding the distribution of each other in that region. A comparison of the simultaneously-measured $CO_2$ and CO vmr profiles by Atmospheric Trace Molecule Spectroscopy Spacelab3 (ATMOS-SL3) and by ISAMS with TIME-GCM model calculations at mid-latitudes in the upper part of the middle atmosphere is presented in section 4. Zonal mean distributions of $CO_2$ and CO as predicted by the TIME-GCM model, showing the effects of the stratosphere/mesosphere coupling, are compared with observation and discussed in section 4.3.

## 2. MEASUREMENTS OF CARBON DIOXIDE

$CO_2$ vmr's have been measured in the upper mesosphere and lower thermosphere by different techniques: in-situ rocket-borne mass spectrometers; infrared radiometers and spectrometers on board rocket and satellites; and occultation instruments on the space shuttle (see Table 1). The profiles are shown in Figs. 1–3 and are discussed in the sections below.

### 2.1. In Situ Rocket-Borne Mass-Spectrometer Measurements

The first $CO_2$ abundance profiles measured in the upper mesosphere and lower thermosphere were carried out by in-situ ion mass spectrometers on board sounding rockets. The first measurements were taken by *Offermann and Grossmann* [1973] in the 120–140 km range at night. They found that $CO_2$ is in diffusive equilibrium in that region, although the errors of their measurements are rather large (30%). *Philbrick et al.* [1973] measured the $CO_2$ volume mixing ratio (vmr, a quantity which is more independent of the atmospheric conditions than the $CO_2$ number density itself) at night in the 86–98 km altitude interval. Their $CO_2$ vmr shows a very oscillating profile (see Fig. 1), which could be due to the presence of turbulent layers

(observed in the experiment) combined with the large errors of the measurements ($\pm 25\%$). Hence, no clear evidence of the separation from the diffusive equilibrium in this region can be drawn from these measurements.

*Trinks and Fricke* [1978] and *Trinks et al.* [1978] measured the $CO_2$ vmr in the sunlit upper mesosphere/lower thermosphere (93–150 km) during the Aladdin 74 campaign (see Fig. 1). The errors in their measurements are also large, 25–30% (they have been plotted in Fig. 1 only at certain altitudes for clarity). By using a simple composition model, they concluded that the observed daytime $[CO_2]/[Ar]$ ratio (a quantity similar to the $CO_2$ vmr) above 100 km decreases more rapidly than expected from diffusive equilibrium, which necessarily implies the presence of a net destruction process for $CO_2$.

*Offermann et al.* [1981] carried out four rocket campaigns measuring four $CO_2$ vmr profiles in the 80–120 km altitude range. The quality of their measurements is similar to other rocket observations: an overall error between $\pm 20\%$ and $\pm 30\%$. The S75-A and S75-B $CO_2$ profiles are systematically different in most of the altitude range, but lie within each other's accuracy limits. During the B2 wintertime flights, the atmosphere was considerably disturbed. Thus, the B2-1 profile shows considerable structure around 100 km, and exhibits rather large values between 85 and 95 km. There is also an unusually strong decrease in the B2-2 profile above 93 km (see Fig. 1).

In summary, all in-situ rocket $CO_2$ vmr profiles are nearly constant up to about 90 km (and perhaps a few km above), with a value of $330\pm 50$ ppmv, and start declining rapidly above this altitude up to about 110-120 km and then more gradually beyond. The profiles are generally rather noisy with errors of the order of 20–30%. Unusual features were observed in some profiles, such as a wave-like structure (B2-1), and a strong decline above 92 km (B2-1), both at northern mid-latitudes at winter. The mean of these profiles starts departing from diffusive equilibrium at around 90 km, although, given the large errors, a departure at 95 km cannot be ruled out.

### 2.2. $CO_2$ Abundances Derived from Emission Measurements

We describe here the $CO_2$ vmr profiles derived from measurements of $CO_2$ infrared emissions. Compared to in situ mass-spectrometer measurements, $CO_2$ vmr is not so directly nor accurately derived from infrared measurements since the radiances depend on the kinetic temperature and on the number density, apart from the $CO_2$ vmr itself. Also, in many cases, the radiometers and spectrometers are affected by large calibration er-

Table 1. Measurements of CO$_2$ abundances

| Type | Name | Date | SZA | Lat/Long | Reference |
|---|---|---|---|---|---|
| Mass-Spect./Rocket | SN-5 | 13 Oct 70 | Night | 40°N/9°E | Offermann and Grossmann, 1973 |
| Mass-Spect./Rocket | NACS | 20 Nov 70 | 98° | Eglin AFB, FLA | Philbrick et al., 1973 |
| Mass-Spect./Rocket | Aladdin74 | 9 Jun 74 | 43° | 38°N/75°W | Trinks and Fricke, 1978; Trinks et al., 1978 |
| Mass-Spect./Rocket | S75-A | 25 Mar 72 | Night | 40°N/9°E | Offermann et al., 1981 |
| Mass-Spect./Rocket | S75-B | 30 Mar 72 | Night | 40°N/9°E | Offermann et al., 1981 |
| Mass-Spect./Rocket | B2-1 | 4 Jan 76 | 87.2° | 37°N/7°W | Offermann et al., 1981 |
| Mass-Spect./Rocket | B2-2 | 21 Jan 76 | 85.4° | 37°N/7°W | Offermann et al., 1981 |
| IR-Emission/Rocket | SPIRE | 28 Sep 77 | Night, 78°-84° | 65°N | Wintersteiner et al., 1992; Nebel et al., 1994 |
| IR-Emission/Rocket | SISSI-3 | 02 Aug 90 | 94 | 67°N, 21°E | Grossmann et al., 1994; Vollmann and Grossmann,1997 |
| IR-Emission/Rocket | SISSI-4 | 09 Apr 91 | ~95 | 67°N, 21°E | Grossmann et al., 1994; Vollmann and Grossmann,1997 |
| IR-Emission/Satellite | SAMS | 78-81 | Day (0-85°) | Global | López-Puertas and Taylor, 1989 |
| IR-Emission/Satellite | ISAMS | 1991-92 | Day (0-85°) | Global | Zaragoza et al., 2000 |
| Occultation/Satellite | EV13 | 2 Dec 83 | Terminator | 35°N/58°E | Vercheval et al., 1986 |
| Occultation/Satellite | EV14 | 3 Dec 83 | Terminator | 31°N/197°E | Girard et al., 1988 |
| Occultation/Satellite | EV18 | 3 Dec 83 | Terminator | 29°N/131°E | Girard et al., 1988 |
| Occultation/Satellite | EV21 | 3 Dec 83 | Terminator | 28°N/86°E | Girard et al., 1988 |
| Occultation/Satellite | SL3-1 | 01 May 85 | Terminator | ~30°N, 300°E | Rinsland et al., 1992 |
| Occultation/Satellite | SL3-2 | 01 May 85 | Terminator | ~48°S, 290°E | Rinsland et al., 1992 |

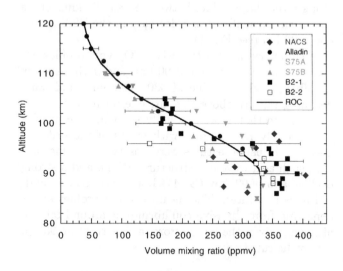

**Figure 1.** In situ ion-mass spectrometer rocket-borne measurements of CO$_2$ volume mixing ratio (vmr) in parts per million by volume (ppmv). For clarity, errors have been plotted only at certain altitudes.

rors. In addition, for many experiments, radiance measurements are affected by uncertainties in the non-LTE populations of the emitting levels.

In a typical wide band measurement near 15 $\mu$m, several CO$_2$ bands (including weak isotopic and hot bands)

contribute to the limb radiance at tangent heights up to about 100 km, the CO$_2$(010) level of the major isotope being the main contributor above this altitude [e.g., *Edwards et al.*, 1993]. The excitation (vibrational) temperature, $T_V$, of the CO$_2$(010) level is close to LTE up to 100 km. Above this height it depends mainly on three quantities: the kinetic temperature, the atomic oxygen abundance, and the CO$_2$(010)–O($^3$P) collisional rate coefficient, $k_{CO_2-O}$. Thus, to derive the CO$_2$ vmr from a wide band emission measurement at 15 $\mu$m, we would ideally require additional simultaneous measurements of pressure, kinetic temperature, and of atomic oxygen, in addition to having an accurate knowledge of the $k_{CO_2-O}$ rate and its temperature dependence.

The analysis of CO$_2$ 4.3 $\mu$m daytime measurements presents fewer unknown quantities than at 15 $\mu$m because the non-LTE daytime populations of the emitting states at 4.3 $\mu$m are known better than those emitting at 15 $\mu$m. The 4.3 $\mu$m levels are pumped by absorption of solar radiation which, provided the solar elevation is known, can be determined accurately. Also, as a consequence of the sunlight absorption, the non-LTE populations are very little dependent on the kinetic temperature (i.e., temperature serves only to build up the hydrostatic pressure); and the O($^3$P) number density is also of little importance. The collisional relaxation rate of the $CO_2(\nu_3)$ levels to $N_2(1)$, $k_v(\nu_3)$, is not completely known to date, but its effect on the total radiance measured by a broad band radiometer (contributed by several fundamental, isotopic and hot bands near 4.3 $\mu$m) is of little effect [*Nebel et al.*, 1994; *López-Puertas et al.*, 1998]. $CO_2$ 4.3 $\mu$m populations are also affected by the excitation from $O(^1D)$ through $N_2(1)$ [*Edwards et al.*, 1996]. The uncertainty introduced by this process due to not measured $O_3$ abundance is however, very small, unless we are close to or under twilight conditions [*Vollmann and Grossmann*, 1997].

*2.2.1. SPIRE.* This instrument measured the $CO_2$ 15 and 4.3 $\mu$m limb radiances at 50-150 km tangent heights. The radiometric calibration of SPIRE is about $\pm25\%$ [*Stair et al.*, 1985]. The analysis of the $CO_2$ 15 $\mu$m (13.0-16.5 $\mu$m) measurements have been presented in a number of papers [e.g. *Caledonia et al.*, 1985; *Sharma and Wintersteiner*, 1990; *Wintersteiner et al.*, 1992]. We have mentioned above the importance of knowing the pressure, kinetic temperature, and atomic oxygen to analyze $CO_2$ 15 $\mu$m measurements, but none of these quantities were measured in the SPIRE experiment. *Wintersteiner et al.* [1992] included the pressure, temperature and atomic oxygen predicted by MSIS-86 in their analysis of the $CO_2$ 15 $\mu$m measurements. They found the best overall fit to the measurements when using a $k_{CO_2-O}$ rate close to $5\times10^{-12}$ cm$^3$s$^{-1}$ at 300 K and the mean $CO_2$ vmr profile obtained from previous rocket measurements (ROC profile in Figs. 1 and 2). Above about 100 km, the $CO_2$ vmr and the $k_{CO_2-O}$ rate affect $CO_2$ radiances in the same sense: for a faster $k_{CO_2-O}$ rate the $CO_2(010)$ $T_V$ is closer to the kinetic and hence larger. Thus, these authors showed that a $CO_2$ vmr profile significantly lower in the 80–110 km region (similar to the EV14-EV21 profiles in Fig. 3) also reproduces the observations very well if a $k_{CO_2-O}$ rate 15% larger, $6\times10^{-12}$ cm$^3$s$^{-1}$ at 300 K, is used. Since this rate has not yet been determined to that accuracy, it seems that the SPIRE 15 $\mu$m measurements are compatible with these two $CO_2$ vmr profiles.

The analysis of the SPIRE 4.3 $\mu$m daytime measurements showed that the measured radiances are generally consistent with the mean rocket $CO_2$ vmr profile (ROC) up to 110 km [*Nebel et al.*, 1994]. Between this altitude and 130 km, the model underestimates the measurements by a factor of $\sim2$. This could be explained by an underestimation of the $CO_2$ abundance (number density or vmr) or by an unknown non-LTE excitation process. The fact that this underprediction does not occur in the 15 $\mu$m measurements taken simultaneously prompts us to suggest that it could possibly be due to a non-LTE process.

Hence, it seems that SPIRE $CO_2$ 15 $\mu$m measurements could be consistent with either the mean rocket profile (ROC) or a substantially depleted profile in the 80–110 km region. On the other hand, the 4.3 $\mu$m measurements are consistent only with the rocket profile, and suggests even larger values between 110 and 130 km. The uncertainties in the non-LTE parameters and in several atmospheric quantities not measured make it very difficult to extract further information (estimates larger than 50%) on the $CO_2$ vmr.

*2.2.2. SISSI.* The Spectroscopic Infrared Structure Signatures Investigation (SISSI) measured the $CO_2$ 15 and 4.3 $\mu$m zenith radiances in four rocket experiments under twilight conditions. The kinetic temperature, pressure, and $O(^3P)$ number density were also simultaneously inferred, although with large errors ($\pm10\%$ for temperature and $\pm30\%$ for atomic oxygen). The radiance measurements themselves presented errors of $\pm30\%$ [*Vollmann and Grossmann*, 1997]. These authors successfully explained the 4.3 $\mu$m radiances below 100 km with the $CO_2$ vmr rocket (ROC) profile but have to multiply it by factors of 2.5 (SISSI-3) and 4 (SISSI-4) above 105 to fit the radiances. This increased $CO_2$ profile also explained the 15 $\mu$m intensities if a $k_{CO_2-O}$ rate coefficient of $\sim1.5\times10^{-12}$ cm$^3$s$^{-1}$ was used (a factor of 4 smaller than that used in SPIRE).

SISSI data can also be interpreted in a different way. The increase in the 4.3 $\mu$m radiances above 105 km (*Nebel et al.* [1994], and *López-Puertas et al.* [1998] found similar enhancements in SPIRE and ISAMS data) may be caused by an unidentified non-LTE source (e.g., $NO^+(\nu_3)$ hot bands emission) instead of by an increase in the $CO_2$ number density. Regarding the 15 $\mu$m radiances, they would also be consistent with the ROC profile if the $k_{CO_2-O}$ rate were between 3 and $5\times10^{-12}$ cm$^3$s$^{-1}$ (a plausible range).

Thus, SISSI inferences of $CO_2$ abundance have rather large errors (an overall error of around $\pm50\%$). Their analysis favours the $CO_2$ ROC profile below about 105 km, but increased in a factor 2–4 above this height. The current uncertainties in the non-LTE parameters, however, do not discard that this factor is close to unity and hence the $CO_2$ vmr ROC profile is also favoured above 105 km.

*2.2.3. SAMS.* The Stratospheric and Mesospheric Sounder (SAMS) on Nimbus 7 measured the $CO_2$ 4.3 $\mu$m limb emission globally in latitude during four years, 1978-1981, using the pressure modulation technique. This technique made SAMS more sensitive to the fundamental than to the hot $CO_2$ 4.3 $\mu$m bands. The daytime measurements were analyzed in detail in *López-Puertas and Taylor* [1989] and recently revised by *López-Puertas*

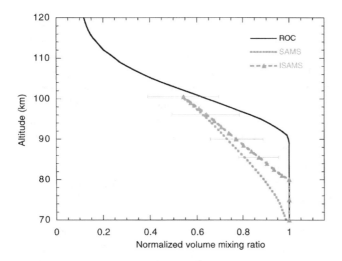

**Figure 2.** CO$_2$ volume mixing ratio profiles derived from infrared emission measurements at 15 and 4.3 $\mu$m. The profiles have been normalized to their respective values in the lower region (330 ppmv for the ROC and SAMS profiles, and 350 ppmv for ISAMS). Labels as in Table 1.

*et al.* [1998]. The difficulties mentioned in section 2.2 about the feasibility of deriving CO$_2$ vmr profiles from 4.3 $\mu$m radiances also apply to SAMS, although in this case the pressure-kinetic temperature field was measured simultaneously up to about 60 km (above this altitude the CIRA 72 climatology was used). The CO$_2$ vmr profile that gave the best overall agreement is shown in Fig. 2. This profile shows significantly lower values than the rocket (ROC) profile between 80 and 95 km. It is also slightly smaller than the ISAMS and ATMOS profiles in the region between around 70 and 80 km. This small difference could be due to a bias in the climatological upper mesospheric and lower thermospheric pressure-temperature used in the SAMS analysis [*López-Puertas et al.*, 1998].

2.2.4. *ISAMS.* ISAMS on board UARS measured the CO and CO$_2$ emissions near 4.6 $\mu$m between approximately 10 and 105 km, globally (in latitude) during six months, covering the entire seasonal cycle. ISAMS used the Pressure Modulation technique which produces two signals: the pressure modulated (PM) and the wide band (WB). The PM signal was used to measure the CO abundances (see section 3.3) and the WB signal, more sensitive to the CO$_2$ bands near 4.3 $\mu$m in the daytime, provided information on the CO$_2(\nu_3)$ non-LTE populations and on the CO$_2$ vmr. ISAMS WB measurements are similar to those of SAMS but differ in several respects. First, the ISAMS measurements are more sensitive to the second major CO$_2$ isotope (636) fundamental and hot bands, while SAMS was more sensitive to the major 626 isotope fundamental band. The

uncertainty in the non-LTE populations of the emitting vibrational levels in SAMS and ISAMS are, however, of a similar magnitude. Secondly, ISAMS had a much better sensitivity than SAMS, with a signal to noise ratio equal one up to about 120 km. Thirdly, simultaneous measurements of the pressure-temperature were taken by ISAMS up to 80 km (about 20 km higher than in SAMS). Radiometric calibration of ISAMS CO WB channel was rather good, with detector noise of $1.8 \times 10^{-5}\bar{B}(290$ K), (18% of the atmospheric signal at 100 km), an 0.38% in the blackbody thermometry, and a mean space-view radiance of $-4 \pm 4 \times 10^{-6}\bar{B}(290$ K) (4% of the atmospheric signal at 100 km) [*Rodgers et al.*, 1996].

The ISAMS WB measurements have been analyzed in detail by *López-Puertas et al.* [1998] and *Zaragoza et al.* [2000]. The CO$_2$ vmr profile that gives the best overall fit to the measurements is plotted in Fig. 2 with the errors varying from ∼10% at 85 km to 30% at 100 km. That analysis has also shown that the CO$_2$ vmr profile does not change (within ISAMS uncertainties) with latitude or season up to 100 km in the daylight atmosphere [*Zaragoza et al.*, 2000].

### 2.3. Solar Occultation Measurements

2.3.1. *Grille.* Carbon dioxide concentrations were measured by the Grille spectrometer on board Spacelab from the mesosphere to the lower thermosphere late in 1983. It used the solar occultation technique in the infrared, taking the sun as a light source in the limb at sunrise and sunset. CO$_2$ abundances were measured in four occultations (events): EV13, analyzed by *Vercheval et al.* [1986], and EV14, EV18 and EV21, reported by *Girard et al.* [1988]. In this experiment the total number density was not measured. The CO$_2$ vmr profiles shown in Fig. 3 were obtained using the MSIS-86 model. This adds another significant uncertainty to the derived CO$_2$ vmr's, not included in the error bars in Fig. 3.

It is noticeable that the EV14, EV-18, and EV-21 profiles are much lower than EV13 and other CO$_2$ profiles in the 90–110 km region (see also Fig. 6). The fact that the total number density was not measured might explain that difference. For example, event EV14 reaches values close to 400 ppmv at 60-70 km (not shown) when the US Standard atmosphere is used, and close to 450 ppmv if MSIS-90 model is taken. These values are hardly credible and hence cast some doubt on the validity of these profiles.

2.3.2. *ATMOS.* The ATMOS instrument on the Spacelab 3 shuttle mission took infrared occultations spectra from the troposphere up to the lower thermosphere from April 29 to May 6, 1985. An important advantage of these measurements is that, in addition to the $CO_2$ mixing ratios, profiles of kinetic temperature and pressure were also derived from the spectra. Furthermore, since they are absorption measurements, they were not affected by non-LTE uncertainties.

*Rinsland et al.* [1992] retrieved $CO_2$ profiles between 70 and 116 km from six occultations. Fig. 3 shows the mean $CO_2$ vmr profiles for the four northern hemisphere occultations (SL3-1) and for the two occultations in the southern hemisphere (SL3-2), together with their estimated 1-$\sigma$ uncertainties. The northern (SL3-1) and southern (SL3-2) ATMOS $CO_2$ mixing ratios average 320 ppmv between 70 and 90 km. A rapid $CO_2$ vmr decrease begins around 90 km in the southern hemisphere data and around 100 km in the northern hemisphere data. The decrease continues in both hemispheres up to 116 km where the $CO_2$ vmr is about 70 ppmv. A comparison of this data with other measurements and model results is presented in section 4.

Occultation spectra covering the $CO_2$ infrared region were measured more recently by ATMOS in the Atlas-1, 2 and 3 missions in 1992, 1993, and 1994, respectively. $CO_2$ profiles were operationally retrieved from these spectra and are available at http://remus.jpl.nasa.gov/-atmos/atftp.html. However, the altitude registration of those profiles is not correct. A dedicated retrieval is being currently undertaken.

# 3. MEASUREMENTS OF CARBON MONOXIDE

## 3.1. Balloon and Rocket Measurements

*Solomon et al.* [1985] compiled and presented a number of balloon-borne infrared techniques and cryogenic sampling experiments which measured CO in the stratosphere. These observations continue nowadays and represent very valuable in-situ measurements, which can be used for correlative studies with the few satellite observations available [*López-Valverde et al.*, 1996]. Detailed descriptions of these two techniques are given by *Erdman and Zipf* [1982] for cryogenic sampling, and by *Turner and Drummond* [1991] for pressure modulated radiometer. The dispersion of the Solomon et al. compilation of measurements is presented in Fig. 4 together with other CO determinations at mid-latitudes, as explained below. These were the only stratospheric data available until recently, since ground observations in the microwave region, which have been made for a long time, are unable to sound stratospheric altitudes very reliably [*Clancy et al.*, 1984].

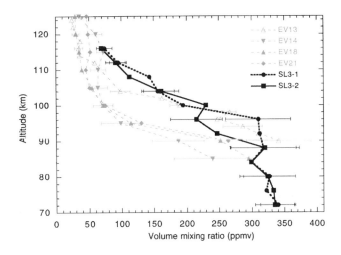

**Figure 3.** $CO_2$ volume mixing ratio profiles in ppmv derived from solar occultation measurements. Labels as in Table 1.

## 3.2. Microwave Measurements from the Ground

Microwave observations of atmospheric CO have been carried out sporadically for some decades (see for example *Waters et al.* [1976]). Analyses of both absorption and emission lines have been used, and CO column as well as vmr profiles have been retrieved in the mesosphere at specific locations and times. They constitute a unique set of CO data, although irregularly distributed in space and time. This dataset was the basis for the detailed photochemical and dynamical study of the CO distribution in the middle atmosphere by *Solomon et al.* [1985]. The variability shown in the CO relative abundance by these early data, as compiled by these authors and pointed out previously by *Clancy et al.* [1984], already showed a seasonal cycle at mesospheric altitudes, which was the most remarkable result of mesospheric CO for almost a decade (until the global sounding by ISAMS/UARS, see below), and which triggered the modeling efforts of Solomon and coworkers.

Briefly, microwave measurements have been made principally using the J=1→0 and 2→1 spectral lines at 115.271 and 230.5 GHz, respectively, and both in absorption against the sun and in emission [*Bevilacqua et al.*, 1985]. Unlike emission measurements, absorption data seem more affected by baseline curvature errors, but are less subject to calibration errors. There seems to be an overall preference of emission measurements for retrieval of CO profiles and of absorption techniques for column density determinations. For profile measurements, the lower altitude of the sounding is determined by the signal to noise ratio, and is located usu-

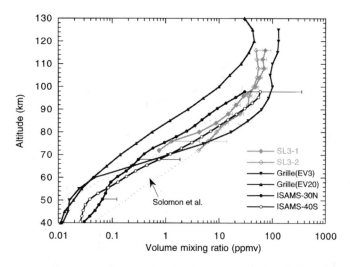

**Figure 4.** A compilation of CO volume mixing ratio profiles in ppmv measured at mid-latitudes. For clarity, 1-$\sigma$ errors are plotted only for ATMOS Spacelab-3 and for ISAMS at selected altitudes. The shaded area represents the variability in the measurements compiled by Solomon et al. [1985]. See text.

ally around 60 km, since CO densities are very small in the stratosphere. The profile upper limit, on the other hand, is usually imposed by the pressure broadening of the spectral line, which is very weak in the mesopause and above [*Bevilacqua et al.*, 1985]. The CO lines mentioned above are frequently much narrower than the spectrometer bandwidth, and about half a dozen points can be taken to retrieve CO at the same number of mesospheric altitudes, with typical spacings of 5 km. The vertical resolution, however, as given by the width of the weighting functions, amounts to 10-12 km [*Aellig et al.*, 1995; *Bevilacqua et al.*, 1985].

As mentioned above, the microwave dataset has shown a large seasonal and latitudinal variability in the CO abundance. Recently, large changes have been observed within a few days at a given location, which have been related to mesospheric temperature inversions observed simultaneously, possibly the result of turbulent diffusion or transient descent of air [*Aellig et al.*, 1995].

*3.3. Space Experiments*

There have been a small number of space instruments devoted to measuring CO in the middle atmosphere. They are SAMS and ISAMS, on board the satellites NIMBUS 7 and UARS, respectively, and the much higher resolution instruments ATMOS and Grille, on the Space Shuttle. Other infrared CO data taken from the Space shuttle include those from the spectrometer CRISTA [*Riese et al.*, 1997] and the Michelson inter-

ferometer CIRRIS-1A [*Dodd et al.*, 1993], but neither of these have been inverted yet to produce CO abundances.

The SAMS instrument measured the strongest vibrational-rotational band of CO in the infrared, at 4.7 $\mu$m, in the stratosphere and mesosphere, with global coverage. Not only is that emission out of LTE in the mesosphere but the noise level was high, and large averages of data had to be used in order to derive indirect information on mesospheric abundances [*Murphy*, 1985]. They confirmed a seasonal/inter-hemispheric variation of CO vmr.

With improved detectors and narrower filters, ISAMS was able to measure the same CO(1-0) band with better geographical coverage and for nearly a whole year. A non-LTE specific model was developed for this emission [*López-Puertas et al.*, 1993] and was incorporated into the retrieval scheme [*López-Valverde et al.*, 1991]. Still, the emission from the night mesosphere was too weak under normal conditions, but during daytime the solar pumped CO atmospheric emission allowed meaningful retrievals from about 30 km up to about 73 km operationally, without averaging, and to higher altitudes in special retrievals [*López-Valverde et al.*, 1998]. The vertical scanning step was 2.5 km but typical widths of the averaging kernels range between 5 and 10 km, which should be considered as the actual vertical resolution. More details of the ISAMS instrument have been presented by *Taylor et al.* [1993], of the validation methodology by *Rodgers et al.* [1996], and of the validation of the CO measurements by *López-Valverde et al.* [1996].

The importance of the ISAMS database of CO data, which is available from the UARS archives, should not be understated. It contains more than 2600 operational profiles of CO vmr per day, during 180 days of measurements from September, 1991, to July, 1992, and is therefore the most extensive set of measurements of CO in the middle atmosphere that has been taken almost continuously and on a global scale. Maps of zonal means of CO vmr from ISAMS have clearly confirmed previous microwave measurements and 2D model predictions about its seasonal and latitudinal distribution. Particularly valuable insights into the polar night have been recorded for the first time [*López-Valverde et al.*, 1996; *Allen et al.*, 1999]. A plot of zonal mean CO abundances from ISAMS for equinox and solstice conditions is shown in Fig. 5, where an average of four days of zonal means from daytime measurements are combined to give a general picture of two very different distributions. Detailed discussions of similar maps of CO are given by *López-Valverde et al.* [1996] and *Allen et al.*

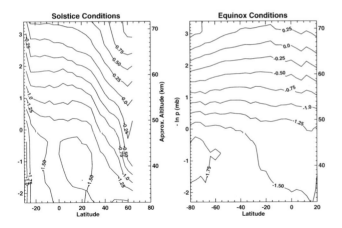

**Figure 5.** Zonal mean of the CO volume mixing ratios measured by ISAMS for solstice and equinox conditions. Four days of ISAMS daytime retrievals are averaged for each plot: 1-4 April 1992 and 1-4 January 1992. Data correspond to the latest version of the ISAMS operational processing (Version 12). Contours (of vmr in ppmv) are log10 spaced at 0.25. The approximate altitude scale is taken from the UARS climatology.

[1999]. In short, at solstice there is a much larger concentration of CO at high latitudes of the winter Hemisphere than at low latitudes. At equinox, the meridional circulation reverses and initiates a net transport of CO in the upper mesosphere in the opposite direction. This reversal is observed in the CO distribution measured by ISAMS during the first four days of April (left panel in Fig. 5).

A number of instruments have flown on the Space Shuttle to measure CO, like ATMOS, Grille, CRISTA and CIRRIS-1A. All of them have much higher spectral resolution than the SAMS and ISAMS radiometers. The first two are absorption experiments, observing the Sun through the limb of the planet, and consequently possess much higher SNR than the emission instruments and are free of non-LTE effects. For these reasons, the CO vmr profiles derived from them extend to higher altitudes than ISAMS and have better vertical resolution, typically that of their altitude scan, about 4 km in the case of the ATMOS data. On the other hand, the number and distribution of the profiles obtained are much more restricted than for ISAMS.

The first detailed ATMOS profiles of CO from the Spacelab 3 flight were analyzed by *Gunson et al.* [1990], and the Grille ones were presented by *Vercheval et al.* [1986] and by *Girard et al.* [1988]. Both experiments were carried out for equinox conditions and found differences between the northern and southern hemispheres. Particularly large is the variation between two of the Grille retrieved profiles, EV3 and EV20 (see Fig. 4).

However, these profiles (particularly EV20) may be affected by assumptions in the mean state of the atmosphere (see section 2.3.1 above). The ATMOS work on CO was re-analyzed by *Rinsland et al.* [1992], after adopting a special retrieval method more appropriate to high altitude soundings (70 to 116 km). This later study made use of corrected and simultaneous pressure, temperature and $CO_2$ retrievals. We consider their results possibly some of the best CO abundances determined at mesospheric and thermospheric altitudes to date but, like the Grille data, they are restricted to two latitude zones and at equinox conditions. These CO profiles, which correspond to 30°N and 48°S, SL3-1 and SL3-2 in Fig. 4, do not show a significant inter-hemispheric variation at these latitudes.

A comparison between the Spacelab ATMOS profiles and the ISAMS data is also shown in Fig. 4. We have taken one full day of ISAMS data (April 1, 1992, typical of equinox conditions) to perform extended retrievals from the daylight measurements, and then have extracted zonal means at 10 degree latitude bands. These have been plotted in Fig. 4 for latitudes 30°N and 40°S. The comparison with ATMOS data, taken 6 years earlier, is very reasonable, although the noise of the ISAMS data (not shown in the figure) is quite large at these altitudes, of the order of the profile-to-profile variability in the latitude boxes selected. If the standard deviation from the mean of these latitude averages (shown at 50 and 96 km for the 40°S profile in Fig. 4) is taken to characterize the ISAMS error, then there is agreement between ISAMS and most of the other data in the figure.

During the ATLAS-1 flight, on March-April 1992, both ATMOS and Grille were flown again, with the advantage of having simultaneous measurements by the ISAMS/UARS mission, although no collocated profiles were taken. An effort to correlate ISAMS and ATMOS data of CO vmr was carried out recently by *López-Valverde et al.* [1998], after computing extended ISAMS retrievals at daytime to obtain CO concentrations up to about 85 km. Unfortunately, no special high-altitude retrievals of the ATMOS measurements are yet available for that flight; only the operational retrievals were available to these authors, which may be affected by the lack of accurate pressure determination above about 70 km (M. Gunson, personal communication). Therefore, the CO data from ATMOS/ATLAS-1 and those comparisons with ISAMS have to be taken with a large degree of caution until a full validation is carried out.

After ISAMS and ATMOS data, our knowledge of the CO abundance has improved dramatically in two major areas, its global distribution in latitude and season,

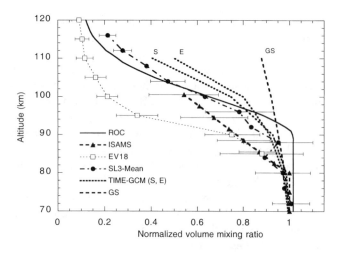

**Figure 6.** Comparison of $CO_2$ vertical profiles observed by rocket-borne mass spectrometers (ROC), ATMOS (SL3-Mean), Grille (EV18) and ISAMS with results from the NCAR TIME-GCM and the Garcia-Solomon two-dimensional model (GS), all of them at mid-latitudes. Two curves from TIME-GCM are shown: (E) for equinox and (S) for solstice conditions.

and its variability in polar regions (this was particularly unexplored a decade ago). It has also been possible to study details of the dynamics of the polar vortex, where CO has been found to be a very useful tracer, and to compare with 3D chemical models of the middle atmosphere [*Allen et al.*, 1999]. An interesting result in these regions is the strong downwelling from the mesosphere to the stratosphere, and the important exchange between these distinct regions. *López-Valverde et al.*, 1996, for example found cases of well defined tongues of enriched-CO air descending down to about 30 km that seemed to coincide with the location of the polar vortex. These structures were not present in the ISAMS data at higher altitudes during the same observation days, but a more zonally symmetric descent occurs at mesospheric altitudes. Similar longitudinally asymmetric descent was also found in ATMOS measurements of CO in the Antarctic stratosphere by *Rinsland et al.* [1999], during the ATLAS 3 mission, flown in November 1994. These authors show that, within the vortex, the descent is clear in the CO vmr down to around 20 km.

It is important to mention a number of features observed in the ISAMS data set and not predicted correctly by models previously, like the local maximum in the tropical upper stratosphere, possibly due to methane oxidation [*Allen et al.*, 1999], or not explained yet, like the strong latitudinal gradient at lower mesospheric altitudes in the winter hemisphere [*López-Valverde et al.*, 1996]. These authors suggested that

this discrepancy may be partially due to biases in zonal means when important displacements of the polar vortex occur, but it seems to be fairly systematic and common to other CO datasets. Most models fail to reproduce the high CO abundances observed in the lower mesosphere, with a typical deficit of a factor of 1.5–3 (see *Allen et al.*, 1999, for a clear discussion of this problem).

## 4. MODELING AND COMPARISON WITH OBSERVATIONS

We present in this section 2D and 3D model calculations and compare them with the observations with the aim of understanding better the $CO_2$ and CO abundances in the middle atmosphere. We address three major points: 1) the role of molecular diffusion on the $CO_2$ abundance; 2) the effect of eddy diffusion on CO and $CO_2$ mixing ratios; and 3) the latitudinal distributions of $CO_2$ and CO and how they are affected by the stratosphere/mesosphere coupling.

### 4.1. Molecular Diffusion of $CO_2$

Predictions of $CO_2$ vmr by one-dimensional models in the past decade [see, e.g., *Offermann et al.*, 1981; *Rodrigo et al.*, 1986, and references therein; *Taylor*, 1988; *López-Puertas and Taylor*, 1989] significantly overestimated the $CO_2$ vmr in the upper mesosphere/lower thermosphere. This was likely due to the neglect or underestimation of diffusive separation below the turbopause. The problem is illustrated in Fig. 6, which compares the measured $CO_2$ profiles compiled in section 2 with calculations carried out with a two-dimensional model [*Garcia and Solomon*, 1985; *Garcia et al.*, 1992; *Garcia and Solomon*, 1994] and with the NCAR TIME-GCM as described by *Roble* [2000; this issue]. The Garcia-Solomon model was designed to study the middle atmosphere below 100 km and does not take into account diffusive separation, since this was thought to be negligible below the turbopause. The TIME-GCM, on the other hand, includes a complete treatment of molecular diffusion, including its seasonal variability. It is immediately clear that the results from the TIME-GCM are in much better agreement with the observations. The result is surprising, since eddy diffusion should be considerably larger than molecular diffusion below the turbopause ($\simeq$ 100 km). Nevertheless, additional model calculations with the TIME-GCM in the next section, bear out this result.

A closer look at Fig. 6 reveals that the TIME-GCM model agrees well with all the measurements within their uncertainties (see section 2), except for three cases

(EV14, EV18, and EV21) of the Grille measurements. They agree particularly well with the ATMOS mean $CO_2$ profile up to 95 km and slightly overestimate the volume mixing ratios above this altitude (the TIME-GCM results lie near their upper error range). TIME-GCM underpredicts the rocket in-situ measurements between 80 and 95 km but, since they have large uncertainties (20-30%), this underestimation is not significant. Above 95 km, as for ATMOS, the measurements are slightly underpredicted. When comparing with the ISAMS profile, the TIME-GCM predictions are generally larger throughout the altitude range but, since ISAMS errors are larger than those of ATMOS, they also lie at the edge of their upper error bars.

To reproduce the very low $CO_2$ vmr values of the Grille occultations taken on 3 Dec 83 (see Table 1) would be very difficult. Although the Grille observations have large uncertainties (see section 2), they cannot be completely ruled out. There is another example, the in-situ B2-2 rocket profile, also taken at winter mid-latitudes under perturbed atmospheric conditions (see Table 1), which also shows an unusually low $CO_2$ vmr (see Figs. 1 and 2). It appears that the mid-latitude winter lower thermosphere may exhibit, under certain perturbed conditions, very low $CO_2$ vmr's. Even so, it is not clear what kind of perturbation could produce mixing ratios of 150 ppmv at 95 km, since normally such values are only found above about 105 km.

### 4.2. Sensitivity to Eddy Diffusion

To better understand the influence of molecular diffusion on the vertical profile of $CO_2$ and CO in the upper mesosphere and lower thermosphere we performed a number of numerical experiments using the global average model of *Roble* [1995] that is updated to be consistent with the aeronomic scheme in the current version of the TIME-GCM as described by *Roble* [2000; this issue]. This model uses an eddy diffusion coefficient given by

$$k_{zz} = H^2 \, k_0 \exp[-A(z - z_0)] \qquad (1)$$

where $H$ is the atmospheric scale height, $z = -\ln(p/p_0)$ is the vertical coordinate, $p$ is the pressure in mb, and $p_0 = 5.0 \times 10^{-7}$. The values of the parameters $k_0$, $A$, and $z_0$ for the altitude range are given in Table 2. This eddy coefficient is consistent with the global average of $k_{zz}$ used in the 3D TIME-GCM (see below).

We investigate the effect of changes in $k_{zz}$ by decreasing (increasing) the entire profile by a factor of 5 (3). Although the larger values are probably more than can be justified by the uncertainty in model estimates, they

Table 2. Parameters for $k_{zz}$

| Range of $z$ | $k_0$, s$^{-1}$ | $A$ | $z_0$ |
|---|---|---|---|
| $z \geq -6$ | $2.5 \times 10^{-6}$ | 1.0 | −6 |
| $-6.4 > z > -9$ | $8.5 \times 10^{-7}$ | 0.43 | −9 |
| $z < -9$ | $5.0 \times 10^{-7}$ | 0.066 | −17 |

serve to illustrate the response of $CO_2$ to changes in diffusive transport. The lower values, on the other hand, are plausible given that the Prandtl number for breaking gravity waves is likely to be in the range 1-10 [*Fritts and Dunkerton*, 1985; *McIntyre*, 1989; *Garcia*, 1989]. It is seen in Fig. 7 that, as the eddy diffusion coefficient is made smaller, the $CO_2$ mixing ratio begins to decrease rapidly at a lower altitude, and vice-versa. The profiles shown in Fig. 7 also demonstrate that the ratio of molecular to eddy diffusion has a pronounced influence on the altitude where the mixing ratio changes from a constant fully mixed atmosphere to one that approaches diffusive equilibrium in the thermosphere.

Fig. 8 shows a comparison of measured and modelled CO and $CO_2$ vmr profiles for mid-latitude conditions during equinox focussed on the upper mesosphere and lower thermosphere, where their interconversion is most noticeable. Simultaneous measurements of both constituents were taken by the ATMOS and ISAMS instruments. ATMOS measurements are possibly the most detailed analysis of the CO and $CO_2$ budget so far carried out at these altitudes. *Rinsland et al.* [1992] show that CO vmr is much smaller than that of $CO_2$ below 100 km, and that they are approximately equal near 116 km. A similar result was found by *Vercheval et al.* [1986] using GRILLE data (EV3 and EV13, respectively). The ISAMS CO and $CO_2$ profiles are in reasonable agreement with ATMOS measurements, although ISAMS CO has rather large errors and could be measured only up to 95 km.

When comparing these measurements with TIME-GCM calculations we see that $CO_2$ vmr's are slightly overestimated above about 95 km, and CO is largely underpredicted, by about a factor of 2-3. This is not very significant for ISAMS measurements, given their large errors, but it is significant in the case of ATMOS data. A faster $CO_2$ photolysis could possibly explain both the $CO_2$ and CO discrepancies, since it would reduce $CO_2$ and increase CO. This, however, does not seem likely in view of our knowledge of the $CO_2$ photoabsorption rates. A slower recombination rate of the three body

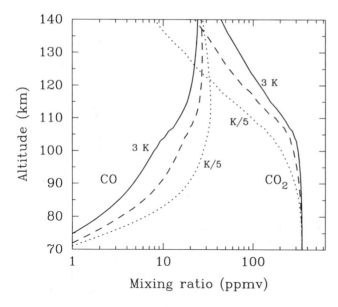

**Figure 7.** 1D global model sensitivity calculations of the CO$_2$ and CO volume mixing ratios (in parts per million, ppmv) to changes by factors 0.2 and 3 in the eddy diffusion coefficient used. Nominal results (dashed lines) are also plotted. See text.

reaction CO+O+M to produce CO$_2$, or an unknown dissociation chemical process are also other possibilities. A less mixed atmosphere would also produce the same effect, see Fig. 7. In this sense, an eddy diffusion coefficient smaller than that used in TIME-GCM and other 2D models by about a factor of 2–3 would give a better agreement both in the CO$_2$ and CO vmr profiles in the upper mesosphere and lower thermosphere. Given the current uncertainties in the eddy diffusion coefficient, the latter might be the most likely explanation, and warrants further investigation.

### 4.3. Latitudinal Distributions of CO$_2$ and CO in the Middle Atmosphere

Distributions of CO$_2$ and CO vmr were obtained from TIME-GCM runs similar to the perpetual solstice and equinox cases described by *Roble* [2000; this issue], except that the results presented here include a full seasonal cycle. The NCAR Thermosphere-Ionosphere-Mesosphere Electrodynamics General Circulation Model (TIME-GCM) is the latest in a series of three-dimensional time-dependent models that have been developed over the past two decades to simulate the circulation, temperature and composition of the upper atmosphere and ionosphere. It combines all of the features of the TGCM [*Dickinson et al.*, 1981, 1984], TIGCM [*Roble et al.*, 1988] and TIE-GCM [*Richmond et al.*, 1992]

that were described previously. The model has been extended downward to 30 km altitude, and aeronomical processes appropriate for the mesosphere and upper stratosphere have been included, as described by *Roble and Ridley* [1994], *Roble et al.* [1987] and *Roble* [1995]. The model has been updated as described by *Roble* [2000; this issue] and the CO$_2$ and CO photochemistry is consistent with that described by *Brasseur and Solomon* [1986] and *Garcia and Solomon* [1985; 1994]. The eddy diffusion is taken from the Lindzen parameterization but adjusted so the zonal average eddy diffusion values are consistent with those given by *Garcia and Solomon* [1985].

The calculated CO$_2$ and CO mixing ratios at solstice (December 21) and equinox (April 10) are shown in Figs. 9a–9d. The CO$_2$ and CO distributions manifest (more clearly that of CO) the stratosphere/mesosphere exchange, where they are very much driven by advection. The CO$_2$ mixing ratio at solstice conditions remains constant up to high altitudes (around 90 km) in regions of upward motion, e.g., in the summer hemisphere, and it is lowered in regions of downward motion, e.g., in the winter hemisphere (see Fig. 9a). Thus, CO$_2$ reflects the stratosphere/mesosphere coupling only above the upper mesosphere in the summer hemisphere

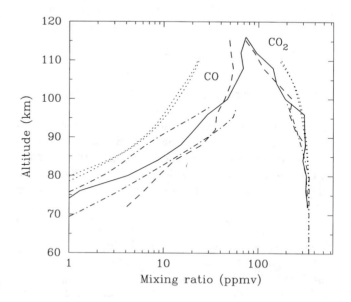

**Figure 8.** A comparison of measured and TIME-GCM calculations of CO$_2$ and CO volume mixing ratios for mid-latitude conditions during equinox. CO$_2$ and CO profiles are from ATMOS SL3 30°N (solid lines); ATMOS SL3 48°S (dashed lines); ISAMS at 30°N (dot-dashed, CO left curve) and 45°S (dot-dashed, CO right curve); and TIME-GCM outputs for 30°N and 45°S (dotted lines). See text.

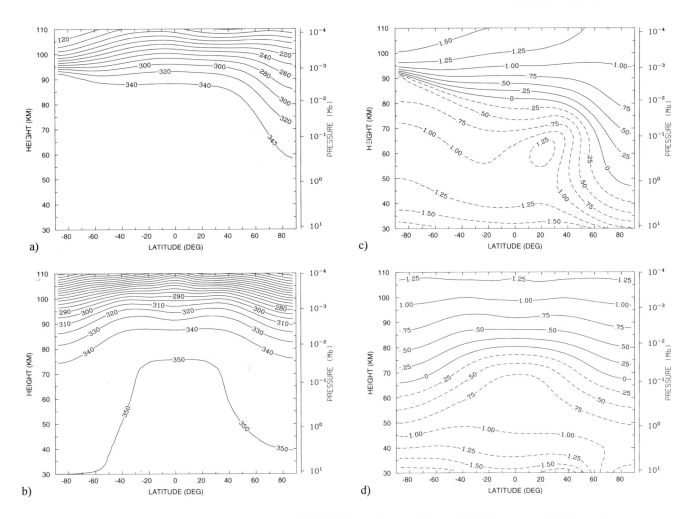

**Figure 9.** Zonal mean cross sections of TIME-GCM results for $CO_2$ and CO for solstice (21 December) and equinox (10 April) conditions. a) $CO_2$ for solstice; b) $CO_2$ for equinox; c) CO for solstice; and d) CO for equinox. Values of vmr are shown with contours linearly spaced at 20 ppmv for $CO_2$ and log10 spaced at 0.25 ppmv for CO.

but exhibits a significant depletion down to the lower mesosphere near the winter pole.

The latitudinal distribution of $CO_2$ shown in Fig. 9a for the daylight atmosphere (latitudes lowermost than 40°N) is consistent with the ISAMS measurements, since ISAMS errors are generally larger than the 5-10% change exhibited in the model. The larger depletion predicted at polar winter latitudes (around 15–20% in the 80–100 km region), however, could not be confirmed because ISAMS could not measure under those conditions. Given that $CO_2$ 15 $\mu$m emissions are routinely used to sound the atmospheric temperature and composition in this region, accurate measurements of $CO_2$ in the lower mesosphere (better than 3% at 60 km, 8% at 70, and than 15% at 80 km) at winter polar regions are urgently needed.

For equinox conditions, $CO_2$ is symmetrically distributed about the equator (see Fig. 9b). ISAMS observations of $CO_2$ and TIME-GCM model predictions are also consistent for these conditions. This model shows latitudinal variations only above about 75 km with an equator-to-pole difference of 5–10% at 80–100 km. These changes, although significant, are smaller than the ISAMS errors.

The CO distribution from the TIME-GCM (Fig. 9c,d) shows all the major features of the ISAMS data in Fig. 5 for both solstice and equinox conditions, although the ISAMS zonal mean cross sections do not extend above approximately 73 km. The model shows a smaller CO in the polar winter region around 70 km than ISAMS, although the data is particularly noisy (errors larger than 70%) at these altitudes. This discrepancy between TIME-GCM and ISAMS is in line with previous model comparisons. These showed a general agreement at stratospheric altitudes, including a reasonable description of the polar vortex as seen by ISAMS, but were unable to simulate the intense vertical gradients of zonal mean CO measured in the mesosphere at the mid-high latitudes of the winter hemisphere [*Allen et al.*, 1999].

Equinox is a time of rapid transition when the CO mixing ratios are strongly influenced by dynamics. TIME-GCM calculations for March 21 (not shown) still present asymmetric distribution of CO remnant from the previous solstice conditions. However, 20 days later, by April 10 they exhibits a symmetrical distribution about the equator (see Fig. 9d). ISAMS could not measure CO at equinox in daytime, but (noisier) nighttime data for the Northern Hemisphere (NH) are available from March 30 to April 10, 1992. These indicate that on April 1 there was a persistent region of high CO at high latitudes of the NH, as a remnant of the previous winter period, consistent with the TIME-GCM results. Examination of the ISAMS data on a day-by-day basis shows that the high latitudes of the SH experienced a rapid increase in CO while the NH side was simultaneously experiencing a steady decline, reaching a symmetrical distribution by April 10; in consonance with TIME-GCM predictions. Even this symmetry surely lasted for only a few days, according to ISAMS data.

As a consequence of the transport of CO following the mean meridional circulation, important exchanges between mesosphere and stratosphere at high latitudes of the winter hemisphere seem to occur, as both model results and data (ISAMS and ATMOS) indicate. As mentioned in section 3, both ISAMS and ATMOS distributions show details of how this occurs, and have been used to study transport associated to the polar vortex [see *López-Valverde et al.*, 1996; *Allen et al.*, 1999; *Rinsland et al.*, 1999].

Regarding the lower mesosphere, detailed comparisons between the Goddard 3-D Chemistry and Transport Model and ISAMS data have been recently carried out by *Allen et al.* [1999]. They found that CO is systematically underestimated above the stratopause at all latitudes in about a factor 1.5–3; a fact that is difficult to explain with the standard dynamics and chemistry. This discrepancy was already found with simpler models and earlier datasets [*Strobel et al.*, 1987]. Since the effect of eddy diffusion on the CO abundance is very small in the lower mesosphere, these authors proposed a chemical solution: a new chemical source ($CO_2+O(^1D)\rightarrow CO+O_2$) and/or a revision of

the main chemical sink ($CO+OH\rightarrow CO_2+H$). The standard photochemistry (JPL97), however, as shown by *Allen et al.*, underpredicts the mesospheric CO. New data from MAHRSI recently available is helping to constrain the OH mesospheric chemistry [*Summers and Conway*, 2000; this issue]. These authors show that the mesospheric OH measured by MAHRSI is about a factor 1.5 smaller than that predicted by the standard photochemistry. This is in line with narrowing the model/measurements differences in CO discussed above, since a lower OH abundance would lead to a slower destruction of CO. This is a topic that should be investigated further in the future and might shed some light on a long-standing problem.

## 5. SUMMARY AND CONCLUSIONS

$CO_2$ mixing ratios have been measured since the 70's by in-situ rocket-borne ion-mass spectrometers, infrared radiometers on board rocket and satellite, and infrared occultation spectrometers on board the space shuttle. These measurements are generally affected by large errors (~30%), with the exception of the ATMOS occultation and ISAMS infrared observations. Most of the profiles have been measured at mid-latitudes, with the exception of ISAMS measurements which were taken globally over the sunlit atmosphere.

The measured profiles shows that $CO_2$ vmr start deviating from the diffusive equilibrium at about 90 km, declining rapidly above this altitude up to about 110-120 km, and then more gradually above. Predictions of $CO_2$ vmr by one-dimensional models in the past decade significantly overestimated the $CO_2$ vmr in the upper mesosphere/lower thermosphere. Model simulations presented in this work with and without diffusive separation (TIME-GCM and Garcia-Solomon model, respectively) show that the effect of diffusive separation begins to become important above about 85 km. This is corroborated by the good agreement between the TIME-GCM predictions and $CO_2$ observations. They suggest that the homopause, the altitude where the atmosphere ceases to be well mixed, lies considerably below 100 km. This result is surprising since the altitude where molecular diffusion becomes larger than eddy diffusion, the turbopause, has usually been estimated to be about 100 km.

TIME-GCM predictions slightly overestimate $CO_2$ observations above about 95 km, lying near the upper error bars of the observations. A possible reason for this small discrepancy is that the eddy diffusion coefficient used in TIME-GCM an other 2D models is overestimated in a factor 2-3. Such a factor appears to be larger than the likely uncertainty in model estimates of the diffusion coefficient, but it must be admitted that rather little is known about the time variability of eddy mixing in the lower thermosphere. Other possibilities, as a slower recombination rate of CO and O, or the presence of an unknown chemical dissociation process cannot be ruled out.

There are a few observations showing very low $CO_2$ mixing ratios in the lower thermosphere (90-110 km) in winter at mid-latitudes under perturbed atmospheric conditions. It is unclear if the observed $CO_2$ vmr's are real or the result of measurement's errors. If real, they do not appear to have a simple dynamical or photochemical explanation.

The TIME-GCM shows a slight latitudinal variation of $CO_2$ vmr above about 80 km for the sunlit atmosphere (equinox conditions and illuminated solstice regions at latitudes equatorwards of 40° in the winter hemisphere). This is consistent with ISAMS observations; this instrument could not detect that small variability given its larger measurements errors. The TIME-GCM model predicts, however, a large depletion at polar winter nighttime latitudes, being significant down to near the stratopause level, which could not be detected by ISAMS, and still remains to be confirmed. Given that $CO_2$ 15 $\mu$m emissions are routinely used to sound the atmospheric temperature and composition in this region, accurate (better than 3% at 60 km, 8% at 70, and than 15% at 80 km) measurements of $CO_2$ in these regions are of high priority.

Understanding the CO distribution in the middle atmosphere has been significantly improved during the last decade using new measurements from space, both in emission on board satellites and in absorption from the space shuttle. The ATMOS data have provided possibly the best vertical profiles of CO up to the lower thermosphere, while the infrared measurements by ISAMS on board UARS, in spite of their large noise and retrieval errors, represent the most extensive dataset of CO in the stratosphere and mesosphere, both in terms of geographical coverage and resolution and in terms of amount of data available. This dataset has been used to show the CO seasonal and latitudinal variability in the mesosphere as well as its distribution in the winter polar regions, including comparison with 2D ad 3D dynamical and chemical models [López-Valverde et al., 1996; Allen et al., 1999].

We have compared zonal means for equinox and solstice conditions from the TIME-GCM with similar data averages from ISAMS, which illustrate a good overall agreement between current model predictions and the general features of the CO distribution. The ISAMS data shows details not reproduced correctly by models thus far. López-Valverde et al. [1996] pointed out that latitudinal gradients observed in the lower mesosphere of the winter hemisphere are much stronger than those simulated by 2D models, and Allen et al. [1999] noted that the chemical transport model they used produced smoother vertical gradients in the lower mesosphere. In addition, a maximum of CO at low-latitudes in the tropics, perhaps of photochemical origin, was not produced by the model.

Comparison of standard models and measurements reveals that CO is underpredicted in the mesosphere at mid-latitudes in a factor 1.5–3. The recent OH observations by the Middle Atmosphere High Resolution Spectrograph Investigation (MAHRSI) might partially solve this long-standing problem.

There are a number of areas of future interest that we would like to highlight, in addition to more detailed comparisons between models and the ISAMS CO data. The UARS orbit does not allow full latitudinal coverage of the planet on a single day, except for certain periods, which do not include equinox. These periods are particularly interesting for the dynamics of the upper mesosphere at low latitudes, since the reversal of the inter-hemispheric circulation takes place then. CO promises to be a very good candidate to study the upper mesosphere under those conditions, and its abundance and distribution must be strongly linked to that of $CO_2$ in the lower thermosphere. Unfortunately, the ISAMS data are very noisy in the upper mesosphere (operational data exist up to 73 km approximately), and particularly during nighttime conditions. Perhaps future experiments like new sun occultation experiments on the space shuttle or emission instruments like MIPAS on the European ENVISAT mission, all of these with much higher spectral resolution than ISAMS, may supply new CO data at these altitudes.

*Acknowledgments.* We thank Peter Wintersteiner for supplying us with a compilation of $CO_2$ rocket measurements, and Martin Mlynczak and David Siskind for very useful comments and discussions. M.L.-P. and M.A.L.-V. have been partially supported by CICYT under contracts ESP97-1798 and ESP97-1773-C03-01. The National Center for Atmospheric Research is sponsored by the National Science Foundation.

## REFERENCES

Aellig, C. P., N. Kampfer, and A. Hauchecorne, The variability of mesospheric CO in the fall and winter as observed with ground-based microwave radiometry at 115 GHz, *J. geophys. Res.*, *100*, 14125–14130, 1995.

Allen, D. R. et al., Observations of Middle Atmosphere CO from the UARS ISAMS During the Early Northern Winter 1991/1992, *J. Atmos. Sci.*, *56*, 563–583, 1999.

Bevilacqua R. M., A. A. Stark, and P. R. Schwartz, The variability of carbon monoxide in the terrestrial mesosphere as determined from ground-based observations of the J=1-0 emission line *J. Geophys. Res.*, *90*, 5777-5782, 1985.

Bischof, W. et al., Increased concentration and vertical distribution of carbon dioxide in the stratosphere, *Nature*, *316*, 708–710, 1985.

Brasseur, G. P. and S. Solomon, Aeronomy of the Middle Atmosphere, D. Reidel Publishing Co., second edition, 452 pp., Dordrecht, Holland, 1986.

Brasseur, G.P. et al., Chapter 5 in *Atmospheric Chemistry and Global Change*, G.P. Brasseur, J.J. Orlando, and G.S. Tyndall eds., Oxford University Press, New York, 1999.

Caledonia, G. E., B. D. Green, and R. M. Nadile, The analysis of SPIRE measurements of atmospheric limb CO₂(ν₂) fluorescence, *J. Geophys. Res.*, *90*, 9783–9788, 1985.

Clancy, R. T., D. O. Muhleman, and M. Allen, Seasonal variability of CO in the terrestrial mesosphere, *J. Geophys. Res.*, *89*, 9673–9676, 1984.

Clerbaux C. et al., Remote sensing of CO, CH₄ and O₃ using a spaceborne nadir viewing interferometer, *J. Geophys. Res.*, *103*, 18999-19013, 1998.

Danilov, A.D., Review of long-term trends in the upper mesosphere, thermosphere and ionosphere, *Adv. Space Res.*, *22*(6), 907–915, 1998.

DeMore, W.B. et al., Chemical Kinetics and Photochemical Data for Use in Stratospheric Modelling, Evaluation No. 12, Publ. 97-4, Jet Propulsion Laboratory, Pasadena, CA, 1997.

Dickinson, R. E., E. C. Ridley, and R. G. Roble, A three-dimensional general circulation model of the thermosphere, *J. Geophys. Res.*, *86*, 1499-1512, 1981.

Dickinson, R. E., E. C. Ridley, and R. G. Roble, Thermospheric general circulation with coupled dynamics and composition, *J. Atmos. Sci.*, *41*, 205-219, 1984.

Dodd, J. A. et al., CIRRIS 1A observation of ¹³C¹⁶O and ¹²C¹⁸O fundamental band radiance in the upper atmosphere, *Geophys. Res. Lett.*, *20*, 2683–2686, 1993.

Edwards, D. P., M. López-Puertas, and M. A. López-Valverde, Non-LTE studies of the 15-μm bands of CO₂ for atmospheric remote sensing, *J. Geophys. Res.*, *98*, 14955-14977, 1993.

Edwards, D.P. et al., Non-local thermodynamic equilibrium limb radiance near 10 μm as measured by CLAES, *J. Geophys. Res.*, *101*, 26577–26588, 1996.

Endemann, M. and H. Fischer, Envisat's high-resolution limb sounder: MIPAS. *ESA bulletin*, *76*, 47-52, 1993.

Erdman, P. W., and E. C. Zipf, A closed-cycle cryogenic pump suitable for sounding rocket, *Rev. Sci. Instrum.*, *53*, 106, 1982.

Fleming E. L. et al., The middle atmospheric response to short and long term solar UV variations: analysis of observations and 2D model results, *J. Atmos. Terr. Phys.*, *57*, 333-365, 1995.

Fritts, D.C. and T.J. Dunkerton, Fluxes of heat and constituents due to convectively unstable gravity waves, *J. Atmos. Sci.*, *42*, 549-556, 1985.

Garcia, R.R. and S. Solomon, The effect of breaking gravity waves on the dynamics and chemical composition of the mesosphere and lower thermosphere, *J. Geophys. Res.*, *90*, 3850- 3868, 1985.

Garcia, R.R., Dynamics, radiation, and photochemistry in the mesosphere: Implications for the formation of noctilucent clouds, *J. Geophys. Res.*, *94*, 14605-14616, 1989.

Garcia, R.R. et al., A new numerical model for the middle atmosphere 1. Dynamics and transport of tropospheric source gases, *J. Geophys. Res.*, *97*, 12967–12991, 1992.

Garcia, R.R., and S. Solomon, A new numerical model of the middle atmosphere 2. Ozone and related species, *J. Geophys. Res.*, *99*, 12937–12951, 1994.

Girard, A. et al., Global results of Grille Spectrometer experiment on board Spacelab 1, *Planet. Space Sci.*, *36*, 291-300, 1988.

Grossmann, K.U., Homann D., and J. Schulz, Lower thermosphere infrared emissions of minor species during high latitude twilight. Part A: Experimental results, *J. Atmos. Terr. Phys.*, *56*, 1885, 1994.

Gunson, M. R. et al., Measurements of CH₄, N₂O, CO, and O₃ in the middle atmosphere by the atmospheric trace molecule spectroscopy experiment on Spacelab 3, *J. Geophys. Res.*, *95*, 13867-13882, 1990.

Hays, P.B., and J. J. Olivero, Carbon dioxide and monoxide above the troposphere, *Planet. Space Sci.*, *18*, 1729–1733, 1970.

Hansen, J.E. et al., Climate forcings in the industrial era, *Proc. of the National Academy of Sciences*, *95*, 12,753-12,758, 1998.

Keeling, C. D., and T. P. Whorf, Atmospheric CO₂ records from sites in the SIO air sampling network, in *Trends: A Compendium of Data on Global Change, Carbon Dioxide Information Analysis Center*, Oak Ridge National Laboratory, Oak Ridge, Tenn., 1998.

López-Puertas, M., and F. W. Taylor, Carbon dioxide 4.3-μm emission in the Earth's atmosphere. A comparison between NIMBUS 7 SAMS measurements and non- LTE radiative transfer calculations, *J. Geophys. Res.*, *94*, 13045-13068, 1989.

López-Puertas, M. et al., Non- local thermodynamic equilibrium populations of the first vibrational excited state of CO in the middle atmosphere, *J. Geophys. Res.*, *98*, 8933-8947, 1993.

López-Puertas, M. et al., Non-LTE atmospheric limb radiances at 4.6 μm as measured by UARS/ISAMS II. Analysis of the daytime radiances, *J. Geophys. Res.*, *103*, 8515–8530, 1998.

López-Valverde, M. A., M. López-Puertas, and C. J. Marks, Non-LTE modelling for the retrieval of CO abundances from ISAMS measurements, in *Technical Digest on Optical Remote Sensing of the Atmosphere*, Vol.18, 3-33, Optical Society of America, Washington, D.C., 1991.

López-Valverde, M. A. et al., Global and seasonal variations in middle atmosphere CO from UARS/ISAMS, *Geophys. Res. Lett.*, *20*, 1247–1250, 1993.

López-Valverde, M.A. et al., Validation of measurements of carbon monoxide from the improved stratospheric and mesospheric sounder, *J. Geophys. Res.*, *101*, 9929-9955, 1996.

López-Valverde, M.A. et al., Correlation between ISAMS and ATMOS measurements of CO in the middle atmosphere, *Adv. Space Res.*, *22*, 1517- 1520, 1998.

McIntyre, M.E., On dynamics and transport near the mesopause in summer, *J. Geophys. Res.*, *94*, 14617-14628, 1989.

Mlynczak, M. G., and J. M. Russell III, An overview of the SABER experiment for the TIMED mission, *Optical Remote Sensing of the Atmosphere*, *2*, OSA Technical Digest Series, Washington, DC, 5-7, 1995.

Mlynczak, M. G., A contemporary assessment of the mesospheric energy budget, in *this monograph*.

Murphy A. K., Satellite measurements of atmospheric trace gases, *Ph.D. Thesis*, Oxford University, Oxford, 1985.

Nebel, H. et al., $CO_2$ non-local thermodynamic equilibrium radiative excitation and infrared dayglow at 4.3 $\mu$m: Application to spectral infrared rocket experiment data, *J. Geophys. Res.*, *99*, 10409-10419, 1994.

Novelli, P. C., K. A. Masarie, and P. M. Lang, Distributions and recent changes of carbon monoxide in the lower troposphere, *J. Geophys. Res.*, *103*, 19015-19033, 1998.

Offermann, D., and K. U. Grossmann, Thermospheric density and compositions as determined by s mass spectrometer with cry-ion source, *J. Geophys. Res.*, *78*, 8296-8304, 1973.

Offermann, D. et al., Neutral gas composition measurements between 80 and 120 km, *Planet. Space Sci.*, *29*, 747-764, 1981.

Philbrick, C.R., G.A. Faucher, and E. Trzcinski, Rocket measurements of mesospheric and lower thermospheric composition,. In *Space Research*, Vol. 13, pp. 255-260. Akademie, Berlin, 1973.

Richmond, A.D., E.C. Ridley, and R.G. Roble, A thermosphere/ionosphere general circulation model with coupled electrodynamics, *Geophys. Res. Lett.*, *19*, 601-604, 1992.

Riese, M., Measurements of trace gases by the cryogenic infrared spectrometers and telescopes for the atmosphere (CRISTA) experiment, *Adv. Space Res.*, *19*, 563-566, 1997.

Rinsland, C.P. et al., Middle and upper atmosphere pressure-temperature profiles and the abundances of $CO_2$ and CO in the upper atmosphere from ATMOS/Spacelab 3 observations, *J. Geophys. Res.*, *97*, 20479-20495, 1992.

Rinsland, C.P. et al., Polar stratospheric descent of $NO_y$ and CO and arctic denitrification during winter 1992-1993, *J. Geophys. Res.*, *104*, 1847-1861, 1999.

Roble, R.G., E.C. Ridley, and R.E. Dickinson, On the global mean structure of the thermosphere, *J. Geophys. Res.*, *92*, 8745-8758, 1987.

Roble, R.G. et al., A coupled thermosphere/ionosphere general circulation model, *Geophys. Res. Lett.*, *15*, 1325-1328, 1988.

Roble R.G., and R.E. Dickinson, How will changes in carbon dioxide and methane modify the mean structure of the mesosphere and thermosphere?, *Geophys. Res. Lett.*, *16*, 1441, 1989.

Roble R.G., "Greenhouse cooling" of the upper atmosphere, *EOS*, Feb. 23, 1993.

Roble, R.G. and E.C. Ridley, A thermosphere-ionosphere-mesosphere electrodynamics general circulation model (TIME-GCM): Equinox solar cycle minimum simulations (30–500 km), *Geophys. Res. Lett.*, *21*, 417-420, 1994.

Roble, R.G., Energetics of the mesosphere and thermosphere, in *The Upper Mesosphere and Lower Thermosphere: A Review of Experiment and Theory*, Geophys. Mono., *87*, 1-21, 1995.

Roble, R.G., On the feasibility of developing a global atmospheric model extending from the ground to the exosphere, in *this monograph*.

Rodgers, C.D. et al., Improved Stratospheric and Mesospheric Sounder validation: General approach and in-flight radiometric calibration, *J. Geophys. Res.*, *101*, 9775-9785, 1996.

Rodrigo, R. et al., Neutral atmospheric composition between 60 and 220 km: A theoretical model for middle latitudes, *Planet. Space Sci.*, *34*, 723-743, 1986.

Sharma, R. D., and P. P. Wintersteiner, Role of carbon dioxide in cooling planetary atmospheres, *Geophys. Res. Lett.*, *17*, 2201-2204, 1990.

Solomon, S. et al., Photochemistry and transport of carbon monoxide in the middle atmosphere, *J. Atmos. Sci.*, *42*, 1072-1083, 1985.

Stair, A.T., Jr. et al., Observations of limb radiance with cryogenic spectral infrared rocket experiment, *J. Geophys. Res.*, *90*, 9763-9775, 1985.

Strobel, D.F. et al., Vertical constituent transport in the mesosphere, *J. Geophys. Res.*, *92*, 6691-6698, 1987.

Summers, M.E., and R.R. Conway, Insights into middle atmospheric hydrogen chemistry from analysis of MAHRSI observations of hydroxyl (OH), in *this monograph*.

Taylor, F. W. et al., Remote sensing of atmospheric structure and composition by pressure modulator radiometry from space: The ISAMS experiment on UARS, *J. Geophys. Res.*, *10*, 10799-10814, 1993.

Taylor, F.W., Studies of planetary atmospheres by optical methods, in *Progress in Atmospheric Physics*, edited by R. Rodrigo, J.J. López-Moreno, M. López-Puertas and A. Molina, pp. 33–45, Kluwer Academic Pub., Dordrecht, 1988.

Trinks, H., and K. H. Fricke, Carbon dioxide in the lower thermosphere, *J. Geophys. Res.*, *83*, 3883-3886, 1978.

Trinks, H. et al., Neutral composition measurements between 90- and 220-km altitude by rocket-borne mass spectrometer, *J. Geophys. Res.*, *83*, 2169-2176, 1978.

Turner D. and J. R. Drummond, Measurements of carbon compounds in the stratosphere using a pressure modulator radiometer, *J. Geophys. Res.*, *96*, 17279-17290, 1991.

Vercheval, J. et al., $CO_2$ and CO distribution in the middle atmosphere and lower thermosphere deduced from infrared spectra, *Ann. Phys.*, *4*, 161-164, 1986.

Vollmann, K., and K.U. Grossmann, Excitation of 4.3 $\mu$m $CO_2$ emissions by $O(^1D)$ during twilight, *Adv. Space Res.*, *20*, 1185–1189, 1997.

Waters J. W., W. J. Wilson, and F. I. Shimabukuro, Microwave measurements of mesospheric carbon monoxide, *Science*, *191*, 1174-1175, 1976.

Wintersteiner, P.P. et al., Line-by-line radiative excitation model for the non-equilibrium atmosphere: application to

CO$_2$ 15 $\mu$m emission, *J. Geophys. Res.*, *97*, 18083-18117, 1992.

Zaragoza, G. et al., Global distribution of CO$_2$ in the upper mesosphere as derived from UARS/ISAMS measurements, *J. Geophys. Res.*, *97*, in press, 2000.

R. R. Garcia, Atmospheric Chemistry Division, National Center for Atmospheric Research, P.O. Box 3000, Boulder, CO 80307-3000. (e-mail: rgarcia@ucar.edu)

M. López-Puertas and M. Á. López-Valverde Instituto de Astrofísica de Andalucía, Apartado Postal 3004, 18080 Granada, Spain. (e-mail: puertas@iaa.es)

R. G. Roble, National Center for Atmospheric Research, High Altitude Observatory, P.O. Box 3000, Boulder, CO 80307-3000. (e-mail: roble@ucar.edu)

# On the Coupling Between Middle and Upper Atmospheric odd Nitrogen

David E. Siskind

*E. O. Hulburt Center for Space Research, Naval Research Laboratory, Washington DC*

This paper reviews observations and models of the descent of mesospheric and thermospheric $NO_x$ into the stratosphere. Data from the Upper Atmosphere Research Satellite (UARS) have validated the suggestion that the large abundances of NO known to exist in the upper atmosphere (z > 90 km) can be transported to the polar stratosphere in winter. Upper atmospheric $NO_x$ is identified in the stratosphere by its association with very low values of $CH_4$, a dynamical tracer. Springtime $NO_x$ enhancements are seen down to 25 km in the Southern Hemisphere and tend to vary from year to year in concert with the previous winter's energetic electron precipitation. Model calculations show that mesospheric transport processes are critical in modulating how much of the NO produced in the thermosphere can be deposited in the stratosphere. Increased horizontal mixing due to breaking planetary waves reduces the net downward transport. One problem is that the model overestimates the $NO_x$ transport, possibly due to an underestimate of NO chemical loss processes or an underestimate of mesospheric horizontal mixing processes. However, the model does qualitatively reproduce the greater enhancement in the Southern Hemisphere relative to the Northern. Outstanding issues for future research are also discussed.

## 1. INTRODUCTION

One of the most interesting questions in aeronomy is the extent to which the two atmospheric odd nitrogen ($NO_y$) layers are coupled. That two such layers exist in the earth's atmosphere has long been recognized. The lower layer, in the stratosphere, first received attention in the context of understanding the stratospheric ozone ($O_3$) budget [*Crutzen*, 1970]. It is produced by the oxidation of nitrous oxide ($N_2O$) and includes such molecules as NO, $NO_2$, $HNO_3$, $N_2O_5$, $ClONO_2$ and other species [*Brasseur and Solomon*, 1986]. One subset of total $NO_y$ which reacts with odd oxygen is $NO_x$ = NO + $NO_2$. Above 30 km, most of the $NO_y$ is in the form of $NO_x$. Typical $NO_x$ mixing ratios are in the range 5-15 ppbv with the peak value located around 40 km. The upper $NO_y$ layer, in the thermosphere, was postulated to exist by *Nicolet* [1945], was first observed by *Barth* [1964] and is reviewed by *Barth* [1992]. It is produced by the dissociation and ionization of $N_2$ by solar photons and energetic electrons. Here, NO and $N(^4S)$ are the dominant $NO_y$ species with peak mixing ratios as high as 10-100 ppmv. In density units, the two $NO_y$ layers are much more comparable; peak values in the thermosphere are about 1-2 $\times 10^8$ while in the stratosphere they are about $1 \times 10^9$ $cm^{-3}$.

Normally the stratospheric and thermospheric $NO_y$ layers are separated by a deep mesospheric minimum [*Siskind et al.], 1998a*. This minimum results from the photolysis of $NO$, i.e.,

$$NO + h\nu \rightarrow N + O. \qquad (1)$$

Atmospheric Science Across the Stratopause
Geophysical Monograph 123
Published in 2000 by the American Geophysical Union

followed by the recombination of atomic nitrogen with NO to form molecular nitrogen

$$N + NO \rightarrow N_2 + O. \qquad (2)$$

Taken together, reactions (1) and (2) represent a net loss of two NO molecules per photolysis event. The lifetime of NO against photolysis in the mesosphere can be as short as several days [*Minschwaner and Siskind*, [1993]. During high latitude winter, however, conditions are dramatically different. Here, photolysis is weak or absent and strong downwelling is known to occur. The combination of these two effects led *Solomon et al.* [1982] to propose that the polar night mesosphere was a region where the thermosphere was coupled to the lower atmosphere. *Frederick and Orsini* [1982] also made this suggestion at about the same time.

For a long time, there were very few observations to support the *Solomon et al.* [1982] suggestion. One was data from the Limb Infrared Monitor of the Stratosphere (LIMS) which showed very high $NO_2$ in the polar upper stratosphere and lower mesosphere [*Russell et al., 1984*]. Unfortunately there were no concurrent NO observations and since most $NO_y$ is in the form of *NO* above 50 km, the net source of $NO_x$ could not be quantified. There was also a report of high NO (over 50 ppbv) by *Frederick and Horvath*, [1985] from a rocket sounding over Alaska. Because of the lack of comprehensive $NO_x$ data, some researchers tried to indirectly infer the existence of high $NO_x$ through its expected effects on $O_3$. Thus *Solomon and Garcia*, [1984] and *Rusch and Clancy*, [1989] looked for springtime reductions in $O_3$ that could be linked to enhanced $NO_x$ which had descended from the thermosphere the previous winter. Both these studies concentrated on the upper stratosphere (0.7 - 4.0 mb) whereas as we shall see, the signature of enhanced springtime $NO_x$ is most evident at much higher pressures. At the same time, other work showed that descent in the winter mesosphere has observational consequences on other trace constituents; for example, bringing down air enriched in CO [*Solomon et al.*, 1985] and depleted in $H_2O$ [*LeTexier et al.*, 1988] and in causing the winter anomaly in the ionospheric D region [*Garcia et al.*, 1987]. In general, however, the topic of odd nitrogen coupling, and the possible relationship with other mesospheric tracers, lagged due to lack of data.

The advent of the Upper Atmospheric Research Satellite (UARS) has drastically changed this situation. The discovery by the Halogen Occultation Experiment (HALOE) of very low (i.e. mesospheric) values of $CH_4$ at altitudes as low as 22 km in the Southern Hemisphere (SH) polar vortex [*Russell et al.*, 1993] provided clear evidence that mesospheric air can penetrate deep into

the stratosphere. Imbedded within these regions of descended mesospheric air are regions of enhanced $NO_x$ that are linked to the upper mesosphere and the lower thermosphere. Supporting evidence for long range polar $NO_x$ descent is provided by the Atmospheric Trace Molecule Spectroscopy (ATMOS) experiment [*Rinsland et al.*, 1999] and the Polar Ozone and Aerosol Measurement (POAM) experiment [*Randall et al.*, 1998]. As a result of these new datasets and associated theoretical advances, the question of enhanced $NO_y$ descent is thus seen to be closely coupled to the more general question of mesosphere/stratosphere coupling in the winter polar vortex.

The odd nitrogen problem remains unique, however, for several reasons. First it has a source at very high altitudes (z > 85 km); to the extent that this NO is brought down into the stratosphere, we may therefore speak more properly of thermosphere/stratosphere coupling, not just mesosphere/stratosphere coupling. This greatly expands the range of atmospheric processes which must be considered to understand the problem. Second, since $NO_x$ chemistry dominates $O_3$ catalytic loss in the mid-stratosphere, it suggests the possibility of a thermospheric influence on stratospheric $O_3$. Finally, since the thermospheric NO source varies quite dramatically in response to variable solar cycle and auroral forcing, it suggests a mechanism whereby solar terrestrial effects could propagate deep into the stratosphere [*Brasseur*, 1993; *Garcia et al.*, 1984]. This last point has been the subject of some controversy. In 1986, *Callis and Natarajan* proposed that the descent of enhanced $NO_x$ was a cause of the then recently discovered polar $O_3$ hole. This theory was quickly disproved because the $O_3$ hole was observed to be depleted of, rather than enhanced with, $NO_y$ [*Solomon*, 1999]; however, [*Callis et al.*, 1998] have since suggested that mesospheric $NO_y$ is a significant contributor to non-polar stratospheric $NO_y$. Furthermore they argue that variability in mesospheric $NO_y$ due to varying amounts of particle precipitation above 80 km could explain variations in the column abundance of $NO_2$ seen by the Stratospheric Aerosol and Gas Experiment (SAGE) II during the 1980s.

This last question, whether mesospheric NO explains the SAGE $NO_2$ variability, reamins unanswered. However, in the last 6 years we have learned much about the complexity of atmospheric $NO_y$ coupling. It involves dynamical and chemical processes in both the stratosphere and mesosphere, and probably also the thermosphere. This paper summarizes the observations which document this coupling, using data mostly from HALOE. In addition I will review and expand upon the results of published two dimensional model calculations of the descent of odd nitrogen from the upper atmosphere into the stratosphere. Finally, areas for future work are outlined.

## 2. OBSERVATIONS OF $NO_x$ ENHANCEMENTS IN THE STRATOSPHERE AND MESOSPHERE

### 2.1. Introduction

The most relevant UARS data for this question are the HALOE NO data which extend from the lower stratosphere to the lower thermosphere [*Gordley et al., 1996*]. This is because HALOE is the first experiment to simultaneously measure the two atmospheric $NO_y$ layers. Below 50 km, the NO data must necessarily be supplemented by the HALOE $NO_2$ since NO + $NO_2$ = $NO_x$ is a long lived family. Another reason the HALOE data is so useful is that it simultaneously measures $CH_4$ which is a long lived tracer that can be used to diagnose atmospheric motions. In this paper, we use data from Version 19 processing.

A sample of this data is shown in Plate 1. The data in the plate are a 5 year (1992-1996) average of SH wintertime conditions (June/July/Aug). The plate shows the large mixing ratios prevalent in the upper atmosphere above 80 km, the mesospheric minimum and then the stratospheric $NO_x$ layer. Although looking at NO in mixing ratio units may appear to overstate the importance of the upper atmosphere, it does clearly demonstrate that the upper atmosphere is enriched in $NO_y$. Furthermore, advection by the mean meridional circulation conserves mixing ratio and the existance of downward sloping isolines in the winter mesosphere, evident in the plate, is a straightforward demonstration that downward transport of NO is occurring. A table of number density values for the mesosphere which correspond to the data presented in Plate 1 is given by *Siskind et al.* [1998a, Tables 3b and 3c].

Plate 1 also clearly illustrates the lack of global sampling associated with the HALOE dataset. Because of the 57° inclination of the UARS orbit and the observational requirement for sunlight, HALOE never samples the polar latitudes in winter. The highest latitude seen by HALOE in winter is 50-55 degrees and HALOE does not observe poleward of 70 degrees until around spring equinox. This is important because as we will show, the largest NO densities are predicted to occur in winter, poleward of where HALOE can see. A summary of the HALOE high latitude measurement periods from 1992 through 1998 is given in Table 1. It shows that there are five periods each year (for a given hemisphere) when mid- to high-latitude observations relevant to the question of polar descent are made and when enhancements

in polar $NO_x$ could, in principle be seen. One of these periods is in late autumn, two are in winter and two are in early spring. Due to orbital precession, each year's sampling is slightly different than the previous. Typically a given sampling pattern occurs several days earlier each year and thus from 1992-1998 a shift of about 1 month is seen. Also the peak latitude often changes as well so that in 1998, the maximum latitude reached during Period 3 dropped from 75 to 60°. The arrows in the table indicate the temporal and latitudinal shifts in sampling from 1992 to 1998. The indicated times correspond to the approximate time when the extrema in latitude occurred. For example, in the SH HALOE sampled the springtime polar vortex at sunset at a peak latitude of 80S in late October 1991 shifting to a peak latitude of 70S in late September 1998.

*Siskind and Russell* [1996] showed how the HALOE data can be used to detect enhanced $NO_x$. $CH_4$ has a source in the troposphere and several chemical sinks which maximize in the upper stratosphere. Thus it exhibits a mixing ratio profile which decreases with altitude. Values of $CH_4 < 0.4$ ppmv are generally representative of air that resides above the stratopause. Such low $CH_4$ values are also seen in the stratospheric high latitude winter regions and are used as a diagnostic of descending air [e.g. *Russell and Pierce*, this volume; see also *Strahan et al.*, 1996; *Bacmeister et al.*, 1995]. At the same time, above about 40 km, $NO_x$ also decreases with increasing altitude. Thus, in the absence of a high altitude source, $NO_x$ should be positively correlated with $CH_4$. We identify cases where the $CH_4$ mixing ratio is $< 0.4$ ppmv and where NO (or $NO_x$ in the stratosphere) and $CH_4$ vary oppositely as regions where there is an extra $NO_x$ source from the upper atmosphere.

### 2.2. Wintertime $NO_x$ and $CH_4$

From Table 1, there are two times in winter and one time in late autumn when HALOE samples as far poleward as 50° and when high latitude NO enhancements are expected. *Siskind and Russell* [1996] presented some preliminary results from one year (1994) for 0.18 mb for an older version of the HALOE processing (V17). *Callis and Lambeth* [1998] presented an analysis of a high latitude enhancement during Period 1. Here we show a five year average (1992-1996) of V19 data for Periods 1 - 3 for both hemispheres at 0.18 mb (about 60 km), binned in 5° increments. Figure 1 shows that for all periods in both hemisphere, the NO begins increasing as one moves poleward of about 30° latitude. If NO were purely of stratospheric origin, the increasing descent rates at higher latitudes would yield lower NO values. The existence of an enhancement is a clear indication of an upper atmospheric source. The enhancements are always larger in the SH and the largest (a factor of 2.5) is during Period 2. Since Period 2 is in early winter, the photodissociation of NO is slowest at this time and this may facilitate the large buildup. During Periods 1 and 3 the N/S differences are smaller (2-3 ppbv or 40%), but nonethelesss greater than the variance of the averaged data (less than 0.5 ppbv). For the total of Periods 1, 2 and 3, there are about 110 daily zonal means which are included in the 45-50° latitude bin and which have mixing ratios greater than 10 ppbv; over 80% of these are in the SH. We thus conclude that the N/S asymmetry is statistically robust.

The relatively greater enhancements in the SH could either be due to greater particle precipitation during the Southern winter or greater downward transport from the upper mesospheric source. Given the limited temporal and spatial coverage of HALOE at high latitudes, its difficult to resolve this question solely from observational considerations. In Section 4 we will show that dynamical factors alone can lead to an interhemispheric difference in our model.

The enhancements seen in Figure 1 occasionally, but not consistently, extend into the upper stratosphere. As in the mesosphere, $NO_x$ enhancements are larger in the SH. Plate 2 shows an example of this at 1 mb from 1994 for Period 2. The plate shows a scatterplot of HALOE $CH_4$ vs $NO_x$. For both the February (NH) and the August (SH) data, very low values of $CH_4$ were observed ($< 0.2$ ppmv) indicating that mesospheric air was being measured. In the August data, a large $NO_x$ enhancement (15-20 ppbv) is evident at the lowest $CH_4$ values; in February, $NO_x$ shows a much smaller increase (5-7 ppbv) at the lowest $CH_4$. This comparison is particularly interesting because *Callis et al.* [1998] have suggested that elevated $NO_2$ values seen by HALOE in the NH in April and May 1994 might be the result of $NO_x$ descent. However, we would expect any enhancement to be more apparent in winter (when sunlight is weaker and descent is stronger); the absence of any enhancement three months earlier, despite the detection of mesospheric air, poses a significant challenge for the *Callis et al.* [1998] suggestion.

### 2.3. Springtime $NO_x$ and $CH_4$

The possibility of $NO_x$ enhancements persisting into the spring (Periods 3 and 4 of Table 1) is of particular interest because the potential for $NO_x$-catalyzed $O_3$ destruction is greater in the stronger sun. Below 40 km, $NO_x$ should be quite long lived since the NO photolysis becomes very slow. To the extent that mesospheric $NO_x$ penetrates this level, it represents a net source to the stratospheric $NO_y$ layer, even after the polar regions become sunlit. Guided by the earlier searches of *Solomon and Garcia* [1984] and *Rusch and Clancy* [1989], *Siskind and Russell* [1996] looked for $NO_x$ enhancements in the HALOE dataset in the upper stratosphere and did not find any. However, *Callis et al.* [1996], also using HALOE, and *Rinsland et al.* [1996], using ATMOS data, showed that enhancements could be observed much lower in the stratosphere, at pressures around 10-20 mb, and, as in winter, only in the SH [*Rinsland et al.*, 1999]. *Randall et al.*, [1998] used POAM data to continuously monitor the descent of a layer of enhanced SH $NO_2$ from 40 km down to 25 km. The observed descent rate from August through October was consistent with theoretical expectations. Interestingly, the enhancement varied from year to year in a manner that qualitatively mirrored the variation of the geomagnetic Ap index, consistent with an upper atmospheric source.

Most recently, *Siskind et al.* [2000], have attempted to quantify the net contribution of mesospheric $NO_x$ to the SH stratosphere for each year from 1992-1996. They assumed that the enhanced $NO_x$ remains well confined within the vortex. To better isolate those measurements that were within the vortex, they used equivalent latitudes (lat$_{eq}$) to regrid the data. *Randall et al.*, [1998] and *Manney et al.* [1999] have illustrated the advantages of the lat$_{eq}$ coordinate system (based upon the area enclosed with a given value of potential vorticity) in regions where planetary waves distort the polar vortex and create zonal asymmetries in the flow pattern.

An example of the usefulness of equivalent latitudes for diagnosing $NO_x$ enhancements is given in Plate 3, taken from *Siskind et al.* [2000]. The plate shows $CH_4$ and $NO_x$ at the 750K potential temperature surface (about 12 mb) as a function of SH equivalent latitude for 1994 (3a) and 1996 (3b). The sharp transition just inside the vortex edge (defined as $(\Delta lat_{eq}) = 0$; thus $\Delta lat_{eq} > 0$ refers to inside the vortex, $< 0$ is outside) is evident. In 1994, the $CH_4$ decrease coincides with increasing $NO_x$ in an identical manner as the 1 mb data from August shown in Plate 2. In 1996, a factor of three less $NO_x$ is seen in the vortex, despite the evidence from the low values of $CH_4$ that the amount of descent in both years was similar. This supports the *Randall et al.* [1998] suggestion that the difference between the two years is due to a difference in the source function at high altitudes, rather than differing meteorological conditions.

The monotonic nature of the $NO_x$ variation when viewed as a function of lat$_{eq}$ supports the argument that this coordinate system is an appropriate one to spatially isolate the geographic regions where $NO_x$ is enhanced. *Siskind et al.* [2000] looked at the year to

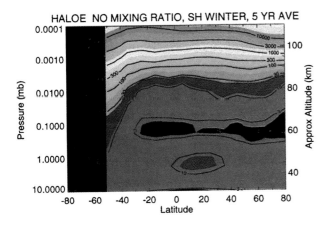

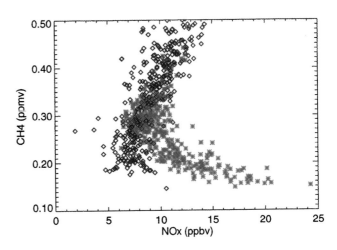

**Plate 1.** Average of HALOE zonal mean nitric oxide mixing ratios (ppbv) for 1992-1996 for June-August. The black region poleward of 50S represents no data.

**Plate 2.** Scatterplot of HALOE $NO_x$ and $CH_4$ for February, 1994 (blue) and August, 1994 (red) at 1mb. Both sets of data are from Period 2 (late winter); the February data is from the NH, the August data is from the SH. Both the SH and NH data show $NO_x$ enhancements at the lowest values of $CH_4$ indicative of a source from high altitudes; however, the SH enhancements are 2-3X greater.

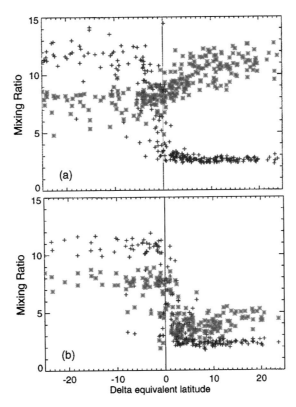

**Plate 3.** $NO_x$ (ppbv, red) and $CH_4$ (ppmv x 10, blue) for $\Theta = 750K$ for (a) 1994 and (b) 1996. The vortex edge is defined as $\Delta(\mathrm{lat}_{eq}) = 0$.

Table 1. Summary of HALOE High Latitude Measurements

| Period | Time of Year (1992 → 1998) | Time of Day | Highest Latitude |
|---|---|---|---|
| 1 | 3 → 6 weeks prior to winter solstice | sunset | 50 |
| 2 | +3 → -1 weeks past winter solstice | sunrise | 50 |
| 3 | +6 → +3 weeks past winter solstice | sunset | 55 → 50 |
| 4 | +2 → -2 weeks past spring equniox | sunrise | 75 → 60 |
| 5 | +3 → 0 weeks past spring equniox | sunset | 80 → 70 |

year variability of the total column inside the vortex. This is shown in Figure 2 for 6 Antarctic springs. Also shown is the geomagnetic Ap index for the previous winter (May-Aug). With the exception of 1992, larger $NO_x$ enhancements correlate with a bigger wintertime Ap index. Using the Ap index as a qualitative measure of the precipitation of energetic electrons at high altitudes (> 80 km), this figure is consistent with wintertime production of $NO_y$, followed by descent all the way down to the mid-stratosphere. Note the very large enhancement in 1991. This supports the earlier suggestions of *Callis et al.* [1996] and *Randall et al.* [1998] that balloon measurements of low $O_3$ in the polar vortex at 25-30 km were due to the descent of enhanced $NO_x$ and not heterogeneous chemistry as first thought [*Hofmann et al.*, 1992]. By integrating the values in Figure 2 over the surface area of the polar vortex, a total amount of enhanced $NO_x$ (relative to the assumed baseline year, 1996) that peaked at around 0.8 - 1.3 gigamoles (GM) in 1991 was estimated. This is on the order of 3-5% of the $N_2O$ source, estimated to be 26-29 GM [*Vitt and Jackman*, 1996].

### 2.4. Discussion

The preceding discussion shows that enhanced $NO_x$ from the upper atmosphere is clearly observed to descend to the SH mid-stratosphere. This is confirmed by the association of enhanced $NO_x$ with $CH_4$ mixing ratios that are typically associated with altitudes above the stratopause. Although we only showed HALOE data, other data from ATMOS and POAM confirm this picture. Inside the vortex, the total column $NO_x$ can vary by a factor of two due to variations in the mesospheric source and at certain altitudes near 10 mb, local enhancements of almost a factor of three are seen. Globally, the observed enhancements can be up to 3-5% of the $NO_y$ source from oxidation of $N_2O$. Finally, these $NO_x$ enhancements appear to be linked to $O_3$ reductions in the vortex at 25-30 km [*Randall et al.*, 1998].

Despite this impressive validation of an almost two decades old theory, important questions regarding the N/S asymmetry and interannual variability remain unanswered. An interesting view of this is seen in Figure 3. This figure shows altitude profiles for the 5 years which

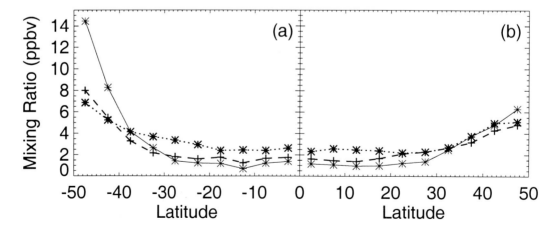

**Figure 1.** Latitudinal variation of zonal mean NO at 0.18 mb (about 60 km). The dashed line in each panel is a 5 year average of the sunset data from Period 1 (see Table 1). The solid line is the sunrise data from Period 2 and the dotted line is for the sunset data from Period 3. (a) SH (b) NH.

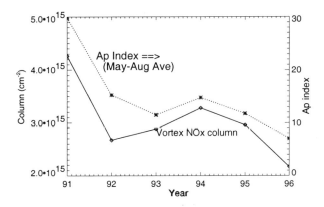

**Figure 2.** Average column $NO_x$ inside the vortex for 1991-1996. The column is bounded on the bottom by the lowest altitude to which mesospheric $CH_4$ is observed (defined as $CH_4 < 0.4$ ppmv, typically about 20-25 km). The top altitude is $\sim 32$ km, above that altitude the vortex has already begun to break down. See Siskind et al., 2000 for details. The 4-month averaged geomagnetic $A_p$ index (dotted line with stars) is given by the right hand axis.

make up the average in Figure 1 (Period 2) at the highest latitudes observed (45-50S for Figure 3a, 45-50N for Figure 3b). At 0.1 mb, in the SH, a factor of 5 variation is seen between the maximum in 1994 and the minimum in 1996 with intermediate values for 1992, 1993 and 1995. This is generally consistent with the variation seen in the mid stratosphere in October. Thus, in Figure 2, the 1992 enhancement appeared anomalous because it had low column $NO_x$ in the stratosphere despite a relatively high $A_p$ index the previous winter. In Figure 3, 1992 is similarly low between 0.3 and 1.0 mbar, but becomes relatively higher than other years above 0.1 mbar. This suggests that the relative timing of the particle production and the $NO_x$ descent should be considered.

Comparing Figure 3a with 3b clearly shows the larger enhancements in the SH below 0.1 mb relative to the NH. Interestingly, above 0.1 mb, the NH NO becomes larger than the SH [e.g. Figure 6, *Russell and Pierce*, 2000]. The model calculations we will discuss later suggest greater mesospheric descent in the SH winter relative to the NH winter. However, it may also be true that the particle precipitation in the NH winter occurred too late in the season and thus the NO could not penetrate into the lower mesosphere without being photodissociated. Finally, we must remember that we are only seeing the edge of the polar enhancements and even these are sampled infrequently. Thus it is premature to say that there is not any transport of mesospheric $NO_x$ to the NH stratosphere; indeed, the LIMS data show there can be. More careful consideration of both the source function from energetic particles and mesospheric trans-

port is required to completely disentangle the relative roles of chemistry and dynamics on the observed variability.

## 3. REVIEW OF TWO DIMENSIONAL MODEL

### 3.1. Introduction

The problem of polar descent has been studied theoretically by a number of authors. *Rosenfield et al.* [1994] used a one dimensional vortex interior model and showed that air initialized at 50-52 km descended about 27-29 km during the winter for both hemispheres. *Bacmeister et al.* [1995] used the Rosenfield et al. heating code in a two dimensional model to study the effects of planetary wave mixing on this descent. For realistic values of wave driving, they were able to simulate the descent of mesospheric air into the polar stratosphere. *Eluszkiewicz et al.* [1995] and *Strahan et al.* [1996] used three dimensional models to study specific aspects of vortex descent, in particular, the relative roles of diabatic descent and horizontal mixing on tracer isopleths,

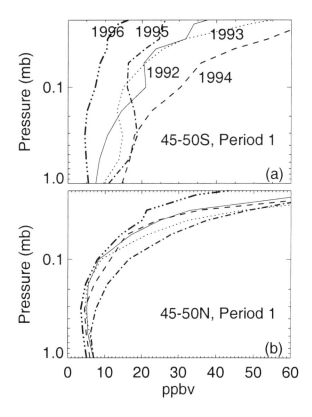

**Figure 3.** Altitude profiles of NO averaged from 45-50S (Figure 6a) and 45-50N (Figure 6b). These latitudes are the highest seen by HALOE each year during mid-winter. The symbols for the curve in 6b are the same as those indicated for 6a.

such as $N_2O$ and $CH_4$. Lagrangian approaches have been adopted by *Fisher et al.* [1993] and *Sutton,* [1994]. They found that by late winter, most of the air in the polar upper stratosphere is of mesospheric origin.

The model we have used to study the descent of enhanced $NO_x$ is a two dimensional (2D), or zonally averaged, chemical transport model. Such models are useful because they are the simplest class of model to incorporate a self-consistent formulation of atmospheric dynamics, radiation, and chemistry. The 2D model we discuss here has been documented by *Siskind et al.* [1997, 1998 and references therein]. In this section, I review several of the salient features of this model that are pertinent to the problem of $NO_y$ coupling.

The dynamical part of the model is based upon the Transformed-Eulerian-Mean (TEM) formulation [e.g. *Brasseur and Solomon*, 1986] which solves for the evolution of zonally averaged angular momentum $(\overline{M})$ and potential temperature ($\Theta$). Previously, the model used the *Prather* [1986] advection scheme; the newest version discussed here uses the piecewise-parabolic-method (PPM) of *Lin and Rood* [1996]. The calculated $\overline{M}$ and $\Theta$ fields are used to derive an elliptic equation for the stream function $\Psi$ from which the residual meridional (v*) and vertical (w*) wind fields are obtained. The vertical domain of our model extends from 0-108 km. Thus our top boundary is near the peak of the thermospheric NO layer and thus we can model both atmospheric NO layers without the need to specify a flux of NO entering the mesosphere. The vertical resolution is 2.66 km and the latitudinal resolution is 4.8°. The dynamics are integrated with a 2 hour time step.

Our model is similar in type to those classified as "2.5-D" or "interactive" by *Hall et al.* [1999]. In other words, as described below, it contains a 3D representation of the propagation of the lowest order planetary wave. The propagation of this wave is governed by the model zonal wind field and in turn, the breaking of this wave affects the zonal wind. Such a formulation readily lends itself to studying the effects of wave propagation and breaking on the transport of chemical tracers.

### 3.2. Forcing Terms in the Model

The transport of $\overline{M}$, $\Theta$ and the various chemical tracer fields is driven by several momentum and thermodynamic forcing terms which are discussed in detail by *Bacmeister et al.* [1998]. The thermodynamic forcing (i.e. the heating and cooling by $O_3$, $CO_2$, $H_2O$ and aerosols) is discussed in depth by Bacmeister et al and will not be discussed here. We will, however, briefly review the momentum forcing because varying several of the free parameters associated with this forcing serves to highlight the processes governing $NO_x$ descent. Our

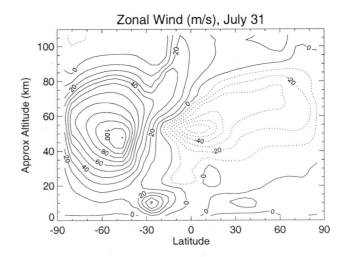

**Figure 4.** Calculated zonal wind from the 2D model for July. Easterly winds are dotted.

overall guidance for "tuning" the model is to give a good overall representation of the zonal wind field and temperature. Figure 4 shows our calculated zonal wind for July 31 (SH winter) and illustrates the strong winter westerlies and summer easterlies seen in the *Randel et al.* [1992] climatology and discussed by *Summers et al.* [1997].

The total momentum forcing, X, in the model can be expressed as

$$X = D_{\mathrm{GW}} - \alpha_{\mathrm{Ray}}\overline{U} + \frac{\partial}{\partial z}(K_{mom}\frac{\partial \overline{U}}{\partial z}) + \nabla \cdot \mathbf{F}_{\mathrm{PW}} \quad (3)$$

The term $D_{\mathrm{GW}}$ represents drag produced by dissipating gravity waves. The model includes separate parameterizations for stationary and non stationary phase speed waves. For stationary waves, a simple cubic-law drag is assumed according to the following

$$(D_{\mathrm{GW}})_{\mathrm{stat}} = -\gamma(z)\overline{U}^3 \quad (4)$$

where the friction coefficient $\gamma(z)$ is prescribed as a simple function of altitude

$$\gamma(z) = \begin{cases} 0 & \text{if } z \leq z_0 \\ \gamma_1(z - z_0)/20. & \text{if } z_0 > z < z_0 + 20 \\ \gamma_1 & \text{if } z \geq z_0 + 20 \end{cases} \quad (5)$$

with $\gamma_1 = (1.5 \text{ days})^{-1}(60 \text{ m s}^{-1})^{-2}$. Thus, for a zonal wind of 60 m s$^{-1}$, the drag timescale is 1.5 days. *Bacmeister et al.* [1995] explored the sensitivity of the zonally averaged temperature to different values of $z_0$. We later found that the strength of the jets in the lower mesosphere is also very sensitive to this parameter

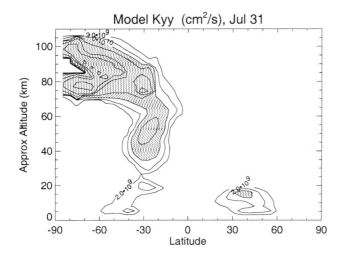

Model Kyy (cm²/s), Jul 31

**Figure 5.** Calculated horizontal eddy diffusion coefficient ($K_{yy}$) derived from the planetary wave parameterization (see text) for July. Regions with values in excess of $1 \times 10^{10}$ cm² s$^{-1}$ are shaded.

[*Summers et al.*, 1997]. A higher value for $z_0$ is analogous to assuming that smaller amplitude waves break at higher altitudes. This leads to less drag at the peak of the zonal jet near the stratosphere (and thus a faster jet), but more drag higher in the mesosphere. Zonal wind climatologies suggest a faster jet in the SH winter than in the NH [*Randel*, 1992] and we have modeled this by using $z_0 = 55$ km in the SH and $z_0 = 45$ km in the NH. The effects of our choice of $z_0$ on NO$_x$ descent will be discussed later.

The model also includes non-zero phase speed waves using a hydrostatic wave packet approach described by *Summers et al.*, [1997]. The model assumes a spectrum of six latitude independent waves which are "launched" at 2.5 km. The amplitudes vary between 15-40 m and the phase speeds vary from $-40 < c_\phi < +40$ m s$^{-1}$ (see table 1 of *Bacmeister et al.*, [1998]).

The second term, $\alpha\overline{U}$, is an ad-hoc Rayleigh damping term. This crudely accounts for momentum deposition due to incompletely specified processes in the model such as high wavenumber planetary waves. We use $\alpha_{\text{Ray}} = 0.01$d$^{-1}$, except $\alpha_{\text{Ray}} = 0.0$ in the tropics ($\pm$ 15 degrees) in order to allow for a more realistic semi-annual oscillation (SAO). [*Summers et al.*, 1997].

The third term, $\partial_z(K_{mom}\partial_z\overline{U})$ represents vertical diffusion redistribution of momentum by turbulent and molecular diffusion. It is generally largest in the upper mesosphere where it is associated mostly with turbulence generated by breaking gravity waves.

The term, $\nabla \cdot \mathbf{F}_{\text{PW}}$ represents EP-flux divergence associated with dissipating or breaking planetary Rossby waves. The planetary wave parameterization used in the model is similar to that of *Garcia* [1991]. We solve a linear, time-dependent potential vorticity equation using zonal winds and temperatures from the TEM model [*Bacmeister et al.*, 1995], considering only the wavenumber $n$=1 stationary mode. The forcing is imposed by a wave geopotential height perturbation at the lowest model level (2.5 km) of the form

$$\Phi(\phi, z_0, t) = \begin{cases} 0 & \phi > 60^o \\ Z_{NH}\sin(3\phi/2) & 0^o < \phi \le 60^o \\ -Z_{SH}\sin(3\phi/2) & 0^o > \phi \ge -60^o \\ 0 & \phi < -60^o \end{cases} \quad (6)$$

A typical value for $Z_{NH}$ is 350m and we also typically assume $Z_{SH} = 0.5 * Z_{NH}$ to approximately account for the greater dynamical activity known to occur in the NH. Later we discuss the effect this has on interhemispheric differences in NO$_x$ descent in the model.

### 3.3. Tracer transport

The evolution of tracers in our model is governed by the continuity relation

$$\frac{\partial\mu}{\partial t} + v^*\frac{\partial\mu}{a\partial\phi} + w^*\frac{\partial\mu}{\partial z} = P - L + \frac{1}{cos}\frac{\partial[cos\phi K_{yy}\frac{\partial\mu}{\partial\phi}]}{\partial\phi} + \frac{1}{\rho}\frac{\partial[\rho K_{zz}\frac{\partial\mu}{\partial z}]}{\partial z} \quad (7)$$

where $\mu$ is the tracer (either $\overline{M}$, $\Theta$ or the mixing ratio of a chemical constituent), P and L the respective production and loss terms and $K_{yy}$ and $K_{zz}$ are the respective horizontal and vertical diffusion coefficients.

$K_{yy}$ is estimated from the planetary wave dissipation by calculating an effective diffusion coefficient for potential vorticity according to *Newman et al.* [1988].

$$K_{yy} = \frac{\overline{v'q'}}{\overline{Q_y}} \quad (8)$$

where $q'$ is the perturbation potential vorticity and $\overline{Q_y}$ is the meriodional gradient of the background zonal mean potential vorticity. The resultant mixing is sensitive to the amplitude of planetary wave forcing which is specified at the lower boundary (eqn (6)), typically 350m for the NH and half that for the SH. The calculated $K_{yy}$ for winter conditions is shown in Figure 5. The figure shows large regions of high $K_{yy}$ both in the upper mesosphere above the peak of the winter jet of Figure 4 and also on the equatorward flank (the so-called "surf zone" of *McIntyre and Palmer*, 1983]. Poleward of the surf zone is a region of very low $K_{yy}$ which is the manifestation of the polar vortex in our model.

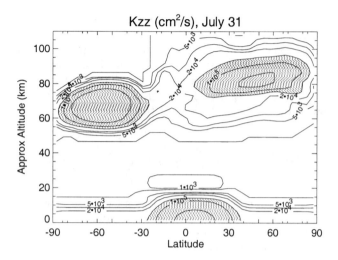

**Figure 6.** Calculated vertical eddy diffusion coefficient ($K_{zz}$) from the gravity wave parameterization (see text) for July. Regions with values in excess of $5 \times 10^4$ cm$^2$ s$^{-1}$ are shaded.

$K_{zz}$ from breaking gravity waves is obtained from the following expression

$$K_{zz,i}(z) = Pr^{-1} \frac{|\overline{U}(z) - c_i|}{\overline{N}^2(z)} D_{GWi}(z) \qquad (9)$$

for the $i^{th}$ wave. $D_{GW}$ is calculated by the gravity wave model for nonzero phase speed waves ($c_i \neq 0$); for zero phase speed waves ($c = 0$), it is given by (4). The Prandtl number (Pr) in (10) represents large uncertainties in the amount and nature of mixing which occurs during gravity wave overturning. Following the arguments summarized in *Summers et al.* [1997], we assume Pr = 3. From equations (4) and (10), it can be seen that $K_{zz}$ depends upon the $4^{th}$ power of the zonal wind. Thus our peak values are associated with the upper mesospheric summer easterlies and the lower mesospheric winter westerlies. This is shown in Figure 6. The winter values are particularly important since, as we will discuss, downward diffusion can supplement the downward advection which occurs in winter.

### 3.4. Chemistry

The chemical part of the model solves for the concentrations of 45 species. This includes all the members of the $O_x$, $NO_y$, $Cl_x$, $HO_x$ and $CH_x$ families s well as longer lived source gases such as $O_2$, $N_2O$, $H_2O$, $H_2$, $CH_4$, CO, $C_2O$ $CH_3Cl$, $CCl_4$ $CFCl_3$ and $CF_2Cl_2$. In addition, the model solves for the distribution of the following species of the ionospheric E region which are important for NO above 90 km: $N(^2D)$, $N_2^+$, $O_2^+$, $NO^+$ and [e-]. The ionization and dissociation from

EUV and soft X ray solar photons is accounted for by empirical fits to pre-calculated g-factors (defined as g = $\int \sigma(E)\phi(z, E)dz$). These fits, in turn, depend upon solar flux, solar zenith angle and overhead column density. For ionization by energetic electrons, we assume an ionization rate profile which peaks at approximately 100 km, is centered at a latitude of 68° and is constant in time. *Siskind et al.* [1997] discuss this in more depth and also give a list of the 13 ionospheric reactions included in the model.

The chemical and dynamical tendencies in (8) are decoupled by using operator splitting [*Summers et al.*, 1997]. We use an iterative Newton-Raphson scheme to solve the coupled chemical production (P) and loss (L) equation using a 24 hour time step. The photolysis calculation is used to account for transport by planetary waves which do not break and would not normally lead to mixing according to (9) [*Siskind et al.*, 1998b]. Non-breaking planetary waves can lead to net transport of chemically active tracers because they can change the solar exposure of an air parcel. An air mass which has an equilibrium latitude in the polar night might nonetheless spend a day at lower latitudes where photodissociation of a given constituent could occur. *Siskind et al.* [1998b] define an effective latitude, $y'$ which oscillates around an equilibrium latitude, $y_0$, according to

$$y' = y_0 \pm \eta' \qquad (10)$$

where $\eta$ is the meridional displacement from the planetary wave. Thus P and L in (8) are defined at $y'$, rather than $y_o$. Inclusion of these "chemical eddies" can have important effects on $NO_x$ partitioning and also on the net deposition of NO into the stratosphere. By providing for greater photolysis of NO, "chemical eddies" lead to about a factor of two less NO deposition.

### 4. COMPARISON OF MODEL WITH DATA

#### 4.1. Overview

Before discussing polar $NO_y$ descent, it is useful to look at the $NO_y$ that results purely from $N_2O$ oxidation. Figure 7 shows the results of two model calculations of equatorial $NO_y$ with a climatology derived from UARS data [*C. Nevison*, private communication, 1999; *Nevison et al.*, 1997]. The $NO_y$ values are plotted as a function of $N_2O$. Since $N_2O$ is presumably the only source of equatorial $NO_y$, using it as an independent variable should minimize any possible errors in model transport and highlight only chemical effects. The upper model curve uses the standard NO photodissociation cross sections [*Minschwaner and Siskind*, 1993]. The lower model curve uses a larger value for the NO photolysis based upon the laboratory study by *Murray*

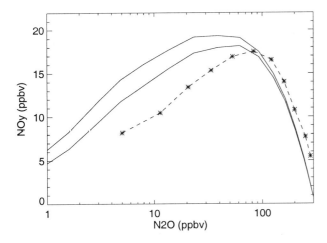

**Figure 7.** Calculated annual average equatorial $NO_y$ as a function of $N_2O$. The upper solid curve uses the NO photolysis cross sections of Minschwaner and Siskind [1993], the lower solid curve uses the cross sections of Murray et al [1994] which are 50% larger. The dashed curve with stars is a climatological average constructed from UARS data.

et al. [1994] of the (0,0) delta band oscillator strength. They reported a value about 50% larger than that used by *Minschwaner and Siskind* [1993]. To date, the Murray et al. value has not yet been utilized for atmospheric studies.

In the upper stratosphere ($N_2O$ < 100 ppbv), it is clear that the models greatly overestimate the data. The discrepancy between our baseline calculation is as large as a factor of 1.7 for values of $N_2O$ around 10 ppbv (which occur near the stratopause). This problem is likely related to the $NO_x$ overestimate reported by *Nevison et al.* [1997]. Figure 7 shows that the higher NO photolysis cross section reduces the discrepancy but does not solve it.

In the discussion that follows, we will link the model overestimate discussed above to the question of $NO_y$ coupling. Table 2 summarizes the 7 model simulations we will be discussing. The first 3 have been discussed in the literature previously [*Siskind et al., 1997; Siskind et al.,* 1998b]. Simulation 4 evaluates the effect of the higher NO cross section. Simulation 5 evaluates the effects of an ad hoc mesospheric mixing term on both the latitudinal distribution of mesospheric $NO_x$ as well as the net deposition in the stratosphere. Simulations 6 and 7 look at the north/south asymmetry in the enhanced $NO_x$ that was previously reported by *Siskind et al.* [1997] and separate out the relative roles of gravity wave drag and planetary mixing in the mesosphere.

### 4.2. Winter Mesosphere

Mesospheric NO serves as a valuable tracer of atmospheric motions. Figure 8 shows calculated NO from the model for July 31 from simulations 1, 3 and 5 compared with latitudinal cross sections of the HALOE climatology presented in Plate 1. Without planetary waves (simulation 1), the model NO remains confined to the polar regions, poleward of where HALOE samples. Inclusion of breaking planetary waves and the associated high $K_{yy}$ values seen in Figure 5 is essential to achieving good agreement between the Model 3 and HALOE at .01 and .04 mb. At .3 mb, even with the breaking planetary waves, the model keeps the NO confined to the polar regions, whereas there is clear evidence from HALOE that enhanced NO is seen as far equatorward as 30°. As seen in Figure 5, there is no mechanism in our model for horizontal transport of polar NO at 0.3 mb because $K_{yy}$ = 0 poleward of 50°(the model polar vortex). Thus Model 5 includes an ad hoc specification of $K_{yy}$ = 3 ×$10^{10}$ cm$^2$ s$^{-1}$ in the winter lower meso-

Table 2. Summary of 2D Model Simulations

| Simulation no. | P wave forcing | $J_{NO}$ | Chemical Eddies | comments |
|---|---|---|---|---|
| 1 | none, Kyy=3×10$^9$ | low | none | see Siskind et al. [1997] |
| 2 | $Z_{NH}$ = 350 m, $Z_{SH}$ = 0.5*$Z_{NH}$ | low | none | ibid |
| 3 | ibid | low | with chem. eddies | [Siskind et al., 1998] |
| 4 | ibid | high | ibid | Murray et al (1994) NO xsec |
| 5 | ibid | high | ibid + Kyy=3e10 in lower mesos. | |
| | *Vary asymmetry between NH and SH in planetary wave (PW) and gravity wave (GW) forcing* | | | |
| 6 | $Z_{NH}$ = 350 m, $Z_{SH}$ = 0.5*$Z_{NH}$ $z_0$(NH) = $z_0$(SH)=50 km | high | with chem. eddies | symmetric GW (Kzz) but asymmetric PW (Kyy) |
| 7 | $Z_{NH}$ = $Z_{SH}$ = 250 m $z_0$(NH) = $z_0$(SH)=50 km | high | with chem. eddies | symmetric GW and PW |

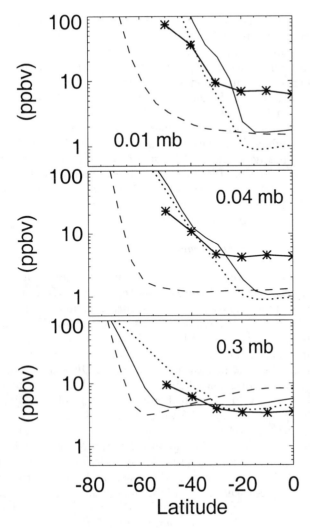

**Figure 8.** Calculated NO for July compared with HALOE (stars). The solid line is simulation 3, the dashed is simulation 1, and the dotted is simulation 5.

sphere. As seen in Figure 8, this added mixing yields good agreement with the data.

The need for an additional horizontal transport in the lower mesosphere is significant from the standpoint of $NO_y$ coupling. In the 50-60 km region, the photodissociation of NO is still rapid enough to cause NO loss on a seasonal time scale [e.g. *Minschwaner and Siskind*, 1993; Figure 3]. As we will discuss below, by bringing more NO out of the polar night region, we reduce the net deposition of upper atmospheric NO into the stratosphere.

### 4.3. Spring Stratosphere

The existence of enhanced $NO_x$ in the spring stratosphere is of interest because it will serve as a net source of odd nitrogen to the stratosphere. The seasonal evolution of SH NO in the model (simulation 3) from days 121-301 is shown in Figure 9a. The figure shows the descent of the 30 ppbv contour from the upper mesosphere into the mid stratosphere during the winter. In early spring, the return of sunlight to the polar regions causes photodissociation of mesospheric, but not stratospheric NO. In addition, the downwelling in the polar mesosphere decreases sharply, thus reducing the downward tracer flux. Thus a region of "cutoff" NO remains.

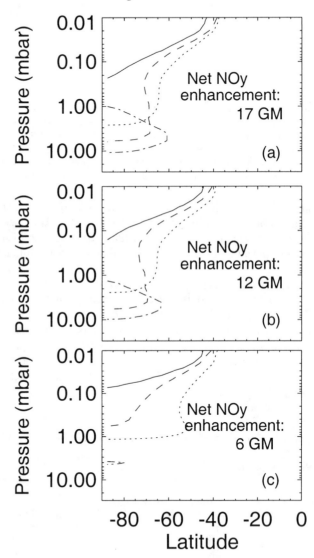

**Figure 9.** Evolution of the 30 ppbv contour of $NO_x$ for 3 model calculations (a) simulation 3, (b) simulation 4 and (c) simulation 5. The 30 ppbv contour serves as a representative measure of high mesospheric $NO_x$ descending into the stratosphere. Each panel shows 4 model days: solid line (day 121), dashed line (day 181), dotted line (day 241) and dot-dashed curve (day 301) The net enhancement in stratospheric $NO_y$ for Day 301 is also indicated. The units are gigamoles (GM).

While the general pattern of descent seen in Figure 9a is consistent with observations, there are some very important differences. First the NO enhancement in the model maximizes near 5 mb whereas the observations shown in Section 2 show the peak to be about a scale height lower in altitude, between 10-20 mb. This likely reflects errors in the specification of drag and mixing in the model. We assume fixed planetary wave and gravity wave forcing which is undoubtedly too simplistic. More serious, the model significantly overestimates the net amount of mesospheric NO which remains in the stratosphere by Day 301. As we have seen, there is no evidence for mixing ratios as large as 30 ppbv anywhere in the springtime stratosphere. In addition, the total amount of $NO_y$ molecules, calculated by performing a parallel "no-thermosphere" simulation and integrating over the difference between the two, is 17 GM which is over 10 times our peak estimates from HALOE. Increasing the NO photolysis cross section (Figure 9b) and then increasing the lower mesospheric mixing (Figure 9c) both act to lower the net deposition in the stratosphere. This can be most clearly seen in the smaller area enclosed by the 30 ppbv contour, most notably on Day 301 (dot-dashed curve in the figure). However, even in Figure 9c, the net source of 6 GM is still at least a factor of 4-5 too large.

### 4.4. North/South Asymmetry

As noted previously, enhanced $NO_x$ is more apparent in the SH relative to the NH in both HALOE and ATMOS data. It is also true in our model [*Siskind et al.*, 1997]; however, the relative roles of planetary wave and gravity wave forcing have not been elucidated. In general, our model yields greater gravity wave drag and mixing in the mid mesosphere in the SH winter relative to the NH. This is shown by the solid lines in Figure 10a (SH winter) and 10b (NH winter) for vertical wind and Figure 10c (SH winter) and d (NH winter) for vertical diffusion velocity. The combination of greater advective and greater diffusive descent is one reason for greater $NO_x$ deposition in the SH. We have previously suggested that this difference explains the observed asymmetry in the annual cycle of HALOE mesospheric $H_2O$ [*Summers et al.*, 1997]; [*Garcia and Pumphrey*, 1999] have made a similar suggestion regarding MLS $H_2O$. The dotted lines in these four panels are from simulation 6 which assumes a hemispherically symmetric gravity wave drag.

In addition because we assume greater planetary forcing amplitudes in the NH, we get greater horizontal mixing in the NH. This is shown by the solid lines in Figures 10e and 10f which plot the latitudinal variation of $K_{yy}$ in our model. The greater horizontal diffusion in the

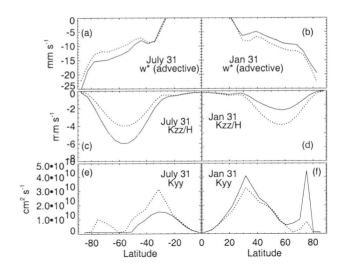

**Figure 10.** Latitudinal variation of model vertical advection (w*, Figures 10a and b), vertical diffusion ($K_{zz}$, Figures 10c and d) and horizontal diffusion ($K_{yy}$, Figures 10e and f) diffusion coefficients for winter at 70 km (.035 mb). The solid line in all six panels is from Simulation 4. The dotted line in all six panels is from Simulation 6 where hemispherically symmetric gravity wave and planetary wave forcings were assumed. The left column panels are for Southern winter, the right column for Northern winter.

NH reduces the descent by increasing the exposure of NO to sunlight and thus photolysis. The dotted lines in Figures 10e and f are from simulation 7 which assumes a hemispherically symmetric planetary wave forcing.

Figure 11 shows the calculated NO for late winter at the stratopause as a function of latitude. The results from simulation 4 (the two solid lines) show a factor of 4 more NO in the SH. This simulation corresponds to that illustrated by the solid curves in Figure 10, i.e asymmetric gravity wave drag and asymmetric planetary wave mixing. The dotted lines in Figure 11 are from simulation 6 (hemispherically symmetric gravity wave drag). Here the weaker descent rate in the SH reduces the peak NO by 50%. The dashed curves are from simulation 7 (symmetric gravity waves and planetary waves). The slightly greater $K_{yy}$ in the SH leads to less NO while the slightly lesser $K_{yy}$ in the NH leads to a small increase in the peak NO. The N/S difference is reduced from a factor of 4 to 1.5 (small N/S differences in the $O_3$ climatology used for the heating and cooling rates as well as the surface boundary conditions still exist and contribute to the residual difference),

The consistency between the above results for NO and those previously seen in $H_2O$ supports the idea that gravity wave drag in the mesosphere is greater in the SH winter relative to the NH winter. This was first suggested by *Shine* [1989] and *Marks* [1989]. As those au-

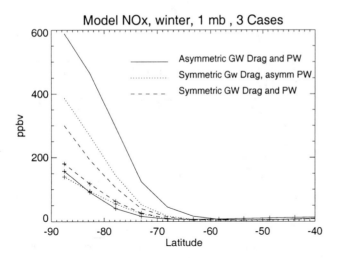

**Figure 11.** Model NO from Simulations 4 (solid), 6 (dotted) and 7 (dashed) at 1 mb as a function of the absolute value of the latitude (i.e., -(lat) for the NH calculation). The three lines without symbols are for July 31 in the SH; the curves with symbols are for January 31 in the NH.

thors noted, its not clear why there is more drag in the SH; the present model is too simplistic to answer that question, rather we are emphasizing the consequences on tracer transport of such drag.

### 4.5. Discussion

From the above discussion, the greatest success of the model is in simulating the N/S differences. The greatest deficiency is that the model significantly overestimates the data. It is likely that both dynamics and chemistry play a part in this problem. From the dynamical perspective, the existence of enhanced NO at mid-latitudes in winter in the 50-60 km region implies a horizontal transport process not in the model. By bringing more NO out of the polar region than in the standard model, the net exposure of NO to UV sunlight is increased, thus enhancing the photodestruction of NO. Is there any other evidence that such transport exists in the atmosphere? The answer appears to be yes. *Callis and Lambeth* [1998] have looked at $NO_2$ enhancements observed by ISAMS in the lower mesosphere for May 1992. They show a region of $NO_2$ which appears to be transported equatorward and suggest that gravity wave mixing is responsible. Also the overly cold polar temperatures in the model may suggest insufficient wave activity at high latitudes [*Siskind et al.*, 1997]. In any event, the sensitivity of polar stratospheric $NO_y$ to horizontal mixing in the mesosphere serves as an important example of atmospheric coupling between the mesosphere and the stratosphere.

From the chemical perspective, our results indicate that the overestimate of the net $NO_y$ deposition is likely related to an overestimate of the tropical $NO_y$ discussed by *Nevision et al.* [1997]. Increasing the photolysis cross section reduces, but does not eliminate both problems. If the tropical model excess were solved, it would further reduce the polar model excess. One complication is that that in the thermosphere, the opposite problem exists, namely a model deficit [*Siskind et al., 1995*]. Yet the chemistry of NO in the thermsophere and in the upper stratosphere is not that dissimilar, i.e. production via atomic nitrogen oxidation and loss via (1) and (2) above. Sensitivity studies that we have done suggest that using the *Murray et al.* [1994] cross sections reduces the calculated thermospheric NO concentrations by about 30%. This might nonetheless be acceptable considering that there are uncertainties in the sources of thermospheric NO which are greater than 30% [*Siskind et al., 1995*]. Also, the temperature regimes of the thermosphere and stratosphere are quite different. It is clear, however, that atmospheric $NO_y$ chemistry needs to be considered as a complete system.

### 5. CONCLUSIONS

#### 5.1. Summary

The descent of upper atmospheric NO into the mid-stratosphere, now a confirmed feature in several satellite datasets, is a fascinating example of long range atmospheric coupling. While its global effects may not be large, in the SH winter vortex, it can cause a significant perturbation to the background $NO_x$ abundance. Although not discussed here, the POAM results suggest significant localized $O_3$ reductions linked to enhanced $NO_x$. In fact, these $O_3$ reductions were likely misinterpreted by *Hofmann et al.* [1992] as an upward extension of the Antarctic $O_3$ hole [*Callis et al.*, 1996; *Randall et al.*, 1998; *Siskind et al.*, 2000].

The descent of upper atmospheric odd nitrogen also ties together several disparate chemical and dynamical atmospheric processes. The coupling of upper and middle atmospheric $NO_x$ is understood to be one facet of the general question of descent into the polar vortex. Polar stratospheric $NO_x$ is a valuable indicator of vertical and horizontal diffusion in the mesosphere. The latitudinal distribution of mesospheric $NO_x$ serves as an important indicator of breaking planetary waves in the mesosphere. In addition, the N/S differences appear to be consistent with N/S differences in $H_2O$ as well as earlier studies of the mesospheric radiative balance. Finally, this topic brings together two scientific specialties which are generally considered separate, namely earth science (dealing with the stratosphere and troposphere) and space science (dealing with the mesosphere, thermosphere and ionosphere).

## 5.2. Outstanding Questions for Future Work

Despite the important validation of the original *Solomon et al.* [1982] hypothesis, many fundamental problems remain. The lack of wintertime $NO_x$ data poleward of 50° continues to limit the strength of our conclusions. Ideally a future instrument with the capabilities of HALOE, in a near polar orbit like POAM, is required to be able to track continuously polar NO from its source region above 80 km down to the stratosphere. From a theoretical standpoint, key questions include:

• Why does the model overestimate the data? Both chemical and dynamical causes likely contribute. More laboratory work to quantify the NO cross section as well as atomic nitrogen reaction rates at middle atmospheric temperatures is needed.

• What is the nature of hypothesized equatorward transport in the lower mesosphere? In the upper stratosphere, it is generally believed that air remains confined inside the vortex and indeed *Siskind et al.* [2000] relied on this assumption to quantify the total SH $NO_y$ enhancement. The validity of this assumption may hinge on the answer to this question.

• What is the nature of the interannual variation of the source due to particle precipitation? A detailed model of particle precipitation in the upper mesosphere and thermosphere throughout the wintertime for every year going back to 1978 (in order to compare with LIMS $NO_2$) is needed.

• What is causing the reported NH SAGE II $NO_2$ variations? The analyses conducted to date do not confirm the suggestion that polar $NO_x$ descent is the explanation. Indeed, given that the NH effects should be much less than the SH, it becomes hard to imagine a significant perturbation to NH $NO_x$; however, the lack of complete seasonal high latitude coverage by HALOE makes a final conclusion premature.

• What are the nature and consequences of the $NO_x$ catalyzed $O_3$ losses in the 25-30 km region? For example, the possible radiative effects of this loss in the vortex should be considered.

Finally, three dimensional studies are clearly called for. The TIME-GCM [*Roble*, this volume] should be well suited for such studies since its bottom boundary is well below the stratopause and it includes a complete calculation of thermospheric NO production. It is clear that odd nitrogen coupling would be an invaluable diagnostic for the ground-to-exosphere model that Roble proposes.

*Acknowledgments.* I thank Charles Barth, Steve Eckermann, Bob Meier, Mike Stevens and the three reviewers for their comments. This work was funded by the Office of Naval Research and the NASA UARS Investigator program.

## REFERENCES

Bacmeister, J. T., et al., Descent of long-lived trace gases in the winter polar vortex, *J. Geophys. Res.*, *100*, 11669, 1995.

Bacmeister, J. T., et al., Age of air in a zonally averaged two-dimensional model, *J. Geophys. Res.*, *103*, 11263, 1998.

Barth, C. A., Rocket measurement of the nitric oxide dayglow, *J. Geophys. Res.*, *69*, 3301, 1964.

Barth, C. A., Nitric oxide in the lower thermosphere, *Planet. Space Sci.*, *15*, 92, 1992.

Brasseur, G., and S. Solomon, Aeronomy of the Middle Atmosphere, D. Reidel Publishing Co., 1986, 441pp.

Brasseur, G. The response of the middle atmosphere to long-term and short-term solar variability: A two-dimensional model, *J. Geophys. Res.*, *98*, 23079, 1993.

Callis, L. B., and M. Natarajan, The Antarctic ozone minimum: Relationship to odd nitrogen, odd chlorine, the final warming, and the 11-year solar cycle, *J. Geophys. Res.*, *91*, 10771, 1986.

Callis, L. B. and J. D. Lambeth, $NO_y$ formed by precipitating electron events in 1991 and 1992: Descent into the stratosphere as observed by ISAMS, *Geophys. Res. Lett.*, *25*, 1875, 1998.

Callis, L. B., et al., Solar atmospheric coupling by electrons (SOLACE) 2. Calculated stratospheric effects of precipitating electrons, 1979-1988, *J. Geophys. Res.*, *103*, 1998.

Crutzen, P. J., The influence of nitrogen oxide on the atmospheric ozone content, *Q.J.R. Meteor. Soc.*, *96*, 320, 1970.

Eluszkiewicz, J, R. A. Plumb, and N. Nakamura, Dynamics of wintertime stratospheric transport in the Geophysical Fluid Dynamics Laboratory SKYHI general circulation model, *J. Geophys. Res.*, *100*, 20883, 1995.

Fisher M., A. O'Neill, and R. Sutton, Rapid descent of mesospheric air into the stratospheric polar vortex, *Geophys. Res. Lett.*, 1267-1270, 1993.

Frederick, J. E., and N. Orsini, The distribution and variability of mesospheric odd nitrogen: a theoretical investigation, *J. Atm. Terr. Physics*, *44*, 479-488, 1982.

Garcia, R.R., Parameterization of planetary wave breaking in the Middle Atmosphere, *J. Atmos. Sci.*, 48, 1405-1419, 1991.

Garcia, R.R. and H.C. Pumphrey, The influence of dynamics on the seasonal variation of water vapor in the middle atmosphere: Modeling and UARS/MLS observations. In Proc. XXV Annual European Meeting on Atmospheric Studies by Optical Methods, M.J. Lopez-Gonzales et al, eds., pp. 156-165. Instituto de Astrofisica de Andalucia, Granada, Spain., 1998.

Garcia, R. R., et al., A numerical response of the middle atmosphere to the 11-year solar cycle, Dynamical and chemical structure of the middle atmosphere, *Planet. Space Sci.*, *32*, 411, 1984.

Garcia, R. R., et al., Transport of nitric oxide and the D-region winter anomaly, *J. Geophys. Res.*, *92*, 977, 1987.

Hall, T. M., et al., Evaluation of transport in stratospheric models, *J. Geophys. Res.*, *104*, 18,815-18,841, 1999.

Hofmann, D.J., et al., Observations and possible causes of new ozone depletion in Antarctica in 1991, *Nature*, *359*, 283-287, 1992.

Horvath, J. J. and J. E. Frederick, In-situ measurements of nitric oxide in the high latitude upper stratosphere, *Geophys. Res. Lett.*, *12*, 495, 1985.

LeTexier, H., S. Solomon, and R. R. Garcia, The role of molecular hydrogen and methane oxidation in the water vapour budget of the stratosphere, *Q. J. R. Meteorol. Soc.*, *114*, 281-295, 1988.

Lin, S-J, and R. B. Rood, Multidimensional flux-form semi-lagrangian transport schemes, *Mon. Wea. Rev.*, *124*, 2046, 1996.

McIntyre, M.E. and T.N. Palmer, The "surf-zone" in the stratosphere, *J. Atmos. Terr. Phys.*, *46*, 825, 1983.

Manney, G., et al., Polar vortex dynamics during spring and fall diagnosed using trace gas observations from the Atmospheric Trace Molecule Spectroscopy instrument, *J. Geophys. Res.*, *104*, 18841, 1999.

Marks, C. J., Some features of the climatology of the middle atmosphere revealed by Nimbus 5 and 6, *J. Atmos. Sci.*, 2485, 1989.

Minschwaner, K., and D. E. Siskind, A new calculation of nitric oxide photolysis in the stratosphere, mesosphere, and lower thermosphere, *J. Geophys. Res.*, *98*, 20401-20412, 1993.

Murray, J. E., et al., Vacuum ultraviolet Fourier transform spectroscopy of the $\delta(0,0)$ and $\beta(7,0)$ bands of NO, *J. Chem. Phys*, *101*, 62, 1994.

Nevison, C. D., S. Solomon and R. R. Garcia, Model overestimates of $NO_y$ in the upper stratosphere, *Geophys. Res. Lett.*, *24*, 803, 1997.

Newman, P. A., et al., Mixing rates calculated from potential vorticity, *J. Geophys. Res.*, *93*, 5221, 1988.

Nicolet, M., Contribution a l'etude de lat structure de l'ionosphere, *Mem. Inst. meteorol. Belg.*, *19*, 38, 1945.

Prather, M.J., Numerical advection by conservation of second order moments, *J. Geophys. Res.*, 91, 6671-6681, 1986.

Randall, C. E., et al., Polar Ozone and Aerosol Measurement (POAM) II stratospheric $NO_2$, 1993-1996., *J. Geophys. Res.*, *103*, 28361-28372, 1998.

Randel W.J., Global atmospheric circulation statistics, 1000-1 mbar, *NCAR Tech. Note NCAR/TN-366+STR*, 256 pp., 1992.

Rinsland, C. P. et al., ATMOS/ATLAS-3 measurements of stratospheric chlorine and reactive nitrogen partitioning inside and outside the November 1994 Antarctic vortex, *Geophys. Res. Lett.*, *23*, 2365-2368, 1996.

Rinsland, C. P., et al., Polar stratospheric descent of $NO_y$ and CO and Arctic denitrification during winter 1992-1993, *J. Geophys. Res.*, *104*, 1847-1861, 1999.

Roble, R. G., On the Feasibility of Developing a Global Atmospheric Model Extending from the Ground to the Exosphere, this volume

Rosenfield, J.E., M.R. Schoeberl, and M.A. Geller, A computation of the stratospheric diabatic circulation using an accurate radiative transfer model, *J. of the Atm. Sci*, *44*, 859-876, 1987.

Rusch, D. W., and R. T. Clancy, A study of the time and spatial dependence of ozone near 1.0 mbar with emphasis on the springtime, *Ozone in the Atmosphere*, proceedings of the 1988 Quadrennial Ozone Symposium, Bojkov and Fabian, eds, 1989.

Russell, J. M. III and R. B. Pierce, Observations of Polar Descent and Coupling in the Thermosphere, Mesosphere, and Stratosphere Provided by HALOE, this volume.

Russell, J. M. III, et al., The variability of stratospheric and mesospheric $NO_2$ in the polar winter night observed by LIMS, *J. Geophys. Res.*, *89*, 7267, 1984.

Russell, J. M. III, et al., The Halogen Occultation Experiment, *J. Geophys. Res.*, *98*, 10777-10797, 1993.

Shine, K., Sources and sinks of zonal momentum in the middle atmosphere diagnosed using the diabatic circulation, *Q.J.R. Meteorol. Soc.*, *115*, 265-292, 1989.

Siskind, D.E. and J.M. Russell III, Coupling between middle and upper atmospheric NO: Constraints from HALOE data, *Geophys. Res. Lett.*, *23*, 137, 1996.

Siskind, D. E., et al., Two-dimensional model calculations of nitric oxide transport in the middle atmosphere and comparison with Halogen Occultation Experiment data, *J. Geophys. Res.*, *102*, 3527, 1997.

Siskind, D. E., et al., The response of thermospheric nitric oxide to an auroral storm, 2. Auroral latitudes, *J. Geophys. Res.*, *94*, 16885-16911, 1989.

Siskind, D. E., C. A. Barth, and J. M. Russell III, A climatology of nitric oxide in the mesosphere and thermosphere, *Adv. Space Res.*, *21*, 1353, 1998a.

Siskind, D. E., J. T. Bacmeister, and M. E. Summers, A new calculation of chemical eddy transport for several middle atmospheric tracers, *J. Geophys. Res.*, *103*, 31,321, 1998b.

Siskind, D. E., et al., An assessment of southern hemisphere stratospheric $NO_x$ enhancements due to transport from the upper atmosphere, *Geophys. Res. Lett.*, 27, 329, 2000,

Solomon, S., Stratospheric ozone depletion: A review of concepts and history, *Rev. Geophys.*, 275, 1999.

Solomon, S., P. J. Crutzen, R. G. Roble, Photochemical coupling between the thermosphere and the lower atmosphere 1. Odd nitrogen from 50 to 120 km, *J. Geophys. Res.*, *87*, 7206-7220, 1982.

Solomon, S., and R. R. Garcia, Transport of thermospheric NO to the upper stratosphere?, *Planet. Space Sci.*, *32*, 399–409, 1984.

Solomon, S., et al., Photochemistry and transport of carbon monoxide in the middle atmosphere, *J. Atmos. Sci.*, *42*, 1072, 1985.

Strahan, S. E., J. E. Nielsen, and M. C. Cerniglia, Long-lived tracer transport in the Antarctic stratosphere, *J. Geophys. Res.*, *101*, 26615, 1996.

Summers, M. E., et al., Seasonal variation of middle atmospheric $CH_4$ and $H_2O$ with a new chemical-dynamical model, *J. Geophys. Res.*, *102*, 3503, 1997a.

Sutton, R., Lagrangian flow in the middle atmosphere, *Q. J. R. Meteorol. Soc.*, *120*, 1299, 1994.

---

D.E. Siskind, Naval Res. Lab, Code 7641, Washington, DC, 20375, siskind@uap2.nrl.navy.mil

# Insights into Middle Atmospheric Hydrogen Chemistry from Analysis of MAHRSI OH Observations

Michael E. Summers[1] and Robert R. Conway

*E. O. Hulburt Center for Space Research, Naval Research Laboratory, Washington, DC*

Observations of mesospheric and upper stratospheric OH made from space by the Middle Atmosphere High Resolution Spectrograph Investigation (MAHRSI) in November 1994 and August 1997 have opened a new window on the study of middle atmospheric hydrogen chemistry. The MAHRSI observations of OH ultraviolet fluorescence have allowed a more detailed testing of models of odd-hydrogen ($HO_x$ = H + OH + $HO_2$) chemistry, and its role in catalytic destruction of ozone, over a wider range of altitudes than ever before. Models of OH using standard photochemical theory overestimate the observed OH abundance by 25 to 35% throughout the mesosphere, yet in the upper stratosphere underestimate OH by nearly 20%. The sharp transition at the stratopause cannot be understood in terms of standard $HO_x$ chemistry or by any previously proposed modifications to that theory. Perhaps the most intriguing result of modeling the OH data is the confirmation of a narrow layer of water vapor near 70 km that had been previously observed by the Halogen Occultation Experiment (HALOE) on the Upper Atmospheric Research Satellite (UARS). The existence of this layer cannot be explained by standard gas phase chemistry at these altitudes and furthermore requires a local source of $H_2O$. This may suggest that heterogeneous chemistry (e.g., surface recombination of $H_2$ and O on meteoric dust to form $H_2O$) plays an important role in the mesosphere. MAHRSI observations of the OH column in the upper stratosphere are in general agreement with both balloon infrared observations extrapolated above 40 km altitude, and high altitude aircraft 2.5 THz observations, and thus resolve an earlier disagreement among OH measurements. Taken together, these results suggest that the standard model of hydrogen chemistry in the middle atmosphere is incomplete.

[1] Also at the Institute of Computational Sciences and Informatics (CSI), Department of Physics and Astronomy, and the Center for Earth Observing and Space Research (CEOSR), George Mason University, Fairfax, VA 22030 (e-mail: msummers@physics.gmu.edu)

Atmospheric Science Across the Stratopause
Geophysical Monograph 123

## 1. INTRODUCTION

The hydroxyl radical, produced from the chemical breakup of water vapor, is one of the most important oxidizing agents in the atmosphere [*Coffey and Brasseur*, 1999]. In the troposphere OH reacts with numerous pollutants and trace gases, and in the middle atmosphere (between ∼15 and 90 km altitude) OH plays a fundamental role as a natural catalyst for the chemical destruction of ozone and atomic oxygen [*Brasseur and Solomon*, 1986]). Yet, due to its high chemical reactivity and consequently extremely low atmospheric abundance, OH has until recently remained one of most poorly measured chemical species.

Most interest in middle atmospheric OH revolves around its role as a catalyst for ozone destruction. Ozone loss in the stratosphere (about 15 to 50 km altitude) is controlled by several coupled catalytic cycles involving nitrogen, chlorine, bromine, and hydrogen radicals which recombine O and $O_3$ (together known as odd-oxygen $O_x = O + O_3$) to form more chemically stable $O_2$ [Brasseur and Solomon, 1986], [Müller and Salawitch, 1999]. OH is also important in the oxidation of both $CH_4$ and HCl. However, in spite of its importance, there are several outstanding issues regarding the role of OH in stratospheric photochemistry that are still not understood. For example, observations of OH and $HO_2$ near 20 km altitude suggest an additional source of $HO_x$ [Wennberg et al., 1999]. Analysis of balloon observations of OH, $HO_2$, and $O_3$ in the middle and upper stratosphere [Jucks et al., 1998] shows that models using standard recommended $HO_x$ photochemistry [DeMore et al., 1997] underpredict $O_3$, i.e., the ozone deficit problem [Müller and Salawitch, 1999]. This long standing model ozone deficit has lead to various suggestions of possible additional sources of $O_3$ production in the atmosphere however none have been shown to be large enough to remove the deficit (see discussion in [Müller and Salawitch, 1999].)

The mesosphere (from about 50 to 95 km altitude) has long been considered to be a much more "pristine" environment for $O_3$ chemistry because ozone loss there is thought to be controlled by odd-hydrogen chemistry alone, and because complexities such as heterogeneous chemistry are generally considered unimportant. However, here standard photochemistry substantially overpredicts the observed OH (see discussion of the MAHRSI observations below) by 25 - 35% throughout the mesosphere [Summers et al., 1996] and [Summers et al., 1997a]. In the models, and presumably in the real atmosphere, lower OH translates to larger $O_3$ abundances. From analysis of coincident MAHRSI OH and CRISTA (Cryogenic Infrared Spectrographs and Telescopes for the Atmosphere) $O_3$ observations at the stratopause (~50 km altitude), [Summers et al., 1997a] found that the observed OH and $O_3$ are mutually consistent with a much reduced ozone deficit. However, the cause of the lower than predicted OH abundances is still uncertain (see below.) Given that OH is lower than predicted, this may partially explain the model underprediction of $O_3$ found in previous mesospheric studies ( [Rusch and Eckman, 1985], [Strobel et al., 1987], [Clancy et al., 1987], [Eluszkiewicz and Allen 1993], [Clancy et al., 1994].) The model ozone deficit

in the lower mesosphere is now much smaller than that found in the earlier studies [Siskind et al., 1995], although still controversial [Crutzen et al., 1995].

Furthermore, photochemical models of the mesosphere generally fail to adequately reproduce the observed variation of $O_3$ on a variety of timescales. The daytime variation of mesospheric ozone appears to be reasonably reproduced by photochemical models, however observed terminator variations [Conner et al., 1994] and night time variations [Zommerfields et al., 1989] remain unexplained. The observed response of mesospheric $O_3$ and temperature due to solar forcing from the 27-day solar rotational variations in $Ly\ \alpha$ are not only unexplained but suggests that between 70 and 80 km altitude, the $O_3$ - temperature dependence is opposite that predicted by standard theory [Summers et al., 1990]. Also, long term trends in $H_2O$ and $O_3$ at the stratopause do not agree with model predictions [Nedoluha et al., 1998].

More recently, as we will discuss below, satellite observations of mesospheric water vapor (source molecule for OH) by HALOE, along with analysis of MAHRSI OH observations, have brought into question the conventional assumptions regarding the chemical sources and sinks of $H_2O$ [Summers et al., 1997a], [Summers and Siskind, 1999]. As discussed by [Summers et al., 1997a], the observed shape of the mesospheric OH profile can only be reproduced by a model that incorporates the HALOE $H_2O$ data which shows a $H_2O$ layer at 70km. However, conventional $HO_x$ chemistry predicts net destruction of $H_2O$ at 70km, not net production as required by the HALOE data.

In addition to the theoretical issues discussed above, there are apparent discrepancies between OH measurements. Whereas recent balloon measurements of OH in the middle and upper stratosphere appear to be in agreement with the standard $HO_x$ model [Jucks et al., 1998], satellite measurements of OH at the stratopause and above [Conway et al., 1996] show OH densities substantially lower than that predicted by the standard model [Summers et al., 1996], [Summers et al., 1997a]. Furthermore, ground-based measurements [Burnett and Burnett, 1996] and [Canty et al., 2000] (this monograph) of middle atmospheric OH column are larger than that implied by both the satellite and balloon measurements.

Resolving these issues and clarifying our understanding of the role of $HO_x$ on the $O_3$ budget in the upper stratosphere and mesosphere are critically important since the contribution to $O_3$ loss due to chlorine radicals, released from anthropogenic CFC's, peaks in that region. Furthermore, the predicted recovery of atmospheric $O_3$ due to the phase out of CFC's mandated by the Montreal Protocol may show it's first manifestation in the upper stratosphere [*Müller and Salawitch*, 1999].

All of the topics we discuss in this paper have cross-stratopause relevance. For example, understanding the source of the mid-mesospheric $H_2O$ layer has implications for quantifying the global middle atmospheric hydrogen budget, as well as for understanding the observed OH distribution in both the mesosphere and upper stratosphere. Similarly, the key chemical reactions which control $OH/HO_2$ partitioning in the mesosphere are thought to also dominate in the upper stratosphere [*Summers et al.*, 1997a].

In this paper we are primarily concerned with what has been learned by studies of MAHRSI OH observations. For discussion of the MAHRSI experiment and a detailed description of the methods used to retrieve mesospheric OH density profiles the reader is directed to the papers by [*Conway et al.*, 1996; 1999]. The CRISTA experiment is described in [*Offermann et al.*, 1999]. The HALOE experiment is discussed in [*Russell et al.*, 1993] and HALOE data validation is addressed in [*Harries et al.*, 1996]. An assessment of current understanding of upper stratospheric ozone chemistry can be found in [*Müller and Salawitch*, 1999]. Discussions of the expected long-term variation of upper stratospheric ozone is given by [*Jucks and Salawitch* 2000] and middle atmospheric global change is presented by [*Brasseur et al.*, 2000] (both in this monograph.)

The plan of this paper is as follows: In Section 2 we provide the reader with a brief $HO_x$ "primer" that covers the conventional photochemical theory of $HO_x$ production and loss and its role in the catalytic destruction of odd-oxygen ($O_x = O + O_2$) in the mesosphere and upper stratosphere. In Section 3 we review the evidence for the mesospheric $H_2O$ layer observed by HALOE and corroborated by modeling of the MAHRSI OH data. In Section 4 we discuss $HO_x$ models and compare their predictions of OH with the MAHRSI OH observations. We consider the possibility that heterogeneous chemistry is the source of the mesospheric $H_2O$ layer in Section 5. In Section 6 we conclude with a discussion of key outstanding questions and suggestions for future research.

## 2. ODD HYDROGEN PHOTOCHEMISTRY - A $HO_x$ "PRIMER"

The ultimate source of odd-hydrogen in the middle atmosphere is the photochemical destruction of water vapor. The production of $HO_x$ occurs by mostly photolysis above about 65 km altitude and primarily by

$$H_2O + O(^1D) \rightarrow OH + OH \qquad (1)$$

below that altitude. In the mesosphere and upper stratosphere $OH/HO_2$ partitioning is controlled by

$$OH + O \rightarrow H + O_2 \qquad (2)$$

$$O + HO_2 \rightarrow OH + O_2 \qquad (3)$$

In the middle and lower stratosphere the reactions

$$HO_2 + NO \rightarrow OH + NO_2 \qquad (4)$$

$$OH + O_3 \rightarrow HO_2 + O_2 \qquad (5)$$

are also important in $HO_x$ partitioning. Ultimately $HO_x$ is lost by

$$HO_2 + OH \rightarrow H_2O + O_2 \qquad (6)$$

$$HO_2 + H \rightarrow H_2 + O_2 \qquad (7)$$

Plate 1 shows the relevant chemical rates for $HO_x$ production and loss.

As is clear from the above chemistry, $HO_x$ radicals are intimately coupled to $O_x$ chemistry. The production of O is via

$$O_2 + h\nu \rightarrow 2O \qquad (8a)$$

$$\rightarrow O + O(^1D) \qquad (8b)$$

and its loss is controlled by the $O_2$ bond-forming reaction (2).

The dominant mechanism for ozone production in the middle atmosphere is the ter-molecular recombination of atomic and molecular oxygen, i.e.,

$$O + O_2 + M \rightarrow O_3 + M \qquad (9)$$

whereas ozone is lost in the mesosphere by

$$O_3 + h\nu \rightarrow O_2 + O \qquad (10a)$$

$$\rightarrow O_2 + O(^1D) \qquad (10b)$$

$$H + O_3 \rightarrow OH + O_2 \qquad (11)$$

$$O + O_3 \rightarrow 2O_2 \qquad (12)$$

Excited oxygen ($O(^1D)$) is lost via

$$O(^1D) + M \rightarrow O + M \qquad (13)$$

where M is a third body non-reactant.

Strong driving of $HO_x$ chemistry by solar UV radiation leads to rapid diurnal variations of OH, $HO_2$ and $O_3$ in the mesosphere. The amplitude of their variations depend largely upon altitude [*Allen et al.*, 1984]. Near the stratopause ($\sim$ 50 km altitude) there is little diurnal variation of $O_3$ but significant variation of

OH and $HO_2$. With increasing altitude the magnitude of the diurnal variation of $O_3$ tends to increase. In the lower mesosphere the diurnal variation of OH and $HO_2$ is fairly symmetric with respect to local solar noon. Above about 65 km altitude the OH (and $HO_2$) shows a peak in early afternoon. This is a consequence of the increasing abundance of $HO_x$ due to cumulative photodissociation of $H_2O$ during daylight hours.

## 3. THE MID-MESOSPHERIC $H_2O$ LAYER

As mentioned above, the source of $HO_x$ is the chemical destruction of $H_2O$. Water molecules are long lived in the middle atmosphere with chemical lifetimes ranging from about a month to over a year, depending upon location [*Brasseur and Solomon*, 1986]. There are two known sources of middle atmospheric water vapor, i.e., the direct transport of $H_2O$ through the cold trap at the tropical tropopause, and the oxidation of stratospheric methane [*LeTexier et al.*, 1988] which is also transported upward from the troposphere. In the stratosphere the $H_2O$ mixing ratio increases with altitude due to the $CH_4$ oxidation. Since almost all $CH_4$ is oxidized below $\sim$65 km altitude, the $H_2O$ mixing ratio is expected to reach its maximum value near that altitude. Above this altitude the destruction of $H_2O$ by solar $Ly\ \alpha$ photolysis will produce a rapid decrease of $H_2O$ abundance with altitude. In recent years, observations from the HALOE experiment [*Russell et al.*, 1993], along with additional ground based measurements [*Nedoluha et al.*, 1997] and modeling studies have confirmed many aspects of this theory [*Summers et al.*, 1997b]. This gives confidence that we have a generally good understanding of the large scale sources and sinks of $H_2O$ and $CH_4$, and the relative roles of chemistry and transport in controlling their distributions.

However, the discovery of a layer of water vapor centered near 68 km altitude by the HALOE experiment, and corroborated by modeling the MAHRSI OH measurements [*Summers et al.*, 1997a], has brought into question the completeness of the above picture of the middle atmospheric water vapor budget. In Plate 2(a), we show an altitude versus latitude plot of OH density as observed by MAHRSI during one typical orbit during the first MAHRSI mission in November 1994. In Plate 2(b) we show the $H_2O$ mixing ratio retrieved from HALOE observations for a time period inclusive of that of the MAHRSI OH observations. The zonal average of the $H_2O$ data shown is generated from daily averages of occultations covering a 38 day time frame. As seen in Plate 2(b), the $H_2O$ mixing ratio is generally constant at $\sim$ 6.0 ppmv between 50 to 63 km altitude at all latitudes. However, a narrow layer of higher water vapor mixing ratio is seen between 64 and 72 km

where it reaches 7.0 ppmv. The similarity between the morphology of the $H_2O$ layer and the coincident layer of OH is striking. Both the $H_2O$ and OH fall off rapidly with altitude above about 75 km as expected due to rapid solar $Ly\ \alpha$ photolysis of $H_2O$. In the region of the $H_2O$ layer the OH chemical lifetime is of order of a minute, whereas the $H_2O$ lifetime is more than a month. Thus OH is in photochemical equilibrium with the $H_2O$ distribution, and this is the cause of the similarity between the observed OH and $H_2O$. (The diurnal variation of OH is also a factor in photochemically mapping the OH distribution to the $H_2O$ distribution.)

The $H_2O$ layer observed by HALOE was generally ignored before 1997 partially because its existence was so grossly contrary to that expected from the standard model of $H_2O$ chemistry, as discussed above. As the analysis of the HALOE data has become more refined in the recent data versions the $H_2O$ layer has become a robust feature of the HALOE retrievals. The most complete discussion of the HALOE $H_2O$ data validation is for version 17 [*Harries et al.*, 1996]. Versions 18 and 19 incorporate a more accurate occultation baseline, improved spectroscopic parameters, and improvements in retrieval algorithms. These three recent versions of the HALOE $H_2O$ data for the time period of the first MAHRSI mission are shown in Plate 3. Also shown in Plate 3 is the location of the HALOE occultation tangent point on November 5, 1994, the first day of the MAHRSI mission which made OH observations over a 9 day period. The black arrows denote the direction of the precession of the HALOE occultation point. Note that the shape of the HALOE $H_2O$ layer is different between sunrise and sunset occultations. Given the different temporal coverage of the HALOE sunrise and sunset data, the $H_2O$ zonal averages shown in Plates 2 and 3 do not represent instantaneous snapshots of the water vapor distribution. Because the dynamical transport time scale for water vapor in the mid and upper mesosphere is less than a season, one expects significant changes in the shape of the layer due to seasonally varying circulation patterns. This may partly explain the differences between HALOE sunrise and sunset zonal averages. Diurnal tides may also play a role in the observed sunrise/sunset differences. It is also possible that there is a systematic instrumental component.

Another reason that the mid-mesospheric $H_2O$ layer seen in the HALOE data was mostly ignored before 1997 was that there were no confirming observations. For the Microwave Limb Sounder (MLS) on UARS, which also measured $H_2O$, the altitude weighting functions for the retrieved mesospheric profile have a width of 6-10 km and are therefore unable to resolve such a thin layer [*Lahoz et al.*, 1996]. This is also true for ground-based water vapor monitors such as WVMS [*Nedoluha et al.*, 1997]. However, the observation of an OH layer by the MAHRSI observations coincident with the $H_2O$ layer is strong confirmation. These two experiments use completely different observational techniques; MAHRSI uses solar ultraviolet resonance fluorescence to measure OH emission, whereas HALOE uses solar infrared occultation to measure $H_2O$ extinction [*Russell et al.*, 1993], [*Conway et al.*, 1999]. Together, the HALOE and MAHRSI observations have been shown to provide a consistent picture of OH and $H_2O$ in the mesosphere and imply near balance in the ozone production and loss at the stratopause [*Summers et al.*, 1997a]. As of this writing there are no candidate mechanisms for producing the observed OH layer other than that of a $H_2O$ layer with the properties of that observed by HALOE.

The $H_2O$ layer is a persistent feature in the HALOE data (from the earliest observations in 1991 to the present) [*Summers and Siskind*, 1999], although it does occasionally disappear for short periods of time. It is typically restricted to latitudes within $\sim30°$ of the equator and to follow the solar declination with a $\sim30$-45 day time lag. It also appears to change shape as one would expect in response to the seasonally changing meridional circulation (i.e., evolving from a single cell circulation near solstices to a two cell system near equinoxes.) Also, the layer occasionally (mostly near solstices) appears to extend to high summer latitudes.

## 4. ODD-HYDROGEN PHOTOCHEMISTRY

In this section we discuss the implications of the MAHRSI OH observations for mesospheric and upper stratospheric $HO_x$ chemistry. Since $HO_x$ chemistry is driven by solar UV photolysis, OH exhibits a strong solar zenith angle (local solar time) dependence. We use the NRL model CHEM1D, which is a time dependent 1-dimensional photochemical model, with model $H_2O$ abundances fixed to values observed by HALOE, to simulate this diurnal variation. The model is the same used in our earlier studies of the MAHRSI data [*Summers et al.*, 1996], [*Summers et al.*, 1997a] but with updated chemical kinetics and cross sections [*DeMore et al.*, 1997]. As validation, our model results for the middle and upper stratosphere reproduce those obtained by [*Jucks et al.*, 1998], when using identical model inputs and constraints on temperature, $H_2O$, $N_2O$, $CH_4$, $O_3$ and total $NO_y$ and $Cl_y$, to better than 5% [*Conway et al.*, 2000].

First we show the OH over-prediction by the model using standard $HO_x$ chemistry, and then discuss the consequences of of the assumed shape of the mesospheric $H_2O$ profile for the shape of the model OH

profile. In Figure 1 we show show several OH calculations using CHEM1D (top panel) along with corresponding model input $H_2O$ abundances (bottom panel). The conditions for this OH model/data comparison are 10N, which is very close to the HALOE occultation on November 8, and a local solar time (LST) of 10AM.

As seen in Figure 1(a) the model OH calculation, using HALOE V.18 $H_2O$ data as input along with standard $HO_x$ chemistry (denoted by JPL97), yields OH densities at all altitudes that are substantially higher than observed by MAHRSI. Based upon analysis of ground-based observations of $HO_2$ and $O_3$, [*Clancy et al.*, 1994] and [*Sandor and Clancy*, 1998], concluded that their observations are more consistent with models in which the rate coefficient for reaction (3) is substantially lower than that recommended [*DeMore et al.*, 1997]. Here we consider a decrease in the rate coefficient for (3) of 50%. This modification dramatically lowers the calculated OH abundance and yields a profile in agreement with the OH data below 70 km but somewhat over-estimates OH above that level. Using a 50% decrease in the rate coefficient of (3), along with the most recent HALOE V.19 $H_2O$ as model input, yields a model OH profile that is in good agreement with the observed OH over most altitudes.

In Figure 1(b) we show a theoretical $H_2O$ profile using standard chemistry [*DeMore et al.*, 1997] obtained from the NRL model CHEM2D, which is a middle atmospheric two dimensional (altitude vs. latitude) chemical-dynamical model [*Summers et al.*, 1997b]. As can be seen in Figure 1(b), the CHEM2D model yields a peak $H_2O$ mixing ratio near 60 km altitude with decreasing $H_2O$ mixing ratio above that level. For comparison, the HALOE $H_2O$ (V.18 and V.19) show more sharply peaked profiles, and larger $H_2O$ mixing ratios near 68 km altitude. The photochemical model OH calculation, using the 2-D $H_2O$ profile as input, yields an OH profile similar to the MAHRSI OH observations everywhere except in the critical 63-73 km altitude region where it seriously underpredicts OH as seen in Figure 1(a). Because the OH abundance is roughly proportional to the square root of the $H_2O$ abundance [*Allen et al.*, 1984] it requires a significant increase in $H_2O$ over that predicted by the 2-D model in order to reproduce the observed enhancement in OH at that altitude. The enhancement in $H_2O$ seen near 70 km in the HALOE experiment is almost exactly that needed in the photochemical model to reproduce the magnitude and shape of the observed OH profile.

It is important to note that the proposed 50% reduction in the rate coefficient for reaction (3) does not provide a unique fit to the OH data [*Summers et al.*, 1997a]; a 20% decrease in (3) plus a 30% in-

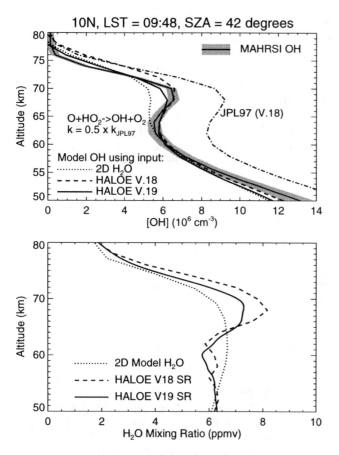

**Figure 1.** (a) A comparison between MAHRSI OH at 10N and local solar time of 9:48 AM with photochemical models. The dot-dashed line is the model using standard chemistry (JPL97) and Version 18 Haloe $H_2O$ data. All other model cases use a 50% reduction in the reaction rate coefficient for (3). (b) Water vapor distributions used in the model cases shown in panel (a).

crease in (6) provides an equally good fit. However, further support for a substantial modification in standard $HO_x$ chemistry comes from a investigation of the MAHRSI OH data and CRISTA $O_3$ data [*Riese et al., 1999*]. CRISTA used three infrared telescopes and spectrometers to measure $O_3$, several other trace gases and temperature. CRISTA and MAHRSI were boresighted and both made limb-scans of trace gas emission. Taken together, the OH and $O_3$ measurements from 1994 CRISTA/MAHRSI mission, along with the 50% change in rate constant for reaction (3), were found to lead to near balance in model $O_3$ production and loss at the stratopause [*Summers et al., 1997a*].

The MAHRSI reflight in August 1997 provided an opportunity to study OH at mid and high northern latitudes. In Plate 4 we show a typical mid latitude sequence (orbit) from this later mission. HALOE was not making observations during the August flight. The nearest HALOE $H_2O$ observations in time (but same location) were made in early September. In Figure 2(a) we show model OH calculations for August 1997 conditions, utilizing HALOE $H_2O$ from September. For LST= 10 am, we see that the photochemical model, incorporating a 50% reduction in (3), reproduces the observed features in the MAHRSI OH data. Thus, under similar daytime conditions (mid morning), the OH observations from the August MAHRSI mission confirm the results from the earlier November mission, i.e., mesospheric OH is substantially lower than that predicted by standard $HO_x$ photochemical theory. Furthermore, the photochemical model with a 50% reduction in the rate coefficient for reaction (3), along with input $H_2O$ from the HALOE experiment, adequately reproduces the observed OH abundance profile in the mesosphere.

A very interesting feature is seen in the midday and afternoon OH observations from the August mission. Whereas the model OH calculations for midday and afternoon predict that the OH layer near 70 km should remain nearly fixed in altitude, the observed OH layer appears to move downward in the morning and then upward in late afternoon (Figure 2(b)&(c)) We speculate that this behavior in the observed OH is due to a diurnal tide and it's effect on OH. The effects of temperature tides on the diurnal variation of $O_3$ has been been previously explored by [*Huang et al., 1998*]. The effects of tides on the diurnal variation of OH as observed in the MAHRSI data will be the topic of a future report.

During its second mission MAHRSI made high quality, low solar zenith angle observations (less than 40°), from which it was possible to retrieve OH abundances down to 40 km altitude. In Figure 3 we show an OH profile that was obtained from an average of mid-day observations over a 5-hour period [*Conway et al., 2000*]. Also shown are model results for upper stratospheric OH that were obtained from CHEM1D simulations utilizing model constraints on long-lived trace species from coincident CRISTA observations ($O_3$, $HNO_3$, and temperature), and HALOE climatologies ($H_2O$, $CH_4$, HCl, $NO_2$, and NO).

The model cases we consider follow the naming conventions of [*Summers et al., 1997a*] and [*Jucks et al., 1998*]. The standard model (A) [*DeMore et al., 1997*] leads to a significant overprediction of OH at all altitudes above 50 km as seen in Figure 3. Model B incorporates a 50% reduction in (3) and yields an OH profile very similar to the MAHRSI OH at altitudes above 50 km. Model C, with a 20% reduction in (3) along with a 30% increase in (6), is also very close to the observed mesospheric OH profile. Here, as in our previous mod-

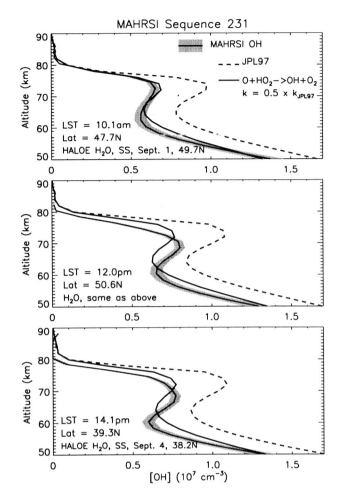

**Figure 2.** A comparison between MAHRSI OH scans and photochemical model for 3 local solar times for sequence 231.

eling studies of OH data from the November MAHRSI flight, we cannot rule out some combination of changes in (3) and (6) as an acceptable solution based upon analysis of mesospheric OH observations alone. Model case D incorporates a 25% reduction in both (3) and (6), as suggested by [*Jucks et al., 1998*]. This produces a mesospheric OH profile almost indistinguishable from case A, i.e. it overpredicts OH above 50 km but underpredicts at the upper stratospheric peak below 45 km.

The most dramatic conclusion that can be drawn by comparing model predictions with the observed OH is that neither standard chemistry nor any of the proposed modifications reproduce the observed OH profile in both the upper stratosphere and mesosphere. Indeed, the OH data appear to show a regime change near the stratopause. However, standard $HO_x$ chemistry is inconsistent with such a regime change because the same reactions, (2) and (3), are responsible for controlling

$OH/HO_2$ partitioning in both the mesosphere and upper stratosphere (see Plate 1(b)). This is clearly seen in Figure 3 where the models of the various $HO_x$ modifications shift the calculated OH profile in magnitude but leave the overall shape mostly unchanged. The OH vertical density profile observed in the upper stratosphere is significantly different, and the observed OH peak near 43 km altitude is ~20% higher, than that predicted by standard $HO_x$ chemistry.

Model/data comparisons from previous $HO_x$ studies in the middle stratosphere reached rather different conclusions than those described above. Several recent balloon-borne experiments used far-infrared limb viewing of atmospheric thermal emission to measure a comprehensive set of abundance profiles of the key radicals along with their precursor (source) molecules [*Chance et al.,* 1996; *Pickett and Peterson,* 1996; *Osterman et al.,* 1997; *Jucks et al.,* 1998]. [*Jucks et al.,* 1998] discussed the most recent observations of upper stratospheric OH and $HO_2$ using the Far Infrared Spectrometer (FIRS-2). They found that their observations of OH and $HO_2$ were in best agreement with Model D described above and shown in Figure 3, which predicts an OH profile nearly identical to that of standard chemistry (Model A). All of the changes in rate coefficients suggested by [*Jucks et al.,* 1998] and implemented in Model D are consistent with the quoted measurement

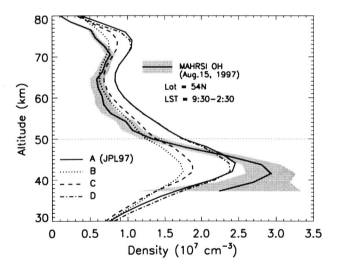

**Figure 3.** MAHRSI observations from Aug. 15, 1997. The shaded area shows the 5-hour average of 34 limb scans for which the local time varied between 0930-1430, the solar zenith angle varied between 32-49 degrees, and latitudes ranged between 42-58N. The photochemical models are (A) Standard chemical kinetics, (B) 50 % reduction for reaction (3) (C) 20 % reduction in rate coefficient for reaction (3) and a 30 % increase for reaction (6), and (D) 25 % reduction in the rate coefficients of (3) and (6).

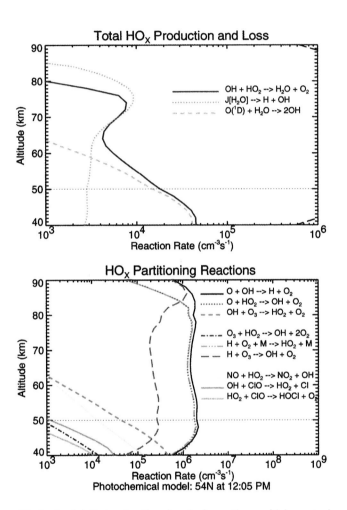

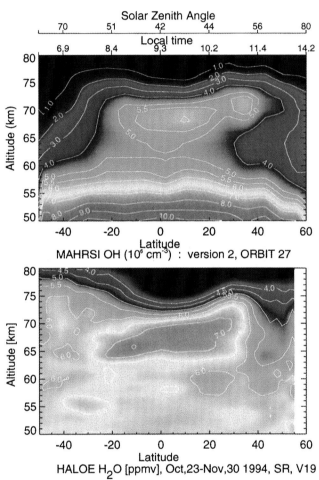

**Plate 1.** (a) Rates for the chemical reactions which control $HO_x$ production and loss, and (b) rates which control $HO_x$ partitioning.

**Plate 2.** (a) Retrieved MAHRSI OH number density as a function of altitude and latitude for orbital sequence 27 on Nov. 5, 1994. Top scale shows the corresponding solar zenith angle and local time of the observations. (b) Zonal average $H_2O$ mixing ratio (ppmv) generated by plotting daily average Version 19 HALOE sunrise (SR) occultation profiles for the time period Oct. 23, through Nov. 30, 1994.

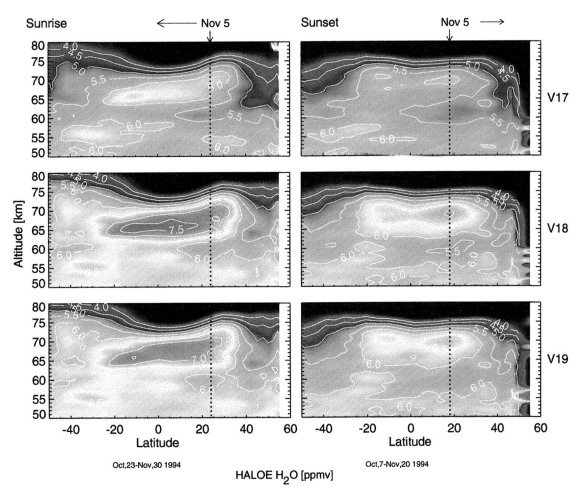

Sunrise     ← Nov 5     Sunset     Nov 5 ⟶

V17

V18

V19

Oct,23-Nov,30 1994     Oct,7-Nov,20 1994

HALOE H$_2$O [ppmv]

**Plate 3.** Comparison of HALOE H2O from the versions 17, 18 and 19, for both sunset and sunrise occultations for same dates as in Plate 2(b).

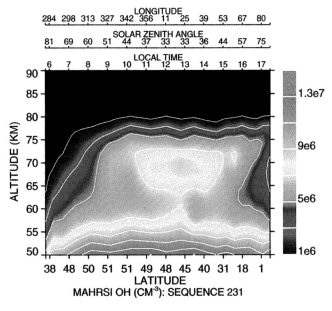

MAHRSI OH (CM$^{-3}$): SEQUENCE 231

**Plate 4.** MAHRSI OH observed on August 11, 1997 for orbit sequence 231.

uncertainties in [*DeMore et al.*, 1997]. Although Model D gave satisfactory agreement between the FIRS-2 OH and HO$_2$, [*Jucks et al.*, 1998] concluded that the model underpredicted upper stratospheric O$_x$ production and thus required an additional, but unknown, source of O$_x$ in the upper stratosphere to resolve the ozone deficit problem (see e.g., [*Siskind et al.*, 1995; *Crutzen et al.*, 1995; *Müller and Salawitch*, 1999]

The balloon data were also used to test standard HO$_x$ chemistry in the upper stratosphere by extending the results above the balloon float height of 38 km using model profiles and the observed vertical column abundance. [*Jucks et al.*, 1998] found good agreement between the extended FIRS-2 OH profiles and Models A and D at 50 km. From a detailed model/model comparison, they also reported good agreement with the models of [*Summers et al.*, 1997a]. Thus [*Müller and Salawitch*, 1999] concluded that the differing results of the MAHRSI and FIRS-2 analyses at 50 km appeared to be due to disagreement between the OH measurements by the two spectrometers. However, the subsequent retrieval of MAHRSI OH observations below the stratopause [*Conway et al.*, 2000] summarized in Figure 3 has provided new information. In particular, the very small scale height of the MAHRSI OH observation below the stratopause and the observed peak density of the upper stratospheric layer result in an apparent OH column density of 2.34 ×10$^{13}$cm$^{-2}$ between 40 and 50 km altitude as compared to a 2.24 ×10$^{13}$cm$^{-2}$ OH column density predicted by Model A. *Jucks and Salawitch* (this monograph) report that the FIRS-2 OH measurements are consistent with the MAHRSI OH measurements between 40 and 50 km and also comment that the FIRS-2 measurements are not sensitive to the mesospheric OH column.

A third independent measurement of middle atmospheric OH is reported by [*Englert et al.*, 2000] (this monograph). The Tera Hertz OH Measurement Atmospheric Sounder (THOMAS) experiment made coordinated measurements from the high-altitude Falcon aircraft during the August 1997 MAHRSI flight. These observations successfully minimized the miss time and distance between the measurements so as to include the same volume of air in the line of sight of each instrument. Comparison of the THOMAS results with OH columns above 40 km derived from MAHRSI profiles that were smoothed by the THOMAS weighting functions show excellent agreement.

Thus the results of the three experiments, MAHRSI, FIRS-2 and THOMAS, each using entirely different techniques to measure the OH column in the upper stratosphere, agree. Furthermore, as noted earlier, the photochemical models used by [*Jucks et al.*, 1998] and

[*Summers et al.*, 1997a] have been found to give almost identical results in both the upper stratosphere and mesosphere. The standing problem now is that both standard HO$_x$ kinetics and all proposed modifications completely fail to account for the observed OH in both the mesosphere and upper stratosphere.

## 5. HETEROGENEOUS CHEMISTRY

The existence of the 70 km water vapor layer is quite surprising from the perspective of standard models of middle atmospheric chemistry and dynamics. First, models of mesospheric photochemistry predict net destruction of H$_2$O above 65 km altitude due to solar *Ly* $\alpha$ photolysis of H$_2$O, and a resulting monotonic decrease of the H$_2$O mixing ratio above that altitude (see e.g., [*Allen et al.*, 1984], [*Brasseur and Solomon*, 1986], [*Le Texier et al.*, 1988], [*Summers et al.*, 1997b].) Although the chemical conversion of CH$_4$ to H$_2$O is a well established part of the conventional picture of middle atmospheric chemistry, an additional yet unknown pathway for CH$_4$ to H$_2$O conversion cannot account for the layer since its presence is also observed in HALOE observations of the sum of "total hydrogen" **D** = 2xCH$_4$ + H$_2$O [*Summers and Siskind*, 1999].

Second, it is impossible to produce the observed H$_2$O layer by dynamical means. Tracer transport in the middle atmosphere occurs by advection due to the general circulation and by mixing which is a consequence of planetary and gravity wave activity. Diffusive separation is important at much higher altitudes, from the turbopause upward into the thermosphere. Advection and mixing cannot account for the existence of the H$_2$O layer since advection preserves the mixing ratio during transport, whereas eddy mixing transports trace gases in a direction opposite to the mixing ratio gradient. So a persistent local maximum in the H$_2$O mixing ratio requires a source in the same location. *Summers and Siskind* [1999] have shown that the H$_2$O layer is a persistent feature of the HALOE observations extending back to 1992.

The H$_2$O source rate required to produce the layer is also inconsistent with an extraterrestrial source of H$_2$O as proposed by [*Frank and Sigwarth*, 1993]. The H$_2$O influx from the small-comet hypothesis is approximately 60 times larger than that required to produce the observed H$_2$O layer [*Siskind and Summers*, 1998]. Such a large influx of H$_2$O as predicted by the small-comet hypothesis would lead to H$_2$O mixing ratios exceeding 20 ppmv throughout much of the mesosphere, and clearly inconsistent with HALOE observations of mesospheric H$_2$O [*Siskind and Summers*, 1998]. Furthermore, models of the combined stratospheric and mesospheric budget of $H_2O$ limit the influx of extraterrestrial $H_2O$ to values about 100 times less than that predicted by the small-comet hypothesis [Hannegan et al., 1998].

If one considers the possible conversion of other hydrogen containing species to $H_2O$, $H_2$ is the only candidate that is present in sufficient abundances. The observed peak $H_2O$ abundance in the layer is approximately 0.6 ppmv larger than that predicted by standard hydrogen chemistry. This corresponds closely to the amount of atmospheric hydrogen available as $H_2$ in the lower mesosphere and upper stratosphere [Pollock et al., 1980], [Fabian et al., 1981], [Dessler et al., 1994]. [Summers and Siskind, 1999] speculated that the reaction

$$O + H_2 \rightarrow H_2O \qquad (17)$$

on the surface of meteoric dust might provide the required conversion of $H_2$ to $H_2O$. The meteoric dust could act as a catalytic agent for the reaction and also help to stabilize the reaction products; however, the mechanism of the surface reaction and the reaction probability depends upon several poorly known factors including the dust composition, sticking probability, surface mobility, size of the particles (these particles may contain only $\sim 100$ molecules and may be non-spherical), and the influence of other species which may also be adsorbed on the dust. Very small ($\sim 1$ nm) meteoric particles formed by the re-condensation of vapors produced during meteor ablation are also thought to be abundant in the region of the $H_2O$ layer [Hunten et al., 1980], [Turco et al., 1981]. Furthermore, [Summers and Siskind, 1999] showed that the tendency of particles as small as 1-10 nm to settle downward through the atmosphere would be countered, at low latitudes, by upward advection, leading to a layer of small meteor dust particles confined between 60-80 km altitude.

The scenario proposed by [Summers and Siskind, 1999] has $H_2$ carried into the dust layer by tropical upwelling. Both O and $H_2$ will diffuse to the particles, react on the surface to form $H_2O$, and then desorb back into the atmosphere. If conversion of $H_2$ to $H_2O$ is occurring in the observed $H_2O$ layer region, then there should exist a region of depleted $H_2$ coincident with the $H_2O$ layer. This source would be consistent with the limited latitudinal extent of the $H_2O$ layer, since at higher latitudes the meteoric dust deposited in the mesosphere will be distributed over a much larger range of altitudes driven by the seasonally varying upward (summer hemisphere) and downward (winter hemisphere) vertical motion [Brasseur and Solomon, 1986]. This will dilute the perturbation of surface chemistry on the background gas phase chemistry outside of low latitudes.

## 6. DIRECTIONS FOR FUTURE RESEARCH

Mesospheric hydrogen chemistry has long been considered to be one of the simplest chemical systems in the middle atmosphere. Although the standard picture of $HO_x$ chemistry appears to explain the general features of mesospheric hydrogen species' distributions, the recent MAHRSI results raise fundamental questions about the completeness of the standard model. The discovery of the narrow layer of high $H_2O$ mixing ratio near 70 km suggests a local source of $H_2O$ which is in conflict with the prediction of net $H_2O$ loss by the standard model. Also, the inability of standard $HO_x$ chemistry (and all proposed modifications) to simultaneously reproduce the OH and $HO_2$ observations in the mesosphere and upper stratosphere suggests the standard model of $HO_x$ partitioning is in error. Below we give some suggestions for future research that will help to clarify our understanding of these issues.

### 6.1. Coordinated Satellite, Balloon and Ground-based $HO_x$ Observations

Although the balloon observations (e.g., FIRS-2) have provided detailed measurements of OH and $HO_2$ profiles in the stratosphere, their extrapolation above the balloon float level ($\sim 38$ km) into the upper stratosphere requires the necessary assumption of a vertical profile shape. The published extrapolations have assumed that their profiles are given by standard $HO_x$ chemistry. However, the MAHRSI OH observations in the upper stratosphere show that the OH vertical profile is dramatically different than that given by the standard model.

Furthermore, the ground-based OH column measurements show a diurnal variation more sharply peaked in early afternoon than that given by standard $HO_x$ models [Burnett and Burnett, 1996]. Since the maximum OH abundance occurs in the upper stratosphere this cautions us against making the assumption that the OH diurnal variation in the upper stratosphere follows that expected from the standard model. It is also possible that the upper stratospheric OH diurnal variation may exhibit a latitudinal dependence through the coupling between $HO_x$ chemical cycles with chlorine and nitrogen cycles.

The ground-based column OH measurements of [Burnett and Burnett, 1996] and discussed by [Canty et al., 2000] (this monograph) have long remained a puzzle. These observations show a long-term trend not easily understood in terms of known solar forcing or changing properties of the atmosphere. They also show higher total column abundances ($\sim 30$-40%) than standard model predictions. Independent OH column measurements would be very useful here in determining whether systematic errors play a role in these results. The preliminary results from the JPL group observing from Table Mountain [*Mills et al.*, 1999] seem to be very promising in this regard.

A coordinated campaign of satellite, balloon, and ground based observations that would provide simultaneous and complete knowledge of OH and $HO_2$ from the lower stratosphere to the upper mesosphere may be necessary in order to solve the $HO_x$ dilemma. Coordination with other measurements including $O_3$, $H_2O$, $CH_4$, and temperature would also be required. Simultaneous observations of OH, $HO_2$, and $O_3$ provide strong constraints on both $HO_x$ partitioning and total $HO_x$. Ground based observations of mesospheric $HO_2$ such as that by [*Sandor and Clancy*, 1998] and OH by [*Canty et al.*, 2000] (this monograph), coordinated with satellite OH and $O_3$ observations, would be especially useful in determining if $HO_x$ partitioning and/or total $HO_x$ is correctly accounted for in photochemical models.

### 6.2. Laboratory Measurements of $OH/HO_2$ Partitioning Reactions

Several independent studies of stratospheric and mesospheric $HO_x$ and ozone suggest that the observations are more consistent with model calculations if the true rate coefficient for (3) were significantly slower than the rate recommended in [*DeMore et al.*, 1997]. However, this rate has been measured in the laboratory on several occasions with mostly consistent results [*DeMore et al.*, 1997]. Additional laboratory determinations of reaction (3), utilizing new innovative measurement techniques, might break this impasse [*Jeffries*, 1999].

### 6.3. Mesospheric $H_2O$ Observations

Although the HALOE $H_2O$ observations have provided a long term record of the mesospheric $H_2O$ layer over the past 8 years, we are still lacking a basic analysis of those observations that would build up a picture of how the layer evolves seasonally and/or in response to middle atmospheric dynamics. It would very useful to know if the layer shows a secular change and to determine if the structure of the layer is correlated with solar UV variations, D and E region electron density, the semi-annual oscillation (SAO), meteor showers, etc., in order to obtain more clues regarding its origin.

The HALOE team has recently utilized the technique of differential absorption between IR channels as a means to remove aerosol contamination in the solar occultation data. This technique makes it possible to remove extinction due to aerosols in the solar occultations. HALOE observations of $H_2O$ at high latitudes will allow a quantitative assessment of the water budget associated with polar mesosphere clouds and a determination of whether the low latitude $H_2O$ layer has a high latitude extension.

Observational and theoretical studies of mesospheric meteor dust deposition and transport might also reveal interesting similarities and/or differences with the observed $H_2O$ layer. Such studies could give important clues to the source of the layer. Perhaps key to understanding the origin of the layer would be coincident observations of mesospheric $H_2$ and $H_2O$ to test the implication that $H_2$ should be depleted in the $H_2O$ layer.

### 6.4. Heterogeneous Chemistry

There is strong theoretical support for the existence of meteoric dust in the mesosphere [*Hunten et al.*, 1980] and the expectation that the distribution of this dust will show strong vertical and horizontal structure [*Fiocco and Grams*, 1971]. In order to determine the role of heterogeneous chemistry on this dust it is essential to know the its composition, structure, and distribution. In situ rocket collection may offer the best promise for studying these particles. A key to understanding the role of heterogeneous chemistry in the mesosphere may lie with laboratory investigations exploring the kinetics of $H_2$, $O_3$, OH, and O on surfaces representing the various types of known meteoric compositions.

The possibility of heterogeneous chemistry on meteoric dust in the mesosphere raises a host of interesting possibilities and questions. Does surface chemistry affect the abundances of other trace gases such as CO, NO, O, and Na containing species? Could downward transport of meteoric dust at higher latitudes influence ozone chemistry in the stratosphere? Does heterogeneous chemistry on polar mesospheric cloud particles influence the $H_2O$ budget there? We must await a better understanding of the distribution of meteoric dust and the kinetic possibilities for surface reactions before we will be in a position to answer such questions.

*Acknowledgments.* We thank Dave Siskind and Ken Jucks for many stimulating discussions about $HO_x$. We are also grateful to Ken for providing the FIRS-2 data and the observational constraints used in his modeling studies. Our appreciation goes to Mike Stevens and Joel Cardon for their careful analysis and inversion of the MAHRSI data, and to Andrew Kochenash for help with the graphics. This work was supported by the Office of Naval Research and the NASA Offices of Earth Science and Space Science.

### REFERENCES

Allen, M., J. I. Lunine, and Y. L. Yung, The vertical distribution of ozone in the mesosphere and lower thermosphere, *J. Geophys. Res.*, 89, 4841-4872, 1984.
Brasseur, G., and S. Solomon, Aeronomy of the Middle Atmosphere, D. Reidel, 2nd Edition, 1986.

Brasseur, G.P. *et al.*, Natural and Human-Induced Perturbations in the Middle Atmosphere: A Short Tutorial, (this monograph).

Burnett, C.R., and E.B. Burnett, The regime of decreased OH vertical column abundances at Fritz Peak Observatory, CO:1991-1995, *Geophys. Res. Lett.*, 23, 1925-1927, 1996.

Canty, T.P. *et al.*, A Critical Review of Hydroxyl in the Middle Atmosphere: Comparison of Measured Vertical Profiles to Model Results and Ground-Based Column Observations, (this monograph).

Chance, K. *et al.*, Simultaneous measurements of stratospheric $HO_x$, $NO_x$, and $Cl_x$: Comparison with a photochemical model, *J. Geophys. Res.*, 101, 9031-9043, 1996.

Clancy, R. T. *et al.*, Model ozone photochemistry on the basis of Solar Mesosphere Explorer mesospheric observations, *J. Geophys. Res.*, 92, 3067-3080, 1987.

Clancy, R. T., B. J. Sandor, and D. W. Rusch, Microwave observations and modeling of $O_3$, $H_2O$, and $HO_2$ in the mesosphere, *J. Geophys. Res.*, 99, 5465-5473, 1994.

Coffey, M. and G. Brasseur, Chapter 14 in *Atmospheric Chemistry and Global Change*, G.P. Brasseur, J.J. Orlando, and G.S. Tyndall eds., Oxford University Press, New York, 1999.

Cohen, R.C. *et al.*, Are models of catalytic removal of $O_3$ by $HO_x$ accurate? Constraints from in situ measurements of the OH to $HO_2$ ratio, *Geophys. Res. Lett.*, 21, 2539 (1994).

Conner B.J. *et al.*, Ground-based microwave observations of ozone in the upper stratosphere and mesosphere, *J. Geophys. Res.*, 99, 16,757-16,770, 1994.

Conway, R.R. *et al.*, Satellite measurements of hydroxyl in the mesosphere, *Geophys. Res. Lett.*, 23, 2093-2096, 1996.

Conway, R.R. *et al.*, The Middle Atmosphere High Resolution Spectrograph Investigations, *J. Geophys. Res.*, 104, 16,327-16,348, 1999.

Conway, R.R. *et al.*, Satellite Observations of Upper Stratospheric and Mesospheric OH: The $HO_x$ Dilemma, *Geophys. Res. Lett.*, in press, 2000.

Crutzen, P.J. *et al.*, A Re-evaluation of the Ozone Budget with HALOE UARS Data: No Evidence for the Ozone Deficit, *Science*, 268, 705-708, 1995.

DeMore, W.B. *et al.*, Chemical Kinetics and Photochemical Data for Use in Stratospheric Modeling, Evaluation No. 12, Publ. 97-4, Jet Propulsion Laboratory, Pasadena, CA, 1997.

Dessler, A.E. *et al.*, An examination of the total hydrogen budget in the lower stratosphere, *Geophys. Res. Lett.*, 21, 2563-2566, 1994.

Dessler A.E. *et al.*, UARS measurements of ClO and $NO_2$ at 40 and 46 km and implications for the model "ozone deficit," *Geophys. Res. Lett.*, 23, 339-342, 1996.

Eluszkiewicz, J., and M. Allen, A global analysis of the ozone deficit in the upper stratosphere and lower mesosphere, *J. Geophys. Res.*, 98, 1069-1082, 1993.

Englert C.R. *et al.*, THOMAS 2.5 THz Measurements of Middle Atmospheric OH: Comparison with MAHRSI Observations and Model Results, (this monograph).

Fabian, P. *et al.*, The Vertical Distribution of Stable Trace Gases at Mid-latitudes, *J. Geophys. Res.*, 86, 5179-5184, 1981.

Fiocco, G. and G. Grams, On the origin of noctilucent clouds: Extraterrestrial dust and trapped water molecules, *J. Atm. Terr. Phys.*, 33, 815-824, 1971.

Frank, L.A. and J. B. Sigwarth, Atmospheric Holes and Small Comets, *Rev. Geophys.*, 31, 1-28, 1993.

Hannegan, G. *et al.*, The dry stratosphere: A limit on cometary water influx, *Geophys. Res. Lett.*, 25, 1649-1652, 1998.

Harries, J.E. *et al.*, Validation of measurements of water vapor from the Halogen Occultation Experiment (HALOE), *J. Geophys. Res.*, 101, 10205-10216, 1996.

Harries, J.E., S. Ruth, and J.M. Russell III, On the distribution of mesospheric molecular hydrogen inferred from HALOE measurements of $H_2O$ and $CH_4$, *Geophys. Res. Lett.*, 23, 297-300, 1996.

Huang, F.G., C.A. Reber, and J. Austin, Ozone diurnal variations observed by UARS and their model simulation, *J. Geophys. Res.*, 102, 12,971-12,986, 1997.

Hunten, D.M, Turco, R.P., and Toon, O.B., Smoke and Dust Particles of Meteoric Origin in the Mesosphere and Stratosphere. *J. Atmos. Sci.*, 37, 1342-1357, 1980.

Jeffries, J.B., Reaction rate coefficients of $HO_2$ + O and $HO_2$ + H, Paper presented at AGU Chapman Conference, Annapolis, MD, 1999.

Jucks, K.W. *et al.*, Observations of OH, $HO_2$, $H_2O$, and $O_3$ in the upper stratosphere: implications for $HO_x$ photochemistry, *Geophys. Res. Lett.*, 25, 3935-3938, 1998.

Jucks, K.W. and R.J. Salawitch, Future changes in upper stratospheric ozone, (this monograph).

Lahoz, W.A. *et al.*, Validation of UARS microwave limb sounder 183 GHz $H_2O$ measurements, *J. Geophys. Res.*, 101, 10,129-10,149, 1996.

LeTexier, H., S. Solomon, and R. R. Garcia, The role of molecular hydrogen and methane oxidation in the water vapour budget of the stratosphere, *Q. J. R. Meteorol. Soc.*, 114, 281-295, 1988.

Mills, F.P. *et al.*, Photochemical Modeling of $HO_x$ in the Terrestrial Mesosphere and Comparison to OH Column Measurements from the JPL Table Mountain Facility, EOS, 80, Number 46, November 16, 1999.

Müller, R., and R.J. Salawitch, Upper Stratospheric Processes, Chapter 6, Scientific Assessment of Ozone Depletion: 1998, World Meteorological Organization - Report No. 44, 1999.

Nedoluha, G.E. *et al.*, A comparative study of mesospheric water vapor measurements from the ground-based water vapor millimeter-wave spectrometer and space-based instruments, *J. Geophys. Res.*, 102, 16,647-16,661, 1997.

Nedoluha, G.E. *et al.*, Changes in upper stratospheric $CH_4$ and $NO_2$ as measured by HALOE and implications for changes in transport, *Geophys. Res. Lett.*, 25, 987-990, 1998.

Offermann, D., et al., Cryogenic Infrared Spectrometers and Telescopes for the Atmosphere (CRISTA) experiment and middle atmospheric variability, *J. Geophys. Res.*, 104, 16311-16326, 1999.

Osterman, G.B., et al., Balloon-borne measurements of stratospheric radicals and their precursors: Implications for the production and loss of ozone, *Geophys. Res. Lett.*, 24, 1107-1110, 1997.

Pickett, H.M. and D.B. Peterson, Comparison of measured

stratospheric OH with prediction, *J. Geophys. Res., 101,* 16,789-16,796, 1996.

Pollock, W., L.E. Heidt, R. Leub and D.H. Enhalt, Measurement of Stratospheric Water Vapor by Cryogenic Collection, *J. Geophys. Res., 85,* 5555-5568, 1980.

Riese, M., et al., Cryogenic Infrared Spectrometers and Telescopes for the Atmosphere (CRISTA) data processing and atmospheric temperature and trace gas retrieval, *J. Geophys. Res., 104,* 16349-13368, 1999.

Rusch, D.W., and R.S. Eckman, Implications of the comparison of ozone abundances measured by the Solar Mesosphere Explorer to model calculations, *J. Geophys. Res.*90, 12,991-12,998, 1985.

Russell, J.M. III, et al., The Halogen Occultation Experiment, *J. Geophys. Res., 98,* 10,777-10,797, 1993.

Sandor, B.J. and R.T. Clancy, Mesospheric HO$_x$ chemistry from diurnal microwave observations of HO$_2$, O$_3$, and H$_2$O, *J. Geophys. Res., 103,* 13,337, 1998.

Siskind, D.E. *et al.*, An intercomparison of model ozone deficits in the upper stratosphere and mesosphere from two data sets, *J. Geophys. Res., 100,* 11,191-11,201, 1995.

Siskind, D.E. and M.E. Summers, Implications of Enhanced Mesospheric Water Vapor Observed by HALOE, *Geophys. Res. Lett., 25,* 2133-2136, 1998.

Strobel, D.F. *et al.*, Vertical constituent transport in the mesosphere, *J. Geophys. Res., 92,* 6691-6698, 1987.

Summers, M.E. *et al.*, A model study of the response of mesospheric ozone to short-term solar ultraviolet flux variations, *J. Geophys. Res., 95,* 22,523-22,538, 1990.

Summers, M.E. *et al.*, Mesospheric HO$_x$ Photochemistry: Constraints from concurrent measurements of OH, H$_2$O, and O$_3$, *Geophys. Res. Lett., 23,* 2097-2100, 1996.

Summers, M.E. *et al.*, Implications of Satellite OH Observations for Middle Atmospheric H$_2$O and Ozone, *Science, 277,* 1967-1970, 1997.

Summers, M.E. *et al.*, The Seasonal Variation of Middle Atmosphere CH$_4$ and H$_2$O with a new chemical-dynamical model, *J. Geophys. Res., 102,* D3, 3503-3526, 1997.

Summers, M.E., and D.E. Siskind, Surface Recombination of O and H$_2$ on Meteoric Dust as a Source of Mesospheric Water Vapor, *Geophys. Res. Lett., 26,* 1837-1840, 1999.

Turco, R.P. *et al.*, Effects of Meteoric Debris on Stratospheric Aerosols and Gases, *J. Geophys. Res., 86,* 1113-1128, 1981.

Wennberg, P.O. *et al.*, Twilight observations suggest unknown sources of HO$_x$, *Geophys. Res. Lett., 26,* 1373-1376, 1999.

Zommerfields, W.C. *et al.*, Diurnal variation of mesospheric ozone obtained by ground based microwave radiometry, *J. Geophys. Res., 94,* 12,819-12,832, 1989.

Michael E. Summers[1] and Robert R. Conway, E. O. Hulburt Center for Space Research, Naval Research Laboratory, Code 7641.5, Washington, D.C. 20375. (e-mail: summers@map.nrl.navy.mil; conway@uap.nrl.navy.mil)

# A Review of Hydroxyl in the Middle Atmosphere: Comparison of Measured and Modeled Vertical Profiles and Ground-Based Column Observations

T. Canty and K. Minschwaner

*Department of Physics, New Mexico Institute of Mining and Technology, Socorro*

K.W. Jucks

*Harvard-Smithsonian Center for Astrophysics, Cambridge, Massachusetts*

A.K. Smith

*National Center for Atmospheric Research, Boulder, Colorado*

We present an intercomparison of previous measurements of hydroxyl (OH) and photochemical model results in the middle atmosphere at midlatitudes. The measurements are categorized as functions of altitude and solar zenith angle (SZA). Hydroxyl vertical profiles obtained under similar conditions are integrated over altitude to estimate a total OH column which is compared to model results and to ground-based observations. Of particular interest is the dependence of OH abundances on SZA, where discrepancies have been identified between the observed total OH column and the column abundance calculated by current photochemical models. OH column abundances in models are generally lower (25%) than observed at small solar zenith angles; however, the agreement generally is good for SZA $>60°$. This suggests that there is a process or processes related to SZA that photochemical models may not be simulating accurately. Currently, there are not enough observations to determine the altitude range where any unknown processes controlling $HO_x$ at high-sun are occurring.

## 1. INTRODUCTION

The hydroxyl radical (OH), is one of the most important trace species for the photochemistry of the middle atmosphere (from about 10 km to 80 km). OH, together with the hydroperoxy ($HO_2$) and atomic hydrogen (H) species, makes up the odd hydrogen ($HO_x$) chemical family. Reactions involving $HO_x$ are important in regulating catalytic processes that destroy stratospheric $O_3$. The current understanding of $HO_x$ photochemistry emphasizes the reaction of water with $O(^1D)$ as the primary production mechanism below 50 km

$$H_2O + O(^1D) \rightarrow 2OH \qquad (1)$$

$HO_x$ is also produced by the oxidation of $CH_4$. Photolysis of $H_2O$ is an important source of $HO_x$ at higher altitudes.

The primary loss mechanism for $HO_x$ is dependent upon altitude. Reactions involving $NO_x$ are thought to be the most important from 10-25 km, for example

$$OH + NO_2 + M \rightarrow HNO_3 + M. \qquad (2)$$

Atmospheric Science Across the Stratopause
Geophysical Monograph 123

If photolysis of $HNO_3$ follows reaction 2, there is no net loss of $HO_x$. However, there is additional loss of $HO_x$ if the $HNO_3$ reacts with OH. In the upper stratosphere and mesosphere the $HO_x$ recombination reaction dominates [*Summers and Conway*, this volume]

$$HO_2 + OH \rightarrow H_2O + O_2. \qquad (3)$$

There are additional reactions involving $NO_x$, $O_3$, and O which control the partitioning of $HO_x$.

There is a longstanding disagreement between modeled and measured column OH [e.g., *Froidevaux et al.*, 1985, *Burnett et al.*, 1988], with photochemical models producing lower ($\sim$25%) column abundances than observed at small solar zenith angles. One difficulty is identifying the altitude range where this discrepancy exists. This study reviews OH measurements at northern midlatitudes, grouped according to solar zenith angle (SZA). By combining existing measurements, we have generated empirical OH vertical profiles and columns which we compare with model results.

## 2. MODEL AND MEASUREMENTS

The model is a modified, 1-D vertical, version of the 3-D Rose model [*Rose and Brasseur*, 1989], employing 24 altitude levels ranging from 10-80 km. The model includes 25 long lived gases and 14 short lived species. It considers 103 gas phase, 32 photolysis, and 5 heterogeneous reactions. Concentrations of short-lived species are determined using a steady state approximation, and long-lived species are calculated using a continuity equation and are updated at the end of each timestep, typically set to 1 hour.

Since $H_2O$, $CH_4$, and $O_3$ are long-lived species believed to be the most important sources of $HO_x$, we constrain their model distributions using monthly and zonally averaged satellite data from the Halogen Occultation Experiment (HALOE) [*Russell et al.* 1993]. These observations are from August 1994, a time frame that represents the best compromise between seasons and years corresponding to available OH measurements. The profiles for $H_2O$, $CH_4$, and $O_3$ are held constant for all SZA. Additional modifications have been made to include effects of vertical diffusion in 1-D mode, and heterogeneous chemical reactions using SAGE II measurements of aerosol surface area [*Yue et al.*, 1994]. Reaction rates are consistent with JPL-97 recommendations [*DeMore et al.*, 1997].

The primary focus of this study is the dependence of OH on SZA; however, the OH data used here were obtained over the course of several different years and seasons, and the derived empirical distributions represent an average over time and longitude. Displayed in Plate 1 (upper left) are the so-

lar zenith angles and altitude ranges for published OH measurements made between 30°N and 40°N. Previously unpublished data is also included from balloon flights of the FIRS-2 instrument [*Johnson et al.* 1995a] from the Harvard Smithsonian Center for Astrophysics. The majority of the observations represented in Plate 1 were made in the afternoon between 20 and 50 km, where there is a relatively wide range of platforms and observing techniques. Conversely, there are very few measurements at small SZA ($<$ 30°) above 40 km. The only published work above 50 km under high-sun conditions has been that of *Torr et al.* [1987]. There are only two satellite-based measurements for this latitude range, one in the early morning [*Conway et al.*, 1996], and one in the late afternoon [*Morgan et al.*, 1993].

## 3. COMPARISON OF OH VERTICAL PROFILES

The measurements were grouped into three solar zenith angle ranges: AM low-sun (60-40° SZA), high sun (SZA $\leq$ 30°), and PM low-sun (60-40° SZA). The corresponding concentrations are presented in Plate 1. Empirically-based OH profiles were determined using a weighted, least squares fitting procedure involving piecewise continuous polynomial functions. The observations were weighted in each case by the inverse of stated uncertainties. Also indicated in Plate 1 are the corresponding range of modeled OH values. Table 1 lists model and empirical OH distributions using the ROSE model altitude grid. The empirical profile from 10 to 20 km (not shown in plate 1) follows airborne in situ measurements by *Wennberg et al.*, [1990,1995].

OH measurements during AM low-sun conditions (Plate 1, upper right) generally show good agreement between experiments, with the exception of a pair of measurements by *Pickett and Peterson* [1996] near 30 km, and a profile measurement between 32 and 36 km by *Heaps and McGee* [1985]. Similarly, the model shows good agreement with measured OH in the 30 to 50 km altitude range, however there is a slight overprediction between 40-45 km that persists throughout the day. There appears to be a significant underprediction below 25 km, although the measurements here are accompanied by a relatively large uncertainty. Above 50 km, the only midlatitude measurement in this range of SZA is from the Middle Atmosphere High Resolution Spectrograph Investigation (MAHRSI) [*Conway et al.*, 1996]. Here, model OH is significantly larger than observed by MAHRSI. This model/measurement discrepancy was noted previously [e.g., *Summers et al.*, 1996] and lent support to earlier suggestions [ *Clancy et al.*, 1994] for a decrease in the currently accepted rate for the reaction

$$HO_2 + O \rightarrow OH + O_2. \qquad (4)$$

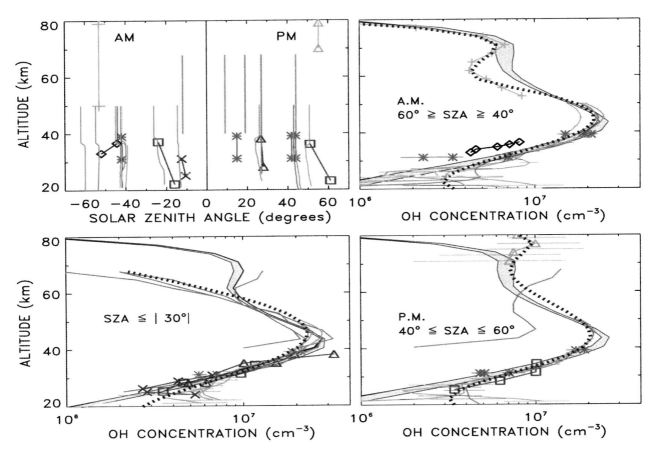

**Plate 1.** Solar zenith angles and altitude ranges for OH measurements from 30 and 40°N (upper left panel). The experiments are: *Carli et al.* [1989] (purple line with triangle), *Heaps and McGee* [1985] (black line with diamond), *Torr et al.* [1987] (blue line), *Stimpfle and Anderson* [1988] (dk. purple line with X), *Stimpfle et al.* [1989, 1990] (dk. blue line with square), *Chance et al.* [1996] (lt. blue line), *Pickett and Peterson* [1996] (red line with asterisk), *Morgan et al.* [1993] (green line with triangle), *Johnson et al.* [1995b] (dk. green line), *Conway et al.* [1996] (lt. orange line with plus), Jucks previously unpublished (orange line). The upper right and two lower panels show measured OH in three SZA ranges, using the same linestyles to indicate experiments. Empirical profiles (black dotted lines) are weighted least squares fits to measurements. Model profiles (grey areas) are calculated for the indicated SZA ranges.

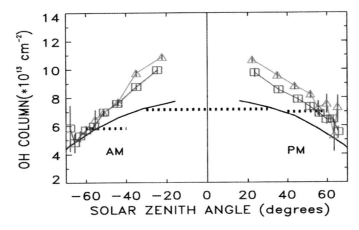

**Plate 2.** Total vertical OH column as a function of solar zenith angle. Ground-based measurements are indicated from New Mexico Tech, NM (red triangles) and Fritz Peak, CO (blue squares). These are mean values for all 1996 data. Error bars are standard deviation of the means. ROSE model results (solid line) are obtained by integration of the calculated OH profile. Empirical OH column (horizontal dotted line) for each SZA range is determined by integration of the empirical profiles in Plate 1.

Table 1. Model and Empirical OH Vertical Profiles

| | AM low-sun | | high-sun | | PM low-sun | |
|---|---|---|---|---|---|---|
| Z (km) | $[OH]_e$ [a] | $[OH]_m$ [b] | $[OH]_e$ | $[OH]_m$ | $[OH]_e$ | $[OH]_m$ |
| 10.0 | — | .4 | 3.5 | .6 | — | .4 |
| 13.0 | — | .2 | 2.5 | .4 | — | .2 |
| 16.1 | — | .4 | 2.3 | .6 | — | .3 |
| 19.1 | 3.8 | .7 | 2.5 | 1.0 | 3.5 | .6 |
| 22.2 | 3.0 | 1.2 | 3.2 | 1.9 | 3.2 | 1.2 |
| 25.2 | 3.4 | 2.3 | 4.3 | 3.4 | 3.7 | 2.3 |
| 28.3 | 4.6 | 4.3 | 6.2 | 6.0 | 5.1 | 4.4 |
| 31.3 | 6.8 | 7.5 | 8.9 | 9.7 | 7.6 | 7.6 |
| 34.3 | 10.1 | 12.0 | 12.4 | 14.7 | 11.2 | 12.0 |
| 37.4 | 14.7 | 18.0 | 16.4 | 21.7 | 15.4 | 18.1 |
| 40.4 | 19.1 | 24.0 | 20.2 | 28.4 | 18.9 | 24.5 |
| 43.5 | 21.8 | 26.2 | 22.7 | 28.7 | 20.6 | 26.3 |
| 46.5 | 21.3 | 22.3 | 23.1 | 23.2 | 20.1 | 22.0 |
| 49.6 | 17.8 | 16.9 | 21.2 | 17.5 | 17.9 | 16.9 |
| 52.6 | 12.8 | 12.8 | 17.6 | 13.5 | 15.0 | 12.9 |
| 55.7 | 7.5 | 10.1 | 13.2 | 11.0 | 12.0 | 10.2 |
| 58.7 | 4.7 | 8.4 | 9.2 | 9.5 | 9.5 | 8.5 |
| 61.7 | 4.4 | 7.5 | 6.0 | 8.8 | 7.7 | 7.5 |
| 64.8 | 4.8 | 7.5 | 3.7 | 9.4 | 6.5 | 7.5 |
| 67.8 | 5.7 | 7.7 | 2.3 | 10.0 | 6.2 | 7.7 |
| 70.9 | 6.0 | 7.6 | — | 9.4 | 6.9 | 7.6 |
| 73.9 | 4.7 | 5.9 | — | 6.9 | 8.4 | 5.9 |
| 77.1 | 2.2 | 2.9 | — | 3.3 | 10.0 | 2.9 |
| 80.0 | .5 | .8 | — | .8 | 7.8 | .8 |

Concentrations are in units of $10^6$ cm$^{-3}$.
[a] Values represent the empirical OH profile obtained from measurements.
[b] Values represent the modeled OH profile at SZA = 43° (low-sun) and SZA = 15° (high-sun)

We decreased the rate of this reaction in the ROSE model by 30% and found improved agreement with the MAHRSI measurements; however, the model in this case significantly underestimated OH between 40 and 50 km. Apparently, it is not possible to reconcile measurements of OH above 50 km with those below 50 km by modification of this reaction rate alone (see also *Jucks et al.* 1998).

For high-sun conditions (Plate 1, lower left), there appears to be satisfactory agreement between experiments, and between modeled and measured OH between 25 and 35 km. Below 25 km, the model underpredicts OH in a pattern similar to the morning results. Above 50 km, there is only one data set [*Torr et al.*, 1987] and no measurements above 65 km. The OH profiles from *Torr et al.* are not consistent with other measurements below 45 km, and the shape of the profiles above 60 km do not suggest a secondary OH maximum between 70 and 75 km as indicated by the model and the morning MAHRSI data. Model OH exceeds the *Torr et al.* measurements by a factor of up to 3 near 65 km.

The PM low-sun measurements shown in Plate 1 (lower right) again demonstrate good agreement among experiments

and a good correspondence with model OH between 25 and 35 km. Consistent also with the morning and midday results, the model predicts less OH than observed below 25 km. Above 40km, there is very poor agreement with observations by *Torr et al.* [1987], whose data do not agree well with any other measurements. There was no explicit uncertainty reported in the *Torr et al.* publication, although the spectral processing and inversion methods required in the analysis suggest that the measurement errors are in excess of 15%. Between 70 and 75 km there is acceptable agreement with the data from *Morgan et al.* [1993].

## 4. COMPARISON OF TOTAL COLUMN

The solar zenith angle dependence of column OH is due to photolytic mechanisms that lead to the production of $HO_x$, primarily $H_2O$ and $O_3$ photodissociation which is regulated by the penetration of solar ultraviolet. This SZA dependence is illustrated in Plate 2, which shows observed OH column abundances as a function of solar zenith angle. The data are averages for 1996 from two locations: Fritz Peak Observatory, Colorado (40°N, 105 °W) and Socorro, New Mexico (34°N, 107°W) [*Burnett and Minschwaner,* 1998]. The Fritz Peak database extends back to 1977, and constitutes one of the most extensive records of a stratospheric trace species that affects $O_3$. The 1996 annual average data indicated here are reasonably consistent with the SZA dependence observed over the previous 15 years, although there have been occasions where the high-sun enhancement was not observed [c.g. *Burnett and Minschwaner* 1998]. In addition, the observations by *Iwagami et al.* [1998] did not show elevated OH abundances at high-sun such as indicated by the Colorado and New Mexico observations used here.

Model results for the OH column are also shown in Plate 2, as well as values obtained by integration of the empirical OH profiles from Plate 1. At altitudes where there are no measurements (c.f. Table 1), model OH concentrations were used to complete the empirical OH column abundance. Below 10 km, both the model and empirical data include an approximate correction for the troposphere, assuming a height and SZA independent value from surface measurements at 40° SZA by *Mount et al.* [1997]. However, inclusion of this data has little impact on the final result, increasing the total OH column by about 7% at high-sun.

It is evident that the ROSE model underestimates the observed OH column for high-sun conditions (small SZA) by about 25%, although there is closer agreement for SZA >60°, particularly in the morning. This behavior is consistent with previous investigations [e.g. *Burnett et al.,* 1988]. The empirical column is also low compared with the ground-based results at small SZA but compares well for SZA >50°. In addition, the empirical column suggests a diurnal asymmetry with higher OH in the afternoon, consistent with the column observations which show an annual average assymetry ($[OH]_{pm}/[OH]_{am}$ -1) of about 15% at 60°. On the other hand, the model predicts very little difference between morning and afternoon OH.

## 5. DISCUSSION AND CONCLUSION

The model deficit in OH column abundances is near zero in the morning for SZA >60°, but grows with decreasing sun angle, approaching $2 \times 10^{13} cm^{-2}$ at overhead sun. The empirical column agrees better with the model than with ground-based observations at high-sun; however, this column is heavily weighted to model results in the upper mesosphere and may also be biased low by the measurements of *Torr et al.* [1987] above 60 km.

Since a uniform discrepancy throughout the column is unlikely, resolution of the problem will require large local perturbations in OH if limited to a particular altitude region. The SZA dependence of the model deficit suggests a pho-

tolytic related process for added production (or reduced loss) which is not captured in the model simulations. The slope of the deficit versus secant(SZA) can be used to constrain the associated values of optical depth; the present results suggest an optical depth of one or more. The corresponding altitude region would be a function of the primary wavelengths involved, ranging from 70 km for 180 nm (or Lyman $\alpha$) radiation, to 20 km for 300 nm radiation [e.g., Figure 1 in *Meier et al.,* 1997].

Comparison of measured OH profiles shows generally good agreement in the mid-stratosphere despite possible annual and seasonal differences. This close correspondence for all three SZA ranges (Plate 1) is convincing evidence that OH is well characterized between 30 and 35 km [see also *Pickett and Peterson,* 1996; *Osterman et al.,* 1997; and *Jucks et al.,* 1998.] In addition, the good agreement between model OH and the empirical profiles would appear to eliminate this region as a suspected source of the model OH column deficit.

Below 25 km, there is a clear underprediction of model OH for all three SZA ranges. These discrepancies may be related to $NO_x$ chemistry and the distribution of $NO_y$ in the model. In regard to the model column deficit, this underprediction is of little consequence due to the lower OH concentrations below 25 km compared to peak values at higher altitudes. This assessment would require revision if the behavior of increasing OH suggested by the FIRS-2 profiles from 25 to 20 km (Plate 1, lower left) persisted down into the lower stratosphere. It would be desirable to obtain more accurate measurements of OH at high sun between 20 and 25 km.

Above 50 km, the OH model/measurement discrepancy at AM low-sun could be important to the deficit in model column OH. Note, however, that revisions to model photochemistry which lower OH concentrations and bring improved agreement with MAHRSI OH may likely worsen the model deficit at high-sun. There is a clear need for future measurements in the 45-80 km altitude region. For example, a modeled increase of $1 \times 10^7 cm^{-3}$ between 45 and 60 km could eliminate most of the column deficit. The spectral regions associated with photolytic processes that have the required SZA dependence would lie from 190 to 200 nm (red tail of the $O_2$ Schumann-Runge bands), or between 230 and 270 nm (peak of the $O_3$ Hartley band). A reexamination of $HO_x$-related photolytic processes in these critical spectral regions could prove useful in resolving the OH column model deficit.

*Acknowledgments.* We thank Clyde Burnett, Bob Conway, Roy Dixon for stimulating discussions and useful suggestions. T.C. thanks Tim Hamlin for technical support. We also acknowledge valuable comments from the two reviewers. Support was provided by the NASA Upper Atmospheric Program # ATM 9730010 and the NSF Atmospheric Chemistry Program # NAG5-4139.

## REFERENCES

Burnett, C.R. and E.B. Burnett, The regime of decrease OH vertical column abundances at Fritz Peak Observatory, CO: 1991-1995, *Geophys. Res. Lett., 23*, 1925-1927, 1996.

Burnett, C. R. and K. Minschwaner, Continuing development in the regime of decreased atmospheric OH at Fritz Peak, Colorado, *Geophys. Res. Lett., 25*, 1313-1316, 1998

Burnett, C.R., K.R. Minschwaner, and E.B. Burnett, Vertical column abundance measurements of atmospheric hydroxyl from 26°, 40°, and 65°N. *J. Geophys. Res., 93*, 5241-5253, 1988.

Carli, B. et al., The mixing ratio of the stratospheric hydroxyl radical from far infrared emission measurements, *J. Geophys. Res., 94*, 11049-11058, 1989.

Chance, K.V. et al., Simultaneous measurements of stratospheric $HO_x$, $NO_x$, and $Cl_x$: Comparison with a photochemical model, *J. Geophys. Res., 101*, 9031-9043, 1996.

Clancy, R.T. et al., Microwave observations and modeling of $O_3$, $H_2O$, and $HO_2$ in the mesosphere, *J. Geophys. Res., 99*, 5465-5473, 1994.

Conway, R.R. et al., Satellite measurements of hydroxyl in the mesosphere, *Geophys. Res. Lett., 16*, 2093-2096, 1996.

DeMore, W.B. et al., Chemical kinetics and photochemical data for use in stratospheric modeling, *JPL Publ. 97-4*, 1997.

Froidevaux, L.A., M. Allen, and Y.L. Yung, A critical analysis of CLO and $O_3$ in the mid-latitude stratosphere, *J. Geophys. Res., 90*, 12,999-13,029, 1985.

Heaps, W.S., and T.J. McGee, Progress in stratospheric hydroxyl measurements by balloon-borne LIDAR, *J. Geophys. Res., 90*, 7913-7921, 1985.

Iwagami, N., S. Inomata, and T. Ogawa, Doppler detection of hydroxyl column abundance in the middle atmosphere: 2. measurement for three years and comparison with a 1D model, *J. of Atmos. Chem. 29*, 195-216, 1998.

Johnson, D.G. et al., Smithsonian stratospheric far-infrared spectrometer and data reduction system, *J. Geophys. Res., 100*, 3091, 1995a.

Johnson, D.G. et al., Estimating the abundance of CLO from simultaneous remote sensing measurements of $HO_2$, OH, and HOCL, *Geophys. Res. Lett., 22*, 1869-1871, 1995b.

Jucks, K.W. et al., Observations of OH, $HO_2$, $H_2O$, and $O_3$ in the upper stratosphere: implications for HOx photochemistry, *Geophys. Res. Lett., 25*, 3935-3938, 1998.

Meier, R.R. et al., Actinic radiation in the terrestrial atmosphere, *J. of Atmos. and Solar-Terr. Physics, 59*, 2111-2157, 1997.

Morgan, M.F., D.G. Torr, and M.R. Torr, Preliminary measurements of mesospheric OH $X^2\Pi$ by ISO on ATLAS 1, *Geophys. Res. Lett., 20*, 511-514, 1993.

Mount, G.H. et al., Measurements of tropospheric OH by long-path laser absorption at Fritz Peak Observatory, Colorado, during the OH photochemistry experiment, fall 1993, *J. Geophys. Res., 102*, 6393-6413, 1997.

Osterman, G.B. et al., Balloon-borne measurements of stratospheric radicals and their precursors: implications for the production and loss of ozone, *Geophys. Res. Lett., 24*, 1107-1111, 1997.

Pickett, H.M. and D.B. Peterson, Comparison of measured stratospheric OH with prediction, *J. Geophys. Res., 101*, 16,789-16,796, 1996.

Rose, K. and G. Brasseur, A three-dimensional model of chemically active trace species in the middle atmosphere during disturbed winter conditions, *J. Geophys. Res., 94*, 16,387-16,403, 1989.

Russell, J.M. et al., The Halogen Occultation Experiment, *J. Geophys. Res., 98*, 10,777-10798, 1993.

Stimpfle, R.M. and J.G. Anderson, In-Situ detection of OH in the lower stratosphere with a balloon borne high repetition rate laser system, *Geophys. Res. Lett., 15*, 1503-1506, 1988.

Stimpfle, R.M. et al., Balloon borne in-situ detection of OH in the stratosphere from 37 to 23 km, *Geophys. Res. Lett., 16*, 1433-1436, 1989.

Stimpfle, R.M. et al., Simultaneous, in-situ measurements of OH and $HO_2$ in the stratosphere, *Geophys. Res. Lett., 17*, 1905-1908, 1990.

Summers, M.E. et al., Mesospheric $HO_x$ photochemistry: Constraints from recent satellite measurements of OH and $HO_2$, *Geophys. Res. Lett., 23*, 2097-2100, 1996.

Summers, M.E. et al., Implications of satellite OH observations for middle atmospheric $H_2O$ and Ozone, *Science, 277*, 1967-1970, 1997.

Summers, M. E. and R. R. Conway, Insights into Middle Atmospheric Hydrogen Chemistry from Analysis of MAHRSI Observations of Hydroxyl (OH), this volume.

Torr, D.G. et al., Measurements of OH $(X^2\Pi)$ in the stratosphere by high resolution UV spectroscopy, *Geophys. Res. Lett., 14*, 937-940, 1987.

Wennberg, P.O. et al., Simultaneous, in situ measurements of OH, $HO_2$, $O_3$, and $H_2O$: A test of modeled stratospheric $HO_x$ chemistry, *Geophys. Res. Lett., 17*, 1909-1912, 1990.

Wennberg, P.O. et al., In situ measurements of OH and $HO_2$ in the upper troposphere and stratosphere, *J. of the Atmos. Sci., 52*, 3413-3420, 1995.

Yue G.K., et al. Stratospheric aerosol acidity, density, and refractive index deduced from SAGE II and NMC temperature data, *J. Geophys. Res., 99*, 3727-3738, 1994.

T. Canty and K. Minschwaner, Dept. of Physics, New Mexico Institute of Mining and Technology, Socorro, NM 87801. (e-mail tcanty@nmt.edu)

K.W. Jucks, Harvard-Smithsonian Center for Astrophysics, 60 Garden Street, Cambridge, MA. 02138

A.K. Smith, Atmospheric Chemistry Division, NCAR, Boulder, CO. 80307

# Energetic Electrons and their Effects on Upper Stratospheric and Mesospheric Ozone in May 1992

W. Dean Pesnell[1,2], Richard A. Goldberg[3], D. L. Chenette[4], E. E. Gaines[4], and Charles H. Jackman[5]

The increased fluxes of precipitating energetic electrons ($E > 1$ MeV) during highly relativistic electron events (HREs) produce ion concentrations in the upper stratosphere and lower mesosphere that exceed the background concentrations. Coupled ion-neutral chemistry models predict that this increased ionization should drive $HO_x$ reactions and deplete mesospheric ozone by up to roughly 25%. As HREs become more intense and frequent during the declining phase of the solar cycle, it was also predicted that mesospheric ozone would show a solar cycle modulation as a result of these events. To calibrate the effect HREs have on mesospheric ozone, we have studied the May 1992 HRE with several instruments on the UARS. Electron fluxes measured with HEPS give the duration and spatial coverage of the HRE. Ozone data from MLS, CLAES, and HRDI were examined for the chemical signature of the HRE, ozone depletions within the magnetic $L$-shell limits of $3 \leq L < 4$. Using the multiple viewing angles of HRDI, we can compare mesospheric ozone at similar local solar times before, during, and after the HRE. This removes some of the ambiguity caused by progressive sampling of the diurnal cycle over a yaw cycle of the satellite. Although we analyzed one of the most intense HREs in the UARS database, we did not find HRE-induced changes in the ozone mixing ratio between altitudes of 55–75 km. Detecting a long-term trend in the ozone driven by precipitating electrons appears to require a substantial increase in the signal-to-noise ratio of the satellite measurements.

## 1. INTRODUCTION

Signatures of solar activity in the terrestrial atmosphere are often simple correlations of some climate measurement with sunspot number or solar radiative output at some particular wavelength. One highly publicized example is the correlation of the length of the sunspot cycle with the increase in average Northern Hemisphere temperature over the last century [*Friis-Christensen and Lassen*, 1991]. Although climate changes are correlated with solar activity, the radiative output of the Sun does not vary with an amplitude sufficient to cause the observed temperature increase.

Fluxes of charged particles emitted by the Sun as the solar wind are also modulated by the solar cycle and could affect the Earth's climate. When these particles enter the Earth's atmosphere they are slowed and stopped by succes-

Atmospheric Science Across the Stratopause
Geophysical Monograph 123
Copyright 2000 by the American Geophysical Union

[1] Nomad Research, Inc., Bowie, Maryland.

[2] Under contract at the Goddard Space Flight Center

[3] Laboratory for Extraterrestrial Physics, NASA Goddard Space Flight Center, Greenbelt, Maryland.

[4] Space Physics Department, Lockheed Martin Advanced Technology Division, Palo Alto, California.

[5] Laboratory for Atmospheres, NASA Goddard Space Flight Center, Greenbelt, Maryland.

sive ionizations of atmospheric gases. The produced ions can alter the chemical composition of the atmosphere by changing the concentration of compounds that serve as catalysts in photo-chemical reactions. Solar proton events are one example of a particle influence on the stratosphere. Energetic protons are able to reach the stratosphere and destroy ozone by enhancing $NO_x$-catalyzed reactions [e.g., *Thomas et al.*, 1983; *McPeters and Jackman*, 1985; *Jackman et al.*, 1990].

More frequent and longer lasting than solar proton events are increases in the flux of electrons with energies in excess of 1 MeV, termed highly-relativistic electron events (HREs) [*Baker et al.*, 1993]. Solar wind electrons are injected into the magnetosphere and then accelerated to these high energies as they diffuse inward toward the Earth. As a result, the highest fluxes are observed at the inner edge of the outer radiation belt, near a magnetic $L$-shell coordinate value of $L = 3$. While solar protons can reach and influence the stratosphere, energetic electrons can penetrate only to the mesosphere and upper stratosphere before depositing their energy. Photochemical models have shown that the enhanced ionization during an HRE should deplete ozone in the mesosphere by changing the concentration of $HO_x$ compounds [*Goldberg et al.*, 1994; 1995a,b; *Aikin and Smith*, 1999]. For the range of electrons in air used in *Goldberg et al.* [1984] electrons with $0.1 \leq E \leq 1$ MeV will produce $HO_x$ radicals between altitudes of about 60 and 80 km and could affect the ozone balance in that region. Unlike the burst-like, geographically-limited nature of low energy electrons linked to auroral activity, electrons with $E > 0.1$ MeV have drift periods less than 0.1 days [*Walt*, 1994]. This means that the HRE electron flux is independent of magnetic local time and is roughly constant along a magnetic $L$-shell [*Vampola and Gorney*, 1983]. Thus, not only are HREs long lasting, they affect a large geographic area while they are active. A typical flux value affecting this region can be derived by averaging all measured precipitating fluxes within a band of $L$.

We describe here an attempt to detect these effects during what we consider to be an optimum energetic electron event, the highly relativistic electron event of May 1992. We first discuss the variation of the energy input into the mesosphere due to energetic electrons as measured by PEM/HEPS. Next, we outline our expectations of how that input would change the ozone in the stratosphere and mesosphere, and finally, the MLS, CLAES, and HRDI measurements that test these expectations.

## 2. HIGHLY RELATIVISTIC ELECTRON MEASUREMENTS FROM UARS

The Particle Environment Monitor (PEM) High Energy Particle Spectrometer (HEPS) instrument on the Upper Atmosphere Research Satellite (UARS) provides pitch-angle-

and energy-resolved measurements of the low-altitude electron flux between 30 keV and 5 MeV [*Winningham et al.*, 1993]. For this analysis the pitch-angle resolution in the measurements is used to distinguish those electron fluxes that are in the bounce loss cone and will directly precipitate into the atmosphere from those that are stably trapped (i.e., they have estimated mirror altitudes above 100 km).

From surveys performed by the PEM/HEPS team, the highly relativistic electron fluxes of May 1992 were the most intense, most energetic, and longest lived of the events seen by UARS from the launch of UARS through the end of 1993 [*Gaines et al.*, 1995]. Within the belt of magnetic $L$-shell $3 \leq L < 4$ (or magnetic latitudes between 55° and 60°), the locally precipitating flux of electrons with both $E > 100$ keV and $E > 1$ MeV reached a large value on May 11 and continued at this value until May 21 [*Pesnell et al.*, 2000]. It then decreased by a factor of about eight in magnitude and continued through May 27. Compared to background effects for magnetically undisturbed times and locations, the ion production rate in the lower mesosphere due to electrons increased at least 100 times during this HRE. Other events measured by HEPS were shorter, less intense, or less energetic.

## 3. PREDICTED ION PRODUCTION RATE AND LOSS OF OZONE DUE TO ELECTRON PRECIPITATION

From the measured precipitating electron energy spectra, we have derived ion-pair production rate ($Q$) height profiles using the techniques described in *Goldberg et al.* [1984]. Figure 1a shows the ion-pair production rate profiles for day-averaged precipitating fluxes in two bands of $L = \bar{L} \pm 1/8$ and for one large burst observed on May 18, 1992. From these profiles it is apparent that large quantities of ions are produced down to 50 km, far deeper than produced during other kinds of electron precipitation events.

Ions produced during an HRE can, through complicated ion chemistry, enhance $HO_x$ compounds and influence the ozone concentration during the particle event [*Solomon et al.*, 1981; 1983]. Approximately two $HO_x$ species are produced from each ion pair up to about 70 km. However, above 70 km, the production of $HO_x$ species from each ion pair has a strong dependence on the ionization rate and the duration of the particle precipitation event [*Solomon et al.*, 1981]. Our model calculations use the production of $HO_x$ constituents per ion pair as a function of altitude from Figure 2 of *Solomon et al.* [1981] for moderate ionization rates. This accounts for the rapid dropoff in ion-related $HO_x$ production with increasing altitude and is within 5% of their rates for the ion production rates in Figure 1a.

We used an assumed continuous source of HRE-related $HO_x$ production from the ion-pair production rate curves of

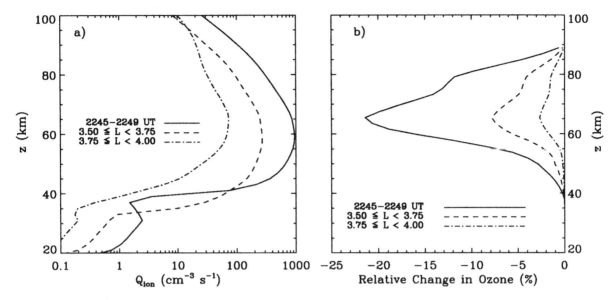

**Figure 1.** On left (a), ion-production rates for the day-averaged electron fluxes of May 18, 1992 for two ranges of $L$ and for a large localized precipitating burst of electron flux that was measured between 2245 and 2249 UT near $L = 3.75$. In (b), the modeled ozone depletion for those ion production rates.

Figure 1a in the GSFC-2D chemistry and transport model of *Jackman et al.* [1990; 1996] at 65°N to predict the ozone change. Our simulated ozone percentages change due to the HREs were computed by comparing a model run that includes the electron-induced ions calculated with the May 18 HRE fluxes shown in Figure 1a with another model run without the HRE flux are presented in Figure 1b. When ion-production rates caused by measured instantaneous electron fluxes are used in both steady-state and time-dependent (*Aikin and Smith* [1999]) photochemical models, ozone depletions up to about 20% are predicted at mesospheric altitudes. Although the ozone data has considerable noise, a depletion of 20% should be observable in the data. We have also displayed the ion-pair production rates and relative ozone depletions of the more conservative daily averaged electron fluxes, which should be a reliable indicator of the impact of the HRE on the mesosphere. The peak computed ozone decrease due to the day-averaged HRE precipitation is about 7.5% at $\sim 65$ km. In addition, the time-dependent model shows that the response of mesospheric ozone to the $HO_x$ radicals produced by the electron flux is greatest in the mid-morning portion of the diurnal cycle.

## 4. OZONE MEASUREMENTS FROM UARS

UARS ozone measurements should be able to quantify the ozone change and determine if the May 1992 HRE was responsible for a significant ozone variation. Although large bursts of electrons are seen in the HEPS data, we were unable to find a corresponding 20% depletion of ozone in the composition measurements. We then searched for reductions in the ozone mixing ratio that would correspond to the day-averaged HRE electron flux. This requires comparing ozone measurements taken within the affected $L$-shells but across times when the electron flux changes rapidly. Due to the way UARS samples the diurnal cycle of material in the mesosphere and upper stratosphere, we must also compare ozone measurements made at similar local solar times as well.

To resolve the diurnal cycle of ozone, we use the precessing orbit of UARS to scan the instruments 24 hours in local solar time (LST). While the cryogenic limb array etalon spectrometer (CLAES) and microwave limb sounder (MLS) measure thermal radiation and can see the entire diurnal cycle, the high resolution doppler imager (HRDI) measures scattered sunlight, and no data is returned from any branch when the sun is absent. At the beginning of a yaw cycle, ozone measurements are made at two values of LST, one in the A.M. branch near noon and one in the P.M. branch near midnight. As the yaw cycle progresses both branches move to earlier values of LST, reaching near midnight and noon, respectively, just before the turning of the satellite that starts the next yaw cycle. Complete coverage of the available portion of the diurnal cycle is available for a given latitude band by combining observations of the A.M. and P.M. branches observed over a yaw cycle. The cold side of UARS was pointed north during May 1992, so all of the ozone measurements are in the northern polar region throughout the event.

Altitude profiles of CLAES and MLS measurements were obtained from the UARS Central Data Handling Facility at the Goddard Space Flight Center. Altitude profiles of HRDI measurements were supplied as 3AT files by the HRDI Sci-

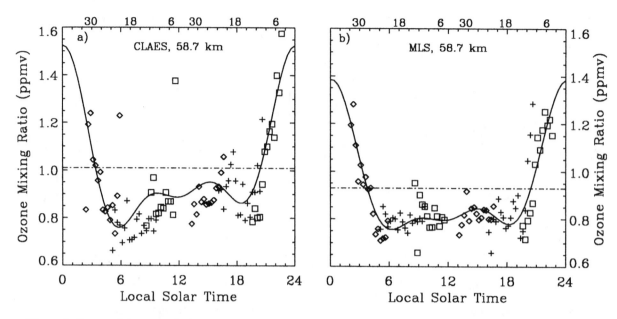

**Figure 2.** Ozone mixing ratios in the upper stratosphere measured by CLAES (a) and MLS (b) between May 6 and May 30, 1992 with $3 \leq L < 4$ are shown. Data taken during the HRE (5/11-5/21) are labeled with a '+' sign, data after 5/21 are labeled with a '◇' and before 5/11 with a '□'. Numbers across the top of each graph show the approximate day in May 1992 when that LST was measured. Solid lines are a low-order Fourier fit to the data and the dot-dashed lines are the average ozone values derived from that fit.

ence Team at the University of Michigan. Each 3AT file contains a day's worth of data sampled at intervals of 65.536 s. Profiles of measured quantities are given on fixed altitude grids. Data from May 1992 with geographic locations within the magnetic $L$-shell band $3 \leq L < 4$, where significant long-lived fluxes of precipitating electrons are seen, were extracted from the data files and analyzed using algorithms described in *Pesnell et al.* [1999; 2000].

In order to average the data, without further mixing undisturbed ozone values with those exposed to the HRE, which would further reduce the strength and significance of the depletion signal, we averaged ozone values that were measured before May 11, after May 21, and from May 11 to May 21 (when the HRE flux is present) within 15 m (CLAES and MLS) or 30 m (HRDI) LST bins. When plotted against LST, we would expect the ozone values measured during the HRE would show a downward shift compared to the ozone when the HRE is not in progress. Ozone mixing ratios from CLAES and MLS are shown as a function of LST in Figures 2a and 2b, where ozone mixing ratio data at 59 km altitude for May 6–30, 1992, with geographic locations in the range $3 \leq L < 4$, are shown. As can be seen in Figures 2a and 2b, there is no consistent downward shift of the ozone mixing ratio in the upper stratosphere when the HRE flux is present. Since these measurements are on the bottomside of a steeply sloped functions (see Figure 1b), measurements in the altitude region of the predicted maximum ozone depletion near 65 km would be a useful extension of the CLAES

and MLS data. Such data is available from the HRDI instrument.

An advantage of the multiple viewing angles of HRDI is the overlapping coverage of the diurnal cycle. As can be seen in Figure 3a, there are four bands of LST values where ozone is measured in two of the three conditions: before, during, or after the May 1992 HRE. We show the LST variation at one altitude and then the altitude variation within the overlapping measurements in the mid-morning. The diurnal cycle in averaged values as observed by HRDI at 69 km agrees with the time-dependent model of *Aikin and Smith* [1999], with a maximum near noon and a minimum near 1500 LST (Figure 3a). In Figure 3a the error bars are drawn to illustrate one standard deviation. As with the MLS and CLAES data, no downward shift is seen in the ozone data taken with the HRE flux compared to that without.

In Figure 3b we compare mid-morning observations (0900 < LST ≤ 1000), where the time-dependent chemistry model of *Aikin and Smith* [1999] predicts the greatest sensitivity to the electrons. At this point in the diurnal cycle the $HO_x$ compounds have increased to a sufficient quantity to have an appreciable effect on the ozone. The profiles in Figure 3b compare data taken across the onset of the HRE, when the precipitating fluxes increased by a factor of 100 at all energies in a short period of time [*Gaines et al.*, 1995; *Pesnell et al.*, 2000]. With the rapid increase in electron flux and the calculated sensitivity to that flux, this is the best opportunity for seeing the predicted effects. However, no consistent de-

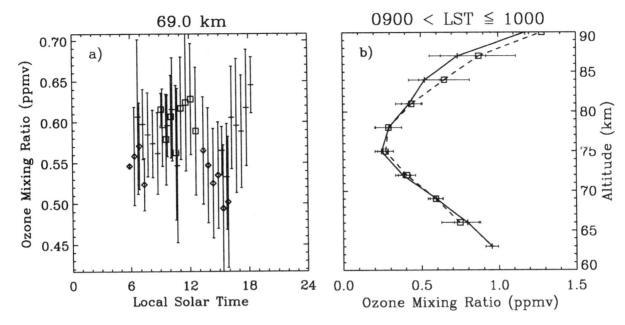

**Figure 3.** a) Ozone data at 69 km measured by HRDI between May 6 and May 30, 1992, with $3 \leq L < 4$, and averaged over 15 m in LST are shown. Data taken during the HRE (5/11-5/21) are labeled with a '+' sign, data after 5/21 are labeled with a '◇' and before 5/11 with a '□'. In (b) are averages of altitude profiles taken between 0900 and 1000 LST for before ('□', connected with a dashed line) and during ('+', connected with a solid line) the HRE.

creases in the ozone altitude profiles are seen when the measured values between 66 and 81 km from before and during the HRE are compared (Figure 3b).

## 5. CONCLUSIONS

According to the two models used to predict ozone depletions, it appears possible that highly relativistic electron events should cause measurable, localized depletions of ozone in the upper stratosphere and mesosphere. Furthermore, the chemical models predict that the depletion amplitude should peak in the altitude region we have examined. Ozone data from MLS, CLAES, and HRDI have been examined for evidence of ozone depletions due to the precipitation of relativistic electrons in May 1992. Ozone data between altitudes of 59 and 81 km were examined to determine if the predicted depletions were visible in UARS measurements. Similar relative depletions should have been seen when comparing ozone measurements made at those LSTs that were observed both with and without the HRE irradiation.

Although we have analyzed one of the most intense and longest lasting HRE event seen during the prime mission period of UARS, we found no measurable effect on ozone in the mesosphere (this work and *Pesnell et al.*, 2000]) or the upper stratosphere [*Pesnell et al.*, 1999]. Without a consistent, measurable effect, these results cast doubt on the long-term significance of the effect that highly relativistic electron events have on the overall ozone budget in the mesosphere.

*Acknowledgments.* This work was supported, in part, by NASA Grants and Purchase Orders. David P. Sletten assisted in the analysis and display of the UARS data.

## REFERENCES

Aikin, A. C., and H. J. P. Smith, Mesospheric constituent variations during electron precipitation events, *J. Geophys. Res.*, *104*, 26,457–26,471, 1999.

Baker, D. N., R. A. Goldberg, F. A. Herrero, J. B. Blake, and L. B. Callis, Satellite and rocket studies of relativistic electrons and their influence on the middle atmosphere, *J. Atmos. Terr. Phys.*, *54*, 1619–1628, 1993.

Friis-Christensen, E., and K. Lassen, Length of the solar cycle: An indicator of solar activity closely associated with climate, *Science*, *254*, 698–700, 1991.

Gaines, E. E., D. L. Chenette, W. L. Imhof, C. H. Jackman, and J. D. Winningham, Relativistic electron fluxes in May 1992 and their effect on the middle atmosphere, *J. Geophys. Res.*, *100*, 1027–1033, 1995.

Goldberg, R. A., C. H. Jackman, J. R. Barcus, and F. Sørass, Nighttime auroral energy deposition in the middle atmosphere, *J. Geophys. Res.*, *89*, 5581–5590, 1984.

Goldberg, R. A., C. H. Jackman, D. N. Baker, and F. A. Herrero, Changes in the concentration of mesospheric $O_3$ and OH during a highly relativistic electron precipitation event, *The Upper Mesosphere and Lower Thermosphere: A Review of Experiment and Theory, Geophysical Monograph 87* (Washington, DC: AGU), 215–223, 1995a.

Goldberg, R. A., D. N. Baker, F. A. Herrero, C. H. Jackman, S.

Kanekal, and P. A. Twigg, Mesospheric heating during highly relativistic electron precipitation events, *J. Geomag. Geoelectr.*, *47*, 1237–1247, 1995b.

Goldberg, R. A., D. N. Baker, F. A. Herrero, S. P. McCarthy, P. A. Twigg, C. L. Croskey, and L. C. Hale, Energy deposition and middle atmosphere electrodynamic response to a highly relativistic electron precipitation event, *J. Geophys. Res.*, *99*, 21,071–21,081, 1994.

Jackman, C. H., A. R. Douglass, R. B. Rood, R. D. McPeters, and P. E. Meade, Effects of solar proton events on the middle atmosphere during the past two solar cycles as computed using a two-dimensional model, *J. Geophys. Res.*, *95*, 7417–7428, 1990.

Jackman, C. H., E. L. Fleming, S. Chandra, D. B. Considine, and J. E. Rosenfield, Past, present, and future modeled ozone trends with comparisons to observed trends, *J. Geophys. Res.*, *101*, 28,753–28,767, 1996.

McPeters, R. D., and C. H. Jackman, The response of ozone to solar proton events during solar cycle 21: The observations, *J. Geophys. Res.*, *90*, 7945–7954, 1985.

Pesnell, W. D., R. A. Goldberg, C. H. Jackman, D. L. Chenette, and E. E. Gaines, A search of UARS data for ozone depletions caused by the highly relativistic electron precipitation events of May 1992, *J. Geophys. Res.*, *104*, 165–175, 1999.

Pesnell, W. D., R. A. Goldberg, C. H. Jackman, D. L. Chenette, and E. E. Gaines, Variation of mesospheric ozone during the highly relativistic electron event in May 1992 as measured by the HRDI instrument on UARS, *J. Geophys. Res.*, submitted, 2000.

Solomon, S., D. W. Rusch, J.-C. Gerard, G. C. Reid, and P. J. Crutzen, The effect of particle precipitation events on the neutral and ion chemistry of the middle atmosphere, 2, Odd hydrogen, *Planet. Space Sci.*, *29*, 885–892, 1981.

Solomon, S., G. C. Reid, D. W. Rusch, and R. J. Thomas, Mesospheric ozone depletion during the solar proton event of July 13, 1982, 2, Comparison between theory and measurements, *Geophys. Res. Lett.*, *10*, 257–260, 1983.

Thomas, R. J., C. A. Barth, G. J. Rottman, D. W. Rusch, G. H. Mount, G. M. Laurence, R. W. Sanders, G. E. Thomas, and L. E. Clemens, Mesospheric ozone depletion during the solar proton event of July 13, 1982, 1, Measurement, *Geophys. Res. Lett.*, *10*, 253–255, 1983.

Vampola, A. L., and D. J. Gorney, Electron energy deposition in the middle atmosphere, *J. Geophys. Res.*, *88*, 6267–6274, 1983.

Walt, M. *Introduction to Geomagnetically Trapped Radiation*, New York, Cambridge University Press, pp. 44–50, 1994.

Winningham, J. D., *et al.*, The UARS Particle Environment Monitor, *J. Geophys. Res.*, *98*, 10,649–10,666, 1993.

C. H. Jackman, Laboratory for Atmospheres, Code 916, NASA/Goddard Space Flight Center, Greenbelt, MD 20771 (e-mail: Charles.Jackman@gsfc.nasa.gov).

D. L. Chenette and E. E. Gaines, Space Physics Department, Lockheed Martin Advanced Technology Center, O/L9-42, Building 255, 3251 Hanover Street, Palo Alto, CA 94304 (e-mail: Chenette@spasci.com; Gaines@spasci.com).

R. A. Goldberg, Laboratory for Extraterrestrial Physics, Code 690, NASA/Goddard Space Flight Center, Greenbelt, MD 20771 (e-mail: Richard.A.Goldberg.2@gsfc.nasa.gov).

W. D. Pesnell, Nomad Research, Inc., 2804 Nomad Court, Bowie, MD 20716 (e-mail: Pesnell@NomadResearch.com).

# Turbulence Dynamics and Mixing due to Gravity Waves in the Lower and Middle Atmosphere

David C. Fritts and Joseph A. Werne

*Colorado Research Associates/a division of Northwest Research Associates*

Internal gravity waves contribute to turbulence generation and mixing throughout the atmosphere due to instability accompanying amplitude growth with altitude and/or reduction of intrinsic phase speeds in shear flows. The instability processes accounting for turbulence generation depend on wave and mean flow structures, but in general exhibit an altitude dependence due to increasing wave scales with increasing altitude. Near the tropopause and away from major sources, characteristic vertical scales are a few km, vertical group velocities are small, and the dominant instability appears to be a Kelvin-Helmholtz (KH) shear instability due to inertia-gravity wave motions. At greater altitudes, characteristic vertical scales increase, causing larger vertical group velocities, larger wave energy fluxes, and a preference for faster convective instabilities of the motion field. This paper will review both the reasons for a change in instability character with altitude and the dynamics accompanying the transition to turbulence via convective wave breaking and shear instability of the motion field. We will also describe the different implications of these wave field instabilities for turbulent mixing and transport of momentum, heat, and constituents.

## 1. INTRODUCTION

Internal gravity waves are ubiquitous in the atmosphere, arise from a variety of sources, and have a number of important effects. Dominant sources at lower altitudes include convection, orography, and wind shear [Fritts and Nastrom, 1992]. Among the more significant effects are forcing of the mean and large-scale circulation and thermal structure, large-scale transport of heat, momentum, and constituents, and turbulent mixing and diffusion of these quantities. The first two topics are the subjects of other papers in this volume by Holton and Alexander [2000] and Garcia [2000]. The

dynamics underlying the latter is the subject of the present paper.

Gravity wave effects vary with altitude for a number of reasons. Most important is the growth of wave amplitudes due to decreasing density with increasing altitude. This causes increases of wave fluxes (per unit mass) with increasing altitude which force the large-scale circulation, structure, and transport. Larger wave amplitudes also increase the tendency for wave instability, leading to higher turbulence intensities and mixing with increasing altitude. The sensitivity of gravity waves to the environments through which they propagate leads to a complex interplay between transport and mixing processes. While the role of gravity waves in large-scale transport processes is becoming better understood, our knowledge of competitive turbulent mixing and diffusion processes is not as advanced. Importantly, such an understanding can only be achieved through detailed

Atmospheric Science Across the Stratopause
Geophysical Monograph 123

assessments of the dynamics of turbulence transitions and the turbulence cascade.

Instability processes accompanying gravity wave propagation in the atmosphere have been studied for many years using a variety of methods. As a result, we now understand in general terms the mechanisms accounting for wave instability and turbulence generation and the factors that govern their relative importance with altitude. The two primary mechanisms for turbulence production are Kelvin-Helmholtz (KH) shear instability and wave breaking. Shear instability is typically more prevalent for lower intrinsic frequencies, and wave breaking (via convective instability) is more prevalent at higher intrinsic frequencies [Fritts, 1984, 1989; Dunkerton, 1984, 1989, 1997; Fritts and Rastogi, 1985; Fritts and Yuan, 1989]. Other instabilities also occur at small and large wave amplitudes, primarily resonant wave-wave interactions [Yeh and Liu, 1981; Thorpe, 1994; Sonmor and Klaassen, 1997]. However, these appear not to contribute directly to the generation of turbulence and mixing and will not be discussed further here.

Despite our earlier understanding of the mechanisms leading to turbulence generation, the details of the instability and turbulence dynamics and their implications for transport and mixing have remained beyond our grasp until recently. Laboratory, theoretical, and numerical studies have made important contributions to our understanding of the initial transition to turbulence and three-dimensional (3D) motions in stratified and/or sheared flows (see Fritts et al., [1998], for a review). Until recently, however, an understanding of the transition to turbulence, and of the dynamics within turbulent flows, has been beyond our grasp. This has changed with the recent availability of very large computational resources. Current simulations now achieve spatial resolutions as high as $\sim 1000^3$, span more than a decade of scales within the "inertial range" of turbulence (i.e., turbulence dynamics for which both density stratification and viscosity are negligible), and promise to provide, with complete analysis, continuing insights into the nature of turbulence dynamics and mixing.

Our objectives in this paper are 1) to argue for a different statistical role of the dominant gravity wave instability processes with altitude and 2) to outline the vorticity dynamics accompanying these instabilities and their implications for turbulent mixing. The first objective is addressed using linear theory and a simple spectral model that conforms closely to observations in Sections 2 and 3. The second objective comprises two parts. We first review what has been learned about the vortex dynamics of turbulence from detailed simu-

lations and analysis of wave breaking in Section 4; we then describe briefly in Section 5 the vortex dynamics accompanying a high-resolution simulation of KH instability employing an incompressible model and comment on the apparent similarity of these two flows. Where appropriate, we also examine the implications of these turbulence sources for mixing, large-scale modeling, and competition with wave-driven transport processes. Our conclusions are presented in Section 6.

## 2. LINEAR THEORY

### 2.1. Dispersion and Polarization Relations

While gravity waves are subject to instabilities at large and small amplitudes [e.g., Müller et al., 1986; Sonmor and Klaassen, 1997], it appears to be only local, as opposed to global, instabilities that lead directly to turbulence. Fortunately, linear theory provides an adequate description of the tendency for such local instability. It is sufficient to assume for our purposes that variations in the mean wind and temperature fields are sufficiently slow on wave scales that wave propagation is described adequately by the WKB approximation [Gill, 1982]. Also assuming an inviscid fluid of scale height $H$ and steady motions of the form $e^{ik(x-ct)}$ (taking wave propagation to be in the $x$ direction without loss of generality), where $\vec{k} = (k, m)$ is the wavenumber vector, we define $\omega = k(c - \bar{u})$ as the intrinsic frequency, with $c$ and $\bar{u}$ the horizontal phase speed and mean wind along the direction of wave propagation. Then the dispersion and polarization relations may be written [Fritts, 1984]

$$m^2 = \frac{k^2(N^2 - \omega^2)}{(\omega^2 - f^2)} - \frac{1}{4H^2}, \qquad (1)$$

$$w' = \frac{-ku'}{m}, \quad v' = \frac{-ifu'}{\omega}, \quad \text{and} \quad \frac{\theta'}{\bar{\theta}_z} = \frac{-iw'}{\omega}, \qquad (2)$$

or in terms of $u'$ and the intrinsic phase speed,

$$\frac{\theta'_z}{\bar{\theta}_z} = \frac{-u'}{(c - \bar{u})} = -a, \qquad (3)$$

where $(u, v, w)$ are component velocities, $\theta$ is potential temperature, $N$ is the buoyancy frequency, $f = 2\Omega \sin \phi$ is the inertial frequency, $\Omega$ is the earth's rotation rate, $\phi$ is latitude, primes and subscripts denote perturbation quantities and derivatives, and $u'$ varies approximately as [Bender and Orszag, 1978]

$$u' \sim m^{1/2} e^{i \int m dz} \qquad (4)$$

in response to gradual variations of $N$ and $\bar{u}$ with altitude. Linear gravity wave dynamics and propagation are described in greater detail by Holton and Alexander [2000] and will not be repeated here.

### 2.2. Gravity Wave Amplitudes and Instability

Equation (3) provides the most intuitive measure of wave amplitude relative to that required for nominal convective instability within the wave field. A gravity wave is convectively stable when $a < 1$ and convectively unstable when $a > 1$, assuming that instability depends only on vertical motions and gradients.

Alternatively, we may ask under what circumstances the wave field is dynamically, as opposed to convectively, unstable. The possibility of a dynamical, i.e., Richardson number based, instability at a lower wave amplitude rather than a corresponding convective instability relies on rotation (nonzero $f$) or additional mean shear. Assuming no mean shear, the nominal amplitude required for a dynamical instability threshold at a Richardson number, $Ri = 1/4$, is [Dunkerton, 1984; Fritts and Rastogi, 1985]

$$a_d = \frac{2(1 - f^2/\omega^2)^{1/2}}{1 + (1 - f^2/\omega^2)^{1/2}}. \tag{5}$$

For $\omega^2 >> f^2$, $a_d \simeq 1$ and the conditions for convective and dynamical instability are essentially identical. However, $a_d$ decreases to zero as $\omega \to f$, reflecting the tendency for potential temperature perturbations to be suppressed relative to perturbation wind shears for lower-frequency wave motions. The tendency for dynamical instability of inertia-gravity waves was assessed by Fritts and Yuan [1989] using linear stability analysis; more recently, LeLong and Dunkerton [1998] evaluated the threshold for dynamical instability numerically. Amplitudes somewhat larger than $a_d$ were found to be necessary to support instability due to propagation of the wave motion. These studies were extended to more general linear instabilities by Dunkerton [1997], while a theoretical link between the instabilities at large wave amplitudes and the resonant triad interactions identified at smaller amplitudes in oceanic and atmospheric contexts was established by Sonmor and Klaassen [1997].

### 2.3. Gravity Wave Energy Density and Flux

To assess the importance of the dominant wave instabilities with altitude, we need to estimate the rate of wave energy dissipation relative to wave energy density with altitude. The ratio of these quantities provides an estimate of the time scale for wave dissipation which

can be compared with the wave period. Presumably, if the wave energy dissipation rate is large, rapid instabilities are favored; if, on the other hand, the wave energy dissipation rate is small, then instabilities occurring at smaller wave amplitudes have a greater potential to impose amplitude constraints.

Defining the total wave energy (per unit mass) to be $E$, the wave energy flux (per unit mass) for a monochromatic gravity wave is given by

$$F_E = c_{gz}E, \tag{6}$$

where $c_{gz}$ is the vertical group velocity, given by

$$c_{gz} = -\frac{\omega}{m}\left(1 - \frac{\omega^2}{N^2} - \frac{f^2}{\omega^2}\right), \tag{7}$$

and where we have neglected terms varying as $1/4H^2$ and $f^2/N^2$.

Following Fritts and VanZandt [1993], we then assume that all of the energy removed from the wave field (whether monochromatic or a broad spectrum) is dissipated and equate the energy dissipation rate, $\epsilon$, to the divergence of wave energy flux (per unit mass)

$$\epsilon = -\frac{1}{\rho}\frac{\partial}{\partial z}[\rho F_E]. \tag{8}$$

Importantly, the altitude dependencies of $E$ and $\epsilon$ vary depending on whether we assume inviscid monochromatic propagation or adhere to the observed statistical description of the wave spectrum. Given our statistical approach to turbulence generation and effects, we will employ the latter in our estimates of wave scales and growth with altitude. However, we note that individual wave packets often account for wave instability and thus consider single-wave instability dynamics in our discussion of turbulence transitions.

### 2.4. The Inertial Range of Turbulence

Turbulence arising due to wave instability processes has often been assumed to have an "inertial range" character at sufficiently small scales. These scales are necessarily smaller than the scales of the waves within which instabilities and turbulence arise. They are also assumed smaller than the scale above which buoyancy plays a strong role, termed the "buoyancy" or "outer" scale [Dougherty, 1962; Ozmidov, 1965], and estimated by $L_B \sim 10(\epsilon/N^3)^{1/2}$ [Weinstock, 1978, 1981]. At smaller scales, the inertial range of turbulence is constrained by viscosity, with a corresponding "inner" scale given by $\ell_0 \sim C\ell_K = C(\nu^3/\epsilon)^{1/4}$, where $\ell_K = (\nu^3/\epsilon)^{1/4}$ is the Kolmogorov scale [Kolmogorov, 1941], $\nu$ is the kinematic viscosity, and $C \sim 7$ to $10$ is a constant which depends on the model employed for the viscous range at scales dominated by dissipation [Heisenberg, 1948; Tatarskii, 1971]. Between these limits ($L_B$ and $\ell_o$), theory and some experiments suggest a 1D turbulence energy spectrum of the form $E(k) \sim k^{-5/3}$. Thus the spatial extent of the inertial range depends on $N$, $\nu$ and $\epsilon$: $L_B/\ell_o = (10/C)(\epsilon/\nu N^2)^{3/4}$. The combination $Re_b = \epsilon/\nu N^2$ is the buoyancy Reynolds number [e.g., Smyth, 1999] In the lower stratosphere, where $\epsilon$ and $\nu$ are small (see below), $\ell_0 \sim 10^{-1}$ m, $L_B \sim 50$ m, and $Re_b \sim 4000$. In the upper mesosphere, where $\epsilon$ and $\nu$ are much larger, $\ell_0 \sim 20$ m, $L_B \sim 300$ m [Lübken, 1997], and $Re_b \sim 40$.

As a result, the anticipated inertial range is broad in the lower stratosphere, spanning $\sim 2$ to 3 decades of scales, while it is much more restricted at higher altitudes, extending $\sim 1$ decade near the mesopause. In all cases, the turbulent mixing and diffusion occurs at scales that are far smaller than the scales of the gravity wave motions themselves. As such, simulations which address turbulent mixing must achieve far higher resolution than required to assess other effects of gravity wave energy and momentum fluxes and wave dissipation.

## 3. VARIATION OF WAVE PARAMETERS WITH ALTITUDE

Given extensive observations over the last few decades, we now have a reasonable understanding of the characteristic scales and amplitudes of gravity waves throughout the lower and middle atmosphere. We also have a fair understanding of the distribution of mean wave energy with observed frequency, but more limited knowledge of intrinsic properties of the wave field. This understanding and a canonical model of the gravity wave spectral variations with altitude are reviewed below. This approach does not imply that there is always a broad spectrum of gravity waves present. Indeed, the wave spectrum is more likely composed of a large number of individual wave packets which may or may not be superposed. The spectral description nevertheless provides a more accurate view of the variation of wave amplitudes and scales with increasing altitude than linear inviscid theory for monochromatic waves.

### 3.1. Observations

Observations or inferences of gravity wave parameters over the last few decades are much too numerous to review here. Thus, we will summarize only the general spectral characteristics of the wave field, its evolution with altitude, and recent evidence for specific wave in-

stability processes at various altitudes. A reader interested in a survey of the earlier observational literature is referred to the reviews by Fritts [1984, 1989].

Frequency spectra of horizontal (or radial) velocity and temperature are defined primarily with radar and lidar data, typically vary as $E(\omega) \sim \omega^{-p}$, and exhibit slopes varying from $p \sim 1$ to 2 for frequencies between $f$ and $N$, with most slopes falling closer to 5/3 [see, e.g., Balsley and Carter, 1982]. These slopes appear in most cases to vary little with altitude, though there are cases in which such variations are observed. Importantly, this implies that there is little or no tendency for filtering or instability processes to alter the distribution of wave frequencies with altitude. At greater altitudes, various tidal structures contribute spectral peaks often extending well above the mean gravity-wave spectral form. Because horizontal wind and temperature variance tend to be concentrated near inertial frequencies, Doppler effects are small and the intrinsic and Doppler-shifted frequency spectra will have nearly the same structures [Fritts and VanZandt, 1987].

Frequency spectra of vertical velocity, on the other hand, exhibit considerable variability with altitude, primarily because these spectra are highly sensitive to Doppler-shifting effects. Assuming that the majority of spectral variance is due to gravity waves, we expect from the dispersion relation that intrinsic frequency spectra of vertial velocities will exhibit a slope of approximately $2-p$, thus having a peak at frequencies just below $N$ for a characteristic value of $p \sim 5/3$. Thus, vertical velocity variance is concentrated near $N$ and is easily Doppler shifted to both higher and lower apparent frequencies, making these spectra perhaps the most sensitive measure of Doppler-shifting effects.

Vertical wavenumber spectra have been measured with many instrument types. Multiple data sets suggest characteristic scales varying from $\sim 2$ to 5 km in the lower stratosphere to $\sim 10$ to 30 km near the mesopause [see Fritts, 1984, 1989]. Spectral slope estimates at higher $m$ also vary, but typical values range from -2.5 to -3 [Dewan et al., 1984]. Importantly, while there are departures from universal amplitudes and slopes, particularly under strongly sheared conditions [Eckermann, 1995], there is remarkable conformity of spectral amplitudes and slopes at high $m$ with the expectations of the various saturation theories that have been offered to date [Dewan and Good, 1986; Allen and Vincent, 1995].

### 3.2. Simple Models

To allow for estimates of energy fluxes and dissipation rates, we describe here a canonical spectrum that appears to have many of the attributes of atmospheric spectra and their evolution with altitude. As noted above, this does not imply that the gravity wave spectrum is broad. Rather, it insures that our estimates are as consistent as possible with observed scales and variations with altitude in the atmosphere. We also note that because of the gravity wave dispersion relation (1), the various spectra are related. Thus, it is necessary only to have accurate characterizations of two spectra, intrinsic frequency and vertical wavenumber, for example, in order to infer the associated gravity wave fluxes of momentum and energy. For this, however, we need to employ the argument above that intrinsic and observed frequency spectra are likely very similar, due to the occurrence of the majority of the variance at low frequencies (for horizontal velocities and temperature).

Spectral models have evolved with time, as we have identified spectral features and constraints more precisely. For example, we now believe the vertical wavenumber spectrum is more highly peaked near the characteristic vertical wavenumber, or scale, than was described initially. Likewise, early models of this spectrum assumed a slope at small $m$ which implied large or infinite vertical wave energy flux. These limitations have now been largely removed, and a spectral description that appears to capture most of the character of the gravity wave field is assumed (for convenience, but with some observational basis) to be separable in frequency and vertical wavenumber and is given by [Fritts and VanZandt, 1993]

$$E(\mu, \omega, \phi) = E_0 A(\mu) B(\omega) \Phi(\phi), \qquad (9)$$

with

$$A(\mu) = A_0 \frac{\mu^s}{1 + \mu^{s+t}} \quad \text{and} \quad B(\omega) = B_0 \omega^{-p}, \qquad (10)$$

where $\mu = m/m_*$, $m_*$ is the characteristic vertical wavenumber, and $A_0$, $B_0$, and $\Phi(\phi)$ are chosen such that each distribution is separately normalized to unity.

For mid-latitudes and $(s, t, p, f/N) = (1, 3, 5/3, 1/200)$ this yields

$$F_E \simeq \frac{N E_0}{18 m_*}. \qquad (11)$$

The factor $N/18m_* \simeq \overline{(1/m)\omega}$ may be interpreted as an effective vertical group velocity $\overline{c}_{gz}$. With $m_* \sim 3$ rad km$^{-1}$ near the tropopause and $\sim 1/3$ rad km$^{-1}$ near the mesopause, $\overline{c}_{gz} \sim 0.5$ and 5 m s$^{-1}$, respectively. With $E_0 \sim 10$ and $10^3$ m$^2$ s$^{-2}$, this implies $F_E \sim 5$ and

$5 \times 10^3$ m$^3$ s$^{-3}$, respectively, near the tropopause and mesopause.

We see from the form of (11) that wave energy density and energy flux vary with both altitude and $N$. Such variations are especially important in regions of rapidly increasing $N$, where the result is increased wave dissipation and flux divergences (VanZandt and Fritts, 1989). For our purposes, however, it is sufficient to assume that $N$ is constant or slowly varying, in which case wave dissipation is dominated by the systematic variations of the wave spectrum with height, yielding

$$\epsilon \simeq \frac{N E_0}{18 m_*} \left( \frac{1}{H} - \frac{3}{2 H_E} \right), \qquad (12)$$

with a dissipation time scale of

$$T_d \sim E_0 / \epsilon \sim \frac{18 m_*}{N} \left( \frac{1}{H} - \frac{3}{2 H_E} \right)^{-1}. \qquad (13)$$

Though there are uncertainties in $E_0$, $m_*$, and the fraction of wave energy that is directly dissipated as a result of wave interaction and instability processes with height, this expression provides a useful estimate of the variation of the dissipation time scale with altitude. More specifically, since $m_*$ varies as $E_0^{-1/2}$, the time scale likewise varies as $E_0^{-1/2}$, causing wave energy to be extracted from the wave field $\sim 10$ times more rapidly near the mesopause than near the tropopause, largely because of the increase of the vertical group velocity (or vertical wave scale) with altitude.

For characteristic wave scales and energies (see above), (13) yields $T_d \sim 100$ hours for the lower stratosphere and $\sim 10$ hours near the mesopause. Importantly, the time scale for wave dissipation in the lower stratosphere is *longer* than all intrinsic wave periods at middle to high latitudes, whereas the time scale near the mesopause is *shorter* than the energy-containing wave periods. The implications are relatively slow and systematic wave dissipation in the lower stratosphere, allowing slower instability processes to be operative, and relatively faster wave dissipation at higher altitudes, often requiring significant energy dissipation within a wave period. Hence, we expect that KH shear instabilities, which are favored at wave amplitudes below the nominal amplitude for convective instability, will be preferred at lower altitudes and away from significant wave sources. On the other hand, convective instabilities appear to be more likely at greater altitudes, where intermittent wave dissipation must shed excess energy more rapidly, and in the vicinity of strong sources of high-frequency waves at lower altitudes [Lilly and Kennedy, 1973; Alexander, 1996].

Alternatively, for measured wave energy densities of $\sim 300$ to $1000$ m$^2$ s$^{-2}$ and corresponding energy dissipation rates of $\sim 10^{-2}$ to $10^{-1}$ W/kg near the mesopause [Lübken, 1997], we obtain a time scale of $T_d \sim 3$ to $10$ hours. In the lower stratosphere, with $E \sim 3$ to $10$ m$^2$ s$^{-2}$ and $\epsilon \sim 10^{-5}$ to $10^{-4}$ W/kg (Lilly and Lester, 1974), we obtain time scale estimates of $T_d \sim 30$ to $100$ hours, generally consistent with our spectral estimates above.

## 4. GRAVITY WAVE BREAKING DYNAMICS

Gravity wave breaking typically occurs accompanying the most energetic (and largest amplitude) components of the wave spectrum. Near the mesopause, typical vertical wavelengths are $\sim 10$ to $30$ km and corresponding unstable layer depths are $\sim 1$ to a few km, though larger scales are occasionally observed. As noted above, unstable layer depths typically constrain the depth and the outer scale, $L_B$, of the resulting turbulence.

Numerical studies of wave breaking over the last few years have defined the dynamics of both the transition to turbulence and the initial turbulence cascade [Andreassen et al., 1998; Fritts et al., 1998]. These studies were constrained by numerical resolution, however, and thus did not address either inertial range dynamics or the implications of wave breaking for mixing. The implications of these results for the dominant instability processes have nevertheless been supported by parallel observational studies [Fritts et al., 1993; Swenson and Mende, 1994; Hecht et al., 1997]. Our purposes in this section are to summarize the dynamics of wave breaking reported elsewhere and to provide a preview of mixing implications based on more recent and higher-resolution studies.

### 4.1. Review of Wave Breaking Dynamics

We will begin by reviewing the dynamics of wave breaking for waves of high intrinsic frequencies, for which convective instability dominates, from a vorticity perspective. This approach was found by Andreassen et al. [1998] and Fritts et al. [1998] to provide considerable insights into both the initial instability processes and the subsequent turbulence dynamics. We will also employ a quantity corresponding approximately to the minimum pressures in the flow in our discussion of the vorticity dynamics, given its ability to characterize vortex structures at many stages of their evolution [see Andreassen et al., 1998, for further details].

The simulation discussed here was performed using a compressible model solving the Navier-Stokes equations in a horizontally-periodic domain, having inflow/outflow boundary conditions in the vertical, subject to a spectral formulation of viscosity and thermal diffusivity, and employing a mean streamwise (along the direction of wave propagation) shear flow to confine wave instability within a high-resolution domain. Additional details of the numerical code and the simulation are provided by Andreassen et al. [1998].

The character of convective wave breaking is illustrated with isosurfaces of potential temperature through the breaking region shortly after initial convective instability and approximately one buoyancy period later in Figure 1. The direction of wave propagation (the streamwise direction) is out of the page and to the right. At early stages, wave breaking is quasi-2D and resembles a wave breaking on a beach. As instabilities evolve, however, the flow becomes increasingly complex. The vorticity field accompanying this evolution is displayed volumetrically in the left panels of Plate 1 for an interval extending for $\sim 1.4$ buoyancy periods beyond initial convective instability and spanning the transition from largely two-dimensional (2D) to fully 3D flow. Views of the vorticity field are from below, with the streamwise direction to the right and the spanwise (across the flow) direction down.

In the absence of rotation, spanwise mean velocity, and 3D noise, the initial vorticity distribution is entirely spanwise and comprises vortex sheets due to both wave and mean velocity shears. As the wave field achieves convective instability, however, the 3D noise seeds counter-rotating, streamwise-aligned convective rolls (see the top two panels on the left in Plate 1). These rolls are confined initially to the convectively unstable phase of the wave field, derive their vorticity (and energy) from baroclinic tendencies, and occupy the full depth of the unstable region of the flow [Andreassen et al., 1998]. For reference, the depth of the unstable region is $\sim 1/5$ of the vertical wavelength of the overturning gravity wave and would typically be $\sim 1$ to a few km near the mesopause [Fritts et al., 1993; Hecht et al., 1997]. The convective rolls immediately decouple from the unstable phase of the gravity wave (the wave experiences phase propagation, while the streamwise vortices undergo mutual advection relative to the 2D flow) and advecting vortex pairs begin to impinge on adjacent spanwise vortex sheets. Divergent spanwise flow ahead of advecting vortex pairs leads to local intensification (via vortex stretching) and thinning and wrapping (via vertical and spanwise advection) of the vortex sheets around the streamwise vortex pairs. The consequence is a succession of intensified spanwise

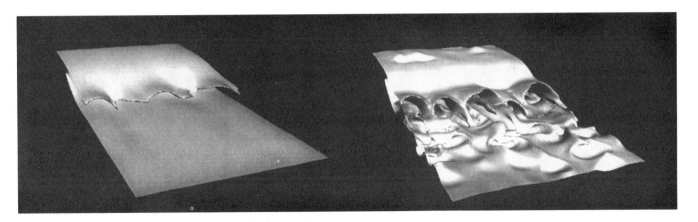

**Figure 1.**   Isosurfaces of potential temperature for the wave breaking simulation shortly after initial convective instability (left) and approximately one buoyancy period later (right). The streamwise direction (wave propagation direction) is out of the page and to the right.

vortex sheets which individually become susceptible to spanwise-localized KH instabilities having largely spanwise vorticity. These secondary KH instabilities lead to spanwise vortex tubes which link to the streamwise rolls accounting for their formation. The result is a succession of intertwined vortex loops at each site of initial spanwise vortex sheet intensification (see the second and third panels on the left in Plate 1). These vortex loops bear a close resemblance to the "horseshoe" and "hairpin" vortices that arise in a variety of turbulent flows and thus may have implications for the evolution of a broad class of flows.

The series of intertwined vortex loops arising from initial convective and secondary dynamical instabilities condition the flow for a rapid subsequent evolution because of the close proximity of adjacent vortices. This further evolution is displayed in the lower four panels on the left in Plate 1. Though the time required for this further evolution spans only $\sim 0.7$ buoyancy periods and the vortex dynamics are somewhat viscously constrained, the increase in flow complexity is dramatic. The vortex loops present in the upper panels on the left in Plate 1 are fragmented into pieces, and new smaller vortex tubes continue to be formed. Additionally, the maximum vorticity and enstrophy (square of the vorticity magnitude) at smaller scales of motion increase throughout the simulation, both because of the initial cascade from larger to smaller scales and because enstrophy is larger at the smaller scales within an inertial range of turbulence.

*4.2. Primary and Secondary Vortex Dynamics*

The processes that appear to govern the evolution of the vortex field and the cascade of energy and en-

strophy to smaller scales of motion within a breaking gravity wave have been discussed in detail by Fritts et al. [1998]. Here, we will simply summarize these findings and illustrate the key processes of these dynamics. A number of these processes can be seen to occur in the latter stages of the full vorticity evolution (Plate 1, left) or more clearly in the closeup images of the vortex field spanning the evolution from Plate 1d to 1f on the left in the expanded images on the right (and compensated for mean advection). Based on our assessments of importance, the dominant processes include 1) orthogonal vortex stretching, 2) vortex sheet and tube formation, and 3) vortex-core oscillations, termed Kelvin waves or twist waves, exhibiting azimuthal wavenumber 0 and 2 structures. Processes that also occur frequently, but appear to be of secondary importance to the turbulence cascade, include 1) vortex reconnection, 2) twist waves exhibiting azimuthal wavenumber 1 structure, and 3) vortex pairing. Because of their lesser role, however, these secondary vortex dynamics will not be discussed further here.

*4.2.1. Orthogonal vortex stretching.* The key element in vortex intensification on short time scales is vortex stretching, either of the ambient shear field by localized vortices or of localized vortices by other localized and closely spaced vortex structures. To be effective, the local component of vorticity, $\omega_i$, and the diagonal component of the strain tensor, $S_{ii}$, must contribute a source, $\omega_i S_{ii}$, which exceeds other baroclinic, compressional, and viscous contributions. This stretching source is especially strong when the vorticity lines of two neighboring vortex structures are nearly orthogonal. Physically, the stretching is due to a flow divergence along the vorticity.

*4.2.2. Vortex sheet and tube formation.* Formation and intensification of vortex sheets and their subsequent roll-up into vortex tubes has been suggested previously as initial steps in the transition to and within turbulent flow. Our wave breaking results and the previous discussion of Plate 1 show that these processes are also of central importance in a stratified fluid. Sheet roll-up occurs where they have intensified due to vortex stretching by the flow of a neighboring orthogonal vortex.

*4.2.3. Mode-zero twist waves.* As observed in the latter evolution displayed in Plate 1, vortex loops arising from secondary KH instability tend to fragment into pieces. One of the causes of this fragmentation is the presence of mode-zero (azimuthal wavenumber 0) twist waves of large amplitude. Examples of mode-zero twist waves can be seen on many of the spanwise-aligned vortices in the left and right panels of Plate 1. In each case, the mode-zero twist waves are initiated through axial stretching and compression of a vortex by a near-orthogonal neighbor. In the case of many of the spanwise vortices, mode-zero twist waves appear to be excited at both ends of the vortex simultaneously (see the top two images on the right of Plate 1), to propagate inward from the sites of excitation, and to jointly cause an axial convergence which weakens the vortex to such an extent that its center is obliterated. The prevalence and consequences of mode-zero twist waves in our simulation suggests this to be a key mechanism for vortex breakup and the evolution toward smaller scales of motion.

The propagation and characteristics of twist waves were first described by Kelvin [1880], while Arendt et al. [1997b] recognized the role of twist waves in the interpretation of our wave breaking vortex dynamics. In general, twist waves are dispersive traveling waves on vortex tubes. The axisymmetric (mode-zero) modes propagate by twisting the vortex lines of the tube, thereby creating an axial flow. The axial flow changes the enstrophy of the tube by axial stretching and compression. This, in turn, changes the rotation rate of the tube which then changes the twist.

*4.2.4. Mode-two twist waves.* A second type of twist wave that plays a major role in the fragmentation of vortex loops is a mode-two (azimuthal wavenumber 2) wave. These waves are typically excited where a vortex tube is perturbed near its end by another vortex structure (either a tube or sheet). The clearest and most long-lived examples of mode-two twist waves in our numerical solution arise due to upstream perturbations of the more axially-uniform streamwise vortex tubes (see the upper left portions of the images in Plate 1). In

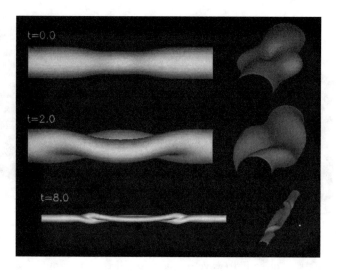

**Figure 2.** Propagation and dispersion of mode-two twist waves excited by a Gaussian pinch. Times are in units of circulation times. Note that modes having opposite helicity propagate in opposite directions.

each case, a mode-two twist wave with a small amplitude is excited on a vortex tube and propagates via self advection, though the propagation of the wave is difficult to quantify because of streamwise flow advection. The waves also grow until they are sufficiently large to unravel the vortex tube on which they propagate into a pair of intertwined helices.

The propagation of a mode-two twist wave excited by a Gaussian pinch is shown from two perspectives at several times (in circulation time units) in Figure 2. Note that the direction of propagation is determined by the direction of the vorticity in the tube and the sense of helicity of the wave packet. In these images, vorticity is to the right and the velocity field due to the vortex tube and its perturbation imposes rightward phase motion for left-handed helicity and leftward phase motion for right-handed helicity. The most conspicuous mode-two twist waves seen at the upper left of the images in Plate 1 have left-handed helicity and positive-$x$ vorticity and right-handed helicity and negative-$x$ vorticity. Therefore both propagate in the positive-$x$ direction. Propagation of mode-two twist waves in the wave breaking simulation is discussed in greater detail by Arendt et al. [1997a, b].

### 4.3. Turbulent Mixing

The results discussed here provide some qualitative insights for mixing due to wave breaking. However, a more complete assessment must await our current

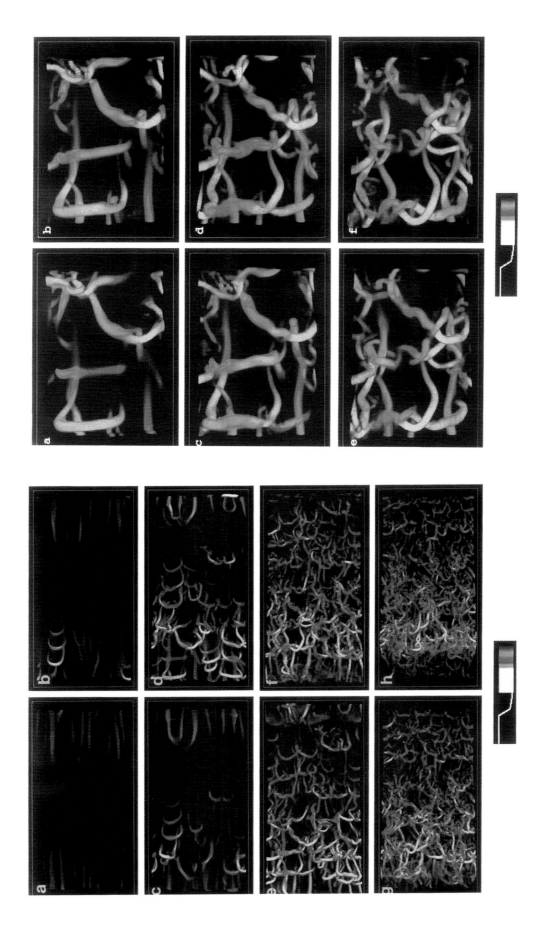

Plate 1. Volume renderings of vortex cores from below with positive $x$ to the right. The left panels span $\sim 1.4$ buoyancy periods following initial convective instability. The right panels display a subdomain at the upper left of the full domain advected with the mean horizontal motion. Color and opacity scales are shown at the bottom.

higher-resolution studies which are addressing much less viscous flows. Though turbulence evolves quickly, with most of the turbulence transition and cascade occupying less than a buoyancy period, the region of active mixing and transport is not confined to the location of initial instability. Instead, the turbulence is advected out of the unstable phase of the wave motion and evolves thereafter in stable stratification. Thus, except at the earliest stages, vertical transport tends to be down gradient and no region of the flow appears to remain well mixed at later stages. This aspect of turbulence due to wave breaking contrasts sharply with the mixing and transport accompanying KH instability discussed below. In this simulation, turbulent diffusion comprises molecular diffusion across large-amplitude, small-scale structures (and shears). The role of vortices in turbulent diffusion is simply to efficiently bring fluids of different character or composition into close proximity, thus greatly enhancing the molecular diffusion operative on otherwise much smaller gradients. While it is too early to advertise the results of higher-resolution simulations, current results appear to support the implications of earlier theoretical efforts suggesting inefficient mixing and a turbulent Prandtl number greater than 1 [Fritts and Dunkerton, 1985; Coy and Fritts, 1988; McIntyre, 1989]. We expect, however, because turbulence does persist after being advected out of the unstable phase of the wave motion, that non-negligible mixing will occur under stably-stratified conditions with largely vertical gradients of heat, momentum, and constituents.

## 5. KH SHEAR-INSTABILITY DYNAMICS

While KH shear instability due to inertia-gravity wave motions in the stratosphere has been inferred from observations and simulated in numerical models [Fritts and Rastogi, 1985; Coulman et al., 1995; LeLong and Dunkerton, 1998], these methods provide few insights into the implications for turbulence and mixing. Instead, such an understanding relies on numerical simulations focussing specifically on KH instability dynamics, due to the enormous demand for computational resources by such simulations.

A great deal has been learned about the dynamics of KH instability in recent years. Theoretical and laboratory studies have contributed significantly to our understanding of the underlying 2D dynamics of unstable shear flows and of the transition to 3D flow in stratified and unstratified fluids (see Thorpe, [1987], for a review). However, it has required high-resolution numerical studies to describe the transition to turbulence

and the resulting mixing more completely. Though a number of authors have addressed the transition to 3D accompanying KH instability, only the study by Werne and Fritts [1999] has captured the emergence of an inertial range of turbulence following instability to date.

The KH simulation described by Werne and Fritts was performed for an environment having uniform stratification and a velocity profile of the form $\bar{u}(z) = u_0 \tanh(z/h)$, with $Ri = N^2 h^2 / u_0^2 = 0.05$, $Re = u_0 h/\nu = 2000$, and $Pr = \nu/\kappa = 1$, where $\nu$ and $\kappa$ are kinematic vicosity and thermal diffusivity. The simulation employed a spectral formulation of the incompressible Navier-Stokes equations solved in a horizontally-periodic domain of wavelength $L = 4\pi h$, corresponding to the most unstable linear eigenfunction of the viscous equations, with domain dimensions $(1, 0.3, 2)$ $L$ and spectral resolution of $(720, 240, 1440)$ for the most vigorous portion of the turbulence evolution. Additional details about the forcing are provided by Werne and Fritts [1999].

The most noteworthy attributes of the simulations presented here are the high spatial resolution and turbulence intensities attained. For example, when the shear layer is fully developed and turbulent, the effective layer Reynolds number defined in terms of the final shear-layer width, $\sim 6h$, the velocity difference, $\sim 2u_0$, and the kinematic viscosity is $Re_{layer} \sim 24,000$. Other measures of the degree of turbulence achieved are the buoyancy Reynolds number, $Re_b \approx 1000$ (discussed above), and the Taylor microscale Reynolds number, $Re_\lambda = \langle u^2 \rangle / \sqrt{\nu \langle u_x u_x \rangle} \approx 230$, where $u$ is the velocity in the $x$ direction, the subscript $x$ denotes differentiation, and $<>$ denotes an average over the layer. All of these measures ($Re_{layer}$, $Re_b$, and $Re_\lambda$) are the highest that have ever been computed for stratified turbulence.

### 5.1. Initial Vorticity Dynamics

The evolution of the flow is shown in Figure 3 with streamwise-vertical cross sections of enstrophy and potential temperature at times of $t = 60, 90, 160$, and 300. Times are in units of $h/u_0$, where $t = 28$ corresponds to a buoyancy period. These panels show the initial instability in the outer billow, the expansion of turbulence throughout the billow core, and the horizontal homogenization and decay of the turbulence layer. The right panels show the corresponding evolution of the thermal field. Note, in particular, the strong thermal gradients within the billow prior to vigorous turbulence and mixing. As turbulence expands, however, thermal gradients within the billow are quickly annihilated, leaving a nearly homogeneous billow core with

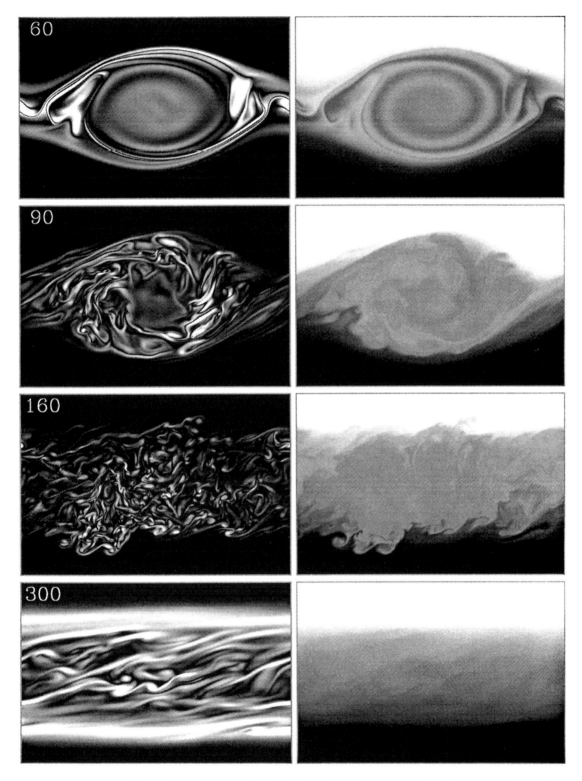

**Figure 3.** Streamwise-vertical cross sections of vorticity magnitude (left) and potential temperature (right) showing the evolution of a KH billow from the onset of 3D instability through the expansion of turbulence, the cascade toward smaller scales, and the restratification of the flow as turbulent motions subside. Times are 60, 90, 160, and 300.

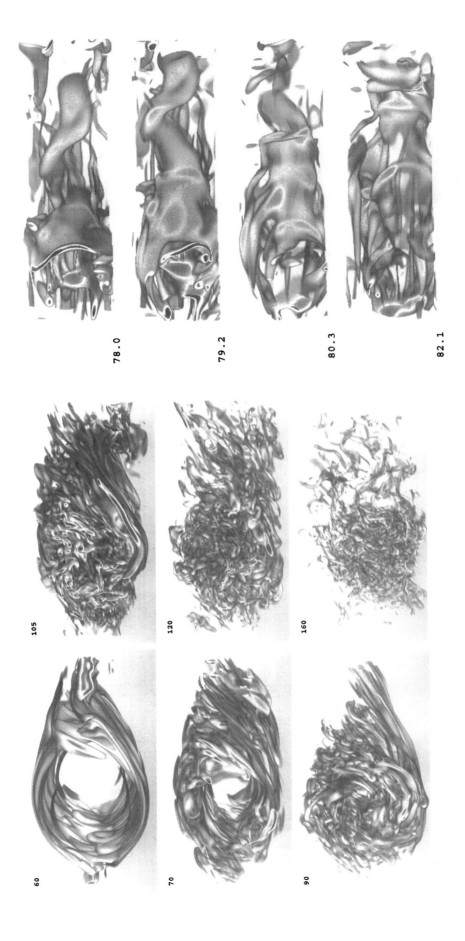

**Plate 2.** Perspective volumetric views of vorticity magnitude throughout the evolution and turbulent breakdown of a KH billow (left and center images). Times are 60, 70, 90, 105, 120, and 160 in units of $h/U_0$, small values of vorticity are transparent or blue and large values of vorticity are yellow or red. The right panels show close-up views of twist waves occurring on a vortex sheet along the bottom of the billow at the times shown.

sharp thermal gradients at the edges of the mixed region. As the turbulence expands horizontally, the mean profile takes on similar structure, again confining large thermal gradients to the edges of the mixed layer.

Vortical structures arising within the KH billow are displayed in the left and center images of Plate 2 in a volumetric fashion at times of 60, 70, 90, 105, 120, and 160 spanning the transition to turbulence and approach to isotropy. Enstrophy is used to display the vorticity field in these plates because it depicts vortex structures having both tube-like and sheet-like character. At early times, the flow is seen to trigger streamwise-aligned (vortex cores along the 2D flow) convective rolls downstream of the entrainment regions of the outer billow (see the initial vortex structures at $t = 60$ in Plate 2), consistent with earlier theoretical predictions and laboratory and numerical studies [Thorpe, 1987].

Following secondary convective instability, the vortex structures become increasingly 3D, largely through the interactions of adjacent vortices. These interactions parallel in important respects those observed in our studies of vortex dynamics accompanying wave breaking discussed above. The dynamics of adjacent convectively unstable and sheared stable layers manifest themselves in two ways. Where counter-rotating streamwise convective rolls are closely spaced, they undergo mutual advection and continued intensification via streamwise stretching. The convective rolls also lead to spanwise divergence at the locations of adjacent spanwise vortex sheets, leading to spanwise-localized thinning and intensification of these sheets. As in wave breaking, the spanwise vortex sheets are wrapped around the streamwise convective rolls accounting for sheet intensification.

Spanwise-localized intensification of the vortex sheets has two effects in our KH simulations. It causes the vortex sheets to become dynamically unstable to secondary (spanwise-localized) KH instability. The vortex sheets in turn perturb the streamwise vortices, resulting in the excitation of twist waves at larger scales. The evolution of a vortex sheet in the lower portion of the KH billow which exhibits several of these responses is shown in the right images of Plate 2. The relative importance of these processes is observed to be very sensitive to initial conditions for the 3D flow. In the entrainment regions, the deformation and roll-up of the spanwise vortex sheets leads to attachment of the emerging spanwise vortices to the streamwise vortices accounting for sheet intensification. Where the vortex structures are more uniform along the streamwise flow, the result is a series of spanwise vortices and the attachment of each to the associated streamwise vortices (see the second and

third images on the left in Plate 2), much as described above for the transition to turbulence in wave breaking. Where streamwise variations are greater, twist waves on the streamwise vortices are more readily excited (see below). In either case, the tendency for vortex sheets to intensify and evolve toward vortex tubes is widespread and operates at many scales.

## 5.2. Dynamics of the Turbulence Cascade

It was proposed previously that twist waves provide the primary means by which energy and enstrophy cascade toward smaller scales within a turbulent flow [Arendt et al., 1997b; Fritts et al., 1998]. The present results provide the first opportunity to test this hypothesis and establish the extent to which the same cascade dynamics prevail when dissipation does not play a primary role on scales at which vortices arise.

In the KH simulation, near-orthogonal vortex alignments arise due to the secondary and tertiary flow instabilities discussed above. Close proximity of adjacent vortices then leads to vortex perturbations which excite various twist wave modes. As shown in Plate 4, twist waves manifest themselves as vortices exhibiting axial variations in vorticity and axial velocity (mode zero) and single or double helices (mode one and mode two). Of these, the mode-one and mode-two helical structures are easiest to identify, though mode-zero twist waves are surely present and may play a greater role in vortex breakdown.

The first clear evidence of twist-wave vortex dynamics in the KH simulation can be seen in the right panels of Plate 2 accompanying vortex-sheet intensification and secondary KH formation, which shows the evolution of the vortex field at the bottom of the billow for times of $t = 78$ and thereafter. A mode-one twist wave having right-handed helicity is seen to emerge on a vortex along the left top of the images from $t = 79.2$ to 80.3. Note that the twist wave amplifies with time, while undergoing an interaction with a neighboring vortex at later stages. A second mode-one twist wave with the same helicity is observed on a much larger scale at the right end of the vortex sheet in the first two images on the right in Plate 2.

Like the vortex dynamics within a breaking wave then, the generation and evolution of turbulence due to KH instability involves a succession of vortex dynamics. Following secondary convective instability and tertiary KH instability of spanwise-localized vortex sheets, the vorticity field is complex, 3D, and composed primarily of vortex tubes. The subsequent vortex interactions

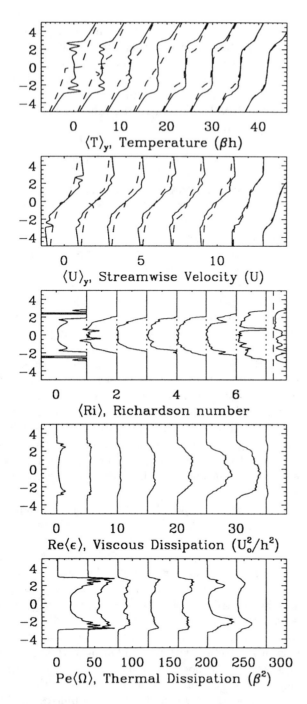

**Figure 4.** Spanwise-averaged vertical profiles of $T(z)$, $u(z)$, and $Ri$ (top to middle) at times of $t = 50, 60, 70, 90, 105, 120, 160,$ and $300$. Profiles of $T(z)$ and $u(z)$ are shown at the streamwise midpoint of the billow (solid lines) and through the braid between adjacent billows (dashed lines). $Ri$ profiles are derived using streamwise and spanwise mean profiles, $\bar{T}(z)$ and $\bar{u}(z)$. Streamwise- and spanwise-averaged profiles of energy and thermal dissipation rates, $\epsilon$ and $\chi$, are shown at the same times in the lower two panels. Successive profiles are offset by a uniform increment in each case.

drive the flow toward ever smaller scales and increasing complexity. Sheets are stretched and roll up to form new tubes; tubes interact and excite twist waves which unravel and fragment the parent tubes until they are arrested by viscous and diffusive effects.

### 5.3. Turbulent Mixing

The effects of momentum and heat transport and turbulent mixing are shown in the top three panels of Figure 4 with spanwise-averaged vertical profiles of $T(z)$, $u(z)$, and $Ri$ at various times. Profiles of $T(z)$ and $u(z)$ are shown at the streamwise midpoint of the billow (solid lines) and through the braid between adjacent billows (dashed lines) in Figure 4a and 4b at $t = 50, 60, 70, 90, 105, 120, 160,$ and $300$. Corresponding profiles of $Ri$ determined from the (streamwise and spanwise) mean profiles, $\bar{T}(z)$ and $\bar{u}(z)$, are shown in Figure 4c. Profiles of mechanical and thermal energy dissipation rates, $\epsilon$ and $\chi$, are displayed at these same times in Figure 4d and 4e.

Mixing accompanies vigorous vortex dynamics, obliterates thermal gradients within the billow core, and drives large-scale thermal and velocity gradients to the edges of the turbulent layer. Mixing at the center of the shear layer is thus fairly complete, but remains confined to the initial shear layer. This is in contrast to mixing due to gravity wave breaking, which largely follows the unstable phase of the wave motion, is thus less complete, but extends over a larger vertical depth.

The edge regions of KH billows and the resulting turbulence layer remain sites of active entrainment dynamics at smaller scales following mixing of the billow cores. They are also the regions of KH billow turbulence most easily observed using radar and acoustic sounding methods in atmospheric measurements of such dynamics because of their high refractive-index gradients [Eaton et al., 1995; Hill et al., 1999; Gibson-Wilde et al., 2000]. However, the mixing accompanying these tertiary instabilities is less complete because of the much smaller energy dissipation rates at these sites. This causes a weakening and spreading of the edge gradients, and of the associated constituent profiles, until the small-scale structures have decayed and the flow has restratified.

The implications for the inertia-gravity wave motions within which KH billows arise are presumably similar, though the simulations to define this have yet to be performed. When KH instability dynamics are vigorous, we expect this to result in a well mixed layer and accompanying momentum, heat, and constituent transport. Because such events are spatially localized, they will alter the inertia-gravity wave structure locally, likely impacting wave amplitude only weakly and causing a readjustment of the flow via radiation of other waves at scales comparable to the event scale. Successive KH mixing events would seem likely accompanying readjustment, though this is speculative at this time.

## 6. SUMMARY AND CONCLUSIONS

Our purposes in this paper were twofold. We first provided evidence that the character of gravity wave instability undergoes a systematic variation with altitude. We then described in some detail the dynamics of the transition to turbulence accompanying wave breaking and KH shear instability and, to the extent presently possible, the implications for turbulent mixing.

The character of the gravity wave spectrum and its variations with altitude suggest that wave energy dissipation increases more rapidly than wave energy density (per unit mass) as altitude increases. This is because wave energy flux varies as the product of wave energy density, $E$, and vertical group velocity, $c_{gz}$, with the vertical group velocity itself increasing as $c_{gz} \sim m_*^{-1/2} \sim E^{1/2}$. Thus, as $E$ increases by $\sim 100$ from the tropopause to the mesopause, the energy flux increases by $\sim 1000$, causing energy to be removed from the wave field $\sim 10$ times more quickly at the higher altitudes. The corresponding time scales for wave energy dissipation (assuming all removed energy cascades to turbulence) are $\sim 30$ to 100 hours near the tropopause and $\sim 3$ to 10 hours near the mesopause.

The implications of these estimates are a tendency for slower, systematic KH shear instabilities due to lower-frequency motions in the lower stratosphere and a tendency for more rapid convective instabilities of higher-frequency wave motions at greater altitudes. Observations in the stratosphere over many years and more recent optical measurements at mesopause altitudes provide supporting data for these inferences.

Our second purpose was to describe the dynamics of the turbulence transition and the turbulence cascade accompanying both wave breaking and KH shear instability. This included a review of previous, relatively low-resolution simulations of wave breaking and more recent, higher-resolution studies of KH instability.

In both wave breaking and KH instability, the transition to 3D structure is triggered by an instability comprising counter-rotating streamwise (or shear-aligned) rolls where the 2D flow is convectively unstable. In both flows, these streamwise vortices occur adjacent to spanwise vortex sheets due to mean and perturbation shears. The interactions of adjacent streamwise vortex tubes and spanwise vortex sheets causes local intensification of the vortex sheets, wrapping of these sheets around pairs of streamwise vortex tubes, and secondary shear (KH) instability of the intensified sheets. This results in additional spanwise vortices which link with the streamwise vortices responsible for their formation, yielding intertwined vortex loops at each site of initial spanwise shear intensification.

The evolution subsequent to intertwined loop formation is rapid and is driven by interactions among vortex loops (and sheets to a lesser extent) in close proximity. These interactions excite perturbations on adjacent vortices that are most naturally described as twist waves which propagate along the vortex tubes and may attain large amplitudes. Twist waves appear to play a large role in the breakup and fragmentation of vortex tubes and the cascade of energy and enstrophy to smaller scales of motion. Mode-zero twist waves introduce large axial flow variations which lead to vortex bursting; mode-two twist waves effectively unravel larger vortices into two intertwined vortices of smaller size. Other vortex interactions, primarily mode-one twist waves, vortex pairing, and reconnection also occur, but are judged to be less significant in the evolution of the vorticity field.

While their turbulence dynamics are similar, the implications of turbulence for mixing and transport due to wave breaking and KH instability are quite different. Though a complete evaluation of the relative roles of these processes in mixing and transport must await higher-resolution studies quantifying the statistical effects of a series of such events, we can at least speculate on their likely roles based on simulations to date. To the extent that mixing by KH instability is fairly complete within the shear layer, and that successive shears tend to occur where stratification is largest, this suggests that such mixing should be fairly efficient relative to the small energy dissipation rates characteristic of the lower stratosphere. Mixing due to wave breaking is likely most vigorous within the unstable phase of the wave motion where vertical gradients are weak, but will also extend with smaller $\epsilon$ (hence smaller intensity and vertical depth) into the stable phase of the wave motion. This suggests less efficient mixing than for KH instability of the same intensity (comparable $\epsilon$) and scale. However, wave breaking dominates turbulence generation at greater altitudes and for larger vertical energy fluxes, group velocities, and vertical scales. Hence, such events are likely capable of mixing and turbulent transport over extended vertical (and horizontal) scales at the larger values of $\epsilon$ anticipated at higher altitudes.

Both $\epsilon$ and mixing efficiency are likely to be highly dependent on the amplitude of the unstable wave, with sporadic, large-amplitude events accounting for more mixing and transport than continuous weaker events yielding the same mean wave energy densities. The tentative implication for parameterization in large-scale models is more efficient mixing (for given $\epsilon$) at lower altitudes where KH instability is predominant, with decreasing efficiency at higher altitudes where wave breaking is predominant.

More quantitative descriptions of turbulent mixing will require further simulations and analysis, but such simulations are now feasible with current computational capabilities. Numerical tracer advection and dispersion studies are anticipated and will offer a potential to compare with similar studies in the laboratory and the atmospheric boundary layer. Together, these capabilities promise significant improvements in our understanding and parameterization of turbulent mixing and diffusion in the near future.

*Acknowledgments:* This research was supported by the NSF under grants ATM-9618004, ATM-9708633, and ATM-9816160 and the AFOSR under contract F4-9620-98-C-0029. Our incompressible simulations were performed with computational resources provided by a DoD HPCMO "Challenge" allocation in support of the Airborne Laser Program. DF provided most of the review material, much of which describes work with various colleagues. JW wrote the spectral code and performed the KH simulations. Both participated in the KH simulation analysis, some of which is presented here for the first time. Images displayed in the plates were created using the volumetric visualization software Viz, written by Per-Øyvind Hvidsten.

## REFERENCES

Alexander, M. J., A simulated spectrum of convectively generated gravity waves: Propagation from the troposphere to the mesopause and effects on the middle atmosphere, *J. Geophys. Res., 101,* 1571-1588, 1996.

Allen, S. J., and R. A. Vincent, Gravity wave activity in the lower atmosphere: Seasonal and latitudinal variations, *J. Geophys. Res., 100,* 1327-1350, 1995.

Andreassen, Ø., et al., Vorticity dynamics in a breaking gravity wave, 1. Initial instability evolution, *J. Fluid Mech., 367,* 27-46, 1998.

Arendt, S., et al., The initial value problem for Kelvin vortex waves, *J. Fluid Mech., 344,* 181-212, 1997a.

Arendt, S., et al., Kelvin twist waves in the transition to turbulence, *Eur. J. Mech. B/Fluids, 17,* 595-604, 1997b.

Balsley, B. B., and D. A. Carter, The spectrum of atmospheric velocity fluctuations at 8 and 86 km, *Geophys. Res. Lett., 9,* 465-468, 1982.

Bender, C. M., and S. A. Orszag, *Advanced Mathematical Methods for Scientists and Engineers,* McGraw-Hill, New York, 593 pp., 1978.

Coulman, C. E., et al., Optical seeing - mechanism of formation of thin turbulent laminae in the atmosphere, *Atmos. Ocean., 34,* 5461-5474, 1995.

Coy, L., and D. C. Fritts, Gravity wave heat fluxes: A Lagrangian approach, *J. Atmos. Sci., 45,* 1770-1780, 1988.

Dewan, E. M., and R. E. Good, Saturation and the "universal" spectrum for vertical profiles of horizontal scalar winds in the atmosphere, *J. Geophys. Res., 91,* 2742-2748, 1986.

Dewan, E. M., et al., Spectral analysis of 10 m resolution scalar velocity profiles in the stratosphere, *Geophys. Res. Lett., 11,* 80-83, 1984.

Dougherty, J. P., The anisotropy of turbulence at the meteor level, *J. Atmos. Terr. Phys., 21,* 210-212, 1962.

Dunkerton, T. J., Inertia-gravity waves in the stratosphere, *J. Atmos. Sci., 41,* 3396-3404, 1984.

Dunkerton, T. J., Theory of internal gravity wave saturation, *Pure Appl. Geophys., 130,* 373-397, 1989.

Dunkerton, T. J., Shear instability of internal inertia-gravity waves, *J. Atmos. Sci., 54,* 1628-1641, 1997.

Eaton, F. D., et al., A new frequency-modulated continuous wave radar for studying planetary boundary layer morphology, *Radio Sci., 30,* 75-88, 1995.

Eckermann, S. D., Effect of background winds on vertical wavenumber spectra of atmospheric gravity waves, *J. Geophys. Res., 100,* 14,097–14,112, 1995.

Fritts, D. C., Gravity wave saturation in the middle atmosphere: A review of theory and observations, *Rev. Geophys. Space Phys., 22,* 275-308, 1984.

Fritts, D. C., A review of gravity wave saturation processes, effects, and variability in the middle atmosphere, *Pure Appl. Geophys., 130,* 343-371, 1989.

Fritts, D. C., et al., Vorticity dynamics in a breaking internal gravity wave, 2. Vortex interactions and transition to turbulence, *J. Fluid Mech., 367,* 47-65, 1998.

Fritts, D. C., and T. J. Dunkerton, Fluxes of heat and constituents due to convectively unstable gravity waves, *J. Atmos. Sci., 42,* 549-556, 1985.

Fritts, D. C., et al., Wave breaking signatures in noctilucent clouds, *Geophys. Res. Lett., 20,* 2039-2042, 1993.

Fritts, D. C., and G. D. Nastrom, Sources of mesoscale variability of gravity waves, II: Frontal, convective, and jet stream excitation, *J. Atmos. Sci., 49,* 111-127, 1992.

Fritts, D. C., and P. K. Rastogi, Convective and dynamical instabilities due to gravity wave motions in the lower and middle atmosphere: Theory and observations, *Radio Sci., 20,* 1247-1277, 1985.

Fritts, D. C., and T. E. VanZandt, Effects of Doppler shifting on the frequency spectra of atmospheric gravity waves, *J. Geophys. Res., 92,* 9723-9732, 1987.

Fritts, D. C., and T. E. VanZandt, Spectral estimates of gravity wave energy and momentum fluxes, I: Energy dissipation, acceleration, and constraints, *J. Atmos. Sci., 50,* 3685-3694, 1993.

Fritts, D. C., and L. Yuan, Stability analysis of inertio-gravity wave structure in the middle atmosphere, *J. Atmos. Sci., 46,* 1738–1745, 1989.

Garcia, R. R., The role of equatorial waves in the semiannual oscillation of the middle atmosphere, this issue, 2000.

Gibson-Wilde, D. E., et al., Direct numerical simulation of VHF radar measurements of turbulence in the mesosphere, *Radio Sci.*, in press, 2000.

Gill, A. E., *Atmosphere-Ocean Dynamics*, Academic Press, New York, 662 pp., 1982.

Hecht, J. H., et al., Wave breaking signatures in OH airglow and sodium densities and temperatures, Part I: Airglow imaging, Na lidar, and MF radar observations, *J. Geophys. Res.*, *102*, 6655-6668, 1997.

Heisenberg, W., Zur statistischen Theorie der Turbulenz, *Z. Phys.*, *124*, 628-657, 1948.

Hill, R. J., et al., Turbulence-induced fluctuations in ionization and application to PMSE, *Earth Planets Space, 51,* 499-513, 1999.

Holton, J. R., and M. J. Alexander, The role of waves in the transport circulation of the middle atmosphere, this issue, 2000.

Kelvin, Lord, Vibrations of a columnar vortex, *Phil. Mag., 10,* 155-168, 1880.

Klaassen, G. P., and W. R. Peltier, The onset of turbulence in finite-amplitude Kelvin-Helmholtz billows, *J. Fluid Mech., 155,* 1-35, 1985.

Kolmogorov, A. N., The local structure of turbulence in incompressible viscous fluid for very large Reynolds number, *Dokl. Akad. Nauk SSSR, 30,* 9-13, 1941a, (reprinted in *Proc. Roy. Soc. Lond. A, 434,* 9-13 (1991)).

LeLong, M.-P., and T. J. Dunkerton, Inertia-gravity wave breaking in three dimensions, 1. Convectively stable waves, *J. Atmos. Sci., 55,* 2473-2488, 1998.

Lilly, D. K., and P. J. Kennedy, Observations of a stationary mountain wave and its associated momentum flux and energy dissipation, *J. Atmos. Sci., 30,* 1135-1152, 1973.

Lilly, D. K., and P. F. Lester, Waves and turbulence in the stratosphere, *J. Atmos. Sci., 31,* 800-812, 1974.

Lübken, F.-J., Seasonal variations of turbulent energy dissipation rates at high latitudes as determined by in situ measurements of neutral density fluctuations, *J. Geophys. Res., 102,* 13,441-13,456, 1997.

McIntyre, M. E., On dynamics and transport near the polar mesopause in summer, *J. Geophys. Res., 94,* 14,617-14,628, 1989.

Müller, P., et al., Nonlinear interactions among internal gravity waves, *Rev. Geophys., 24,* 493-536, 1986.

Ozmidov, R. V., On the turbulent exchange in a stably stratified ocean'Izvestiya, *Acad. Sci. U.S.S.R., Atmos. & Ocean Phys., 1,* 861-871, 1965.

Smyth, W. D., Dissipation-range geometry and scalar spectra in sheared stratified turbulence, *J. Fluid Mech., 401,* 209-242, 1999.

Sonmor, L. J., and G. P. Klaassen, Toward a unified theory of gravity wave stability, *J. Atmos. Sci., 54,* 2655-2680, 1997.

Swenson, G. R., and S. B. Mende, OH emission and gravity waves (including a breaking wave) in all-sky imagery from Bear Lake, UT, *Geophys. Res. Lett., 21,* 2239-2242, 1994.

Tatarskii, V. I., *The Effects of the Turbulent Atmosphere on Wave Propagation*, Israel Program for Scientific Translations, Jerusalem, 1971.

Thorpe, S. A., Transitional phenomena and the development of turbulence in stratified fluids: A review, *J. Geophys. Res., 92,* 5231-5248, 1987.

Thorpe, S. A., Observations of parametric instability and breaking waves in an oscillating tilted tube., *J. Fluid Mech., 261,* 33-45, 1994.

VanZandt, T. E., and D. C. Fritts, A theory of enhanced saturation of the gravity wave spectrum due to increases in atmospheric stability, *Pure Appl. Geophys., 130,* 399-420, 1989.

Weinstock, J., Vertical turbulent diffusion in a stably stratified fluid, *J. Atmos. Sci., 35,* 1022-1027, 1978.

Weinstock, J., Energy dissipation rates of turbulence in the stable free atmosphere, *J. Atmos. Sci., 38,* 880-883, 1981.

Werne, J. A., and D. C. Fritts, Stratified shear turbulence: Evolution and statistics, *Geophys. Res. Lett., 26,* 439-442, 1999.

Yeh, K. C., and C. H. Liu, The instability of atmospheric gravity waves through wave-wave interactions, *J. Geophys. Res., 86,* 9722-9728, 1981.

D.C. Fritts and J.A. Werne, Colorado Research Associates 3380 Mitchell Lane, Boulder, CO 80301. (e-mail: dave@co-ra.com; werne@co-ra.com)

# The Role of Equatorial Waves in the Semiannual Oscillation of the Middle Atmosphere

Rolando R. Garcia

*National Center for Atmospheric Research, Boulder, Colorado*

A review of the observed features of the semiannual oscillation (SAO) of the middle atmosphere is presented, together with numerical simulations that reproduce much of the observed behavior. It is argued that latent heat release by deep convection excites equatorial Kelvin and inertia-gravity waves that play a major role in driving the SAO. The waves in question have scales ranging from planetary ($k = 1 - 3$) to intermediate ($k \simeq 4 - 25$), and periods less than 5 days. In the calculations presented here, planetary-scale Kelvin waves provide over half of the driving for the westerly phase of the stratospheric SAO, while the remainder is supplied by intermediate scale waves; in the mesosphere, the westerly phase is driven by a combination of planetary and intermediate-scale Kelvin waves, while inertia-gravity waves of diurnal period contribute most of the forcing for the easterly phase. The inertia-gravity waves responsible for forcing the mesospheric easterly phase are particularly affected by dissipation during the easterly phase of the quasibiennial oscillation (QBO); this may account for the apparent modulation of the mesospheric SAO by the QBO seen in satellite observations.

## 1. INTRODUCTION

The semiannual oscillation (SAO) of mean zonal winds in the tropical stratosphere and mesosphere was first documented by *Reed* [1962, 1965, 1966]; its main characteristics have since been confirmed by both ground-based [e.g., *Angell and Korshover*, 1970; *Groves*, 1972; *Belmont el al*, 1974, 1975; *Cole and Cantor*, 1975; *Hirota*, 1978, 1980; *Hamilton*, 1982] and space-borne observations [*Hitchman and Leovy*, 1986; *Ortland et al*, 1996]. The SAO is also apparent in zonal mean temperatures; the temperature oscillation is in thermal wind balance with the zonal mean wind [*Andrews et al*, 1987; *Delisi and Dunkerton*, 1988a; *Garcia and Clancy*, 1990].

The SAO is actually two linked oscillations approximately out of phase with each other, one peaking near the stratopause and the other near the mesopause, as shown by *Hirota* [1978]. Recent measurements made by the High-Resolution Doppler Imager (HRDI) onboard the Upper Atmosphere Research Satellite (UARS) have extended the range of observations into the lower thermosphere and documented an apparent coupling between the stratospheric quasibiennial oscillation (QBO) and the mesopause SAO [*Burrage et al*, 1996]. *Garcia et al* [1997] have summarized existing observations and discussed the behavior of the mesopause SAO, including its possible link to the stratospheric QBO.

Even though the mean structure of the SAO is well documented and there is a growing body of data on its variability, the physical mechanism that gives rise to the oscillation is not completely understood. It is generally agreed that the easterly phase in the stratosphere is due to advection of zonal mean easterly mo-

Atmospheric Science Across the Stratopause
Geophysical Monograph 123

mentum by the meridional circulation, with a contribution from planetary Rossby waves propagating into the tropics from the winter hemisphere [*Hopkins*, 1975; *Holton and Wehrbein*, 1980; *Takahashi*, 1984; *Hamilton*, 1986; *Ray et al*, 1998]. (The fact that both advection and Rossby wave forcing are tied to the seasonal cycle, producing easterly accelerations at the solstices, explains the semiannual periodicity of the oscillation). The stratospheric westerly phase is known to be driven, at least in part, by planetary-scale Kelvin waves [*Hirota*, 1980; *Hitchman and Leovy*, 1988]. However, *Hitchman and Leovy* have also shown that large-scale Kelvin waves alone cannot account for the observed westerly acceleration (they estimate their contribution to be between 30 and 70%). This conclusion is consistent with the results of General Circulation Models (GCMs) [e.g., *Sassi et al*, 1993], which do not produce a completely realistic stratospheric SAO, although the computed planetary-scale Kelvin wave amplitudes are comparable or even larger than those observed.

Rather less is known about the mechanism of the mesospheric SAO, except that both the westerly and easterly phases appear to be wave-driven since mean advection at mesopause altitudes cannot account for the observed easterly accelerations. A number of hypotheses have been put forward to explain the mesospheric oscillation; these have generally relied on momentum transport by a spectrum of waves propagating from the lower atmosphere and filtered by the wind system of the stratospheric oscillation. That is, zonal mean westerlies in the stratosphere suppress waves that transport westerly momentum but allow propagation of those that transport easterly momentum, and vice-versa. The result is a mesospheric oscillation out of phase with the stratospheric one, as observed. Although this aspect of the mechanism is almost certainly correct, the nature of the waves involved continues to be the subject of speculation. In *Dunkerton's* [1982] pioneering study of the mesopause SAO it was assumed that momentum was transported by planetary-scale Kelvin waves and small-scale gravity waves (zonal wavelength < 1000 km). A similar scheme was used in a model of the zonally-averaged circulation of middle atmosphere by *Garcia et al* [1992], who dispensed with the planetary-scale Kelvin waves. More recently, *Mengel et al* [1995] have used a parameterization of small-scale gravity waves to model both the quasi-biennial and the semiannual oscillations. Although these mechanistic models have succeeded in producing more or less realistic oscillations in both the stratosphere and mesosphere, the mesospheric oscillation has never been simulated with a GCM [e.g., *Sassi et al*, 1993; *Hamilton et al*, 1995].

The state of theoretical understanding of the SAO can thus be described as mixed. On the one hand, there is little doubt that the westerly phase in the stratosphere, and both phases in the mesosphere, are wave-driven; and that filtering of vertically propagating waves gives rise to the out-of-phase relationship between the stratospheric and mesospheric winds. On the other hand, there is scant evidence about the nature of the waves that drive the SAO, aside from the fact that planetary-scale Kelvin waves are responsible for a substantial fraction of the westerly forcing in the stratosphere.

The purpose of this paper is to review certain observational and modeling studies that shed light on the waves that force the SAO. The studies have been motivated by the work of *Salby et al* [1991] and *Bergman and Salby* [1994], who have analyzed the behavior of tropical deep convection using space-borne OLR observations. In particular, Bergman and Salby have argued that heat release by deep convection can force a broad spectrum of equatorial waves, both in wavenumber and frequency; their findings are consistent with several recent studies [*Manzini and Hamilton*, 1993; *Hayashi and Golder*, 1994; *Sato et al*, 1994; *Sato and Dunkerton*, 1997; *Tsuda et al*, 1994a,b]. It is shown that the upward component of the Eliassen-Palm (EP) flux implied by convective forcing has substantial contributions from Kelvin and inertia-gravity waves at intermediate scales (zonal wavenumbers $k = 4-25$) and short periods (< 2 days); it is also argued from the results of model simulations, that these waves play a major role in driving the SAO, especially in the mesosphere.

It should be noted that wave excitation inferred from OLR observations excludes the very high-frequency, mesoscale gravity waves invoked in gravity-wave parameterizations such as those used in the models of *Dunkerton*, *Garcia et al*, and *Mengel et al*, mentioned above. The possible role of mesoscale waves in forcing the SAO is discussed in Section 5.

## 2. THE OBSERVED SAO

Before discussing the nature and forcing of the waves that drive the SAO, it is convenient to review briefly the main features of the oscillation.

Fig. 1 shows the annual cycle of zonal mean winds obtained from rocketsonde observations at Kwajalein Island (8.7°N, 167.7°W) and Ascension Island (7.6°S, 14.4°W) [see *Garcia et al*, 1997]. The semiannual oscillation is apparent throughout the middle atmosphere above about 40 km. At the solstices, easterly zonal winds are present in the stratosphere, while in the mesosphere the winds are westerly. The mesospheric west-

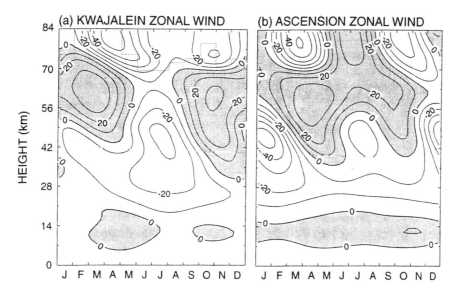

**Figure 1.** Composite seasonal cycle of the semiannual oscillation at Kwajalein (8.7°N, 167.7°W) and Ascension (7.6°S, 14.4°W) islands. Contour intervals are 10 m s$^{-1}$. Data for altitudes/times demarcated by the dashed line in panel ($a$) are extrapolated from lower levels and interpolated between available times. From Garcia et al [1997].

erlies descend gradually until they replace the stratospheric easterlies at the equinoxes, while the mesospheric winds become easterly. These features are seen most clearly in the data from Ascension Island; at Kwajalein, observations are scarce above 70-75 km over much of the year, so the apparent lack of a mesospheric westerly phase at that station should be discounted.

There is a tendency for the first cycle of the SAO in each calendar year at the equator to be stronger than the second [*Delisi and Dunkerton*, 1988*b*]. This can be appreciated in Fig. 2, which shows the seasonal evolution of the zonal wind as a function of latitude at 40 and 80 km. Both easterly and westerly wind maxima are stronger in the half-year comprising northern hemisphere winter and spring than they are northern summer and fall. *Delisi and Dunkerton* have argued that this seasonal asymmetry is caused by a stronger mean meridional circulation in northern winter, which –through advection of mean easterly momentum from the summer hemisphere– produces stronger tropical easterlies near the stratopause. This stratospheric easterly phase then sets the stage for a strong westerly phase because it enhances the vertical propagation of waves that force the westerly phase.

Recent satellite observations made by the HRDI instrument onboard UARS have documented the interannual behavior of the SAO throughout most of the stratosphere and mesosphere. Fig. 3 shows slightly more than 3 years of HRDI observations, and reveals an interesting relationship between the strength of the easterly phase of the mesospheric SAO and the phase of the QBO in the lower stratosphere. As first noted by *Burrage et al* [1996], the mesospheric easterly phase is strong when the stratospheric QBO winds are westerly (e.g., in April, 1993 and March, 1995), and relatively weak when the QBO winds are easterly (e.g., in November, 1993 and May, 1994). On the other hand, the westerly phase of the mesospheric SAO appears to be unaffected by the QBO.

The observations just discussed are part of a growing body of data on the SAO [see, *Garcia et al*, 1997 for a comprehensive review]. These data document the annual cycle of the SAO throughout the entire middle atmosphere, and begin to illustrate its interannual variability. It is shown below that many of the features displayed in the data can be simulated with numerical models provided that these take into account the forcing produced by the full spectrum of equatorial waves excited by tropical deep convection.

## 3.   FORCING OF EQUATORIAL WAVES BY DEEP CONVECTION

### 3.1.   Theoretical Formulation

*Bergman and Salby* [1994] have used the theoretical formulation of *Salby and Garcia* [1987] to express the geopotential response to deep convective heating in the tropical troposphere. The result is derived for a spherical atmosphere at rest (or in the present of a constant angular momentum background wind) which, as argued

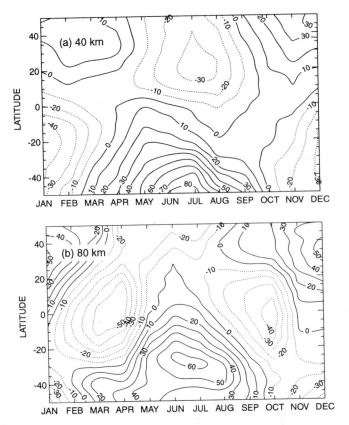

**Figure 2.** Composite seasonal cycle of the semiannual oscillation at 40 and 80 km derived from UARS/HRDI observations. From Garcia et al [1997].

by *Salby and Garcia*, is a reasonable approximation of conditions in the "near field", i.e., immediately above the forcing region.

The method consists of writing the solution in the form $\phi' = \Phi_k^\omega(\varphi, z) \, e^{i(k\lambda - \omega t)}$ and combining the primitive equations into a separable partial differential equation. Separation then yields a horizontal structure equation, Laplace's equation, whose solutions are the "Hough modes" [*Longuet-Higgins*, 1968], and a vertical structure equation, which is solved using Green's functions. The Hough functions are the normal modes of a shallow, rotating atmosphere; they consist of three manifolds: eastward-propagating inertia-gravity waves (including Kelvin waves), westward-propagating inertia-gravity waves, and Rossby waves. Rossby-gravity waves are included under either the eastward gravity wave manifold or the Rossby manifold, depending on their direction of propagation.

Details of the derivation are given by *Salby and Garcia* [1987]; the result for the geopotential field above

the forcing region, for the zonal wavenumber/frequency pair $(k, \omega)$ is:

$$\Phi_{kn}^\omega(z) = i \, \frac{\kappa Q_{kn}^\omega}{\omega \Gamma \, H} \, P_m^- e^{(1/2H - im)z} \qquad (1)$$

where $Q_{kn}^\omega$ is the projection of the $(k, \omega)$ component of the heating onto the Hough mode of meridional index $n$; and $P_m^-$, the projection onto vertical wavenumber $m$. $H$ is the atmospheric scale height; $\Gamma = (1 + i\alpha/\omega)$; $\alpha$ is a Newtonian cooling coefficient; $\kappa = R/c_p$; $R$ is the gas constant for air; and $c_p$, the specific heat at constant pressure.

According to (1), the response $\Phi_{kn}^\omega(z)$ to convective forcing depends upon the wavenumber-frequency spectrum of the heating, and its projection onto the merid-

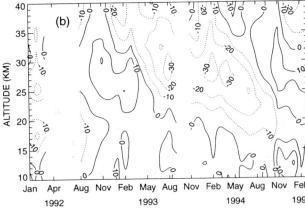

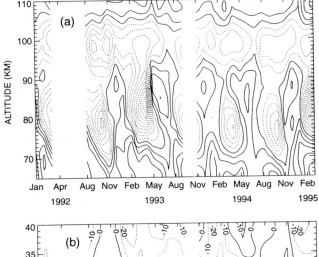

**Figure 3.** Evolution of the mean zonal wind in (a) the mesosphere and (b) the stratosphere, as observed by UARS/HRDI. The quasibiennial oscillation is apparent in the stratosphere below 35 km. Note that the easterly phase of the mesospheric SAO is much stronger during the QBO westerly phase. From Garcia et al [1997].

ional structure of the Hough modes and the vertical structure of the forced wave. An important implication of this result is that, if the heating has a characteristic depth, $D$, then the vertical projection $P_m^-$ behaves as shown in Fig. 4a, i.e., it has a primary maximum for waves of vertical wavelength twice the depth of the forcing; a secondary maximum is associated with waves of vertical wavelength approximately two-thirds the depth of the forcing. This mechanism of vertical scale selection was first noted by *Holton* [1972, 1973], and has been discussed in detail by *Salby and Garcia* [1987].

The EP flux associated with each class of wave forced by convective heating can be expressed in terms of the geopotential if thermal damping is neglected, which is a reasonable assumption in the lowermost stratosphere, immediately above the convective forcing region. *Horinouchi and Yoden* [1998] have shown that, with this assumption, the vertical component of EP flux can be written as:

$$\{F_z\}_{kn}^\omega = -\frac{k}{aN^2} Re\{m\} |\Phi_{kn}^\omega(z)|^2. \quad (2)$$

Note that $F_z$ behaves as $|\Phi_{kn}^\omega(z)|^2$ multiplied by the vertical wavenumber, $m$. Thus, relative to the geopotential response, the vertical component of EP flux is emphasized for waves of smaller vertical wavelength (larger $m$), as shown in Fig. 4b.

### 3.2. Inferences From Tropical OLR Observations

The formalism just described can be applied to observations of convection to deduce the geopotential response (1) and the EP flux (2) associated with equatorial waves excited by convective heating. Estimates of the heating can be obtained from outgoing longwave radiation (OLR) measurements made by the International Satellite Cloud Climatology Program. These data have been processed by Salby and co-workers to create a global dataset of OLR, the Global Cloud Imagery (GCI) [*Tanaka et al*, 1991; *Salby et al*, 1991]; the data are available every three hours, on a $512 \times 512$ latitude/longitude grid. The GCI data can be converted into equivalent brightness temperatures, $T_{br}$, whence an index of convective activity can be defined by selecting suitably low $T_{br}$, which are associated with deep convective clouds in the tropics [*Richards and Arkin*, 1981; *Hendon and Woodberry*, 1993; *Elbert et al*, 1996]. Further, as shown by *Ricciardulli and Garcia* [2000], the convective activity index can be related to precipitation, and therefore to the latent heat released by convection.

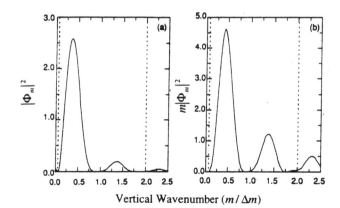

**Figure 4.** Vertical projection response for (a) the geopotential squared and (b) the vertical component of EP flux, as functions of the vertical wavenumber $m$. $\Delta m = 2\pi/D$, where $D$ is the characteristic depth of the convective forcing, so maximum response occurs at $(m/\Delta m) = 0.5$, i.e., at $\lambda_z = 2D$. The vertical component of EP flux behaves as $m$ times the geopotential squared, emphasizing the secondary response at $(m/\Delta m) \simeq 1.3$ ($\lambda_z \simeq 2D/3$). Adapted from *Bergman and Salby* [1994].

The studies of *Richards and Arkin* and *Elbert et al* have shown that the correlation between tropical rainfall rates and low $T_{br}$ is quite good over a broad range of spatial and temporal scales, so the GCI data are expected to reproduce accurately the shape of the spectrum of convective heating over the space and time scales resolved by the OLR observations. Of course, if there is significant variability on scales smaller than the resolution of the GCI dataset, this will not be captured by the observations and in fact will alias onto resolved wavenumbers and frequencies. Individual convective cells –or even convective complexes smaller than about 100 km– are not resolved by the GCI data, and neither is convective variability that occurs at periods shorter than 6 hours. In particular, heating variability associated with oscillations of convective turrets in mesoscale systems [*Fovell et al*, 1992; *Alexander et al*, 1995] is not resolvable by the GCI.

Another source of uncertainty in the estimation of convective heating rates from GCI data is the proportionality factor relating $T_{br}$ to heating rate. *Ricciardulli and Garcia* [2000] show how $T_{br}$ can be related to heating, but the relationship is predicated on time-mean values [*Janowiak and Arkin*, 1991] and assumed to hold over all time scales. To the extent that this assumption is inaccurate, it will introduce errors into the magnitude of the heating rates inferred from the GCI data.

**Power Spectrum of Heating (GCI)**

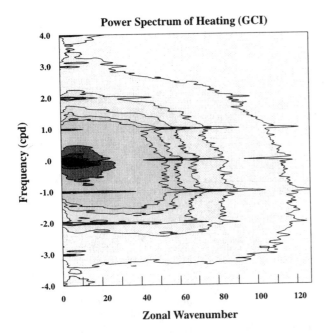

**Figure 5.** Normalized wavenumber-frequency power spectrum of convective heating derived from observations of outgoing longwave radiation. See text for details.

Finally, the calculation of wave excitation from the GCI data requires the assumption of a vertical profile for the heating, since this is not available from the OLR observations. *Salby and Garcia* [1987] have shown that the vertical projection of the heating, $P_m^-$ in Eq. (1), depends strongly on the depth of the heating distribution and less so on the details of the vertical profile. For the calculations presented in this paper, a sinusoidal profile of depth 12 km is chosen for simplicity.

Fig. 5 shows the wavenumber-frequency spectrum of convective heating derived from the GCI dataset for the period 14 November – 20 March, 1984. Although the spectrum is red in both wavenumber and frequency, a striking feature is the slow fall-off of variance at large wavenumbers and frequencies; in particular, the contributions of the diurnal frequency and its harmonics are clearly seen.

Fig. 6 shows the wavenumber-frequency distribution of the vertical component of EP-flux obtained from Eqs. (1)-(2) when the heating $\mathcal{Q}_{kn}^{\omega}$ is derived from the GCI spectrum shown in Fig. 5. Apart from the large amplitude near the origin, the distribution displays a two-lobe structure at higher frequencies and wavenumbers, with largest amplitude occurring along lines corresponding to phase speeds $\omega/k \simeq 40$ and 15 m s$^{-1}$. These preferred values of phase speed arise from the vertical scale selection discussed earlier. For Kelvin waves

(and high-frequency inertia-gravity waves), the vertical wavenumber $m$ is related to the phase speed according to $c = N/m$, where $N$ is the buoyancy frequency; the preferred phase speeds of 40 and 15 m s$^{-1}$ correspond to vertical wavelengths, $\lambda_z = 2\pi/m$ approximately twice and two-thirds the depth of the heating $D$ which, as mentioned above, is taken to be 12 km.

Although the distribution of $F_z$ is dominated by the response at very low frequency, this is associated with waves that do not propagate readily into the middle atmosphere [*Bergman and Salby*, 1994; *Ricciardulli and Garcia*, 2000]. Elsewhere, significant amplitude is found at frequencies well beyond the diurnal, and at wavenumbers up to about $k = 30$ in the 40 m s$^{-1}$ lobe and up to $k = 70 - 80$ in the 15 m s$^{-1}$ lobe. *Ricciardulli and Garcia* have argued that waves associated with the higher phase speed lobe are likely to play a role in the semiannual oscillation; those associated with the slower phase speed lobe, which have smaller intrinsic frequency and so are likely to be dissipated in the lower stratosphere, may be important for the QBO.

*Ricciardulli and Garcia* have compared wave momentum fluxes inferred from GCI data with estimates obtained by *Sato and Dunkerton* [1997]. *Sato and Dunker-*

**Vertical Component of EP Flux (GCI)**

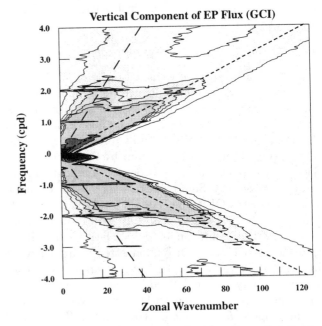

**Figure 6.** Normalized wavenumber-frequency distribution of the vertical component of EP flux, derived from the convective heating spectrum shown in Fig. 5. In order to enhance readability the quantity plotted is the absolute value, $|F_z|$; note that $F_z$ is actually negative at positive (eastward) frequencies.

*ton* find that (1) momentum fluxes at tropopause level are about evenly divided between eastward and westward propagating components, with magnitudes in the range $0.7 - 6 \times 10^{-2}$ m$^2$ s$^{-2}$, and (2) long-period (5-20 days) waves carry no more than 25% of the total momentum flux of any one sign. These results are broadly consistent with those of *Ricciardulli and Garcia*: momentum fluxes inferred from GCI data are also evenly divided between eastward and westward-propagating components, with typical magnitudes at tropopause level of $1.3 - 1.4 \times 10^{-2}$ m$^2$ s$^{-2}$, and a contribution from long-period waves of about 20%.

Although not explicitly stated by *Sato and Dunkerton*, long period waves are synonimous with large-scale Kelvin waves, since inertia-gravity waves generally occur at periods shorter than 2 days. For example, Kelvin waves of $k = 1$ and 2, and phase speed 40 m s$^{-1}$ (corresponding to the high phase speed lobe in Fig. 6) have periods of about 15.6 and 7.8 days, respectively. This implies that the fraction of $F_z$ carried by long-period, large-scale Kelvin waves inferred from GCI data is not unrealistic.

Fig. 7 shows $F_z(\omega)$ for Kelvin and inertia-gravity waves of phase speed $c > 30$ m s$^{-1}$; this includes the 40 m s$^{-1}$ lobe and therefore selects the faster waves that are expected to be important for forcing the SAO. $F_z(\omega)$ is obtained from the distribution in wavenumber and frequency by summing over all wavenumbers $k < \omega/c$, where $c = 30$ m s$^{-1}$, including the Kelvin and gravity wave manifolds. The Rossby manifold is ignored since those waves occur at low frequencies and do not propagate efficiently into the middle atmosphere. The top panel of Fig. 7 shows $F_z$ derived from the same GCI data as Fig. 6, while the bottom panel shows results obtained from version 2 of the NCAR Community Climate Model (CCM2) [*Hack*, 1994; *Hack et al*, 1994; *Boville*, 1995.] Several striking features are worth noting: The $F_z(\omega)$ distribution derived from GCI data has large amplitude at frequencies $|\omega| > 0.5$ cycles per day (cpd), with prominent diurnal and semidiurnal harmonics; at positive (eastward) frequencies contributions to $F_z$ are fairly evenly divided between Kelvin and inertia-gravity waves, while at negative (westward) frequencies $F_z$ is due solely to inertia-gravity waves. By contrast, the CCM2 results exhibit much smaller contributions by either eastward or westward inertia-gravity waves, although the Kelvin wave contribution is comparable to that of the GCI. The implications of this result are touched upon in Section 5.

The results shown in Figs. 6 and 7 suggest that high frequency Kelvin and inertia-gravity waves are likely to

VERTICAL COMPONENT OF EP FLUX, C > 30 m/s

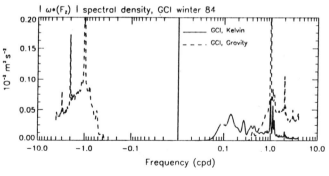

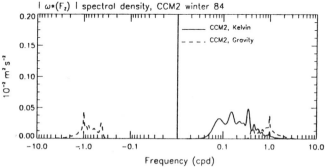

**Figure 7.** Frequency-weighted distribution of the magnitude of the vertical component of EP flux, for equatorial waves of phase speed $|c| > 30$ m s$^{-1}$. The top panel shows results derived from observational estimates of convective heating; the bottom panel shows results obtained from the convective heating calculated by the NCAR Community Climate Model. See text for details.

play a role in the SAO, and that the forcing of gravity waves is severely underestimated in some GCMs (possibly accounting for the fact that such models do not produce a realistic SAO). In what follows, the results of some model calculations that illustrate and support these ideas are presented.

## 4. MODELING THE SAO

### 4.1. Numerical Model

The role of high-frequency equatorial waves in forcing the SAO can be tested by using estimates of convective heating to force a numerical model of the tropical middle atmosphere. The model used here is a three-dimensional, beta-plane model that solves the momentum and temperature equations for the zonal mean and zonal wavenumber $k = 1 - 15$. The model is quasi-nonlinear in that the waves interact individually with the zonal mean state, but wave-wave interactions are neglected. The model domain extends from the surface to 120 km, with 1 km vertical resolution, and spans the latitude range ±40°, with 3° horizontal resoltion. The model and the method of solution are described in detail by *Sassi and Garcia* [1997].

Waves are excited in the model by including a latent heating term in the thermodynamic equation. As explained by *Sassi and Garcia* the spatial and temporal behavior of the heating are specified so that its wavenumber-frequency spectrum resembles that of Fig. 5.

Because the model is formulated on a beta-plane, it cannot simulate the global mean meridional circulation or its seasonal variation (which is principally responsible for producing easterly winds near the stratopause at the solstices, as discussed in the Introduction). To simulate this effect, an easterly drag is imposed that maximizes at the solstices and produces stratospheric easterlies similar to those observed. The easterly drag during northern hemisphere winter is specified to be somewhat stronger than in northern summer, to mimic the seasonally-asymmetric effect of mean advection by the meridional circulation discussed in Section 2.

It should be borne in mind that some of the simplifications inherent in the beta-plane model may affect the results of the calculations, and therefore the generality of the conclusions derived from them. The uncertainties associated with inferring convective heating rates from OLR observations have been discussed above. Since the heating distribution used in the calculations is specified on the basis of the GCI data, the excitation of waves in the model is subject to those uncertainties. In addition, the background zonal wind in the lower stratosphere influences the propagation of waves emanating from the troposphere [*Bergman and Salby*, 1994]. In the simulations presented below, the zonal wind in the lower stratosphere is weak ($|\bar{u}| < 5$ m s$^{-1}$). Thus, the results of the calculations could be biased insofar as the actual winds in the region may be stronger than this value. However, as will be seen presently, the waves that drive the SAO in the model have phase speeds $|c| > 30$ m s$^{-1}$, so they are unlikely to be affected strongly by lower stratospheric winds, even if the magnitude of latter was in the 10–15 m s$^{-1}$ range.

### 4.2. Simulation of the SAO

Fig. 8 shows the evolution of the zonal wind at the equator obtained with the model just described. The evolution compares well with observations of the SAO (e.g., Fig. 1), as regards both amplitude and phasing of the easterly and westerly wind systems. In the mesosphere ($\sim 60-75$ km) the zonal wind is in superrotation throughout most of the annual cycle, consistent with

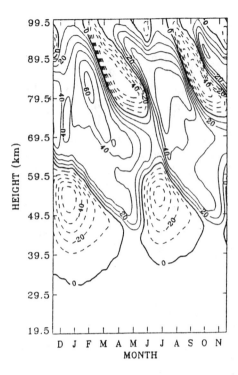

**Figure 8.** Simulation of the semiannual oscillation produced by the numerical model of Sassi and Garcia [1997]. Adapted from Sassi and Garcia [1997].

Fig. 1 [see also *Dunkerton*, 1982; *Garcia et al*, 1997]. A seasonal asymmetry in the strength of the modelled SAO, similar to that observed, is also apparent. The stronger easterly phase in the stratosphere is, of course, a direct result of the stronger easterly drag imposed in that season, but the subsequent strong westerly phase results from enhanced wave transmission, as hypothesized by *Delisi and Dunkerton* [1988b].

An unrealistic feature of the simulation is that the mesospheric easterly phase is weaker and the westerly phase stronger than observed. A possible reason for this behavior is that momentum deposition by the solar diurnal tide is neglected in the model. *Lieberman and Hays* [1994] have shown that the solar diurnal tide can induce accelerations of as much as $-5$ to $-10$ m s$^{-1}$ day$^{-1}$ near 90 km. Easterly forcing by the diurnal tide would introduce an easterly bias in the zonal wind at and above mesopause level.

Fig. 9 shows the seasonal evolution of wave driving as a function of zonal wavenumber near the stratopause and mesopause. At the stratopause, the westerly phase at the equinoxes is driven mainly by planetary-scale waves ($k = 1 - 3$), with a smaller but significant contribution from intermediate wavenumbers. The stratopause easterly phase is not wave driven, but depends on the

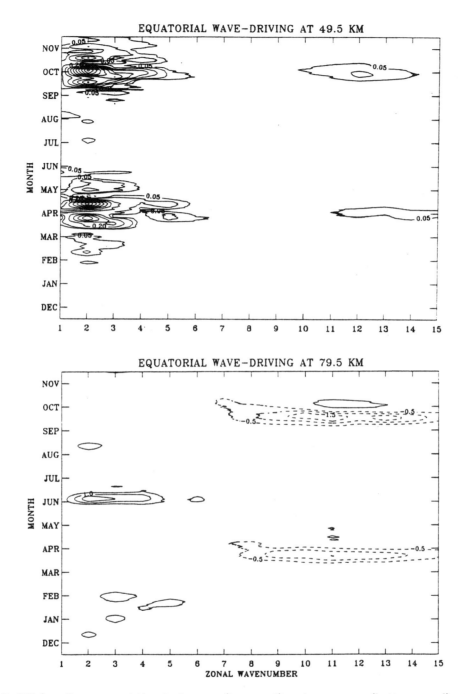

**Figure 9.** EP flux divergence at the stratopause (top panel) and mesopause (bottom panel) calculated with the numerical model of Sassi and Garcia [1997]. The contour intervals are 0.1 m s$^{-1}$day$^{-1}$ in the top panel and 0.5 m s$^{-1}$day$^{-1}$ in the bottom panel. Adapted from Sassi and Garcia [1997].

specified easterly forcing discussed above. As shown by *Sassi and Garcia* [1997], these solsticial easterlies are a crucial part of the SAO mechanism in that they set the "clock" for the oscillation, thus producing its semi-annual periodicity. At mesopause altitudes, westerly wave driving occurs at the solstices, and is associated

with wavenumbers $k = 1 - 6$, but with a fairly uniform contribution from all wavenumbers in this range. By contrast, easterly wave driving at the equinoxes is provided almost exclusively by wavenumbers $k \geq 7$.

*Sassi and Garcia* [1997] have analyzed in detail the wave field and identified the waves that drive the SAO in

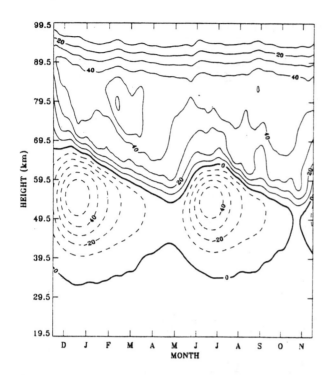

**Figure 10.** As in Fig. 8, but for a calcualtion wherein wave driving is restricted to planetary-scale waves ($k = 1 - 3$).

this numerical model. The planetary-scale waves that provide westerly forcing at both the stratopause and mesopause are Kelvin waves; at higher wavenumbers, westerly forcing is due to both Kelvin and eastward-propagating inertia-gravity waves. Easterly forcing at wavenumbers $k = 7 - 15$ is due entirely to westward-propagating inertia-gravity waves. Thus, the modeled SAO depends on the presence of equatorial waves at intermediate scales, i.e., scales smaller than those of the planetary waves ($k = 1 - 3$), but substantially larger than small-scale gravity waves (zonal wavelength < 1000 km). In the model, these intermediate scales include wavenumbers $k = 4 - 15$, the latter being the truncation limit of the calculations. However, it is likely that the range of intermediate-scale wavenumbers that contribute significantly to the SAO extends well beyond $k = 15$, perhaps to $k = 20 - 30$, as suggested by the $F_z$ distribution of Fig. 6.

### 4.3. Role of Intermediate-scale Waves

To illustrate the role of intermediate-scale waves in the modeled SAO, the results of a simulation that includes only planetary-scale waves ($k = 1 - 3$) are shown in Fig. 10. The evolution of the zonal wind at the equator bears little resemblance to the results obtained with the full model or to the observations. A semian-

nual oscillation is still present in the stratosphere, but the westerly phase fails to descend below 50 km, indicating that the intermediate-scale waves are necessary to produce a realistic westerly phase in the model. This is consistent with *Hitchman and Leovy's* [1986] finding that planetary-scale Kelvin waves do not account in full for the observed accelerations during the stratopause westerly phase.

In the mesosphere there is no longer a semiannual oscillation; instead, a broad layer of westerlies is present above about 60 km. These westerlies are produced by planetary-scale Kelvin waves, which now dominate wave driving at all altitudes. Although planetary-scale, westward-propagating inertia-gravity waves are present in the simulation and provide some easterly wave driving, this is negligible (cf. Fig. 9) and is overwhelmed by the westerly driving due to the Kelvin waves.

Comparison of Figs. 8 and 10 suggests that the westerly wind layer that persists throughout the course of the year in much of the mesosphere ($\sim 60 - 75$ km) is due to forcing by planetary-scale Kelvin waves. When the model is truncated at $k = 3$, the layer extends beyond the mesopause because the intermediate-scale waves that would provide easterly forcing in the mesosphere are absent.

### 4.4. Role of Diurnal Forcing

It was noted in connection with Figs. 6 and 7 that the EP flux generated by tropical convection exhibits prominent maxima at the diurnal frequency and its harmonics. A strong, spatially coherent diurnal cycle is characteristic of convective forcing over the tropical landmasses [*Salby et al,* 1991]. To assess the importance of the diurnal cycle, its contribution to the convective heating field can be suppressed in model simulations [see *Sassi and Garcia,* 1997, for details]; Fig. 11 shows the results of a model run wherein the diurnal cycle of convection has been eliminated. The stratospheric SAO is comparable to that obtained with the full model (Fig. 8), although the westerlies do not descend as deeply into the stratosphere at equinox. In the mesosphere, however, the zonal wind evolution resembles that obtained with the truncated ($k = 1 - 3$) version of the model, suggesting that removal of the diurnal cycle of convection has resulted in much weaker excitation of intermediate-scale inertia-gravity waves. As a consequence, the westerly accelerations produced by planetary-scale Kelvin waves dominate the mometum budget throughout the mesosphere and produce time mean westerly winds up to about 90 km; only above this altitude is a (weak) semiannual oscillation present.

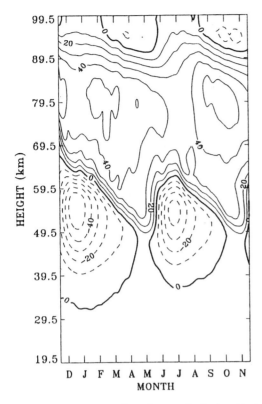

**Figure 11.** As in Fig. 8, but for a calculation that omits forcing by the diurnal cycle of convection.

*Sassi and Garcia* [1997] have argued that the importance of diurnal forcing for inertia-gravity waves arises from the differences between the dispersion relationship of these waves and that for the Kelvin waves. Fig. 12 shows these dispersion relationships. Since the spectrum of convective heating is red (Fig. 5), the large amount of variance at low frequency and wavenumber will project strongly upon the planetary-scale Kelvin waves, but not upon the inertia-gravity waves since, even for the gravest meridional mode, these do not occur at periods smaller than about a day. Thus, forcing of inertia-gravity waves depends sensitively on the amount of power available near the diurnal frequency; when the diurnal cycle of convection is removed, forcing of these waves is considerably diminished.

### 4.5. Interaction With the QBO

The importance of diurnal excitation for intermediate-scale inertia-gravity waves has some interesting consequences for the forcing of the mesospheric SAO, which may help explain the apparent correlation observed between the latter and the stratospheric QBO. *Burrage et al* [1996] noted that the easterly (but not the westerly) phase of the mesospheric SAO is substantially stronger

when QBO westerlies prevail in the stratosphere than it is during the QBO easterly phase (see also Fig. 3 above).

*Garcia and Sassi* [1999] have recently shown how such a modulation can come about through filtering of the upward-propagating wave spectrum by the QBO winds. Fig. 13 displays the results of a simulation carried out with a version of the beta-plane model described earlier, wherein a quasibiennial variation of the zonal wind in the lower stratosphere has been introduced. This is done by relaxing the model zonal wind to observed values. Just under four years of the simulation are shown in the figure, but it is clear that the calculated mesospheric winds exhibit the same relationship with the QBO as do the HRDI observations shown in Fig. 3. For example, the two mesospheric easterly phases of year 2 (when stratospheric winds are easterly) are considerably weaker than the first easterly phase of year 1 or the second easterly phase of year 3 (when stratospheric winds are westerly). During the weak easterly phases, the mesospheric zonal winds are about $-30$ m s$^{-1}$, but during the strong phases they reach between $-40$ and $-45$ m s$^{-1}$. The variation, although considerable, underestimates that seen in the observations of Fig. 3, where the change can be as large as a factor of two ($-40$ vs. $-80$ m s$^{-1}$), possibly because the $k = 15$

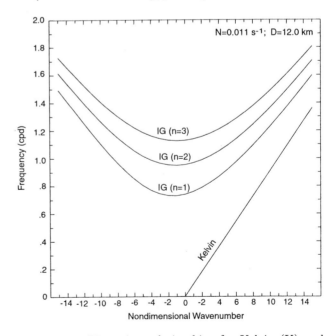

**Figure 12.** Dispersion relationships for Kelvin (K) and inertia-gravity (IG) waves forced by latent heat release distribution of depth $D = 12$ km. The numbers labeling the inertia-gravity wave dispersion curves denote the meridional indices of the modes.

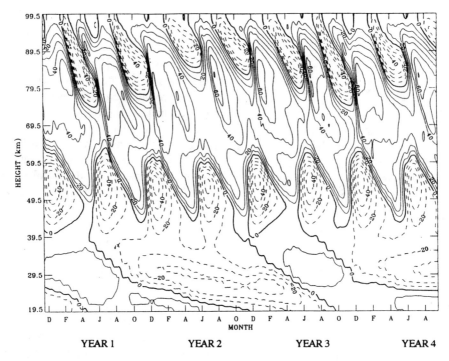

**Figure 13.** Interannual variability of the semiannual oscillation produced by the model of Sassi and Garcia [1997] when QBO wind variability is introduced in the stratosphere. From Garcia and Sassi [1999].

truncation used in the model underestimates the forcing due to inertia-gravity waves [*Garcia and Sassi*, 1999].

The peculiar modulation of only the easterly phase of the mesospheric SAO can be understood by noting (1) that the stratospheric QBO is asymmetric, i.e., winds are stronger in the easterly phase ($\sim -30$ m s$^{-1}$) than they are during the westerly phase ($\sim 15$ m s$^{-1}$), and (2) that, as a result, inertia-gravity waves are dissipated preferentially during the easterly phase of the QBO. The asymmetry of the QBO winds can be appreciated from Fig. 13; to illustrate the effect of those winds on the vertically-propagating waves, the power spectrum of eddy temperature at 60 km calculated with the model for both phases of the QBO is shown in Fig. 14.

Variance due to inertia-gravity waves is concentrated at constant frequencies of $\pm 1$ cpd. These correspond to westward- and eastward-propagating inertia-gravity waves excited by the diurnal cycle of convection. Note that there are marked differences in the spectra between the two phases of the QBO: diurnal variance at negative frequencies is substantially smaller at wavenumbers $k > 10$ during the easterly phase of the QBO than during the westerly phase. The waves in question have phase velocities comparable to the QBO easterlies, which reduces their intrinsic frequency and thus

enhances their rate of dissipation. For example, if $\omega_d$ denotes the diurnal frequency, then $c = -\omega_d/k = -42$ m s$^{-1}$ at $k = 11$ and $-31$ m s$^{-1}$ at $k = 15$). Clearly, these waves will be subject to strong dissipation as they propagate through QBO easterly wind maxima, such as those occurring during the two (weak) SAO easterly phases of year 2 in Fig. 13. The weakness of these SAO phases is then attributable to decreased easterly forcing due to dissipation in the stratosphere of part of the inertia-gravity wave spectrum at negative frequencies.

The westerly phase of the mesospheric SAO, on the other hand, is driven mainly by Kelvin waves. As shown in Fig. 13, Kelvin wave variance lies along a line of constant phase speed $c \simeq 54$ m s$^{-1}$. This is higher than the 40 m s$-1$ lobe of preferred phase speed seen in the EP flux spectrum of Fig. 6; the shift to higher phase speed is due to dissipation of the slower waves between the source region and the lower mesosphere (60 km), where the spectrum of Fig. 14 is calculated. Such selective damping of the slower components of the wave spectrum has been widely documented in the past [*Holton*, 1973; *Hayashi et al*, 1984; *Garcia and Salby*, 1987, *Boville and Randel*, 1992]. Propagation of the Kelvin waves is largely unaffected by the QBO westerlies, which at 15 m s$^{-1}$ are much slower than the phase speed of the waves.

z = 60 Km

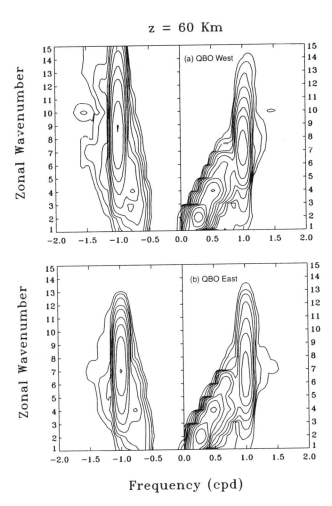

**Figure 14.** Normalized wavenumber-frequency spectrum of temperature for the model calculation of Fig. 13 during (a) the westerly and (b) the easterly phase of the QBO. Easterly wave power (negative frequencies) is reduced at $k \geq$ 11 during the easterly phase of the QBO. Adapted from Garcia and Sassi [1999].

As a consequence, modulation of the mesospheric westerly phase of the SAO does not occur [see *Garcia and Sassi*, 1999, for details].

## 5. SUMMARY AND DISCUSSION

It has been argued that the salient features of the semiannual oscillation can be understood in terms of selective propagation and dissipation of a spectrum of vertically-propagating equatorial waves forced by deep convection. The waves in question include the familar planetary-scale Kelvin waves, which provide over half of the forcing for the westerly phase of the stratospheric SAO, and intermediate-scale ($k = 4 - 25$) Kelvin and

inertia-gravity waves, which are particularly important for the mesospheric SAO.

A series of numerical simulations of the SAO wherein tropical waves are forced by a distribution of convective heating that resembles that observed yields the following picture of the oscillation:

1. The SAO "clock" is set by the occurrence of equatorial easterly winds at the solstices. These are the result of the seasonal cycle of the global mean meridional circulation, which advects easterly momentum across the equator at the solstices. Forcing by planetary Rossby waves propagating from the winter hemisphere may also contribute to the generation of equatorial easterlies at the solstices.

2. Planetary-scale Kelvin waves provide more than half of the forcing for the westerly phase of the stratospheric SAO, the remainder is supplied by intermediate-scale Kelvin and inertia-gravity waves. The westerly phase descends from the mesosphere as stratospheric easterlies disappear at the equinoxes.

3. Kelvin wave forcing is responsible for the westerly wind layer in the mesosphere (60-75 km). Around the midpoint of this layer the SAO is weakest: it lies above the altitude range influenced by the seasonal cycle of easterly mean advection, and below the altitude range where forcing by inertia-gravity waves can give rise to mesospheric easterlies.

4. In the upper mesosphere, intermediate-scale waves become increasingly important in the momentum budget; wave driving of the easterly phase in the mesosphere is provided almost exclusively by inertia-gravity waves of intermediate scale. Calculations that exclude wavenumbers $k > 3$ fail to produce the mesospheric SAO (or a wholly realistic stratospheric SAO).

5. Because inertia-gravity waves do not occur at frequencies much smaller than 1 cpd, the diurnal cycle of convective heating is a very important excitation mechanism for these waves. Removal of diurnal excitation results in a simulated mesospheric SAO that is much too weak and occurs at altitudes much higher than observed.

6. The inertia-gravity waves that force the easterly phase of the mesospheric SAO are susceptible to strong dissipation in the lower stratosphere during the easterly phase of the QBO. This may account for the apparent modulation by the QBO of the easterly (but not the westerly) phase of the mesospheric SAO.

The foregoing picture is consistent with many of the available observations, but it must be borne in mind that the calculations presented in this paper have been carried out with a mechanistic, beta-plane model rather than a comprehensive GCM. Although the model is forced with a heating distribution whose wavenumber-frequency spectrum resembles that inferred from OLR observations, the estimates of heating from OLR data are themselves subject to uncertainties, as discussed in Section 3.2. It should be noted, however, that comparison of momentum fluxes inferred from OLR data with the recent estimates of *Sato and Dunkerton* [1997] reveals broad agreement, both as regards the magnitude of the fluxes and the fraction of the total carried by long-period, planetary-scale Kelvin waves (see Section 3.2).

The waves that drive the SAO in the numerical model have spatial scales that are well resolved by GCMs; nevertheless, the mesospheric SAO has never been simulated in a GCM, while the amplitude of the stratospheric oscillation and its penetration into the middle stratosphere are usually underestimated. A possible reason for this state of affairs is that the higher frequency, intermediate-scale waves that play a crucial role in the calculations discussed here are not forced efficiently in GCMs. *Ricciardulli and Garcia* [2000] have argued that, at least in the NCAR Community Climate Model (CCM), the problem arises from the failure to reproduce the observed variability of convective heating. Convection operates on small scales that cannot be resolved with current GCMs and must therefore be parameterized; the convective schemes used in recent versions of the NCAR CCM reproduce quite well the observed time-mean heating distribution, but underestimate its variability, sometimes severely. Reduced heating variance at higher frequencies then results in unrealistically weak excitation of inertia-gravity waves; an example of this behavior, for CCM2, was shown in Fig. 7.

Small-scale gravity waves (which for the most part are not resolvable by GCMs) may also play a role in the SAO, especially away from the equator, where they could help explain the latitudinal asymmetry of the stratospheric oscillation [*Hamilton*, 1986; *Garcia et al*, 1997; *Ray et al*, 1998]. If small-scale waves are indeed important in forcing the SAO, they must be excited at very high frequencies, since otherwise their phase speeds would be too small to allow propagation into the upper stratosphere and mesosphere. For example, a wave of wavelength $\lambda_x = 1000$ km and phase speed $c = 30$ m s$^{-1}$ has a frequency $\omega = 2.6$ cpd; for a more typical mesoscale gravity wave ($\lambda_x \sim 100$ km), $\omega \sim 26$ cpd, so the period of such a wave is shorter than one hour.

Recent observations [e.g., *Sato*, 1993; *Alexander and Pfister*, 1995; *Sato et al*, 1995; *Dewan et al*, 1998] do indicate that convection excites small-scale gravity waves at very high frequencies. The forcing mechanism may be related to the oscillation of convective turrets, which takes place on time scales comparable to the buoyancy period [*Fovell et al*, 1992; *Alexander et al*, 1995]. On the other hand, the diurnal cycle of convection, which plays a central role in forcing the intermediate-scale waves discussed in this paper, is not a viable excitation mechanism for small-scale gravity waves able to propagate to great heights. Thus, insofar as the diurnal cycle excites small-scale waves, these waves are more likely to play a role in driving the QBO than the SAO.

It is difficult to confirm observationally the importance of the intermediate-scale waves discussed here (or of small-scale gravity waves) because of the lack of global, synoptic data. Observations from polar-orbiting satellites can resolve unambiguously zonal wavenumbers $k \leq 6$ and periods $\geq 2$ days; ground-based observations have much better frequency resolution but provide only local information. Coordinated use of satellite observations from NASA's forthcoming TIMED mission (Thermosphere, Ionosphere and Mesosphere Energetics and Dynamics) and existing ground-based stations may help shed light on the presence of smaller-scale, higher-frequency waves and their role in the SAO.

*Acknowledgments.* The author wishes to thank W.J. Randel, A.K. Smith, and two anonymous reviewers for their comments on the original manuscript. This work was supported in part by a grant from the National Aeronautics and Space Administration. The National Center for Atmospheric Research is sponsored by the National Science Foundation.

## REFERENCES

Alexander, J.A., J.R. Holton, and D.R. Durran, The gravity wave response above deep convection in squall line simulations, *J. Atmos. Sci.*, 52, 2212-2226, 1995.

Alexander, J.A. and L. Pfister, gravity wave momentum flux in the lower stratosphere over convection,, *Geophys. Res. Lett.*, 22, 2029-2032, 1995.

Andrews, D.G., J.R. Holton, and C.B. Leovy, *Middle Atmosphere Dynamics*, Academic Press, 489pp, 1987.

Angell, J.K. and J. Korshover, Quasi-biennial, annual and semiannual zonal wind and temprature harmonic amplitudes and phases in the stratosphere and low mesosphere of the northern hemisphere, *J. Geophys. Res.*, 75, 543-550, 1970.

Belmont, A.D., D.G. Dartt, and G.D. Nastrom, Periodic variations in stratospheric zonal wind from 20-65 km, at 80N to 70S, *Q. J. R. Meteor. Soc.*, 100, 505-530, 1974.

Belmont, A.D., D.G. Dartt, and G.D. Nastrom, Variations in stratospheric zonal winds, 20-65 km, 1961-1971, *J. Appl. Meteor., 14*, 585-594, 1975.

Bergman, J.W. and M.L. Salby, Equatorial wave activity derived from fluctuations in observed convection, *J. Atmos. Sci., 51*, 3791-3806, 1994.

Boville, B.A., Middle atmosphere version of CCM2 (MACCM2): Annual cycle and interannual variability, *J. Geophys. Res., 100*, 9017-9039, 1995.

Boville, B.A. and Randel, W.J., Equatorial waves in a stratospheric GCM: Effects of vertical resolution, *J. Atmos. Sci., 49*, 785-801, 1992.

Burrage, M. D., R. A. Vincent, H. G. Mayr, W. R. Skinner, N. F. Arnold, and P. B. Hays, Long term variability in the equatorial mesosphere and lower thermosphere zonal winds, *J. Geophys. Res., 101*, 12847-12854, 1996.

Cole, A.E. and A.J. Cantor, *Tropical Atmospheres, 0 to 90 km*, Project 8624, AFCRL-TR-75-0527, Aeronomy Laboratory, Air Force Cambridge Research Laboratories, Hanscom, Massachussetts, 1975.

Delisi, D.P. and T.J. Dunkerton, Equatorial semiannual oscillation in zonally averaged temperature observed by the Nimbus 7 SAMS and LIMS, *J. Geophys. Res., 93*, 3899-3904, 1988*a*.

Delisi, D.P. and T.J. Dunkerton, Seasonal variation of the semiannual oscillation, J. Atmos. Sci., 45, 2772-2787, 1988*b*.

Dewan, E.M., R.H. Picard, R.R. O'Neill, H.A. Gardiner, J. Gibson, J.D. Mill, E. Richards, M. Kendra, and W.O. Gallery, MSX satellite observations of thunderstorm generated gravity waves in mid-wave infrared images of the upper stratosphere. *Geophys. Res. Lett., 25*, 939-942, 1998.

Dunkerton, T.J.. Theory of the mesopause semiannual oscillation, *J. Atmos. Sci., 39*, 2681-2690, 1982.

Elbert, E.E.. M.J. Manton, P.A. Arkin, R.J. Allam, G.E. Holpin, and A. Gruber, Results from the GPCP algorithm intercomparison programme, *Bull. Amer. Meteor. Soc., 77*, 2875-2887. 1996.

Fovell, R.G., D.R. Durran, and J.R. Holton, Numerical simulation of convectively generated stratospheric gravity waves, *J. Atmos. Sci., 49*, 1427-1442.

Garcia, R.R. and M.L. Salby, Transient response to localized episodic heating in the tropics. Part II: Far-field behavior, *J. Atmos. Sci., 44*, 499-530, 1987.

Garcia, R.R. and R.T. Clancy, Seasonal variation in equatorial mesospheric temperatures observed by SME, *J. Atmos. Sci., 47*, 1666-1673, 1990.

Garcia, R.R., F. Stordal, S. Solomon, and J.T. Kiehl, A new numerical model of the middle atmosphere. 1. Dynamics and transport of tropospheric source gases, *J. Geophys. Res., 97*, 12967-12991, 1992.

Garcia, R.R., T.J. Dunkerton, R.S. Lieberman, and R.A. Vincent, Climatology of the semiannual oscillation of the tropical middle atmosphere, *J. Geophys. Res., 102*, 26019-26032, 1997.

Garcia, R.R. and F. Sassi, Modulation of the mesospheric semiannual oscillation by the quasibiennial oscillation, *Earth Planets Space, 51*. 563-570, 1999.

Groves, G.V., Annual and semiannual zonal wind components and corresponding temperature and density variations, 60-130 km, *Planet. Space Phys., 20*, 2099-2112, 1972.

Hack, J.J., Parameterization of mosit convection in the NCAR Community Climate Model CCM2, *J. Geophys. Res.,99*, 5551-5568, 1994.

Hack, J.J., B.A. Boville, J.T. Kiehl, P.J. Rasch, and D.L. Williamson, Climate statistics from the National Center for Atmospheric Research Community Climate Model CCM2, *J. Geophys. Res.,99*, 20785-20813, 1994.

Hayashi, Y., D. Golder, and J.D. Mahlman, Stratospheric and mesospheric Kelvin waves simulated by the GFDL "SKYHI" general circulation model, *J. Atmos. Sci., 41*, 1971-1984, 1984.

Hayashi, Y., Kelvin and mixed Rossby-gravity waves appearing in the GFDL "SKYHI" model and the FGGE dataset: Implications for their generation mechanism and role in the QBO. *J. Meteor. Soc. Japan, 72*, 901-935, 1994.

Hamilton, K., Rocketsonde observations of the mesospheric semiannual oscillation at Kwajalein, *Atmos. Ocean, 20*, 281-286, 1982.

Hamilton, K., Dynamics of the stratospheric semiannual oscillation, *J. Meteor. Soc. Japan, 64*, 227-244, 1986.

Hamilton, K., R.J. Wilson, J.D. Mahlman, and L.J. Umscheid, Climatology of the SKYHI troposphere-stratosphere-mesosphere general circulation model. *J. Atmos. Sci., 52*, 5-43, 1995.

Hendon, H.H. and K. Woodberry, The diurnal cycle of tropical convection, *J. Geophys. Res., 98*, 16623-16637, 1993.

Hirota, I., Equatorial waves in the upper stratosphere and mesosphere in relation to the semiannual oscillation of the zonal mean wind, *J. Atmos. Sci., 35*, 714-722, 1978.

Hirota, I., Observational evidence of the semiannual oscillation in the tropical middle atmosphere: A review, *Pure Appl. Geophys., 118*, 217-238, 1980.

Hitchman, M.H. and C.B. Leovy, Evolution of the zonal mean state in the equatorial middle atmosphere during October 1978-May 1979, *J. Atmos. Sci., 43*, 3159-3176, 1986.

Hitchman, M.H. and C.B. Leovy, Estimation of Kelvin wave contribution to the Semiannual Oscillation, *J. Atmos. Sci., 45*, 1462-1475, 1988.

Holton, J.R., Waves in the equatorial stratosphere generated by tropospheric heat sources, *J. Atmos. Sci., 29*, 368-375, 1972.

Holton, J.R., On the frequency distribution of atmospheric Kelvin waves, *J. Atmos. Sci., 30*, 499-501, 1973.

Holton, J.R. and W.M. Wehrbein, The role of planetary-waves in the annual cycle of the zonal-mean circulation of the middle atmosphere, *J. Atmos. Sci., 37*, 1968-1983, 1980.

Hopkins, R.H., Evidence of polar-tropical coupling in upper stratosphere zonal wind anomalies, *J. Atmos. Sci., 32*, 712-719, 1975.

Horinouchi, T. and S. Yoden, The excitation of transient waves by localized episodic heating in the tropics and their propagation into the middle atmosphere, *J. Meteor. Soc. Japan, 74*, 189-210, 1998.

Janowiak, J.E. and P.A. Arkin, Rainfall variations in the

tropics during 1986-1989 as estimated from observations of cloud top temperature, *J. Geophys. Res.*, *96*, 3359-3373.

Lieberman, R.S. and P.B. Hays, An estimate of the momentum deposition in the lower thermosphere by the observed diurnal tide, *J. Atmos. Sci.*, *51*, 3094-3105, 1994.

Longuet-Higgins, M., The eigenfunctions of Laplace's tidal equation over a sphere, *Phil. Trans. Roy. Soc. London*, *A262*, 511-607, 1968.

Manzini, E. and K.P. Hamilton, 1993: Middle atmospheric travelling waves forced by latent and convective heating, *J. Atmos. Sci.*, *50*, 2180-2200, 1993.

Mengel, J.H., H.G. Mayr, K.L. Chan, C.O. Hines, C.A. Reddy, N.F. Arnold and H.S. Potter, Equatorial oscillations in the middle atmosphere generated by small-scale gravity waves, *Geophys. Res. Lett.*, *22*, 3027-3033, 1995.

Ortland, D.A., W.R. Skinner, P.B. Hays, M.D. Burrage, R.S. Lieberman, A.R. Marshall and D.A. Gell, Measurements of stratospheric winds by the High Resolution Doppler Imager, *J. Geophys. Res.*, *101*, 10351-10363, 1996.

Ray, E.A., M.J. Alexander, and J.R. Holton, An analysis of the structure and forcing of the equatorial semiannual oscillation in zonal wind, *J. Geophys. Res.*, *103*, 1759-1774, 1998.

Reed, R.J., Some features of the annual temperature regime in the tropical stratosphere, *Mon. Weather Rev.*, *90*, 211-215, 1962.

Reed, R.J., The quasi-biennial oscillation of the atmosphere between 30 and 50 km over Ascension Island, *J. Atmos. Sci.*, *22*, 331-333, 1965.

Reed, R.J., Zonal wind behavior in the equatorial stratosphere and lower mesosphere, *J. Geophys. Res.*, **71**, 4223-4233, 1966.

Ricciardulli, L. and R.R. Garcia, The excitation of equatorial waves by deep convection in the NCAR Community Climate Model (CCM3), *J. Atmos. Sci.*, in press, 2000.

Richards, F. and P.A. Arkin, On the relationship between satellite-observed cloud cover and precipitation, *Mon. Wea. Rev.*, *109*, 1081-1093, 1981.

Salby, M.L. and R.R. Garcia, Transient response to localized episodic heating in the tropics. Part I: Excitation and short-time, near-field behavior, *J. Atmos. Sci.*, *44*, 458-498, 1987.

Salby, M.L. H.H. Hendon, K. Woodberry, and K. Tanaka, Analysis of global cloud imagery from multiple satellites, *Bull. Amer. Meteor. Soc.*, *72*, 467-480, 1991.

Sassi, F., R.R. Garcia, and B.A. Boville, The stratopause semiannual oscillation in the NCAR Community Climate Model, *J. Atmos. Sci.*, *50*, 3608-3624, 1993.

Sassi, F. and R.R. Garcia, The role of equatorial waves forced by convection in the tropical semiannual oscillation, *J. Atmos. Sci.*, *54*, 1925-1942, 1997.

Sato, K., Small-scale wind disturbances observed by the MU radar during the passage of typhoon Kelly, *J. Atmos. Sci.*, *50*, 518-537, 1993.

Sato, K., F. Hasegawa, and I. Hirota, Short period disturbances in the equatorial lower stratosphere, *J. Meteor. Soc. Japan*, *72*, 859-872, 1994.

Sato, K., H. Hashiguchi, and S. Fukao, Gravity waves and turbulence associated with cumulus convection observed with the UHF/VHF clear-air Doppler radars, *J. Geophys. Res.*, *100*, 7111-7119, 1995.

Sato, K. and T.J. Dunkerton, Estimates of momentum flux associated with Kelvin and gravity waves, *J. Geophys. Res.*, *102*, 26247-26262, 1997.

Takahashi, M., A numerical model of the semiannual oscillation, *J. Meteor. Soc. Japan*, *62*, 52-68, 1984.

Tanaka, K., K. Woodberry, H.H. Hendon, and M.L. Salby, Assimilation of global cloud imagery from multiple satellites, *J. Atmos. Ocean. Tech.*, *8*, 613-626, 1991.

Tsuda, T., Y. Maruyama, H. Wiryosumarto, S.W.B. Harijono, and S. Kato, Radiosonde observations of equatorial atmosphere dynamics over Indonesia. Part I: Equatorial waves and diurnal tides, *J. Geophys. Res.*, *99*, 10491-10505, 1994a.

Tsuda, T., Y. Maruyama, H. Wiryosumarto, S.W.B. Harijono, and S. Kato, Radiosonde observations of equatorial atmosphere dynamics over Indonesia. Part II: Characteristics of gravity waves, *J. Geophys. Res.*, *99*, 10507-10516, 1994b.

R.R. Garcia, National Center for Atmospheric Research, P.O. Box 3000, Boulder, CO 80307-3000 (e-mail: rgarcia@ncar.ucar.edu)

# Modeling Atmospheric Tidal Propagation Across the Stratopause

M. E. Hagan

*High Altitude Observatory, National Center for Atmospheric Research, Boulder, Colorado*

This review of solar atmospheric tides in the middle atmosphere focuses on the achievements in modeling tidal phenomena during the past decade with particular attention to advances in diurnal tidal modeling. Modeling achievements with steady-state linearized tidal models, linear and non-linear time-dependent mechanistic models, and first principles general circulation models are highlighted along with the insight obtained with the recent emergence of assimilative tidal schemes. The review includes references to select measurement diagnostics in order to elucidate the impact of some modeling achievements. On-going investigations of unresolved issues in tidal modeling research are also highlighted including the plausible role of gravity wave and tidal interactions in the formation of mesospheric inversion layers, the underlying physics governing the semiannual oscillation of diurnal tide in the MLT, and whether gravity wave-tide interactions result in tidal acceleration or deceleration.

## INTRODUCTION

Atmospheric tides are persistent and global variations in wind, temperature, pressure, density, and geopotential height with periods that are harmonics of a solar or lunar day. Since their restoring force is gravity, tides are a special class of gravity waves characterized by their sources, their global scale, and the aforementioned periodicities. But, unlike other gravity waves, atmospheric tides are affected by the Earth's rotation and sphericity because of their distinguishing horizontal scale and periodicities. Solar atmospheric tides can be excited by the absorption of solar radiation, by large-scale latent heat release associated with deep convective activity in the troposphere or by the gravitational pull of the Sun. The solar diurnal (semidiurnal) tide has a period of 24 (12) hours. Migrating tidal components propagate with the apparent motion of the Sun, so the migrating diurnal (semidiurnal) tide is a westward propagating $s = -1(s = -2)$ perturbation, where $s$ represents zonal wavenumber. So-called non-migrating solar diurnal (semidiurnal) tidal components are $s \neq -1(s \neq -2)$ global perturbations with periods of 24 (12) hours, so these waves can be standing (i.e., $s = 0$), or they can propagate either eastward (i.e., $s > 0$) or westward (i.e., $s < 0$).

The Moon's gravity forces the lunar atmospheric tide. Atmospheric tides that are thermally excited are significantly stronger than either solar or lunar gravitational tides and are therefore the focus of this report. They can be observed throughout the Earth's atmosphere, but those components that are excited in the troposphere and stratosphere and propagate upward deserve particular attention. Tides that propagate across the stratopause into the MLT profoundly affect the large-scale dynamics of these regions. Like other gravity waves these tidal components grow in amplitude with increasing altitude since atmospheric density decreases and energy must be conserved. Recent advances in modeling upward propagating tidal components are highlighted in this report. While there is a significant body of research that reports on tidal propagation from

Atmospheric Science Across the Stratopause
Geophysical Monograph 123
Copyright 2000 by the American Geophysical Union

the mesosphere into the thermosphere, that work is beyond the scope of this report. Likewise, reports on tidal excitation and propagation between the troposphere and stratosphere are also excluded. Herein, the focus is on tidal coupling between multiple atmospheric regimes that include at minimum the stratosphere, mesosphere, and lower thermosphere. This topic is worthy of particular attention because tides in the upper mesosphere and lower thermosphere are the most profound and persistent feature of the large scale dynamics therein.

Nearly 20 years ago *Forbes and Garrett* [1979] comprehensively reviewed the advances in linear tidal modeling that occurred since *Chapman and Lindzen* [1970] authored their monograph on the subject. *Kato* [1980], *Volland* [1988], and *Forbes* [1995] also provide tutorial reviews of classical tidal theory and describe modifications to those solutions that are associated with non-classical effects including those attributable to zonal mean zonal winds, latitudinal temperature gradients, and dissipation. Notably, it is critical to include all of the aforementioned characteristics and processes in a realistic tidal model of the Earth's middle and upper atmosphere.

*Vial* [1989] and *Vial and Forbes* [1989] describe the achievements in tidal modeling during the 1980s. *Vial* [1989] points to *Forbes'* [1984] interim review and celebrates increasingly more precise semidiurnal tidal predictions, but characterizes the initial numerical assessments of the role of dissipation and mean winds on diurnal tidal structure [*Vial and Teitelbaum*, 1984; *Vial*, 1986; *Aso et al.*, 1987; *Forbes and Hagan*, 1988] as the most significant tidal modeling advance of the decade. *Vial* [1989] further identifies interhemispheric asymmetries in tidal structure, more refined investigations of the diurnal tide, and the effects of non-migrating tides as important unresolved subjects for future tidal modeling studies. In his review of non-migrating tides *Kato* [1989] concurs with Vial's third recommendation. Therein, Kato describes then current knowledge of non-migrating tides above the stratosphere as rudimentary and recommends further study with both mechanistic and general circulation models (GCMs). During the 1990s tidal modelers reported on research that addressed the issues raised by *Kato* [1989] and *Vial* [1989]. Beyond these challenges modelers were also motivated by the rich tidal diagnostics that emerged from Upper Atmosphere Research Satellite (UARS) measurement analyses and were accompanied by a largess of ground-based lidar and radar tidal determinations. Specific tidal modeling advances are discussed herein after the qualitative description of the various numerical models employed in on-going tidal research which follows.

## NUMERICAL TIDAL MODELS

Numerical predictions of atmospheric tides have long been made by solving the linearized Navier-Stokes equations for perturbation fields with characteristic zonal wavenumbers and periodicities that are assumed a priori along with the zonal mean background atmosphere. Mean zonal winds are included, but mean meridional winds are neglected in linearized tidal model formulations. This approach is an outgrowth of the classical analytic tidal theory. Linearized tidal models may be steady-state or time-dependent. The steady-state models calculate perturbation fields that are affected by the background atmosphere. Some time-dependent tidal models are actually quasi-linearized models. That is, they solve for evolving perturbation fields as well as an evolving zonal mean atmosphere which are coupled by tide-mean flow interaction terms. Like their steady-state counterparts, time-dependent quasi-linearized tidal models are run for an assumed tidal frequency and zonal wavenumber. These classes of tidal models are sometimes called mechanistic tidal models because they are uniquely designed to study tidal and occasionally planetary wave dynamics. As described in the following sections several mechanistic tidal models emerged during the past decade. Some middle and upper atmospheric GCMs are also discussed, since they too emerged as important tools that are increasingly used in numerical investigations of atmospheric tides. In characterizing all of the aforementioned tidal modeling tools distinguishing features of forcing and dissipative schemes are paid particular attention because they have a significant impact on the tidal predictions.

### Mechanistic Tidal Modeling Across the Stratopause During the 1990s

State-of-the-art mechanistic tidal models are 2-dimensional and extend from the ground into the thermosphere and from pole to pole. The zonal wavenumber that is specified in each tidal calculation quantifies the longitudinal behavior of the wave. The steady-state linearized subset of these models are primarily used to successfully predict tidal climatologies and to carry out numerical experiments that quantify how tides are affected by variable forcing, dissipation processes, or zonal mean atmospheric states. Several such models emerged during the 1990s. The global-scale wave model (GSWM) is described by *Hagan et al.* [1995, 1997a, 1999a] (hereafter GSWM-95, GSWM-97, GSWM-98, respectively). *Wood and Andrews* [1997a, b, c], *Ekanayake et al.* [1997], and *Zhu et al.* [1999] (hereafter WA97,

EAM97, and Z+99, respectively) also report the development of linearized steady-state tidal models. The respective systems of equations and hierarchical structure of these models are all described in their reports. In brief, all account for equivalent physical processes, but in ways that, as highlighted below, differ in detail.

WA97 and Z+99 employ a spectral scheme, while GSWM and EAM97 use finite difference algorithms to resolve atmospheric tidal signatures in the lower, middle, and upper atmosphere. All routinely employ the best available empirical models [e.g., *Fleming et al.*, 1990; *Hedin,* 1991] to specify their zonal mean winds, temperatures, and densities. There are consequential differences in their tidal forcing functions, however. These differences are most pronounced at tropospheric altitudes and partly reflect how difficult it is to parameterize the radiative budget of the troposphere. The differences also speak to the uncertainties associated with this problem. WA97 and GSWM employ *Groves'* [1982a] parameterization of tropospheric tidal forcing due to absorption of solar radiation. Z+99 derive solar heating rates with water vapor climatologies from National Center for Environmental Prediction (NCEP) reanalyses [*Kalnay et al.*, 1996]. Throughout the entire model regime EAM97 force tidal harmonics with heating rates derived from those calculated in the Kyushu University GCM (KU-GCM) [*Miyahara et al.*, 1993]. In the troposphere, these rates include both the latent and dry convection heat sources in addition to solar radiative forcing. GSWM can also be forced with tropospheric latent heating rates that are determined from global cloud imagery data analyses [e.g., *Williams and Avery*, 1996; *Forbes et al.*, 1997]. In the middle atmosphere ozone absorbs solar ultraviolet radiation which in turn forces additional tidal components. WA97 use *Groves'* [1982b] ozone heating rates, while Z+99 derive strato-mesospheric tidal heating using UARS ozone climatologies. GSWM middle atmospheric heating rates are based either on COSPAR International Reference Atmosphere (CIRA) ozone climatologies [*Keating et al.*, 1990] or on ozone climatologies derived from UARS data.

There are additional and important distinctions in the way the aforementioned modelers account for dissipation processes in the upper atmosphere. Most notable are the assumptions that they invoke to parameterize the considerable effects of the divergence of heat and momentum flux due to molecular and eddy diffusion. As detailed by *Forbes and Garrett* [1979] it is critical to include these effects in realistic tidal models that extend into the MLT. The mathematical formulation of the relevant terms in the horizontal momentum and thermal energy equations may take the form:

$$\frac{1}{\rho_0}\frac{\partial}{\partial z}(\mu_0 + \rho_0\nu_{eddy})\frac{\partial}{\partial z}\mathbf{U}' \quad \text{and}$$

$$\left(\frac{\gamma-1}{R\rho_0}\right)\frac{\partial}{\partial z}(K_0 + \rho_0 K_{eddy})\frac{\partial}{\partial z}\delta T$$

respectively, for $\mathbf{U}' = (u', \nu')$, the zonal ($u'$) and meridional ($\nu'$) tidal fields; $\delta T$, the tidal temperature; $\rho_0$, zonal mean density; $z$, altitude; $\mu_0$, the dynamic molecular viscosity; $\nu_{eddy}$, the kinematic eddy viscosity; $R$, the gas constant; $\gamma$, the ratio of heat capacity at constant pressure to heat capacity at constant volume; $K_0$, the molecular thermal conductivity; and $K_{eddy}$, the eddy thermal conductivity. If these processes are explicitly included in the system of tidal equations, it is of eighth-order (i.e., four coupled second-order partial differential equations). However, if the aforementioned effects are parameterized by Rayleigh friction, $\nu_R$, and Newtonian cooling, $\alpha_N$, coefficients, the aforementioned terms reduce to, $\nu_R\mathbf{U}'$ and $\alpha_N\delta T$, respectively, and the system of linearized tidal equations may be written as a single second-order differential equation. Both WA97 and Z+99 invoke this computationally attractive so-called inviscid approximation. Their effective diffusion coefficients are dependent upon the background atmosphere they assume and on the vertical scale of various harmonic components that they resolve. In their diurnal tidal calculations $\nu_R$ are of order 100–300 m$^2$/s in the mesopause region. In contrast, EAM97 and GSWM are viscid models because they explicitly solve for the aforementioned dissipative effects. GSWM assumes latitude dependent eddy diffusion coefficient profiles which are slightly (up to 20%) smaller than the latitude independent profile of EAM97 which peaks at $\sim$120 m$^2$/s just below 100 km. In diurnal tidal calculations GSWM includes an additional dissipative term to parameterize the effects of gravity wave drag on the diurnal tide. This addition is based upon the results of a numerical calculation performed by *Miyahara and Forbes* [1991] (hereafter MF91) and takes the form of an effective Rayleigh friction term (i.e., $\nu_R\mathbf{U}'$) in GSWM. The work of MF91 is discussed further in the paragraphs below.

Before turning to a discussion of the more sophisticated mechanistic models that contributed to tidal research during the past decade, it is important to acknowledge one of the exceptional achievements in recent tidal modeling, the development of a linearized steady-state tidal model hybrid known as the tuned mechanistic tidal model (TMTM). TMTM is an assimilative scheme initially formulated by *Khattatov et al.* [1997] and extended by *Yudin et al.* [1997]. Their assimilative approach is invaluable for correlative analyses of ground-based and satellite-borne measurements and will serve as a viable interpretive and investigative tool well into the future. TMTM was initially developed to quantify diurnal tidal signatures in UARS high resolution doppler interferometer (HRDI) wind measurements made at MLT heights. TMTM is an iterative scheme that solves the linearized steady-state system of equations for an assumed period and zonal wavenumber that the aforementioned tidal modelers resolve, except the coupled diurnal tidal perturbation wind, temperature, geopotential, pressure, and density fields are constrained to be consistent with the amplitudes and phases of the meridional tidal wind component observed by HRDI. More recently, TMTM was extended to assimilate ground-based tidal diagnostics and to quantify the behavior of the semidiurnal as well as the diurnal harmonic [*Yudin et al.*, 1997]. TMTM is also a valuable diagnostic tool. It was used to evaluate current parameterizations of tidal dissipation in the MLT which have substantial uncertainties. Initially, *Khattatov et al.* [1997] used an inviscid approximation in TMTM to resolve the $\nu_R$ and $\alpha_N$ that dissipate the migrating (zonal wavenumber 1) diurnal tidal harmonic when the perturbation fields were constrained by the HRDI data and the diurnal forcing was parameterized in the model. They subsequently inferred pseudo-eddy diffusion coefficients that account for both molecular and eddy processes from the implied dissipation rates upon consideration of the observed diurnal tidal vertical wavelength and wavenumber. *Yudin et al.* [1997] report on the viscid extension of the TMTM scheme which allows direct determinations of eddy dissipation coefficients ($K_{zz}$) when the model is tuned with HRDI measurements. Inherent in these ($K_{zz}$) determinations is the fundamental assumption that MLT tidal dissipation is attributable to diffusive processes as opposed to wave-wave interactions, for example.

*Vial et al.* [1991] report on a very simple inviscid time-dependent tidal model which excludes zonal mean wind effects as well as latitude gradients in the background temperature and density fields. They use this model to investigate the time it takes for a tide of lower atmospheric origin to propagate into the MLT and reach steady-state, and to explore the transient nature of upward propagating tides. *Aso* [1993] also investigates tidal transience with a windless viscid model. Further, he resolves semidiurnal tidal signatures that arise from diurnal tidal forcing by including non-linear advection terms in the time-dependent horizontal wind and temperature perturbation equations. *McLandress*

[1997] reports the development of a simple inviscid time-dependent tidal model too, although he did not investigate transience. Rather, he uses the simple model to aid the interpretation of tidal signatures in a GCM known as the Canadian middle atmosphere model (CMAM).

*McLandress and Ward* [1994] (hereafter MW94) report on a time-dependent quasi-geostrophic model that extends from ~16 to 140 km and simulates the zonal mean circulation as well as the first three zonal wavenumbers of the large-scale circulation. Monthly mean geopotential observations define their model lower boundary. This model is incapable of generating tidal motions, so linearized tidal model structures [*Forbes*, 1982a, b] are included a priori throughout the model regime. There is no direct coupling between the model atmosphere and the tidal fields, but the latter modulate GW propagation and dissipation in their model regime. This type of model is readily used to quantify the effects of tidal modulated GW drag on the zonal mean atmosphere, on stationary planetary waves, or as a secondary tidal source.

*Miyahara et al.* [1991] developed a more complex quasi-linearized viscid mechanistic model and used it to provide valuable insight into the dynamical interactions between tides and the zonal mean atmosphere. This model iteratively resolves the coupled and evolving characteristics of both the zonal mean atmosphere and a specified tidal harmonic. Not only does the background atmosphere affect the tide, the tide also affects the zonal mean atmosphere in this type of model. The mathematical formulation that quantifies the impact of the zonal mean on the tide is wholly comparable to the steady-state formulation. *Miyahara et al.* [1991] formulate their model in a log-pressure coordinate system which is typical of most time-dependent tidal models. If $z_p = -H \ln(p/p_s)$ represents this height coordinate and $p_s$ is the surface pressure, the eddy flux terms $\overline{F_u}$, $\overline{F_\nu}$, and $\overline{F_T}$ that appear in the zonal mean horizontal momentum and energy equations and quantify tidal effects on the zonal mean take the form:

$$\overline{F_u} = -\frac{1}{a}\frac{\partial \overline{u'\nu'}}{\partial \theta} - \frac{2\cot\theta}{a}\overline{u'\nu'} - \frac{1}{p_0}\frac{\partial p_0 \overline{u'w'}}{\partial z_p}, \quad (1)$$

$$\overline{F_\nu} = -\frac{1}{a}\frac{\partial \overline{\nu'\nu'}}{\partial \theta} + \frac{\cot\theta}{a}\overline{u'u'}$$
$$- \frac{\cot\theta}{a}\overline{\nu'\nu'} - \frac{1}{p_0}\frac{\partial p_0 \overline{\nu'w'}}{\partial z_p}, \quad (2)$$

$$\overline{F_T} = -\frac{1}{a}\frac{\partial \overline{\nu'\delta T}}{\partial \theta} - \frac{\cot\theta}{a}\overline{\nu'\delta T} - \frac{1}{p_0}\frac{\partial p_0 \overline{\delta T w'}}{\partial z_p}, \quad (3)$$

respectively, where $a$ is the Earth's radius, $\theta$ is colatitude, $w'$ is the vertical velocity tidal field, the overbars represent zonal means and $p_0$ is zonal mean pressure. The initial conditions for the zonal mean are given by CIRA [*Miyahara et al.*, 1991]. Lower and middle atmospheric thermotidal forcing is from *Groves* [1982a, b] and gravity wave (GW) momentum deposition forcing of the zonal mean winds is parameterized by the *Matsuno* [1982] scheme.

*GCMs that Model Tidal Propagation Across the Stratopause*

GCMs are increasingly used to investigate tidal propagation across the stratosphere and tidal effects in the atmospheric regions aloft. In this section GCMs that contributed to tidal research during the past decade are briefly characterized. All of these models employ GW parameterization schemes to quantify the sub-grid scale effects of eddy diffusion and of GW momentum and energy deposition which significantly impact tidal propagation and dissipation. Some of these models calculate radiative transfer and atmospheric chemistry from first principles while other models resolve the dynamics, but parameterize relevant chemical and radiative effects. A few distinguishing characteristics are noted in the following descriptions along with cursory remarks about each model domain. The remaining details of these GCM formulations are contained in the relevant references cited herein.

*Hunt* [1990] describes updates to a dynamical grid-point GCM (hereafter H-GCM) that extends from the ground to 100 km and inherently resolves tidal motions when the radiative calculation is carried out every 2 hours. The extended UK Universities Global Atmospheric Modelling Programme (UGAMP) GCM (EUGCM) [*Jackson*, 1995; *Norton and Thuburn*, 1996, 1997] includes the same atmospheric regimes, but solves the governing equations using spectral techniques. H-GCM and EUGCM have been successfully used to quantify the effects of the diurnal cycle in the upper mesosphere and to investigate the effects of GW parameterization on tidal motions as they propagate across the stratopause and into the mesosphere. However, spurious effects associated with the upper boundaries preclude detailed studies of tides in the mesopause region with these models. *McLandress* [1997] showed that when a sponge layer with a top near ~100 km is included in such GCMs, it will mitigate the spurious effects and provide useful information about tides in the mesosphere. The spectral mesosphere/lower thermosphere model (SMLTM) extends to ~220 km and is also suitable for such investigations [*Akmaev et al.*, 1996, 1997]. Although this dynamical model includes realis-

tic parameterizations of radiative transfer, the SMLTM lower boundary is at ~15 km and requires an additional parameterization of the tropospheric tidal source. This forcing is introduced by including GSWM geopotential height perturbations at the SMLTM lower boundary.

The KU-GCM [*Miyahara et al.*, 1993], the CMAM [*de Grandpre et al.*, 1997], and the National Center for Atmospheric Research thermosphere-ionosphere-mesosphere-electrodynamics GCM (TIME-GCM) [*Roble and Ridley*, 1994] all include self-consistent calculations of atmospheric chemistry in addition to dynamics. Both the KU-GCM and the CMAM are spectral models that extend from the ground to the lower thermosphere with upper boundaries at 165 km and 95 km, respectively. TIME-GCM is a grid point model with a lower boundary at ~30 km that extends to ~500 km in the upper thermosphere. Upward propagating atmospheric tides are calculated self-consistently in both KU-GCM and CMAM, but TIME-GCM, like SMLTM, must account for tidal forcing below the lower boundary. This is done either by including GSWM perturbations as a boundary condition, or by simultaneously running the TIME-GCM with the NCAR tropospheric Community Climate Model (CCM) and passing fields between the two models via a flux-coupler.

## SCIENTIFIC ACHIEVEMENTS IN TIDAL MODELING DURING THE 1990s

An EUGCM numerical experiment recently reconfirmed that realistic mesospheric models must include solar atmospheric tides. In keeping with previous KU-GCM [*Miyahara*, 1984; *Wu et al.*, 1989], H-GCM [*Hunt*, 1990], and mechanistic model [*Miyahara et al.*, 1991] findings, zonal mean winds are unrealistically large near the equatorial mesopause when EUGCM is run without the diurnal cycle [*Jackson*, 1995]. This result illustrates one of many important effects that tides of lower atmospheric origin have on the MLT dynamics and chemistry. Recent numerical investigations that characterize additional effects are highlighted in this section along with studies of tidal excitation, propagation, and dissipation.

*Tidal Signatures and Climatologies*

As described above realistic representations of the middle and upper atmosphere must include solar tides, particularly the diurnal tide. Figure 1 contains additional evidence of this claim. It illustrates horizontal winds at noon during equinox as observed by UARS [after *McLandress et al.*, 1996] and modeled with TIME-GCM [after *Shepherd et al.*, 1998]. Diurnal tidal signatures are particularly evident in the meridional com-

**Neutral Wind Climatology; Equinox 12LT**

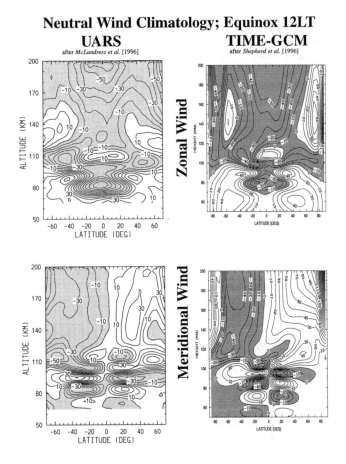

**Figure 1.** Eastward (top) and Northward (bottom) wind signatures during equinox at 12 hours local solar time as observed by WINDII and HRDI (left) [after *McLandress et al.* 1996] and modeled by TIME-GCM (right) [after *Shepherd et al.*, 1998]. There are diurnal tidal signatures below ~120 km. TIME-GCM predictions extend to polar latitudes beyond the UARS diagnostic regime where there is evidence of auroral forcing above ~130 km.

ponent at latitudes equatorward of ~40° and altitudes below ~120 km. Zonal mean meridional winds are small throughout this regime, so a wave with a vertical wavelength of order 25 km which is asymmetric about the equator is apparent in the wind field. Similar though considerably less pronounced and symmetric features are arguably evident in the zonal winds. They are seen in a more confined area (20–40°) near the low latitude mesopause (80–100 km) which is also where the zonal diurnal tide is comparatively strong and zonal mean zonal winds are weak. The contours of eastward and westward wind centered on the equator near 70–120 km are not direct evidence of the diurnal tide. Rather, they are probably associated with acceleration of the zonal mean flow by breaking and dissipating eastward

and westward propagating waves, including the diurnal tide.

Figure 2 provides a different perspective on diurnal tidal signatures. This illustration of diurnal tidal wind and temperature amplitude and phase as a function of latitude and altitude typifies illustrations of mechanistic model results. Salient tidal features are obvious, since only the perturbation fields are illustrated. As discussed above horizontal wind amplitudes peak at low and middle latitudes near 100–105 km, but measurable signatures are also evident in the stratosphere. Zonal (meridional) phases are symmetric (anti-symmetric) about the

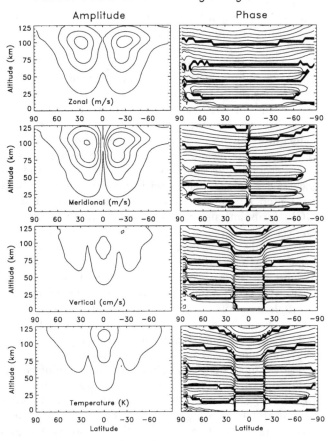

**Figure 2.** Another representation of the salient features of diurnal tidal wind and temperature amplitude (left column) and phase (right column). Tidal phase generally refers to the local solar hour of maximum. Amplitudes increase with increasing altitude into the lower thermosphere. Zonal (1st row) and meridional (second row) contour intervals may range between 5 and 15 m/s, while vertical velocity (third row) and temperature (bottom row) amplitudes may range between 2.5 and 10, cm/s and degrees K, respectively. Minimum (maximum) values of contour intervals are characteristic of solstice (equinox). See text for details.

equator. The distance between signatures of phase wrap (contour buildup) are indicative of a vertical wavelength of 25 or so km. Some additional features also emerge from this perspective. Like the meridional counterpart the zonal wind component of the diurnal tide is small near the equator which is where the diurnal temperature and vertical velocity perturbations peak. There are also secondary maxima in diurnal temperature and vertical winds which occur near 30–40° and are out of phase with the equatorial maxima. Finally, there is evidence of interhemispheric asymmetry in the aggregate diurnal tidal response.

The salient features of the diurnal tide that are described above persist, but amplitudes in particular may evolve significantly on interannual, seasonal, and even on intraseasonal time scales. *Khattatov et al.* [1997], *McLandress* [1997], WA97, *Yudin et al.* [1997], and Z+99 all describe monthly variations in diurnal tidal perturbations based upon their climatological calculations. Strong seasonal variability in the MLT is evident in these results with amplitude maxima (minima) during equinox (solstice) (e.g., Figure 2). *Akmaev et al.* [1997], *Geller et al.* [1997], *Roble and Shepherd* [1997], and GSWM-98, also report seasonal variability in their diurnal tidal estimates which capture the diurnal wind signatures found in both space-borne [e.g., *Burrage et al.*, 1995; *McLandress et al.*, 1996] and ground-based [e.g., *Vincent et al.*, 1989] measurements. Notably, as evidenced in Figures 3 and 4 [after *Wu et al.*, 1998] respectively, CMAM tidal temperature climatologies are consistent with the seasonally variable diurnal tidal determinations inferred from the Microwave Limb Sounder (MLS) instrument on UARS. Thus, during the past decade the seasonal variability of the MLT diurnal tide was firmly established. Despite significant efforts to understand the underlying physical processes, the source(s) of this variability remain unresolved. Highlights of recent efforts to investigate plausible sources including tidal forcing and dissipation are discussed in the next several paragraphs.

*Migrating Tidal Sources*

As described in the introduction migrating solar tides propagate with the apparent motion of the sun, so the migrating diurnal (semidiurnal) tide is a westward propagating wavenumber one (two) oscillation with a period of 24 (12) hours. For many years it has been known that migrating solar tides are primarily excited by the absorption of solar radiation throughout the atmosphere and that these waves are the dominant solar atmospheric tidal components. In independent numerical experiments with Z+99 and GSWM [*Hagan*, 1996]

the tidal component excited by the absorption of infrared radiation in the troposphere is contrasted with that forced by the absorption of ultraviolet radiation in the middle atmosphere. These reports suggest that the diurnal response is dominated by the former and modulated by the latter at middle and low latitudes where the diurnal response is large. However, there is a complicated interplay between the semidiurnal responses to these two sources and neither one systematically dominates the aggregate response aloft. The GSWM was also used to explore migrating tidal excitation associated with another tropospheric tidal source in a series of numerical investigations wherein GSWM was forced with increasingly sophisticated estimates of latent heating rates. The migrating tidal responses [*Hagan*, 1996; GSWM-97; *Forbes et al.*, 1997] illustrate that the components excited by this additional tropospheric source may further affect tidal signatures in the MLT. *McLandress* [1997] quantifies the seasonal variability of CMAM diurnal heating rates due to both solar radiative and deep convective sources and confirms that the tropospheric radiative source dominates the diurnal tidal response in the MLT. He also reports significant variability in tropospheric tidal forcing, but that it is insufficient to account for the variability in the MLT response. *Burrage et al.* [1995] and *Geller et al.* [1997] reach the same conclusion in their related GSWM and TMTM studies of seasonal variability, although they omit the assessment of latent heating effects.

*Braswell and Lindzen* [1998] use a classical tidal model to calculate the diurnal surface pressure response to increases in migrating tropospheric tidal forcing due to anomalous absorption of solar radiation by water vapor in clear air and in the presence of clouds. The latter calculation produced equatorial surface pressure increases which are consistent with observations, but the observed response at low to middle latitudes remained underestimated by their results. Further, they do not explore whether or how the middle or upper atmospheric diurnal tide responds to such tropospheric radiative heating increases.

Two particularly noteworthy results emerge from the aforementioned investigations of migrating tidal sources. First, no lower atmospheric diurnal heat source can be neglected in realistic tidal models. Second, it is critical to include realistic tidal signatures at the lower boundaries of middle and upper atmospheric GCMs that do not extend down to the surface.

*Non-migrating Tidal Sources*

Non-migrating tides have different longitudinal characteristics than their migrating counterparts. As dis-

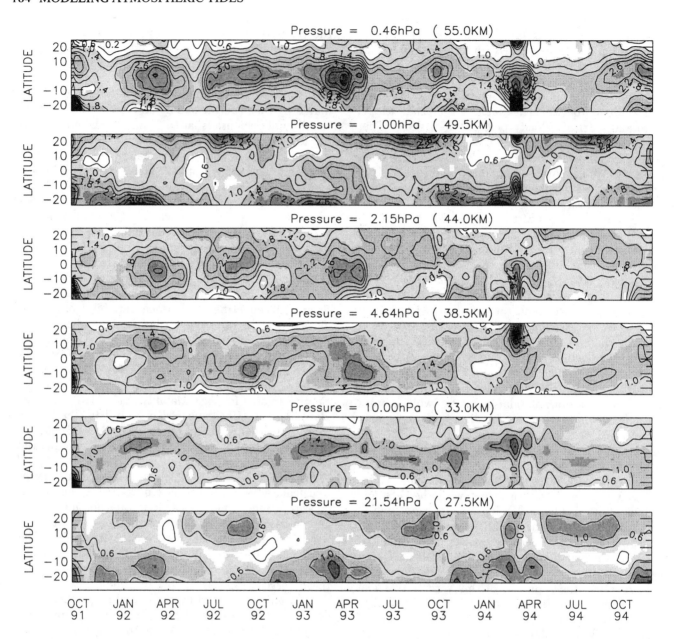

**Figure 3.** Contours of diurnal temperature amplitude versus time and latitude at 6 stratospheric and mesospheric altitudes as measures by the MLS instrument on UARS [after *Wu et al.*, 1998].

cussed in the introduction non-migrating tides can be standing global waves or they may propagate either eastward or westward as long as their characteristic zonal wavenumbers are not equal to the frequency (per day) of the wave. MW94 identify localized GW drag which occurs as a result of GW-tide interactions as a plausible non-migrating tidal source. A series of additional investigations focused on non-migrating tidal excitation due to tropospheric latent heat release associated with deep convective activity in the tropics. GSWM-97 and *Hagan et al.* [1997b] describe latent

heating as an important non-migrating tidal driver and quantify its non-negligible effects on middle and upper atmospheric dynamics. EAM97 include all radiative and latent heat sources and show that upward propagating non-migrating diurnal tides posses significant amplitudes at MLT low and middle latitudes. Eastward propagating components dominate all but the migrating diurnal response during southern hemisphere summer. The GSWM-97 and EAM97 results also confirm the important effects of mean winds on non-migrating tidal propagation that were initially reported by *Miya-*

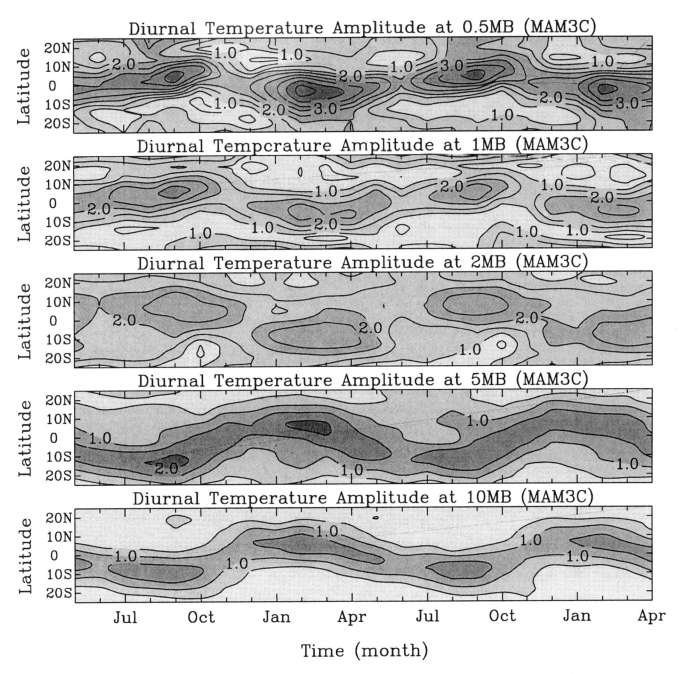

**Figure 4.** Same as Figure 3 except temperatures are modeled by CMAM [after *Wu et al.*, 1998].

*hara et al.* [1993]. Specifically, the Doppler shifted frequency $\tilde{\sigma} = \sigma + sU_0/a\cos\theta$ of a given tidal component (wavenumber s) becomes comparatively higher when tidal propagation is opposite to the direction of the strato-mesospheric zonal mean winds ($U_0$). When this occurs tidal components are able to propagate into MLT middle and upper latitudes. Note, the Doppler effect increases with increasing *s*. It is the comparatively large wavenumber components that propagate most efficiently to high latitudes. Since the migrating diurnal tide is small at these locations (e.g., Figure 2), non-migrating signatures may dominate the diurnal response there.

*Khattatov et al.* [1996] and *Hagan et al.* [1997b] present evidence of longitudinal variability in diurnal tidal signatures deduced from UARS HRDI and WIND Imaging Interferometer (WINDII) measurements, respectively. This variability is consistent with numerical predictions of non-migrating tidal signatures. Specifically, the interplay of diurnal components with multiple zonal wavenumbers leads to pronounced longitudinal amplitude variations but negligible phase effects, because diurnal phases track the dominant longitudinally invariant (i.e., migrating diurnal tidal) response (e.g., *Hagan et al.* [1996]). The calculations that quantified the importance of non-migrating effects in the MLT and prompted observational confirmation represent some of the important tidal modeling achievements of this decade.

*Tidal Dissipation*

The focus in this section is the parameterization of GW effects in numerical tidal modeling efforts. Without exception every middle and upper atmospheric tidal modeler addresses this issue because (1) no global model explicitly resolves all GWs, (2) eddy diffusion attributable to the turbulence associated with GWs as they become unstable and finally break affects MLT tides, and (3) momentum deposition due to GW breaking decelerates or accelerates the diurnal tide as well as the zonal mean flow at MLT altitudes. Although other dissipative processes including radiative damping, molecular viscosity and conductivity, and ion drag are important to accurate MLT tidal modeling, recent numerical investigations of tidal dissipation primarily focus on GW effects.

MF91 used a time-dependent model and a GW parameterization scheme based on *Lindzen* [1981] in a solstice calculation which focused on GW-tide interaction. They describe regions where MLT diurnal tidal amplitudes are suppressed by GW drag and derive an effective $\nu_R$ based upon their model results. MW94 confirm this result with a comparable numerical calculation. The MF91 $\nu_R$ are the basis of the standard GSWM GW drag formulation. The absence of any seasonal variability in the MF91 drag estimates led to an unrealistically small equinoctal diurnal tide in GSWM-95, but this shortcoming was corrected in GSWM-98 based in part on insight provided by a series of numerical experiments [e.g., *Burrage et al.*, 1995; *Geller et al.*, 1997].

In a tutorial paper on the parameterization of unresolved GWs in GCMs *McLandress* [1998] describes several GW schemes in detail. These include the parameterizations of *Lindzen* [1981], *McFarlane* [1987], *Fritts and Lu* [1993], *Medvedev and Klaassen* [1995], and *Hines* [1997a, b]. *McLandress* [1998] also reports the effects of including various schemes in CMAM, but he focuses on the zonal mean circulation. All parameterizations produce comparatively more eddy dissipation during solstice than equinox. To investigate GW effects on the diurnal tide *McLandress* [1998] employs a

windless linearized time-dependent viscid model. Like MF91 he reports diurnal phase advances when any GW drag is included in the model, but the aggregate impact on diurnal tidal amplitude is not uniform. A reduction in diurnal amplitude occurs with the inclusion of the monochromatic [after *Lindzen*, 1981] and *Fritts and Lu* [1993] GW schemes, while the Doppler spread parameterization [*Hines*, 1997a, b] enhances the diurnal tidal amplitudes. The contrasting impacts that *McLandress* [1998] reports are consistent with the findings of *Mayr et al.* [1998] and *Meyer* [1999]. They are particularly noteworthy in that they point to one of the major outstanding problems in diurnal tidal modeling in the MLT. Specifically, do GW-tide interactions reduce or enhance diurnal amplitudes?

The most innovative estimates of GW effects in recent years emerged from the assimilative work of *Khattatov et al.* [1997] and *Yudin et al.* [1997]. TMTM calculates monthly estimates of $K_{zz}$ characterized by pronounced seasonal and interhemispheric variability. These TMTM results introduce a new and valuable observational constraint on realistic numerical modeling. The $K_{zz}$ deduced from the TMTM are weakest at equinox and comparatively stronger in winter than in summer. This dissipative variation is exactly opposite to the observed MLT diurnal amplitude response which suggests that variable dissipation may explain the semiannual tidal variation. *Yudin et al.* (this volume) extend this argument to explain the interannual variability of diurnal tidal amplitudes during March. They deduce $K_{zz}$ that are modulated by the quasi-biennial oscillation of the strato-mesospheric zonal mean winds that periodically filter upward propagating GW.

Even though there are significant differences in the details of the state of the art formulations most numerical modelers concur with the findings of the TMTM group and ascribe the largest signatures of seasonal variability in the MLT diurnal tide to the variations that they include in their dissipative parameterizations. For example, after conducting a series of special GSWM calculations that invoked a hybrid *Lindzen* [1981]/*Matsuno* [1982] GW parameterization scheme *Meyer* [1999] concludes that annual variations in the zonal mean atmosphere result in similar variations in and the diurnal harmonic of momentum flux divergence produced by breaking GWs. The seasonal variations in these parameters are significant enough to produce a realistic estimate of the semiannual variation in MLT tidal amplitude. *McLandress* [1997] offers a rather different perspective, however. In his comparative analysis of tidal fields calculated with CMAM and with a linear model, he finds good (poor) agreement in the MLT diurnal response for April (July) when the models are identically forced. He correlates the weaker GCM response during July with large positive values of Eliassen-Palm flux divergence and concludes that non-linear effects due to resolved motions in the model play a role in damping the diurnal tide and causing the observed semiannual variation [*McLandress*, 1997]. Further, *McLandress* [1998] compares the differences in the seasonal variability of the diurnal tidal response from two CMAM simulations, one that excluded non-orographic gravity waves and one that included them. He reports no qualitative difference in the semiannual tidal variation in these CMAM results and concludes that unresolved gravity waves do not account for the seasonal variation of the diurnal tide. The divergent conclusions that emerge from these recent studies of GW effects suggest that the underlying cause(s) of the observed seasonal variability of the diurnal tide in the MLT remains an open research question.

EAM97 and *Hagan et al.* [1999b] reexamine the effects of realistic tidal dissipation on the diurnal tide in the presence of mean winds. They note that diurnal tidal amplitudes are larger in atmospheric regions where the Doppler shifted frequency of the tide is comparatively large. They attribute this behavior to the aforementioned Doppler effect and note that it is harder (easier) to dissipate a higher (lower) frequency wave for vertical wavelengths that are of comparable order. These results imply that the westward propagating migrating diurnal tide should be comparatively large where the wave undergoes dissipation in MLT regions of large westward zonal mean zonal winds. Thus, tidal dissipation in the presence of mean winds may explain at least part of the interhemispheric asymmetry in diurnal tidal perturbations commonly reported in the literature.

*Simulations of Chemical Dynamical Coupling due to Solar Atmospheric Tides*

*Roble and Shepherd* [1997], *Yee et al.* [1997], *Yudin et al.* [1998], and *Shepherd et al.* [1998] describe the tidal influence on airglow at equatorial and middle latitudes from global perspectives calculated with coupled chemical dynamical models. In sum, tidal forcing may enhance the emission rate of the atomic oxygen green line, $O(^1S)$, during the evening hours and cause it to descend. The emission then weakens drastically near midnight owing to quenching and the reduced supply of atomic oxygen. Thereafter, the emission recovers when downward tidal winds refresh the supply of atomic oxygen from above. The diurnal variation in $O(^1S)$ can be particularly large (factor of 2) at the equator where, as illustrated in Figure 1, opposing meridional diurnal tidal winds converge at ~95 km [*Shepherd et al.*, 1998]. The $O_2(^1\Sigma)$ and OH emissions often exhibit

similar behavior, but the depletion is shifted with local time in keeping with the phase progression of the diurnal tide and the altitudes of these emissions. Similar local time shifts occur with latitude in all airglow and the behavior at middle latitudes exhibits signatures that are suggestive of semidiurnal tidal forcing [*Shepherd et al.*, 1998]. All of the aforementioned modelers conclude that the pronounced seasonal variations in the magnitude and/or occurrence of the airglow signatures are attributable to the competing effects of dissipation and the magnitude of the tidal driver at MLT altitudes.

## CONCLUDING REMARKS

Numerical models of atmospheric tides evolved considerably during the past 30 years. The results of the early investigations of the 1970s led modelers to understand the need to account for zonal mean zonal winds, latitudinal temperature gradients, and realistic tidal forcing and dissipation in order to accurately model tidal propagation from the lower and middle atmosphere into the upper atmosphere. Nonetheless, it has taken the better part of two decades to adequately refine numerical tidal models so that they can produce realistic MLT migrating tidal climatologies that vary seasonally and include interhemispheric asymmetries in their latitudinal structure. Since model evaluations and updates largely rely on tidal observations, the time line for tidal model refinements is intimately connected to the availability of such data. During the 1980s modelers principally relied on ground-based observations at middle latitudes. These data fostered increasingly accurate predictions of the semidiurnal tide. The emergence of UARS and correlative low latitude diagnostics during the past decade shifted the focus to the diurnal tide. The increased availability of computational resources also contributed to this shift in focus. Diurnal tidal calculations require comparatively higher model resolution than comparable semidiurnal calculations due to the differences in the characteristic vertical wavelengths and latitudinal structures of the harmonic components. Diurnal tidal calculations that were previously difficult or impossible are now routine. The accessibility of computer resources also facilitated the recent emergence of a significant number of new models. Many are GCMs with regimes that extend over multiple atmospheric layers. These models proved particularly suited for investigations of non-linear interactions and chemical dynamical processes involving tides. Accessibility to computer resources will continue to expand. Roble (this volume) describes the feasibility of the next generation GCM that will extend from the ground to the exosphere. Linearized steady-state tidal models as well as quasi-linear time-dependent tidal models remain valuable numerical tools, however. They not only aid the interpretation of the more complicated GCM results, they continue to provide comparatively inexpensive but reliable climatological predictions of atmospheric tides.

As detailed in this review recent model assessments led to a variety of numerical investigations with a focus on the diurnal tide. The related topics that were central to these investigations include tropospheric and stratospheric migrating and non-migrating tidal forcing, tidal propagation across the stratopause and into the MLT, wave-tide and tide-mean flow interactions, chemical and dynamical effects of MLT tides, and diurnal tidal dissipation. Despite significant advances several issues remain unresolved. There are lingering uncertainties about tropospheric tidal forcing and MLT tidal dissipation that must be addressed. The semiannual variation in the MLT diurnal tide can be better understood through further numerical investigations of the comparative importance of the seasonal variability in migrating as well as non-migrating tidal forcing, dissipation, and the non-linear effects of GW-tide-mean flow interactions. New insight obtained by correlative investigations with time-dependent mechanistic and first principles models will be particularly critical to further advances. Studies of secondary waves generated by tides interacting with other global scale waves [e.g., *Norton and Thuburn*, 1996; *Palo et al.*, 1998, 1999] will quantify how important the secondary waves are to MLT chemistry and dynamics including their impact on the primary waves, whether they propagate further into the upper atmosphere, and whether they affect the ionosphere. Additional numerical studies of GW-tide interactions will also be important since they will address the fundamental issue of whether and how GWs accelerate or decelerate the diurnal tide.

Other plausible effects of GW-tide interactions that can be investigated with mechanistic models include the formation and evolution of mesospheric inversion layers (MILs). MILs are large temperature perturbations that exhibit downward phase propagation consistent with the diurnal tide, but whose amplitudes are an order of magnitude larger than tidal predictions [e.g., *Dao et al.*, 1995; *Leblanc and Hauchecorne*, 1997; *Meriwether et al.*, 1998]. *Liu and Hagan* [1998] recently postulated that GW stability is significantly affected by the diurnal tide. They quantify enhancements in local dynamical cooling and turbulent heating due to GW breaking in the presence of the diurnal tide and suggest that this type of GW-tide interaction produces the observed MILs. However, their work is hardly conclusive and lacks a global perspective. Much additional research is needed to understand MILs in the middle and upper atmosphere.

Finally, the correlative perspective provided by ground-based and satellite borne tidal diagnostics must continue to be explored. The assimilative techniques developed by *Khattatov et al.* [1997] and *Yudin et al.* [1997] along with those which will extend this formulation will prove particularly invaluable in the very near term. The thermosphere-ionosphere-mesosphere energetics and dynamics (TIMED) satellite scheduled for launch in May 2000 will sample the MLT with almost complete horizontal coverage. However, like other satellites that orbit at high inclination angles, TIMED will have a slow precession rate precluding a data stream with high local time resolution. Conversely, a number of ground-based observers will conduct correlative campaigns under the auspices of the NSF coupling, energetics, and dynamics of atmospheric regions (CEDAR) program and measure the MLT with high temporal resolution and low spatial coverage. Data assimilation into MLT models will provide optimum use of TIMED and CEDAR data samples and facilitate the deconvolution of zonal mean, tidal and planetary wave signatures from these diagnostics. The results will not only reconfirm many of the tidal features deduced from the UARS data during the past decade. They will also provide a global perspective on tidal signatures at higher altitudes and higher latitudes and undoubtedly reveal new phenomena that the modelers will attempt to unravel.

*Acknowledgments.* The author thanks J. M. Forbes and C. K. Meyer for comments on the initial draft of this report and to the reviewers for their suggestions. The National Center for Atmospheric Research (NCAR) is sponsored by NSF. The efforts of M. E. Hagan are supported in part by the NSF CEDAR program and by NASA grant S-10105X to NCAR.

## REFERENCES

Akmaev, R. A., J. M. Forbes, and M. E. Hagan, Simulation of tides with a spectral mesosphere/lower thermosphere model, *Geophys. Res. Lett.*, *23*, 2173-2176, 1996.

Akmaev, R. A., V. A. Yudin, and D. A. Ortland, SMLTM simulations of the diurnal tide: Comparison with UARS observations, *Ann. Geophys.*, *15*, 1187-1197, 1997.

Aso, T., Time-dependent numerical modelling of tides in the middle atmosphere, *J. Geomag. Geoelectr.*, *45*, 41-63, 1993.

Aso, T., S. Ito, and S. Kato, Background wind effect on the diurnal tide in the middle atmosphere, *J. Geomag. Geoelectr.*, *39*, 297-305, 1987.

Braswell, W. D., and R. S. Lindzen, Anomalous short-wave absorption and atmospheric tides, *Geophys. Res. Lett.*, *25*, 1293-1296, 1998.

Burrage, M. D., et al., Long-term variability in the solar diurnal tide observed by HRDI and simulated by the GSWM, *Geophys. Res. Lett.*, *22*, 2641-2644, 1995.

Chapman, S., and R. S. Lindzen, *Atmospheric Tides*, 201 pp., D. Reidel, Dordrecht, Holland, 1970.

Dao, P., et al., Lidar observations of the temperature profile between 25 and 103 km: Evidence of strong tidal perturbation, *Geophys. Res. Lett.*, *22*, 2825-2828, 1995.

de Grandpre, J. J., et al., Canadian middle atmosphere model: Preliminary results from the chemical transport module, *Atmosphere-Ocean*, *35*, 385-431, 1997.

Ekanayake, E. M. P., T. Aso, and S. Miyahara, Background wind effect on propagation of nonmigrating diurnal tides in the middle atmosphere, *J. Atmos. Solar-Terr. Phys.*, *59*, 401-429, 1997.

Fleming, E. L., et al., Zonal mean temperature, pressure, zonal wind and geopotential height as functions of latitude. COSPAR International Reference Atmosphere 1986. Part II: Middle atmosphere models, *Adv. Space Res.*, *10*, 11-62, 1990.

Forbes, J. M., Atmospheric tides, 1, Model description and results for the solar diurnal component, *J. Geophys. Res.*, *87*, 5222-5240, 1982a.

Forbes, J. M., Atmospheric tides, 2, The solar and lunar semidiurnal components, *J. Geophys. Res.*, *87*, 5241-5252, 1982b.

Forbes, J. M., Middle atmosphere tides, *J. Atmos. Terr. Phys.*, *46*, 1049-1067, 1984.

Forbes, J. M., Tidal and planetary waves, in *The Upper Mesosphere and Lower Thermosphere: A Review of Experiment and Theory*, 67-87, edited by R. M. Johnson and T. L. Killeen, American Geophysical Union, Washington, DC, 1995.

Forbes, J. M., and H. B. Garrett, Theoretical studies of atmospheric tides, *Rev. Geophys.*, *17*, 1951, 1979.

Forbes, J. M., and M. E. Hagan, Diurnal propagating tide in the presence of mean winds and dissipation: A numerical investigation, *Planet. Space Sci.*, *36*, 579-590, 1988.

Forbes, J. M., et al., Upper atmospheric tidal oscillations due to latent heat release in the tropical troposphere, *Ann. Geophys.*, *15*, 1165-1175, 1997.

Fritts, D. C., and W. Lu, Spectral estimates of gravity wave energy and momentum fluxes, Part II: Parameterization of wave forcing and variability, *J. Atmos. Sci.*, *50*, 3695-3713, 1993.

Geller, M. A., et al., Modeling the diurnal tide with dissipation derived from UARS/HRDI measurements, *Ann. Geophys.*, *15*, 1198-1204, 1997.

Groves, G. V., Hough components of water vapor heating, *J. Atmos. Terr. Phys.*, *44*, 281-290, 1982a.

Groves, G. V., Hough components of ozone heating, *J. Atmos. Terr. Phys.*, *44*, 111-121, 1982b.

Hagan, M. E., Comparative effects of migrating solar sources on tidal signatures in the middle and upper atmosphere, *J. Geophys. Res.*, *101*, 21,213-21,222, 1996.

Hagan, M. E., J. M. Forbes, and F. Vial, On modeling migrating solar tides, *Geophys. Res. Lett.*, *22*, 893-896, 1995.

Hagan, M. E., J. L. Chang, and S. K. Avery, GSWM estimates of non-migrating tidal effects, *J. Geophys. Res.*, *102*, 16,439-16,452, 1997a.

Hagan, M. E., C. McLandress, and J. M. Forbes, Diurnal tidal variability in the upper mesosphere and lower thermosphere, *Ann. Geophys.*, *15*, 1176-1186, 1997b.

Hagan, M. E., et al., GSWM-98: Results for migrating solar tides, *J. Geophys. Res.*, *104*, 6813-6828, 1999a.

Hagan, M. E., et al., QBO effects on the diurnal tide in the upper atmosphere, *Earth, Planets, and Space*, *51*, 6571-578, 1999b.

Hedin, A. E., Extension of the MSIS thermosphere model into the middle and lower atmosphere, *J. Geophys. Res.*, *96*, 1159-1172, 1991.

Hines, C. O., Doppler-spread parameterization of gravity wave momentum deposition in the middle atmosphere, 1, Basic formulation, *J. Atmos. Terr. Phys.*, *59*, 371-396, 1997a.

Hines, C. O., Doppler-spread parameterization of gravity wave momentum deposition in the middle atmosphere, 2, Broad and quasi-monochromatic spectra and implementation, *J. Atmos. Terr. Phys.*, *59*, 387-400, 1997b.

Hunt, B. G., A simulation of the gravity wave characteristics and interactions in a diurnally varying model atmosphere, *J. Meteor. Soc. Japan*, *68*, 145-161, 1990.

Jackson, D. R., Tides in the extended UGAMP general circulation model, *Q. J. R. Meteorol. Soc.*, *121*, 1589-1611, 1995.

Kalnay, E., et al., The NCEP/NCAR 40-year reanalysis project, *Bull. Amer. Meteorol. Soc.*, *77*, 437-471, 1996.

Kato, S., Dynamics of the upper atmosphere, *Center for Academic Publications*, Tokyo, Japan, 1980.

Kato, S., Non-migrating tides, *J. Atmos. Terr. Phys.*, *51*, 673-682, 1989.

Keating, G. M., M. C. Pitts, and C. Chen, Improved reference models for middle atmosphere ozone, *Adv. Space Res.*, *10*, (6)37-(6)49, 1990.

Khattatov, B. V., et al., Dynamics of the mesosphere and lower thermosphere as seen by MF radars and the high-resolution Doppler imager/UARS, *J. Geophys. Res.*, *101*, 10,393-10,404, 1996.

Khattatov, B. V., et al., Diurnal migrating tide as seen by HRDI/UARS, 2, Monthly mean global zonal and vertical velocities, pressure, temperature, and inferred dissipation, *J. Geophys. Res.*, *102*, 4423-4435, 1997.

Leblanc, T., and A. Hauchecorne, Recent observations of the mesospheric temperature inversions, *J. Geophys. Res.*, *102*, 19,471-19,482, 1997.

Lindzen, R. S., Turbulence and stress due to gravity wave and tidal breakdown, *J. Geophys. Res.*, *86*, 9707-9714, 1981.

Liu, H.-L., and M. E. Hagan, Local heating/cooling of the mesosphere due to gravity wave and tidal coupling, *Geophys. Res. Lett.*, *25*, 2941-2944, 1998.

Matsuno, T., A quasi one-dimensional model of the middle atmosphere circulation interacting with internal gravity waves, *J. Meteorol. Soc. Japan.*, *60*, 215-226, 1982.

Mayr, H. G., et al., Seasonal variations of the diurnal tide induced by gravity wave filtering, *Geophys. Res. Lett.*, *25*, 943-946, 1998.

McFarlane, N. A., The effect of orographically excited gravity wave drag on the general circulation of the lower stratosphere and troposphere, *J. Atmos. Sci.*, *44*, 1775-1800, 1987.

McLandress, C., Seasonal variability of the diurnal tide: Results from the Canadian middle atmosphere general circulation model, *J. Geophys. Res.*, *102*, 29,749-29,764, 1997.

McLandress, C., On the importance of gravity waves in the middle atmosphere and their parameterization in general circulation models, *J. Atmos. Solar-Terr. Phys.*, *60*, 1357-1383, 1998.

McLandress, C., and W. E. Ward, Tidal/gravity wave interactions and their influence on the large-scale dynamics of the middle atmosphere: Model results, *J. Geophys. Res., 99*, 8139-8155, 1994.

McLandress, C., et al., Combined mesosphere/thermosphere winds using WINDII and HRDI data from the Upper Atmosphere Research Satellite, *J. Geophys. Res., 101*, 10,441-10,453, 1996.

Medvedev, A. S., and G. P. Klaassen, Vertical evolution of gravity wave spectra and the parameterization of associated wave drag, *J. Geophys. Res., 100*, 25,841-25,853, 1995.

Meriwether, J. W., et al., Observed coupling of the mesosphere inversion layer to the thermal tidal structure, *Geophys. Res. Lett., 25*, 1479-1482, 1998.

Meyer, C. K., Gravity wave interactions with the diurnal propagating tide, *J. Geophys. Res., 104*, 4223-4239, 1999.

Miyahara, S., A numerical simulation of the zonal mean circulation of the middle atmosphere including effects of solar diurnal tidal waves and internal gravity waves: Solstice conditions, in *Dynamics of the Middle Atmosphere*, edited by J. R. Holton and T. Matsuno, 271-287, D. Reidel, Dordrecht, Holland, 1984.

Miyahara, S., and J. M. Forbes, Interaction between gravity waves and the diurnal tide in the mesosphere and lower thermosphere, *J. Meteor. Soc. Japan, 69*, 523-531, 1991.

Miyahara, S., et al., Mean zonal acceleration and heating of the 70- to 100-km region, *J. Geophys. Res., 96*, 1225-1238, 1991.

Miyahara, S., Y. Yoshida, and Y. Miyoshi, Dynamical coupling between the lower and upper atmosphere by tides and gravity waves, *J. Atmos. Terr. Phys., 55*, 1039-1053, 1993.

Norton, W. A., and J. Thuburn, The two-day wave in a middle atmosphere GCM, *Geophys. Res. Lett., 23*, 2113-2116, 1996.

Norton, W. A., and J. Thuburn, The mesosphere in the extended UGAMP GCM, in *Gravity Wave Processes and their Parameterization in Global Climate Models*, edited by K. Hamilton, 383-401, Springer-Verlag, New York, 1997.

Palo, S. E., R. G. Roble, and M. E. Hagan, TIME-GCM results for the quasi-two-day wave, *Geophys. Res. Lett., 25*, 3783-3786, 1998.

Palo, S. E., R. G. Roble, and M. E. Hagan, Middle atmosphere effects of the quasi-two-day wave determined from a general circulation model, *Earth, Planets, and Space, 51*, 629-647, 1999.

Roble, R. G., On the feasibility of developing a global atmospheric model extending from the ground to the exosphere, this volume, 2000.

Roble, R. G., and E. C. Ridley, A thermosphere-ionosphere-mesosphere-electrodynamics general circulation model (TIME-GCM): Equinox solar cycle minimum simulations (30–500 km), *Geophys. Res. Lett., 21*, 417-420, 1994.

Roble, R. G., and G. G. Shepherd, An analysis of WINDII observations of O($^1$S) equatorial emission rates using the TIME-GCM, *J. Geophys. Res., 102*, 2467-2474, 1997.

Shepherd, G. G., et al., Tidal influence on midlatitude airglow: Comparison of satellite and ground-based observations with TIME-GCM predictions, *J. Geophys. Res., 103*, 14,741-14,751, 1998.

Vial, F., Numerical simulations of atmospheric tides for solstice conditions, *J. Geophys. Res., 91*, 8955-8969, 1986.

Vial, F., Tides in the middle atmosphere, *J. Atmos. Terr. Phys., 51*, 3-17, 1989.

Vial, F., and H. Teitelbaum, Some consequences of turbulent dissipative effects on the diurnal thermal tide, *Planet. Space Sci., 32*, 1559-1565, 1984.

Vial, F., and J. M. Forbes, Recent progress in tidal modeling, *J. Atmos. Terr. Phys., 51*, 663-671, 1989.

Vial, F., J. M. Forbes, and S. Miyahara, Some transient aspects of tidal propagation, *J. Geophys. Res., 96*, 1215-1224, 1991.

Vincent, R. A., T. Tsuda, and S. Kato, Asymmetries in mesospheric tidal structure, *J. Atmos. Terr. Phys., 51*, 609-616, 1989.

Volland, H., *Atmospheric Tidal and Planetary Waves*, Kluwer Academic Publishers, Norwell, Massachusetts, 1988.

Williams, C. R., and S. K. Avery, Non-migrating diurnal tides forced by deep convective clouds, *J. Geophys. Res., 101*, 4079-4091, 1996.

Wood, A. R., and D. G. Andrews, A spectral model for simulation of tides in the middle atmosphere. I: Formulation, *J. Atmos. Solar-Terr. Phys., 59*, 31-51, 1997a.

Wood, A. R., and D. G. Andrews, A spectral model for simulation of tides in the middle atmosphere. II: Results for the diurnal tide, *J. Atmos. Solar-Terr. Phys., 59*, 53-77, 1997b.

Wood, A. R., and D. G. Andrews, A spectral model for simulation of tides in the middle atmosphere. III: Results for the semidiurnal tide, *J. Atmos. Solar-Terr. Phys., 59*, 79-97, 1997c.

Wu, D.-H., S. Miyahara, and Y. Miyoshi, A non-linear simulation of the thermal diurnal tide, *J. Atmos. Terr. Phys., 51*, 1017-1030, 1989.

Wu, D. L., et al., Equatorial diurnal variations observed in UARS Microwave Limb Sounder temperature during 1991-1994 and simulated by the Canadian Middle Atmosphere Model, *J. Geophys. Res., 103*, 8909-8917, 1998.

Yee, J.-H., et al., Global simulations and observations of O($^1$S), O$_2$($^1$S), and OH mesospheric nightglow emissions, *J. Geophys. Res., 102*, 19,949-19,968, 1997.

Yudin, V. A., M. A. Geller, and L. Wang, Interannual variability of diurnal tide in the low-latitude mesosphere and lower thermosphere during equinoxes: Mechanistic model interpretation of the 1992-96 HRDI measurements, this volume, 2000.

Yudin, V. A., et al., Thermal tides and studies to tune the mechanistic tidal model using UARS observations, *Ann. Geophys., 15*, 1205-1220, 1997.

Yudin, V. A., et al., TMTM simulations of tides: Comparisons with UARS observations, *Geophys. Res. Lett., 25*, 221-224, 1998.

Zhu, X., et al., On the numerical modeling of middle atmospheric tides, *Q. J. R. Meteorol. Soc., 125*, 1825-1857, 1999.

M. E. Hagan, High Altitude Observatory, National Center for Atmospheric Research, 3450 Mitchell Lane, Boulder, CO 80301. (e-mail: hagan@ucar.edu)

# Observations of Southern Polar Descent and Coupling in the Thermosphere, Mesosphere and Stratosphere Provided by HALOE

James M. Russell III

*Hampton University, Center for Atmospheric Sciences, Hampton, VA 23668*

R. Bradley Pierce

*Atmospheric Sciences, NASA, Langley Research Center, Hampton, VA 23681*

The most prominent indicators of vertical descent in HALOE data are changes in the vertical profiles of $CH_4$, HF, $H_2O$, HCl and NO. In this paper, we use long-term 3-D trajectories initialized at HALOE occultation points, to examine the 1993 seasonal evolution of $CH_4$, Lagrangian mean descent, and meridional transport in the middle stratosphere at $60°$-$70°$S and within the Antarctic vortex as a whole. At $60°$-$70°$S, poleward and downward transport accounts for the mid-stratosphere seasonal cycle in HALOE $CH_4$. Within this band, tropical and sub-tropical air observed by HALOE near 52 km in May descends to 40 km by July and mid-latitude air observed near 45 km in May, descends to 30 km by August. Lagrangian descent rates vary significantly during the winter. The mean rates are -5.2 km/month at 40 km and -3.0 km/month at 30 km. These values are consistent with rates published in a number of studies using HALOE data when viewed in the context of the Lagrangian analysis. The data show that the effects of descent can reach as low as 25 km, that the descent rate varies interannually, and the signature of descent is most clear at the center of the vortex. Occurrence of descent is obvious in both the northern and southern polar regions. Southern Hemisphere $CH_4$ pressure versus longitude plots demonstrate coupling of the Antarctic polar mesosphere to the subtropical lower stratosphere. Mid-latitude time series of HALOE NO and the F10.7 cm flux demonstrate solar and dynamical coupling of the thermosphere and lower mesosphere.

## 1. INTRODUCTION

The HALogen Occultation Experiment (HALOE) was launched on board the Upper Atmosphere Research Satellite (UARS) on September 12, 1991 by the Space Shuttle Dis-

Atmospheric Science Across the Stratopause
Geophysical Monograph 123

covery into a 585 km 57° inclined orbit. The experiment has operated essentially without flaw from turn-on October 11, 1991 up to the present collecting global scale data on the chemistry and dynamics of the middle atmosphere. HALOE uses the instrument techniques of gas filter and broadband radiometry and the solar occultation measurement approach to provide vertical profiles of HCl, HF, $CH_4$, NO, $H_2O$, $O_3$, $NO_2$, aerosols and temperature versus pressure, the latter inferred from broadband absorption measurements made in a band of $CO_2$. A complete experiment description including geographic coverage and early measurement results was presented by *Russell et al.*, [1993a]. The HALOE retrieval algorithm has been continually improved and we use the third public release of the data, referred to as V19, in this paper.

The purpose of this paper is to describe HALOE observations of descent in the polar regions and evidence of coupling from the polar mesosphere to the near tropical lower stratosphere. Decent in the polar winter night has been studied extensively and is now a well known phenomenon [see e.g. *Brasseur*, 1982, *Solomon et al.*, 1982; *Garcia and Solomon*, 1983 and *Russell et al.*, 1984]. More recently with the launch of UARS and the abundance of data which followed, a series of papers were published describing polar descent including *Russell et al.*, [1993b]; *Fisher* [1993]; *Rosenfield* et al., [1994]; *Schoeberl et al.*, [1995]; *Bacmeister et al.*, [1995]; *Randel et al.*, [1998]; *Kawamoto and Shiotani*, [2000] and *Strahan et al.* [1996]. Analysis of ATMOS measurements of long-lived trace gases *by Abrams et al*, [1996] and *Manney et al.* [1999] and balloon measurements [*Baue, et al.*, 1994] have also contributed to our understanding of polar descent processes. Results from these studies will be discussed later in this paper. A number of the parameters measured by HALOE are well suited for use in identifying and studying descent processes. The long-lived gases HF, HCl and $H_2O$ for example all have vertical stratospheric gradients opposite to that of $CH_4$ which allows for independent confirmation of descent. Maximum values of HF, HCl and $H_2O$ are reached near the stratopause or above while $CH_4$ has its maximum in the troposphere. Therefore the influence of descent will tend to raise the "normal" stratospheric mixing ratio values of the former molecules and lower the $CH_4$ mixing ratios. All of this behavior has been observed in the HALOE polar data and has been analyzed and reported in the more recent references cited. The experiment has been operating for over eight years now collecting data in nine southern polar late winter/springs and eight northern polar winter/spring periods. First we will review HALOE observations and phenomena that have formed the basis for past studies in order to clearly describe the effects of the descent on mixing ratio observations. Next we will show results of our current trajectory analysis to quantitatively estimate the magnitude of the descent observed and then compare these results with past analyses. We will then show HALOE nitric oxide data and describe effects of solar coupling in the thermosphere and dynamical coupling with the mesosphere. Finally we will discuss and summarize all of these findings.

## 2. OBSERVATIONS

The most unambiguous evidence of descent effects on trace gases is for $CH_4$. Plate 1 shows a HALOE sunset zonal mean $CH_4$ pressure versus latitude cross section for the period August 31 to October 6, 1994. Note that because of the occultation sampling pattern, it takes from late August to early October for the measurement location to progress from ~ 75°N to ~73°S. So in this sense, the figure represents a kind of mean state of the atmosphere over this time period. The main features of the $CH_4$ distribution are upwelling in the tropics followed by poleward transport toward both poles with the strongest transport indicated for the fall hemisphere. The effects of strong descent are obvious in the Antarctic region with very low values of $CH_4$ extending down to about 25 km altitude. This effect is also seen in the UARS CLAES data [*Kumer et al.*, 1993]. While the influence of descent in Plate 1 is clear, simple zonal averaging tends to combine extra vortex and interior vortex data thereby masking some of the descent effect. If the data are averaged along potential vorticity (PV) contours [e.g. *Schoeberl et al.*, 1992] and then plotted as a function of equivalent latitude (i.e. a latitude coordinate based on PV contours arranged symmetrically about the pole) so that the interior and exterior vortex air masses are separated, the influence of descent becomes even more pronounced [see e.g. *Randel et al.*, 1998, Fig 1].

The first reported observation of the effect of descent on vertical $CH_4$ profiles is shown in Fig. 1 taken from *Russell et al.*, [1993b]. The solid profile for 247.5°E shows a constant mixing ratio of ~0.2 ppmv from 0.1 mb (~65 km) down to 25 mb (~25 km). These authors observed that the lowest mixing ratios and the constant mixing ratio profiles occurred near the point of minimum winds or vortex center and they interpreted the observation to be a consequence of unmixed vertical descent i.e. descent in the absence of horizontal mixing. The other profiles in the figure were measured at longitudes, which were closer to the vortex edge as indicated by both the wind field and the PV contours. It appears that either vortex edge or extra vortex $CH_4$ air masses were being observed in parts of the profile resulting in the very unusual shapes; i.e. the high values near 10 mb are reflective of extra vortex air while the low values near 22 mb are points measured inside the vortex. The authors also presented data clearly showing descent effects on the HF profile and noted that these effects were apparent in the $H_2O$ and HCl profiles as well. An examination of the entire data record shows that the most sure indicator of the vortex center is the geographic region that contains nearly constant mixing ratio $CH_4$ profiles. This is a common occurrence in the Antarctic spring (e.g. *Siskind* [2000]). Similar behavior has been observed in the northern polar region, but the vortex is not usually as sustained as in the south and while effects of descent are obvious, they are not as clearly displayed. Because of this, the remainder of this paper will keep the focus on the Antarctic region. Fig. 2 shows $CH_4$, HF, $H_2O$, and HCl averages of 121 interior Antarctic vortex

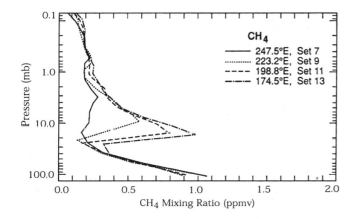

**Figure 1.** HALOE pressure versus $CH_4$ mixing ratio for sunset at 71° S on October 22, 1991 (after Russell et al., 1993b).

profiles along with extra vortex profiles for these gases. The averages were constructed using September and October data measured over the five year period from 1991 to 1995. Selection of profiles inside the vortex was done by finding all $CH_4$ profiles with a nearly constant mixing ratio over the 0.1 mb to 25 mb range. Once events were selected for $CH_4$, the same occultations were used to construct the averages for the other gases. The effect of descent can be seen for all gases and the internal consistency of the profiles as discussed above is apparent.

The plots in Fig. 2 suggest that the occurrence of descent is a regular feature of the polar winter. This can be seen in more detail by examining the $CH_4$ zonal mean 60°S to 70°S average time series. This latitude range was selected to give the most occultation samples at the highest possible latitudes so that a meaningful average could be obtained in the geographic region where descent is most clearly manifested. The time series for 60 km and 50 km is shown in Fig. 3. Also shown is a multiple linear regression fit to the data, which includes mean, linear, second order and annual terms [*John Anderson, 1999, Hampton University, personal communication*]. A reasonable fit is obtained for the entire period HALOE is collecting data in this range (i.e. from ~ September to February) with the exception of late 1996 and early 1997 at 60 km which appears to be an anomalous period. For the most part, $CH_4$ is regularly decreasing with time at each altitude shown from September to February. The 50 km time series is phase shifted relative to 60 km with the decline from the maximum mixing ratio starting later and the minimum occurring later. The decline at 30 km (not shown) is shifted further and does not start until January. This is expected if the changes are occurring because of descending air. The increased $CH_4$ existing at the start of each cycle is due to transport from lower latitudes. We also show the second order term fitted to the data, which indicates that $CH_4$ reaches a minimum in the stratopause region

in mid 1996. We only note this in passing because the methane minimum, concurrent $H_2O$ maximum and trend reversals in other parameters not shown here, are not understood at present and are the subject of intense investigation [see e.g. *Randel et al.*, 1999]. The $CH_4$ changes in Fig. 3 and similar changes at other altitudes contain fundamental information on the vertical descent rate. However extraction of this quantity is complicated because while the signature of descent is most obvious at the vortex core, it is not clear what is happening near the vortex edge or how these regions interact. *Pierce et al.*, [1994] used trajectory mapping of HALOE data and showed that mixing of interior and exterior air occurs over a 5° latitude range either side of the jet maximum which is one way to define the vortex edge. We next apply three dimensional trajectory mapping techniques to HALOE data and use these results to estimate vertical descent rates.

## 3. TRAJECTORY MAPPING AND DESCENT RATE ANALYSIS

We conducted a three dimensional HALOE airmass simulation for 1993 to examine the evolution of polar $CH_4$ mixing ratios and descent during the Antarctic winter. The HALOE airmass simulation predicts the trajectories of airmasses observed by HALOE, allowing us to examine the Lagrangian movement of stratospheric airmasses. Calculations of descent based on this approach combine two of the techniques that were recently used to estimate stratospheric descent rates, i.e. those based on descent rates of observed mixing ratio isopleths [*Schoeberl, et al.*, 1995; *Kawamoto and Shiotani*, 2000; *Abrams et al.*, 1996; *Bauer et al.*, 1994], and modeling studies (trajectory, 2D and 3D) [*Fisher, et al.*, 1993; *Bacmeister, et al.*, 1995; *Rosenfield et al.*, 1994; *Manney et al.*, 1999; *Strahan et al.*, 1996]. By combining these approaches we gain some distinct advantages. First, by mapping the HALOE $CH_4$ observations forward in time, we can predict $CH_4$ mixing ratios during times when HALOE was not observing particular latitude. Second, comparisons with subsequent HALOE observations provide a means of verifying the accuracy of the computed airmass trajectories. The model used in this study is the NASA Langley three-dimensional photochemical/trajectory model which uses potential temperature as a vertical coordinate [*Pierce, et al.*, 1999 and references therein]. Since we are focusing on HALOE $CH_4$ which has a long photochemical lifetime in the stratosphere [Brasseur and Solomon, 1986], photochemical production and loss processes and small-scale mixing were ignored for the simulation. Sensitivity tests of the effects of small scale mixing, which is parameterized in the LaRC trajectory model as an n-member mixing process between neighboring parcels [*Fairlie et al.*,

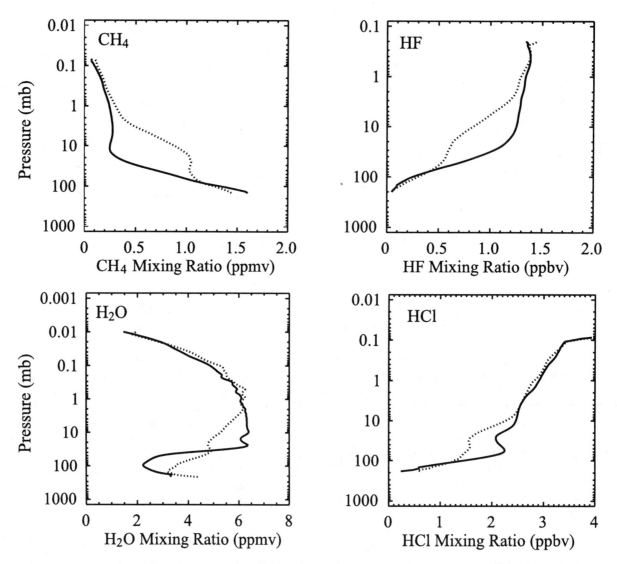

**Figure 2.** Average of 121 CH4, HF, H2O, and HCl Antarctic interior and extra vortex profiles in the months of September and October over the period 1991 to 1995. Solid curves are interior.

1999], indicate that the ensemble mean results presented here are not sensitive to mixing processes. Methane oxidation is slow enough in the middle stratosphere that it can also be ignored for these simulations. Trajectories were initialized beginning January 1 on 35 potential temperature surfaces for all HALOE observations during 1993. Surfaces extended from the lower mesosphere (2400K) to the lowest HALOE measurement altitude for all occultations. The United Kingdom Meteorological Office (UKMO) assimilation for the Upper Atmospheric Research Satellite (UARS) [*Swinbank and O'Neill, 1994*] is used to provide winds and temperatures for the trajectory simulation. Radiative heating rates, which constitute the vertical velocity component in the isentropic framework, are computed using the MIDRAD scheme [*Shine, 1987*] which incorporates the absorption of

solar UV radiation by ozone and molecular $O_2$, and the infrared contributions due to $CO_2$, $H_2O$, and $O_3$. Diurnally averaged heating rates are used in the trajectory calculations.

Systematic biases in the globally averaged heating rates can have a significant impact on long-term trajectory simulations. To remove any bias in the estimated heating rates we determined time-averaged global mean heating residuals on constant pressure surfaces and removed this residual from the daily heating rates. Previous studies have shown that the impact of random errors in the trajectory position associated with long-term trajectory integrations can be effectively mitigated by computing ensemble means for a large number of trajectories [*Pierce et al., 1999*]. We use this approach to examine the seasonal evolution of $CH_4$ in the Antarctic vortex during 1993. We first look at predictions of the zonal

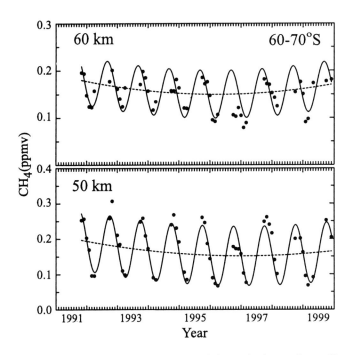

**Figure 3.** Monthly zonal mean $CH_4$ mixing ratio time series at 60 km and 50 km averaged over the 60° - 70° S range for the period October, 1991 to February, 1999. Annual and second order terms of a multiple linear regression fit are shown.

mean $CH_4$ mixing ratios at 60°–70°S. Ensemble mean $CH_4$ predictions at 60°-70° S are constructed by averaging $CH_4$ mixing ratios for all parcels within this latitude band in ± 1 km bins over a range of altitudes in 5-day intervals throughout the simulation. Only those parcels with $CH_4$ signal-to-noise ratios (determined from the uncertainty estimates for the initial HALOE sounding) of greater than 30% are used in constructing the daily ensemble means. The HALOE airmass ensemble mean predictions at 60°–70°S provide insight into the behavior of $CH_4$ during the period from April to August, 1993 when HALOE does not observe these latitudes (cf Fig. 3).

Figure 4a shows the daily predicted ensemble mean $CH_4$ mixing ratio time series at 30 km. The monthly averaged ensemble mean predictions and monthly averaged observed $CH_4$ mixing ratios are also shown. The scatter in the daily ensemble means is relatively low, indicating that ensemble averaging has effectively reduced random errors in the trajectory calculations. The ensemble mean time series fills the gap between March and September, when HALOE does not observe this latitude band. The fact that the ensemble mean time series agrees with the September observations indicates that the trajectory model prediction is consistent with HALOE observations. However, during the period from October to November the trajectory simulation significantly underestimates the observed monthly mean $CH_4$ mixing ratios at 30 km. The underestimate during October

and November may be due to a sampling bias of the zonal mean at 60° –70°S in the ensemble mean trajectory predictions. As will be shown, the ensemble mean trajectory predictions are composed primarily of vortex parcels during this period, since many parcels from higher altitudes have accumulated within the vortex at 30 km. HALOE, on the other hand, observes both inside and outside the vortex at 60° –70°S during October and November because the vortex is significantly distorted due to dynamical activity. Consequently, the HALOE zonal averages include more middle latitude $CH_4$ mixing ratios than are included in the ensemble means. Since middle latitude $CH_4$ mixing ratios are generally higher than those within the vortex, the ensemble mean estimate is low relative to the observed zonally averaged $CH_4$.

The reductions in the observed and predicted $CH_4$ mixing ratios during the Antarctic winter must arise due to descent of lower $CH_4$ mixing ratios from aloft. In the isentropic framework, descent occurs through net radiative cooling. The generally good agreement in predicted and observed $CH_4$ mixing ratios at 30 km indicates that the assimilated winds and computed heating rates used in the trajectory calculations are a good representation of the actual winds and radiative heating rates at and above these altitudes.

Figures 4b-c show time series of the ensemble mean initial altitude and latitude of the parcels which were used to construct the 30 km $CH_4$ time series (i.e. the ensemble mean initial altitude and latitude of parcels observed by HALOE which where subsequently found within 60°–70°S at 30 km). The figures show that air which was initially observed at 45 km in Southern Hemisphere middle latitudes in May (determined from the ensemble mean age, which is the length of time since the trajectory was initialized), is found at 30 km within the 60°–70°S band by August. Similar results are found at 40 km, where air which was initially observed by HALOE near 52 km in the tropics during May is found at 40 km and within the 60°–70°S latitude band by July (not shown). The initial altitude and latitude time series show that poleward and downward transport accounts for the seasonal cycle in the ensemble mean HALOE $CH_4$ prediction within the 60°–70°S latitude band at 30km. Ensemble means of the initial parcel altitude, combined with the trajectory age, can be used to estimate the ensemble mean Lagrangian descent rate D(z,t), that is shown in Fig 4d. It is defined as: $D(z,t)=(z-Z_0)/A$ where z is the altitude (30 km), $Z_0$ is the ensemble mean initial altitude, and A is the ensemble mean age of the trajectories. The ensemble mean Lagrangian descent rate reflects the mean descent over the lifetime of the ensemble of parcels and is therefore strongly influenced by the higher descent rates found at higher altitudes during the ensemble's history. The Lagrangian descent rates remain relatively low until after mid May because the parcels within

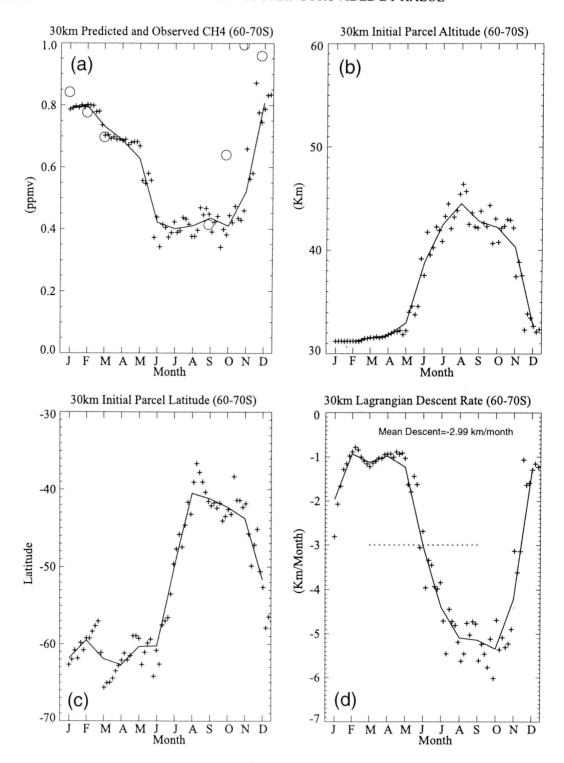

**Figure 4.** Time series at 30 km during 1993 for 60°-70°S of (a) the daily predicted (pluses), monthly mean predicted (line), and monthly mean observed (open circles) $CH_4$ mixing ratios; (b) ensemble mean initial altitudes of trajectories used to construct the $CH_4$ time series in (a); (c) ensemble mean initial latitudes; and (d) ensemble Mean Lagrangian decent rates. The horizontal dashed line in (d) indicates the period used to determine the average winter descent rates.

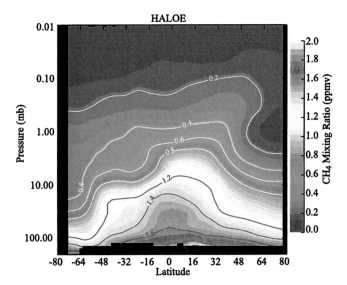

**Plate 1.** HALOE sunset zonal mean CH$_4$ pressure versus latitude cross section for the period August 31 to October 6, 1994.

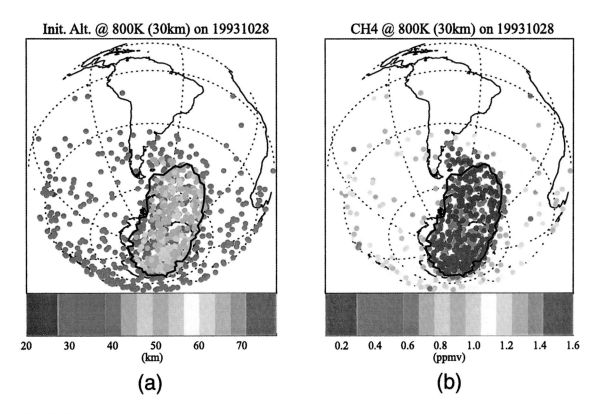

**Plate 2.** Synoptic maps of the parcels used to construct the time series shown in Fig. 4 on the 800 K surface (~ 30 km) for October 28, 1993. (a) shows the parcel's initial altitude and (b) shows the parcel's CH4 mixing ratio. The black line denotes the vortex edge.

the ensemble prior to May were initialized near 30 km (see Fig 4b), where diabatic heating rates (and thus descent rates) are relatively low. *Manney et al.*, [1999] used clustered 32-day back trajectories to assess Lagrangian mean descent rates for airmasses observed by ATMOS during the 1994 southern hemisphere spring (Nov 3-12). This study found maximum Lagrangian mean diabatic descent rates of near 10K/day between 840K and 1200K during this time period. Lagrangian descent rates rapidly increase to near -5 km/month by August as parcels which were originally observed at higher altitudes in Southern Hemisphere middle latitudes arrive within the $60°-70°S$ latitude band at 30km. These results are broadly consistent with trajectory based estimates of middle and upper stratospheric Antarctic descent rates by *Fisher et al.*, [1993] who found descent rates of $-3$ km/month in the Antarctic middle stratosphere and -12 km/month in the Antarctic upper stratosphere. These findings are also generally consistent with the results of *Rosenfield et al.*, [1994] who used a one-dimensional vortex interior descent model, combined with radiative heating estimates, to show that air which originated at 52km in the southern hemisphere on March 1, 1987 descended to 26-29 km by October 31, 1987, resulting in a mean Antarctic winter descent rate of -3.65 km/month for these airmasses. Rosenfield found that air which originated at 52 km on July 1, 1987 descended much more rapidly, falling 14km in the month of July, 1987. The results from the one-dimensional vortex interior descent model were compared with 3-D trajectories for 1992, where similar descent rates were found. The Lagrangian descent rates predicted during August 1993 are averages of the high descent rates encountered by the parcels in the upper stratosphere and the lower descent rates encountered in the middle stratosphere as the parcels descend to 30 km. The average Antarctic winter (March through September) ensemble mean Lagrangian descent rate of HALOE airmasses found within the $60°-70°S$ latitude band at 30 km during 1993 is -3.0 km/month. The average Antarctic winter ensemble mean Lagrangian descent rates in this latitude band are -5.2 km/month at 40km and -0.7 km/month at 20 km (not shown).

Plates 2a and b show synoptic maps of the parcels used in constructing the ensemble mean data for October 28, 1993 at 800K (near 30km). The parcels in Plates 2a and b are colored by their initial altitude and $CH_4$ mixing ratio, respectively. The black line indicates an estimate of the position of the vortex edge. This estimate is based on the relative contributions of rotation and strain (Q) integrated around potential vorticity (PV) contours [*Fairlie, et al.*, 1999]. When Q is positive, strain dominates and fluid elements are stretched; where Q is negative, rotation dominates and fluid elements remain intact. *Haynes* [1990], found that closed circulations (such as the polar vortex) are character-

ized by negative Q, while positive Q is associated with regions of "planetary-wave breaking" [*McIntyre and Palmer*, 1983]. The vortex edge in Plate 2 is defined as the PV contour where the integrated Q is equal to zero. The highest parcel density is found within the polar vortex, which is elongated towards South America. As discussed earlier, the high density of parcels within the polar vortex leads to the low ensemble mean $CH_4$ mixing ratios at $60° -70°S$ shown in Fig. 4a during this period. There is a very sharp gradient in the parcel initial altitudes and $CH_4$ mixing ratios across the vortex edge. Parcels within the vortex originated from between 40 km and 50 km and have mixing ratios between 0.2 and 0.4 ppmv. This sharp gradient reflects a kinematic barrier, which generally keeps air within the vortex isolated from middle latitudes. However, Rossby wave breaking events can disturb the kinematic barrier. The low $CH_4$/high initial altitude air which is equatorward of the kinematic barrier near the southern tip of South America is an example of this process. A large amplitude Rossby wave breaking event is just beginning to draw vortex air from within the kinematic barrier and mix it into middle latitudes. This trajectory-based prediction of Rossby wave induced exchange between the polar vortex and middle latitudes will be confirmed in later sections when we examine daily HALOE observations during this period.

A number of studies have shown that descent rates are larger near the edge of the Antarctic vortex than in the center in the lower stratosphere. [*Pierce et al.*, 1994, *Manney et al.*, 1994, 1999, and *Eluszkiewicz et al.*, 1995]. The initial parcel altitudes shown in Plate 2a do not appear to support these earlier studies. This discrepancy is most likely due to accumulated errors in the individual trajectory positions. The year long trajectory simulation is able to retain the sharp boundary between interior and exterior vortex airmass histories and tracer mixing ratios but is apparently not reliable for assessing the detailed structure within the polar vortex itself. It is obvious from Plate 2 that the strongest Antarctic descent occurs within the polar vortex. Consequently, the $60°-70°S$ ensemble mean Lagrangian descent estimates presented in Fig. 4, while useful in interpreting the seasonal evolution of zonal mean $CH_4$, most likely underestimate the actual descent rate within the polar vortex. To explore this possibility we have computed the vortex averaged descent during the 1993 Antarctic winter in two ways. The first method follows the technique developed by *Schoeberl et al.*, [1995] who used the vertical displacement of the HALOE 0.25 ppmv $CH_4$ mixing ratio isopleth to infer a net descent during the 1992 Antarctic winter. The second method uses D(z,t), as defined earlier.

To follow the descent of $CH_4$ mixing ratio isopleths within the Antarctic vortex we compute the daily ensemble mean altitude for parcels which were within the Antarctic vortex (as defined by negative integrated Q) and within $CH_4$ bins centered around 0.2, 0.3 and 0.4 ppmv. A bin width of ± 5% of the respective $CH_4$ mixing ratio was used. Because the Q diagnostic is sensitive to relatively weak rotational circulations we are able to define the proto-vortex during the Antarctic fall, as the wintertime westerlies are just beginning to develop. At least five vortex parcels must be found within each $CH_4$ bin to define an ensemble mean. In practice, many more than five parcels were used in the ensemble means, except during the early winter. Figure 5a shows the daily predicted vortex ensemble mean altitude time series for the 0.3ppmv $CH_4$ mixing ratio isopleth. The solid line denotes a three-point boxcar average of the daily ensemble means. The scatter in the daily vortex ensemble means is considerably lower than found in the zonal ensemble means, indicating that random errors in the trajectory calculations have been reduced even further by vortex ensemble averaging. The 0.3 ppmv $CH_4$ vortex mixing ratio isopleth begins at 40 km during mid March. During the first part of the Antarctic winter it descends at a relatively constant rate, reaching 31-32km by mid June. During July, the altitude of the isopleth appears to ascend. This anomaly is a consequence of the relatively uniform vertical profile of $CH_4$ within the vortex (see Fig 2a) which leads to the inclusion of upper stratospheric parcels with $CH_4$ mixing ratios near 0.3 ppmv in the ensemble mean. By September, the 0.3 ppmv $CH_4$ isopleth has descended to 29-30 km. The results for the 0.2 ppmv and 0.4 ppmv $CH_4$ isopleths are similar. The 0.2 (0.4) ppmv $CH_4$ isopleth descends from 42km (36km) in March to near 30km (29km) in September. Net winter descent rates, obtained from the difference between the vortex ensemble mean altitudes of the $CH_4$ isopleth for March and September, 1993, are -1.9 km/month, -1.6 km/month, and -1.2 km/month, for $CH_4$ mixing ratios of 0.2 ppmv, 0.3 ppmv, and 0.4 ppmv respectively. *Schoeberl et al,.* [1995] compared $CH_4$ profiles for the 1992 Antarctic fall (February-March) and Antarctic spring (September-October) to infer a vortex mean descent rate of -1.8 km/month for the 0.25 ppmv $CH_4$ isopleth, which is in very good agreement with our estimates for 1993.

*Kawamoto and Shiotani* [2000] used Antarctic HALOE observations to determine the vortex averaged descent rates of $CH_4$ isopleths between 0.3 and 0.9 ppmv for 1992-1997. They found significant year-to-year variations in the Antarctic vortex descent rates which were associated with the Quasi-biennial Oscillation. Their estimate for the 1993 0.3 ppmv isopleth was near -1.7 km/month, while the 1992 0.3 ppmv isopleth descent rate was near -1.9 km/month. Our estimated descent rate for the 0.4 ppmv $CH_4$ isopleth during 1993 is considerably less than Kawamoto and Shiotani obtained (-1.65 km/month) indicating that our radiative cooling rates may be less than the actual diabatic descent below 30 km. This is not surprising since the MIDRAD radiative code

is designed for use in the middle stratosphere and is less accurate in the lower stratosphere since it assumes a climatological upwelling longwave flux from the troposphere to determine the longwave component of the net heating.

Figures 5b and c show the ensemble mean initial latitude and Lagrangian descent rates (D(z,t)) for the 0.3ppmv $CH_4$ vortex parcels. During the early Antarctic winter, the 0.3 ppmv $CH_4$ vortex ensemble mean is composed of parcels which were initially observed by HALOE at polar latitudes, while during late winter and early spring, the 0.3 ppmv $CH_4$ vortex ensemble mean is composed of parcels which were initially observed by HALOE at middle latitudes of the Southern Hemisphere. The ensemble mean Lagrangian descent rates, computed from D(z,t), range from -1 km/month to -3 km/month during March through June. During June and July the ensemble mean Lagrangian descent rates increase rapidly, reaching -7 km/month by late winter and early spring. This rapid increase in descent corresponds to the transition from polar parcels to middle latitude parcels in the 0.3ppmv $CH_4$ vortex ensemble mean. The average Antarctic winter (March through September) Lagrangian mean descent rate of HALOE vortex airmasses with $CH_4$ mixing ratios of 0.3 ppmv during 1993 is -4.9 km/month. The average winter Lagrangian descent rates for the 0.2 ppmv and 0.4 ppmv $CH_4$ mixing ratio isopleths are -5.7 km/month and -3.5 km/month. As anticipated, the vortex averaged Lagrangian descent rates are larger than the Lagrangian descent rates obtained within the 60°-70°S latitude band. However, the overall pattern of poleward and downward transport is similar.

The apparent discrepancy between the net winter descent based on $CH_4$ isopleth displacements and the vortex ensemble mean Lagrangian descent rate is due to the fact that the latter estimate is an ensemble mean of a true Lagrangian descent rate, while the former is not. For the isopleth displacement to be a true Lagrangian estimate of the descent rate, the air which comprised the upper stratospheric 0.3 ppmv $CH_4$ isopleth in the Antarctic fall would have to be the same air which comprised the middle stratospheric vortex 0.3 ppmv $CH_4$ isopleth in the Antarctic spring. From Fig. 5b we see that this is not the case. Instead, the late winter/early spring Antarctic vortex ensemble mean is composed of air which originated in middle latitudes of the Southern Hemisphere, and, as seen from Plate 1, the 0.3 ppmv $CH_4$ isopleth is found at higher altitudes in middle latitudes. Consequently, the vortex ensemble mean Lagrangian descent rate is higher than the descent rate obtained by following the descent of a $CH_4$ isopleth.

## 4. COUPLNG BETWEEN ALTITUDES AND REGIONS

It is clear from Plate 1 and the HALOE airmass simulations that very low levels of $CH_4$ are transported down from

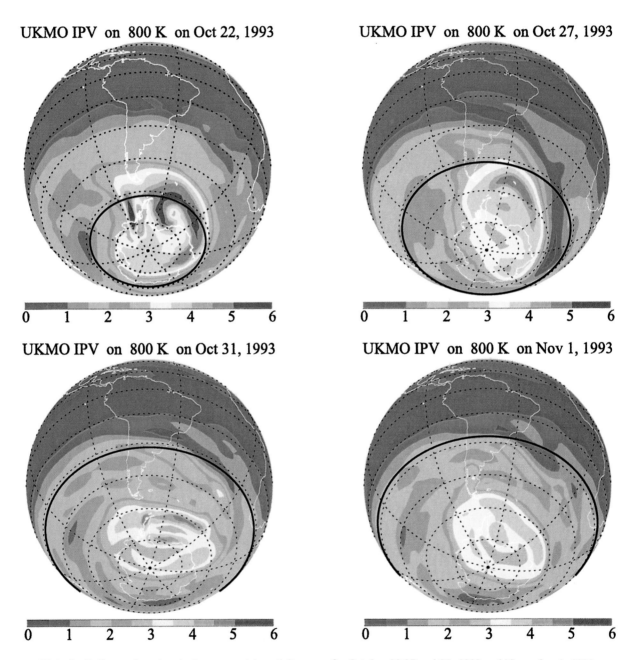

**Plate 3.** Daily southern hemisphere potential vorticity maps for October 22,27 and 31, 1993 and November 1, 1993 showing the influence of vortex air on the mid-latitudes and subtropics. The black circle indicates the latitude of HALOE observations.

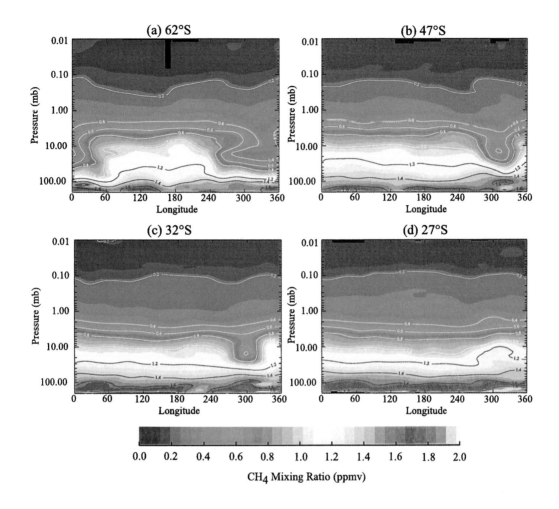

**Plate 4.** Sunset CH₄ pressure versus longitude cross sections at 62° S, 47° S, 32° S and 27° S for October 22,27 and 31, 1993 and November 1, 1993 respectively.

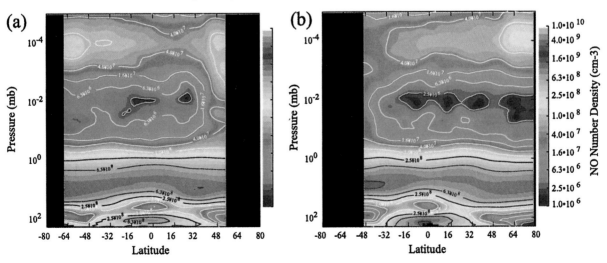

**Plate 5.** Sunset NO pressure versus latitude cross sections for December 31, 1992 to February 7, 1993 (a) and April 11 to May 28, 1993 (b).

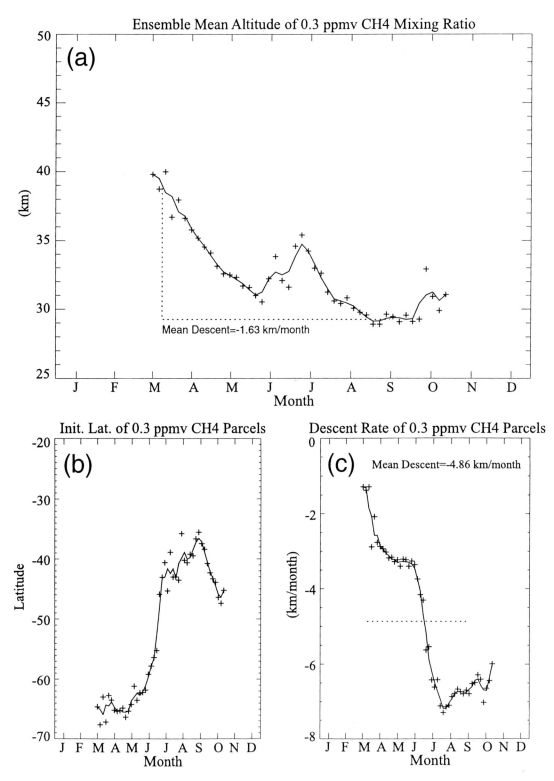

**Figure 5.** Daily predicted vortex ensemble mean time series for the CH₄ 0.3 ppmv isopleth (a), the ensemble mean initial latitude (b) and the Langrangian descent rates inside the vortex for the 0.3 ppmv CH₄ parcels (c).

the mesosphere to the lower stratosphere. While the upper stratospheric air appears to remain within the vortex during the winter, once brought down it will eventually be dispersed into middle latitudes during the Southern Hemisphere final warming, as illustrated by Plate 2. *Bithell et al.*, 1994 performed a study with HALOE data showing evidence of vortex edge air being torn off by anticyclonic flow that carried the low values of $CH_4$ to lower latitudes. This can be seen in HALOE polar stereographic projections on various pressure surfaces but as in Plate 1, it takes about three weeks to a month time period for the solar occultation sampling to cover a sufficiently large latitude range to produce the plots. HALOE polar stereographic projections therefore provide time mean pictures and tend to wash out daily variations in the position of the polar vortex. Rigorous interpretation requires examination of daily PV maps.

Plate 3 shows UKMO-UARS PV maps on the 800 K surface (~ 30 km) for four days in the period October $22^{nd}$ to November 1, 1993. The black circles indicate which latitudes HALOE was observing on each day. On October $22^{nd}$, a large amplitude anticyclone is evident southeast of New Zealand. By October $27^{th}$ this anticyclone has displaced the polar vortex off the pole and a filament of vortex air is just beginning to be peeled off of the vortex near the southern tip of South America. This vortex filament event is also evident in Plate 2, which shows that the filament is composed of air, which originated in the upper stratosphere and has low $CH_4$ mixing ratios. By October $31^{st}$ the filament has broken off of the vortex and the air within the filament is then mixed into the surrounding middle latitude air within the Southern Hemisphere surf zone. Daily $CH_4$ pressure versus longitude cross sections can be used to confirm this wave breaking event because each cross section is a snapshot for a given narrow latitude band of only 2 to 3°.

Plate 4 shows pressure versus longitude cross sections at 62°S, 47°S, 32°S and 27°S on the 22nd, 27th and 31st of October and the 1st of November respectively. It can be seen in Plate 4a that the 62° S cross section includes measurement samples made inside the vortex for the 0°-30° and 240°- 360° longitude ranges and outside the vortex for 30°-240°. Values in the latter range are much higher at the 10 mb level than within the vortex and are associated with air within the middle latitude anticyclone. Of even greater interest are the cross section plots at 47° S and 32° S (Plates 4b,c). Note the very low $CH_4$ values that exist in a narrow longitude band centered at about 300° longitude and extending down to about the 25 mb level. This feature exists for both latitudes and is coincident with the vortex filament discussed earlier. These results show that horizontal transport of low $CH_4$ air from the Antarctic vortex region has extended all the way into the subtropics. Horizontal transport associated with Rossby wave breaking is the only plau-

sible explanation and the PV plots of Plate 3 confirm this. The last panel, Plate 4d, shows no effect of the vortex air. These results provide clear evidence of coupling between the polar mesosphere and the subtropical lower stratosphere.

## 5.  SOLAR AND DYNAMICAL COUPLING

An example of coupling between altitudes, atmospheric regions and solar effects, is seen in HALOE nitric oxide data. We start first with NO pressure versus latitude cross sections for the winter periods of each hemisphere (Plates 5a,b). In the ~60 km to 80 km range, the Northern Hemisphere winter cross section (Plate 5a) shows uniformly low NO number densities from 68° S to 40° N but increased values poleward of this range. Similarly, the Southern Hemisphere winter (Plate 5b) shows low NO from 78°N to 40°S with larger values poleward. Note that the winter latitude coverage only extends to about 52° because of the occultation viewing geometry. We assert that the larger NO values in the high winter latitudes are due to NO vertical descent in the polar night. Figure 6 shows two NO time series extending from the beginning of the UARS mission to January, 1999 at 70 km altitude for the latitude range 40° N to 60°N (Fig. 6a) and 40°S to 60°S (Fig. 6b). Note the periodicity of the NO variations in each hemisphere which are out of phase corresponding to the opposite seasons. Note also the nearly invariant low levels of NO number density that occur in summer for each hemisphere. These plots dramatically illustrate the regular "pulsing" of the atmosphere that occurs due to winter descent bringing high levels of NO from near 110 km where the number density is a maximum down to 70 km. Although we do not show it, effects of the descent can be seen in the NO time series down to about 60 km but below this altitude very little evidence of the phenomenon is present in the data. The very large and anomalous NO density which occurs in the north in January, 1994 is due to solar activity which resulted in significant energetic electron precipitation that caused NO formation [*Callis et al.*, 1997].

The curves represented by the series of closely connected circles in the figures show the variation in the F10.7 cm solar flux which is an indicator of solar output. Further, the F10.7 cm signal is believed to be a good gage of the solar soft X-ray flux that produces energetic electrons leading to NO formation. This connection is born out by the high degree of correlation in simultaneous measurements of the actual soft X-ray flux and thermospheric NO measured by the Student Nitric Oxide Explorer (SNOE) [*Barth et al.*, 1999]. Although we do not show it here, variations in the equatorial NO time series observed by HALOE at 110 km and the F10.7 cm flux are highly correlated over the part of the solar cycle covered by UARS. There is a striking similarity in Fig. 6 between the envelope of the NO time series

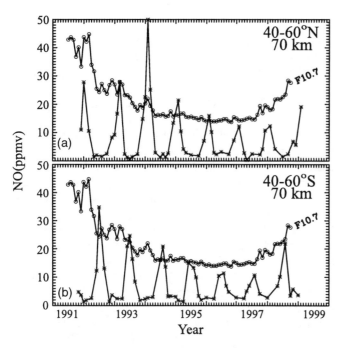

**Figure 6.** HALOE monthly zonal mean NO mixing ratio time series at 70 km averaged over the range from (a) 40° N to 60° N and (b) 40° S to 60° S with the F10.7 cm flux superimposed.

peaks in HALOE data and the variation in the F10.7 cm flux especially in the Southern Hemisphere. These results show clear evidence of solar coupling in the thermosphere where maximum NO production occurs and dynamical coupling with the mesosphere through vertical descent down to 70 km (or even weak coupling to 60 km). Also, HALOE and other data show ample evidence of even deeper penetration of NO from the thermosphere into the lower stratosphere in the southern polar vortex [*Siskind and Russell*, 1996; *Callis et al.*, 1996; *Siskind et al.*, 1997; *Randall et al.*, 1998; *Rinsland et al.*, 1999]. *Siskind et al.* [2000] did a quantitative analysis of HALOE NO data over six Antarctic spring periods using $CH_4$ as a tracer to evaluate the contribution of upper atmospheric NO to the stratospheric polar vortex. Using equivalent latitude coordinates they were able to correlate years of high thermospheric NO with high stratospheric NO and using the $CH_4$ data they concluded that the origin of lower stratospheric enhancement was due to descent from aloft. Strong NO enhancement occurred in four of the six springs they studied (note that the NO lifetime is infinite in winter and only days in summer). A strong correlation also exists in the time series of average column abundance of HALOE $NO_x$ ($NO + NO_2$) for the vortex core and the $A_p$ index (an indicator of geomagnetic activity used to predict electron flux variations). This suggests that particle precipitation and NO formation in the thermosphere followed by descent is the main cause of stratospheric vortex

$NO_x$ variability. More details and additional analysis of this is presented in another paper in this monograph by *Siskind, 2000.*

## 6. DISCUSSION AND SUMMARY

HALOE data have shown the occurrence of polar descent in a variety of ways. The first and most obvious indication of this phenomenon is the occurrence of a nearly constant $CH_4$ mixing ratio profile over the ~ 0.1 mb to ~ 25 mb range in the Antarctic spring. This is in sharp contrast to the almost uniformly increasing profile with decreasing altitude that occurs outside the vortex. Perturbations to other mixing ratio profiles observed by HALOE can also be seen including HF, $H_2O$, HCl and NO. Pressure versus latitude cross sections show the effects of descent as either deep troughs or regions of enhancement in the polar regions depending on whether the parameter mixing ratio is either decreasing or increasing with altitude. The effects of descent displayed in pressure versus longitude cross sections are mixing ratio depressions or enhancements on what is normally a near uniform distribution. While we did not show any profiles or cross sections for the Northern Hemisphere in this paper, the effects are as obvious as they are in the south. The main difference is the southern vortex usually remains intact over the course of winter and therefore, the data are more amenable for study of descent. Time series in the southern polar region over the more than eight years of HALOE observations show that descent is a regular occurrence in winter. Studies have further shown that the strength of descent varies interannually (*Kawamoto and Shiotani*, 2000). Synoptic maps of $CH_4$ developed using HALOE initialized trajectory studies and direct HALOE observations show that the magnitude of the descent is greatest inside the vortex with the clearest signal of descent occurring at the vortex center as indicated by the constant $CH_4$ mixing ratio profiles.

We performed detailed analyses of the HALOE data in 1993 using three-dimensional trajectories developed using UKMO wind and temperature fields. The great advantage of this approach is it fills in data during the deep polar winter when occultation provides no measurements because of polar night. The further advantage is that the trajectories are initialized using HALOE data and can then be checked for accuracy at a later time using subsequent HALOE observations. The trajectory calculations and $CH_4$ observations were in excellent agreement at 30 km and higher altitudes and they were in acceptable agreement at 20 km but the accuracy of the calculations on this surface was not as good because of limitations with the radiation code. Time series of initial altitudes, latitudes and Lagrangian descent rates were calculated at altitudes of 40 km, 30 km and 20 km in order to estimate the amount of poleward and downward transport that occurred. The results showed that air, which was at 45 km and mid latitudes in May, moved poleward and downward to 60°-70°S and 30 km in August. The 60°-70°S Lagrangian descent rate at 30 km varied over the course of the year from ~ -1 km/month in February to a maximum of ~ -5.5 km/Month in September. The March to October average Lagrangian descent rate was ~ -5.2 km/month at 40 km, ~ -3 km/month at 30 km and ~ -0.7 km/month at 20 km. These results are generally consistent with previous Lagrangian modeling studies.

Net descent rates estimated from the vertical displacement of $CH_4$ mixing ratio isopleths were compared to Lagrangian mean descent rates for parcels within the Antarctic vortex. Net winter descent rates, obtained from the difference between the vortex ensemble mean altitudes of $CH_4$ isopleths for March and September 1993, were found to be -1.9 km/month, -1.6 km/month, and -1.2 km/month, for $CH_4$ mixing ratios of 0.2 ppmv, 0.3 ppmv, and 0.4 ppmv. These results are in very good agreement with previous analyses, which used HALOE observations alone. However, the average Antarctic winter Lagrangian mean descent rate of HALOE vortex airmasses with $CH_4$ mixing ratios of 0.2 ppmv, 0.3 ppmv, and 0.4 ppmv were considerably larger; ~-5.7 km/month, ~ -4.9 km/month, and ~  -3.5 km/month, respectively. This discrepancy was shown to be due to the fact that the isopleth estimate of net descent is not a true Lagrangian descent rate, since the air which comprised the upper stratospheric polar $CH_4$ isopleths in the Antarctic fall was not the same air which comprised the middle stratospheric polar $CH_4$ isopleths in the Antarctic spring. Instead, the late winter early/spring Antarctic vortex ensemble mean was composed of air, which originated in middle latitudes of the Southern Hemisphere. Therefore, since the middle latitude $CH_4$ isopleths are found at higher altitudes than the polar isopleths, the vortex ensemble mean Lagrangian descent rate is higher than the descent rate obtained by following the descent of a $CH_4$ isopleth. These results apply to the ensemble means of airmasses observed by HALOE, which does not sample high southern latitudes during the Antarctic winter. Consequently, the contribution due to direct polar descent during mid winter can not be assessed using HALOE observations. However, the good agreement between the ensemble mean and observed time series, and the very good agreement with previous isopleth estimates of descent, indicate that the results presented here capture the main features of the Antarctic wintertime Lagrangian circulation.

The Antarctic wintertime Lagrangian circulation begins with air from polar and middle latitudes in the upper stratosphere, which appears to be continuously drawn into the polar vortex at these altitudes. Once within the vortex, the air initially descends rapidly and then slows as it approaches the middle to lower stratosphere. The vortex air remains highly isolated from middle latitude stratospheric air during this descent process. Synoptic maps of the HALOE airmass simulation during late Southern Hemisphere spring show that there is a very sharp gradient in the parcel initial altitudes and $CH_4$ mixing ratios across the vortex edge which is a consequence of this isolation. Parcels within the vortex originate from between 40 km and 50 km, consistent with the Antarctic wintertime Lagrangian circulation. Rossby wave breaking events during late winter and early spring peel air off the edge of the vortex, transporting it into middle latitudes of the middle and lower stratosphere. This mixing of vortex air with middle to low latitude air in the lower stratosphere completes the Antarctic Lagrangian circulation for the winter season and couples air, which has descended from the polar mesosphere to the near subtropical lower stratosphere.

Nitric oxide pressure versus latitude cross sections in the 60 km to 90 km range show the effects of vertical descent as a winter time filling in of the normally very low NO values in this altitude interval. Time series at 40°-60°N and S for 70 km show uniformly low NO in the summer months and a regular NO "pulsing" of the atmosphere every winter in each hemisphere. The envelopes of the NO peaks in each hemisphere are strongly correlated with the solar F10.7 cm flux. The 70 km enhancements taken together with the F10.7 cm flux changes, which are indicative of NO formation in the thermosphere, show the presence of solar coupling in the thermosphere and dynamical coupling to the mesosphere.

Acknowledgements. We acknowledge the helpful contributions of Patrick N. Purcell and Janet Daniels of SAIC, Inc. and John Anderson and Scott Bailey of the Hampton University Center for Atmospheric Sciences who provided some of the figures and consultation for this paper. We also express appreciation to David E. Siskind of NRL who encouraged us to write the paper and to the entire HALOE Team for their careful data processing activities. The Hampton University effort was supported in part by NASA Grants NAG5-7001 and NAS1-97042. Support for Pierce was provided by the NASA OES Atmospheric Chemistry Modeling and Analysis Program (ACMAP).

## REFERENCES

Abrams, M. C. et al., ATMOS/ATLAS-3 observations of long-lived tracers and descent in the Antarctic vortex in November 1994, *Geophys. Res. Lett.*, 23, 2341-2344, 1996.

Bauer, R. A. et al., Monitoring the vertical structure of the Arctic polar vortex over northern Scandinavia during EASOE: Regular N2O profile observations, *Geophys. Res. Lett.*, 21, 1211-1214, 1994.

Bacmeister Julio T., et al., Descent of long-lived trace gases in the winter polar vortex, *J. Geophys. Res.*, 100, No. D6, 11,669-11,684, June 20, 1995.

Barth, Charles A., Scott M. Bailey and Stanley C. Solomon, Solar-terrestrial coupling: Solar soft x-rays and thermospheric nitric oxide, *Geophys. Res. Lett.*, Vol. 26, No. 9, 1251-1254, May 1, 1999.

Bithell, M., et al., Synoptic Interpretation of Measurements from HALOE, *J. Atm. Sci.*, 51, No. 20, 2942-2956, October 15, 1994.

Brasseur, G., Physique et Chemie de l'Atmosphere Moyene, Mason, Paris, 1982.

Brasseur, G. and S. Solomon, Aeronomy of the Middle Atmosphere, 2nd Ed., D. Reidel, Norwell, Mass, 1986.

Callis, L. B. et al., A 2D model simulation of downward transport of $NO_y$ in the stratopsphere: Effect on the austral spring $O_3$ and $NO_y$, *Geophys. Res. Lett.*, Vol. 23, 1905-1908, 1996.

Eluszkiewicz, J., R. A. Plumb, and N. Nakamura, Dynamics of wintertime stratospheric transport in the geophysical fluid dynamics laboratory SKYHI general circulation model, *J. Geophys. Res.*, 100, 20,883-20,9000,1995.

Fairlie, et al., The contribution of mixing in Lagrangian photochemical predictions of polar ozone loss over the Arctic in summer 1997, *J. Geophys. Res.*, 104, D21, 26,597-26,609, 1999.

Fisher, M., A. O'Neil and R. Sutton, Rapid descent of mesospheric air into the stratospheric polar vortex, *Geophys. Res. Lett.*, Vol. 20, 1267-1270, 1993.

Garcia R. and S. Solomon, A numerical model of the zonally averaged dynamical and chemical structure of the middle atmosphere, *J. Geophys. Res.*, 88, 1,379, 1983.

Haynes, P.H., High-resolution three-dimensional modeling of stratospheric flows: Quasi-two dimensional turbulence dominated by a single vortex, in *Topological Fluid Mechanics*, edited by H.K. Moffat and A. Tsinober, Cambridge University Press, 1990.

Kawamoto, Nozomi and Masato Shiotani, Interannual variability of the vertical descent rate in the Antarctic polar vortex, *J.Geophys. Res.*, In-press, 2000.

Kumer, J. B., L. Mergenthaler and A. E. Roche: CLAES $CH_4$, $N_2O$, and $CF_2CCl_2$ ($F_{12}$) global data, *Geophys. Res. Lett.*, Vol. 20, 1239-1242, 1993.

Manney, G. L., et al., On the motion of air through the stratospheric polar vortex, *J. Atmos. Sci.*, 51, 2973-2994, 1994.

Manney, G. L., et al., Polar vortex dynamics during spring and fall diagnosed using trace gas observations from the Atmospheric Trace Molecule Spectroscopy instrument, *J. Geophys. Res.*, 104, 18,841-18,866, 1999.

McIntyre, M.E. and T.N. Palmer, Breaking planetary waves in the stratosphere, *Nature*, 305, 593-600, 1983.

Norton, W.A., Breaking Rossby Waves in a model stratosphere diagnosed by a vortex following coordinate system and a technique for advecting material contours. *J. Atmos. Sci.*, Vol. 51, pp. 654-673, 1994.

Pierce, R. Bradley, et al., Evolution of southern hemisphere spring air masses observed by HALOE, *Geophys. Res. Letters*, Vol. 21, No. 3. 213-216, February 1, 1994.

Pierce, R. B., et al., Mixing processes within the polar night jet, *J. Atmos. Sci.*, 51, 2957-2972, 1994b.

Pierce, R. B. et al.,Large-scale stratospheric ozone photochemistry and transport during the POLARIS Campaign, *J. Geophys. Res.*, 104, D21, 26,525-26,545, 1999.

Randall, C. E. et al., Polar Ozone and Aerosol Measurements (POAM) II stratospheric $NO_2$, 1993-1996, *J.Geophys. Res.*, Vol. 103, 28361 – 28372, 1998.

Randel, William J., et al., Seasonal cycles and QBO variations in stratospheric $CH_4$ and $H_2O$ observed in UARS HALOE data. *J. Atmos. Sci.*, Vol. 55, pp. 163-185, January 15, 1998.

Randel, William J., et al., Space-time patterns of trends in stratospheric constituents derived from UARS measurements, *J. Geophys. Res.*, 104, No. D3, 3711-3727, February 20, 1999.

Rinsland, C. P. et al., Polar stratospheric descent of $NO_y$ and CO and Arctic denitrification during winter 1992-1993, *J. Geophys. Res.*, 104, 1847-1861, 1999.

Rosenfield , Joan E., Paul A. Newman and Mark R. Schoeberl, Computations of diabatic descent in the stratospheric polar vortex, *J. Geophys. Res.*, 99, No. D8, 16,667-16,689, August 20, 1994.

Russell, J. M., et al., The Variability of Stratospheric and Mesospheric NO2 in the Polar Winter Night Observed by LIMS, *J. Geophys. Res.*, 89, No. C8, August 23, 1984.

Russell, et al., The Halogen Occultation Experiment, *J. Geophys. Res.*, 98, No. D6, 10,777-10,797, June 20, 1983a.

Russell, J. M. III et al., HALOE Antarctic Observations in the Spring of 1991. *Geophys. Res. Lett.*, Vol. 20, No. 8, 719-722, April 23, 1983b.

Schoeberl, M. R et al., The Structure of the Polar Vortex, *J. Geophys. Res.*, 97, 7859-7882, 1992.

Schoeberl, Mark R., Mingzhao Luo and Joan R. Rosenfield, An anlaysis of the Antarctic Halogen Occultation Experiment trace gas observations, *J. Geophys. Res.*, 100, No. D3, 5159-5172, March 20, 1995.

Shine, K. P, The middle atmosphere in the absence of dynamical heat fluxes, *Q. J. R. Met. Soc.*, 113, 605-633, 1987.

Siskind, D. E. and J. M. Russell III: Coupling Between Middle and Upper Atmospheric NO: Constraints from HALOE Observations, *Geophys. Res. Lett.*, Vol. 23, No. 2, 137-140, January 15, 1996.

Siskind, D. E. et al., Two dimensional model calculations of nitric oxide transport in the middle atmosphere and comparison with HALOE data, *J. Geophys. Res.*, 102, 3527. 1997.

Siskind, D. E., et al., A quantitative assessment of stratospheric $NO_x$ enhancements due to transport from the upper atmosphere, In-press, *Geophys. Res. Lett.*, January, 2000.

Siskind, D. E., On the coupling between middle and upper atmospheric odd nitrogen, *AGU Monograph*, 2000, this issue.

Strahan, S. E., J. E. Nielsen, and M. C. Cerniglia, Long-lived tracer transport in the Antarctic stratosphere, *J. Geophys. Res.*, 101, 26,615-26,629, 1996.

Solomon, S., P.J. Crutzen and R.G. Roble, Photochemical coupling between the thermosphere and the lower atmosphere, 1, Odd nitrogen from 50 to 120 km, *J. Geophys. Res.*, 87, No. C9, 7221, 1982.

Swinbank, R. and A. O'Neill, A Stratosphere-Troposphere Data Assimilation System, *Mon. Wea. Rev.*, 122, 686-702, 1994.

# Sudden Stratospheric and Stratopause Warmings: Observations of Temperatures in the Middle Atmosphere above Eureka

Thomas J. Duck [1]

*Department of Physics and Astronomy, York University, Toronto, Canada*

James A. Whiteway

*Department of Physics, University of Wales, Aberystwyth, UK*

Allan I. Carswell

*Department of Physics and Astronomy, York University, Toronto, Canada*

Wintertime observations of strato-mesospheric temperatures have been obtained in the High Arctic at Eureka (80 °N, 86 °W) since 1993 by using a lidar and meteorological balloons. Several minor sudden stratospheric warmings (i.e., large-scale stratospheric circulation disturbances) have been observed since that time. With respect to the measurements obtained at Eureka, the disturbances are associated with lower stratospheric warming and stratopause cooling. In contrast, when the circulation is relatively undisturbed, lower stratospheric cooling and stratopause warming are observed above Eureka. These temperature changes are associated primarily with movements of the wintertime stratospheric vortex, which implies that single station measurements are best examined against the backdrop of the actual three-dimensional circulation rather than the more traditional sudden warming diagnostics. Furthermore, because warming at the stratopause is observed when the circulation is relatively undisturbed, a contribution in addition to planetary wave drag and vortex-vortex interactions is likely important for the dynamics of the upper stratosphere.

## 1. INTRODUCTION

The term "sudden stratospheric warming" can readily confuse discussions regarding wintertime stratospheric temperature observations at single measurement stations. Do observations of stratospheric temperature increases at any level constitute a "sudden stratospheric warming"? Traditionally, the phrase is used to imply a large-scale midwinter event that induces changes the zonal mean thermal structure of the high-latitude stratosphere so that below 10 hPa the polar temperatures warm relative to midlatitudes (see *Schoeberl* [1978] or *Labitzke* [1981] for reviews). The cause of sudden stratospheric warming is usually attributed to the nonlinear growth of planetary waves [*Matsuno*, 1971; *McIntyre and Palmer*, 1983] and vortex-vortex interactions [*O'Neill and Pope*, 1988]. It is of interest to consider how temperature structures change during these intrinsically large-scale events.

---

[1] Now at Haystack Observatory, Massachusetts Institute of Technology

Atmospheric Science Across the Stratopause
Geophysical Monograph 123

A difficulty when treating "sudden stratospheric warming" phenomena is that terminology can quickly complicate the discussion. For example, *Labitzke* [1981] noted that the word 'warming' is a "misnomer" because temperature changes opposite in sign can occur at different levels. For the sake of clarity, we will call the circulation "disturbed" during those intrinsically large-scale flow disruptions previously referred to as sudden warmings. Conversely, a circulation that has polar temperatures below 10 hPa that are significantly colder than the mean at midlatitudes will be called "relatively undisturbed". The words "warming" and "cooling" will be reserved exclusively to describe changes in temperature profiles. For example, an observed temperature increase near the stratopause above a single measurement station will be referred to as a stratopause warming.

In this study, temperature profiles obtained with a lidar in the High Arctic at Eureka (80 °N, 86 °W) are presented. Measurements taken during several stratospheric disturbances are compared with the profiles obtained in relatively undisturbed conditions. Global analyses of height are used to provide detail that is lost in the zonal mean analysis. The temperature profiles obtained during stratospheric disturbances are vastly different from those measured when the flow is relatively undisturbed, and are quite naturally associated with particular meteorological features. Furthermore, the warmest temperatures in the upper stratosphere are found when the flow is relatively undisturbed, which has some implications for our understanding of upper stratospheric dynamics.

## 2. MEASUREMENT AND ANALYSIS TECHNIQUE

### 2.1. The UV-Lidar at Eureka

Measurements of strato-mesospheric temperatures, gravity wave activity, ozone and aerosols have been obtained by using a lidar (laser radar) at Eureka each winter since 1992/93. Technical data concerning the lidar are given by *Carswell et al.* [1993] and *Duck* [1999]. The instrument is used to measure temperatures in the middle atmosphere by assuming hydrostatic balance in a perfect gas, i.e., by employing the standard Rayleigh lidar technique discussed by *Hauchecorne and Chanin* [1980]. The profiles considered here are nightly averages, although shorter duration observations are routinely made (see, for example, *Whiteway and Carswell* [1994] and *Whiteway et al.* [1997]). The temperature measurements typically extend from 25 to above 70 km in altitude, and are complemented by meteorological balloon (radiosonde) measurements below this interval. Lidar measurements obtained during the six wintertime campaigns between 1992/93 and 1997/98 are considered here.

### 2.2. NCEP Global Analyses

Global analyses of temperature and height on eighteen pressure surfaces ranging from 1000 to 0.4 hPa (i.e., from near the ground to approximately 50 km in altitude) were supplied by NCEP for the purposes of this study. Maps of wind were constructed from the height maps by assuming geostrophic balance; profiles of wind sampled from the analyses above Eureka were found to be in good agreement with the radiosonde observations.

The temperature analyses were used to calculate the difference in zonal mean temperatures on the 30 hPa pressure surface between 90 and 60 °N, $\Delta T_{90\text{-}60°N}$, for each day. The 30 hPa level was chosen so as to be consistent with the review by *Labitzke* [1981], and because the differences in $\Delta T_{90\text{-}60°N}$ between disturbed and undisturbed conditions is usually larger than at 10 hPa.

Because distinct temperature changes occur in the Arctic in late December [*Duck et al.*, 1998] and the fact that the polar sunrise certainly changes the radiative environment there come March, only the interval during January and February is considered hereafter. Six distinct stratospheric circulation disturbances, i.e., where $\Delta T_{90\text{-}60°N}$ reached above 0 K, were evident during the measurement campaigns considered. In order to provide a comparison, the six days of coldest $\Delta T_{90\text{-}60°N}$, one day taken from each campaign, were selected.

## 3. OBSERVATIONS

### 3.1. Night-average Temperature Profiles

The temperature profiles obtained during the six days of stratospheric disturbance are presented in Figure 1a. The CIRA reference atmosphere temperatures for February at 80 °N [*Fleming*, 1990] are given for comparison; the CIRA profile gives the approximate form of the average February temperature profile observed at Eureka.

Referring to Figure 1a, the temperatures above Eureka during stratospheric disturbances are relatively warm in the lower stratosphere, but cool in the upper stratosphere/lower mesosphere. At times, the temperature structure during a stratospheric disturbance is such that the profiles become almost isothermal throughout the middle atmosphere (or even have temperature decreasing with altitude), with no discernable stratopause. These isothermal temperature profiles are sometimes observed for several days after the disturbance peak.

The observations taken during the six days of relatively undisturbed circulation are given in Figure 1b. In contrast with the profiles presented in Figure 1a, the temperatures in Figure 1b are relatively cool in the lower stratosphere and warmer in the upper stratosphere / lower mesosphere. The peak of the upper level warming is near the stratopause, and

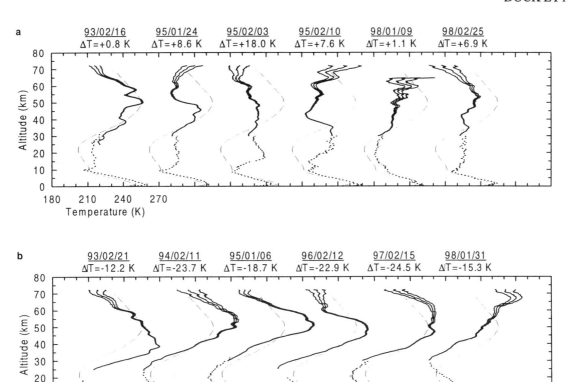

**Figure 1.** Temperature profiles obtained at Eureka during days of peak stratospheric disturbance (**a**) and in relatively undisturbed conditions (**b**). Temperature profiles are given by the solid black lines and are separated by 60 K; the surrounding gray lines represent the uncertainty in each measurement. The dotted profiles are the radiosonde observations, and the dashed profile is used as a reference. The dates for each measurement are listed above each profile, and the diagnostic $\Delta T_{90\text{-}60°N}$ is also provided.

so periods of relatively undisturbed large-scale conditions can also be referred to as stratopause warmings above Eureka. Thus, the temperatures obtained during relatively undisturbed conditions show deviations from the standard profile that are opposite in sense to those measured during the stratospheric disturbances.

It should be noted that the measurement obtained on 31 January is anomalous, since the temperature profile shown is cool throughout the stratosphere and then extremely warm above 55 km in altitude. In fact, the thermal evolution of the vortex during January of 1998 was quite different from that of the preceding five years, an issue that will be discussed in a future communication.

### 3.2. The Large-scale Stratospheric Circulation

More insight into the nature of disturbed / undisturbed flow and the corresponding temperature profiles obtained at Eureka can be gained by examining maps of the large-scale stratospheric circulation. Figure 2 provides maps of geopotential height on the 10 hPa surface during each

stratospheric disturbance. Evident in each map is a deep low that is representative of the Arctic stratospheric vortex, an enormous cyclone that develops in the stratosphere each winter in response to differential meridional solar heating. The height contours in each map serve as streamlines of wind, with the circulation directed counter-clockwise around the vortex low (and directed clockwise around any highs).

Referring to the height maps in Figure 2, it can be seen that each stratospheric disturbance was characterized by the polar vortex having been displaced from its quiescent position over the pole by a strong high. In general, the vortex was moved so that Eureka was positioned outside of the vortex and within the high, or on the very outside edge of the vortex jet (with the one exception being during the anomalous period of January 1998).

Figure 3 provides maps of geopotential height during each day of relatively undisturbed flow. As seen in Figure 3, these conditions corresponded to times when Eureka was positioned within the core of the vortex or on the inside

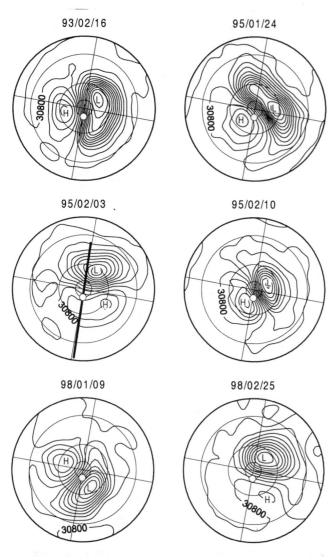

**Figure 2.** Polar stereographic maps of geopotential height (m) on the 10 hPa surface during each day of stratospheric disturbance presented in Figure 1a. The contour interval is 200 m, and the letters L and H are used to mark lows and highs respectively. The Greenwich meridian is to the right on each map, and the concentric circles starting at the outside edge indicate latitudes of 15, 30, 60 and 80 °N. Eureka is marked on each map by a white dot. The solid line running through Eureka on 3 February 1995 is a great circle taken perpendicular to the wind at 30 km in altitude above Eureka (referred to in Figure 4b).

edge of the jet (again, with the one exception being during January 1998).

Comparing the temperature profiles in Figure 1 with the corresponding maps of height in Figures 2 and 3, it appears that relatively cool temperatures in the lower stratosphere and warm temperatures in the upper stratosphere are associated with the core of the wintertime cyclonic vortex, whereas the opposite thermal structure is associated with the stratospheric anticyclone. This is exactly the temperature structure that is required to drive each type of vortex, as can

be readily understood from thermal wind considerations (see, for example, *Holton* [1992]). It should be noted that the unusually warm temperatures in the lower stratosphere of each stratospheric high are undoubtedly the result of adiabatic compressions known to occur during stratospheric disturbances.

### 3.3. Cross Sections of Wind and Temperature

In order to understand the temperature changes that occur due to vortex movements, it is instructive to examine cross sections of winds and temperatures that may be created from the meteorological analyses. Two such cross sections, taken along the thick solid lines that are drawn on the height maps for 3 February 1995 (shown in Figure 2)

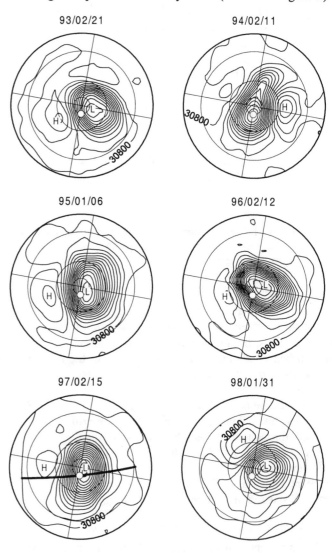

**Figure 3.** Polar stereographic maps of geopotential height (m) on the 10 hPa surface during each day of relatively undisturbed conditions presented in Figure 1b. The solid line running through Eureka on 15 February 1997 is a great circle taken perpendicular to the wind at 30 km in altitude above Eureka (referred to in Figure 4a).

and for 15 February 1997 (shown in Figure 3) are given in Figure 4. The polar vortex is evident in each map as the pair of jets with absolute maximum wind speeds near 30 km in height. In Figure 4b, another jet due to an anticyclone is apparent at –3000 km from Eureka and with peak wind speeds near 35 km.

As shown in Figure 4a, the thermal structure on a day with a strong and undisturbed Eureka-centred vortex is such that cold temperatures exist in the lower stratosphere and warm temperatures are found in the upper stratosphere. The vortex core upper stratosphere is in fact warmer than anywhere else in the stratosphere poleward of the subtropics. Figure 4b shows the case where an anticyclone has pushed the vortex away from Eureka. Notice that, in general, the temperature structure of the vortex is similar to that in Figure 4a: cool temperatures characterize the lower stratosphere and warm temperatures characterize the upper stratosphere of the vortex core. In contrast, the anticyclone has relatively warm temperatures in the lower stratosphere and cool temperatures in the upper stratosphere of its centre, which is again consistent with thermal wind requirements for a closed anticyclone. The warm temperatures found in the lower stratosphere of the anticyclone (and even in the shared jet) are largely dynamical in origin.

## 4. DISCUSSION

The temperature profiles obtained at Eureka during the six days of stratospheric disturbance showed lower stratospheric warming and stratopause cooling. These same days corresponded to times when the vortex was displaced away from Eureka by a strong high. Conversely, during relatively undisturbed conditions, the observations at Eureka showed lower stratospheric cooling in conjunction with strong stratopause warming; these measurements corresponded to times when Eureka was near the centre of the vortex. Therefore, *from the perspective of Eureka*, stratopause warmings are unrelated to large-scale stratospheric disturbances (i.e., "sudden stratospheric warmings"). It should be remarked, however, that compressional warming throughout stratosphere can certainly result from large-scale mass transports apparent during a stratospheric disturbance. For example, dynamically induced warming has been shown to occur in the jet between an anticyclone and the stratospheric vortex [*Fairlie and O'Neill*, 1988; *Fairlie et al.*, 1990].

By viewing cross-sections of thermal and dynamical structure that bisect the vortex, it is clear that changes in strato-mesospheric temperatures above Eureka during disturbed / undisturbed conditions are largely a natural consequence of vortex/anticyclone movements (we suppose here that the anticyclones do not form over Eureka, but rather are created elsewhere and then displace the vortex).

When a minor stratospheric disturbance does occur, the relatively cool (warm) temperatures in the lower (upper) stratosphere of the vortex core do not disappear; they simply shift away from Eureka's point of view. Nevertheless, those same temperatures would be observed over a different locale, yielding temperature changes opposite in sense to those observed above Eureka. That warm temperatures are always apparent in the vortex core upper stratosphere, regardless of whether the stratosphere is disturbed or undisturbed, suggests that those warm

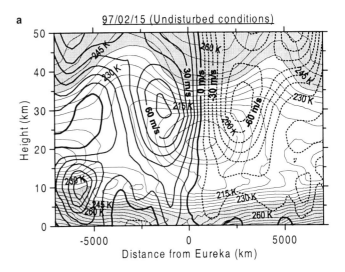

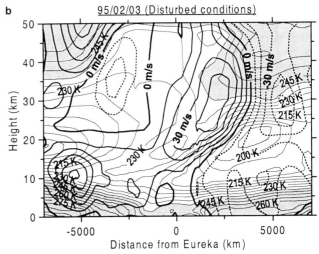

**Figure 4.** Cross sections of the thermal and dynamical structure on **a**, 15 February 1997 (relatively undisturbed conditions), and **b**, 3 February 1995 (disturbed conditions). The cross sections were taken along the great circle paths marked on the corresponding maps of Figures 2 and 3. The wind contours (10 m/s interval) represent the wind magnitude perpendicular to the cross section; positive (negative) values indicate winds directed out of (into) the page and are denoted by the thick solid (dashed) lines. The temperature contours (5 K interval) are given by the thin solid lines; regions where the temperatures are greater (less) than 240 K (200 K) are darkly (lightly) shaded.

temperatures are not forced by a planetary wave driven circulation.

Occasionally a major stratospheric disturbance results in the complete destruction of the polar vortex. In such a case, the temperatures everywhere over the Arctic might look more like those over Eureka during a minor stratospheric disturbance (i.e., like the temperature structure found at the centre of a stratospheric high). It is interesting to note that in earlier studies, major stratospheric disturbances were reported to occur every few years (see, for example, *Schoeberl* [1978] and *Labitzke* [1981]). However, their occurrence rate seems to have diminished during the 1990s, and not one major event occurred during the six years considered here.

The interpretation of single station measurements in the context of zonal mean diagnostics alone presents considerable challenges. However, these difficulties can be alleviated through careful consideration of the large-scale circulation. By separately examining measurements obtained in different vortex regions, it is possible to detect significant warming induced by adiabatic compressions. For example, *Duck et al.* [1998] used the Eureka data to show that the upper stratosphere of the polar vortex core warms annually in late December. The study communicated here indicates that Eureka normally views the vortex core in relatively undisturbed conditions. It is therefore likely that something other than planetary waves or vortex-vortex interactions forces the annual late-December vortex core warming. *Duck et al.* [1998] gave evidence that gravity waves might be important in this regard.

*Acknowledgments.* The authors wish to express their appreciation to J. Bird, D. Donovan, and D. Velkov for their participation in the Eureka lidar measurement programme. The temperature and height global analyses were supplied by the National Centers for Environmental Prediction (NCEP). This work was carried out as part of the research programme of the Centre for Research in Earth and Space Technology (CRESTech) at York University. The Eureka Weather Station and stratospheric observatory are operated by the Atmospheric Environment Service (AES) of Canada. Financial support was provided by the AES and the Natural Sciences and Engineering Research Council of Canada (NSERC).

## REFERENCES

Carswell, A. I., A. Ulitsky, and D. I. Wardle, Lidar measurements of the Arctic stratosphere, *SPIE.*, *2049*, 9-23, 1993.

Duck, T. J., J. A. Whiteway, and A. I. Carswell, Lidar observations of gravity wave activity and Arctic stratospheric vortex core warming, *Geophys. Res. Lett.*, *25*, 2813-2816, 1998.

Duck, T. J., High Arctic observations of strato-mesospheric temperatures and gravity wave activity, *Ph.D. thesis*, York University, Toronto, Canada, April 1999.

Fairlie, T. D. A., and A. O'Neill, The stratospheric major warming of winter 1984/85: Observations and dynamical inferences, *Q. J. R. Meteorol. Soc.*, *114*, 557-578, 1988.

Fairlie, T. D. A., M. Fisher, and A. O'Neill, The development of narrow baroclinic zones and other small-scale structure in the stratosphere during simulated major warmings, *Q. J. R. Meteorol. Soc.*, *116*, 287-315, 1990.

Fleming, E. L., S. Chandra, J. J. Barnett, and M. Corney, Zonal mean temperature, pressure, zonal wind and geopotential height as functions of latitude, *Adv. Space Res.*, *10*, (12)11-(12)59, 1990.

Hauchecorne, A., and M. L. Chanin, Density and temperature profiles obtained by lidar between 35 and 70 km, *Geophys. Res. Lett.*, *7*, 565-568, 1980.

Holton, J. R., *An introduction to dynamic meteorology*, Academic Press, 507 pp., 1992.

Labitzke, K., Stratospheric-mesospheric midwinter disturbances: A summary of observed characteristics, *J. Geophys. Res.*, *86*, 9665-9678, 1981.

Matsuno, T., A dynamical model of the sudden stratospheric warming, *J. Atmos. Sci.*, *28*, 1479-1494, 1971.

McIntyre, M. E., and T. N. Palmer, Breaking planetary waves in the stratosphere, *Nature, 305*, 593-600, 1983.

O'Neill, A., and V. D. Pope, Simulations of linear and nonlinear disturbances in the stratosphere, *Q. J. R. Meteorol. Soc.*, *114*, 1063-1110, 1988.

Schoeberl, M. R., Stratospheric warmings: Observations and theory, *Revs. Geophys. Space Phys.*, *16*, 521-538, 1978.

Whiteway, J. A., and A. I. Carswell, Rayleigh lidar observations of thermal structure and gravity wave activity in the High Arctic during a stratospheric warming, *J. Atmos. Sci.*, *51*, 3122-3136, 1994.

Whiteway, J. A., T. J. Duck, D. P. Donovan, J. C. Bird, S. R. Pal, and A. I. Carswell, Measurements of gravity wave activity within and around the Arctic stratospheric vortex, *Geophys. Res. Lett.*, *24*, 1387-1390, 1997.

A. I. Carswell, Dept. of Physics and Astronomy, York University, 4700 Keele Street, Toronto, Canada, M3J 1P3. (e-mail: carswell@lidar.crestech.ca)

T. J. Duck, Haystack Observatory, Massachusetts Institute of Technology, Route 40, Westford, MA 01886, USA (e-mail: tomduck@haystack.mit.edu)

J. A. Whiteway, Dept. of Physics, University of Wales, Aberystwyth, SY23 3BZ, UK. (e-mail: jjw@aber.ac.uk)

# Year-round Temperature and Wave Measurements of the Arctic Middle Atmosphere for 1995-1998

Andrew J. Gerrard and Timothy J. Kane

Andrew J. Gerrard and Timothy J. Kane

*Department of Electrical Engineering, The Pennsylvania State University, University Park, PA*

Jeffrey P. Thayer

*Radio Science and Engineering Division, SRI International, Menlo Park, CA*

The first near year-round Rayleigh Lidar temperature and wave activity measurements of the Arctic upper stratosphere and lower mesosphere are presented. The data were obtained from Sondrestrom, Greenland (67° N, 51° W) throughout 1995-1998. The relatively continuous, high frequency measurements of vertical thermal profiles, root-mean-square (RMS) atmospheric relative-density perturbations, and their associated variability over the four years at one geographic site complement previously published Arctic climatologies that were based on various mixed data sets. The nightly, monthly, and yearly variability of the values was seen to be much larger in the winter periods than in the summer. This is attributed to both the strong influence of the polar vortex and its associated variability as well as the increased potential for atmospheric wave activity in the middle atmosphere during these periods. Winter temperatures from empirical models resemble measured temperatures only after a long-term average of the observations, suggesting constraints/limitations on both short term observations as well as model comparisons/applications. Model temperatures more closely and consistently resemble observed temperatures obtained from late spring through late fall. RMS values display a strong annual trend, with a maximum in the winter and a minimum in the summer. The need for estimates of geophysical variability in both model outputs and model/data comparisons, as well as the need for multi-site observations, is also discussed.

## 1. INTRODUCTION

Climatologies based on experimental measurements of temperatures of the Arctic middle atmosphere have been made for some time, including data/papers presented in the Middle Atmosphere Program Handbook 16 (hereafter MAP

16) [e.g., *Barnett and Corney*, 1985] and various other papers [e.g., *Clancy and Rusch*, 1989; *Lübken and von Zahn*, 1991; *Clancy et al.*, 1994; *Lübken* 1999]. Experimentally-measured atmospheric wave climatologies at high latitudes have been based primarily on rocket measurements [e.g., *Hirota*, 1984; *Eckermann et al.*, 1995] and more recently satellite measurements [e.g., *Wu and Waters* 1996a,b]. However, much of the data presented in a number of these past studies have come from mixed sources (i.e., from different experimental techniques, different years, varying geographic locations, or some combination thereof) with potentially limited sampling frequencies. For example,

Atmospheric Science Across the Stratopause
Geophysical Monograph 123

both the Committee on Space Programs and Research (COSPAR) International Reference Atmosphere (hereafter CIRA-86) [*Fleming et al.,* 1990] and the mass spectrometer and incoherent scatter model (hereafter MSISE-90) [*Hedin,* 1991] were partially based on such diverse data sets (especially below ~80 km, where a good deal of the temperature data in both models was based on MAP 16). The number of experimental/measurement-based studies at high latitudes with year-round, high frequency sampling of the middle atmosphere at one geographic location and high resolution is noticeably small.

The purpose of this paper is to add to the climatological database of Arctic middle atmospheric temperature profiles and upper stratospheric wave activities, as well as to raise awareness of the nightly, monthly, and yearly variability of such observations. We do this by presenting Rayleigh Lidar temperature profiles of the 30 km to 64 km altitudes and atmospheric relative-density perturbation measurements in the 30 km to 50 km altitudes over Sondrestrom, Greenland (67° N, 51° W) from 1995 through 1998. The data presented here are meant to show the overall climatological structure of the Arctic middle atmosphere, leaving the smaller-scale/seasonal features for a later study. The results of this paper nicely complement the aforementioned papers (discussing similar scales), and can be used with those papers to give a good overview of the "background" structure and associated variability through a large portion of the Arctic middle atmosphere.

## 2. INSTRUMENT DESCRIPTION, DATA COVERAGE, AND ANALYSIS TECHNIQUE

The Sondrestrom Rayleigh Lidar is described in detail in *Thayer et al.* [1997]. Data used for this study were postbinned to 5 minutes temporal (approximately the Brunt-Väisälä period) and 192 m vertical resolution. Atmospheric relative-density profiles were then computed and adjusted to account for lower atmospheric attenuation and system fluctuations. Further calculations utilized standard error propagation assuming a stationary process.

Figure 1 depicts the data coverage for the temperature measurements made for this study. Data coverage for the relative-density perturbation measurements are similar to Figure 1, except that some of the data sets were not included or were shortened due to insufficient length (discussed below) or insufficient data quality. Data sets during summer periods tended to be temporally shorter due to the high solar background (which makes this region difficult for optical observations), while other data gaps can be attributed to cloudy conditions or system upgrades. Biases due to undersampling of any long period structure or use of data sets of varying temporal length were addressed by reprocessing the data with a common temporal data window

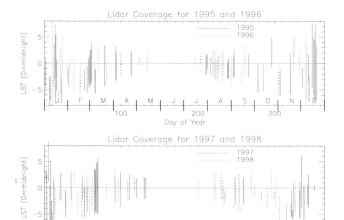

**Figure 1.** Rayleigh Lidar data coverage used for the temperature measurements at Sondrestrom. Vertical bars represent the local time coverage (in hours) of Lidar data used for this study. A 0 represents local midnight, with negative times representing premidnight time periods.

(a four hour window centered on local midnight), and the results match those obtained with the full data set fairly well.

The analysis used for the extractions of mean nightly temperature profiles follows that of *Hauchecorne and Chanin* [1980], and modified as in *Thayer et al.* [1997]. Details of this method involved:

- Prior to calculating the temperature profile, the mean relative-density profile from the data set was smoothed using a running 7 km (constant in height) Blackman window.

- A top altitude for the profile was chosen at a point where the relative error in the filtered mean nightly relative-density profile was at 5%. This top altitude was not allowed to go above ~74 km.

- The temperature at this top altitude was chosen from MSISE-90. The standard deviation on the initial estimate was taken as 20 K. The first 10 km below the initial start altitude was removed from the resultant temperature profiles. It has been shown that the initial temperature estimate yields little impact below this range [*Hauchecorne and Chanin,* 1980].

- If a given month had >10 hours of total collected data (and thus composed of at least 2 sets of data), the individual temperature profiles were averaged together and weighted by their duration (statistical standard deviation), to form the month's "mean" temperature profile.

Standard deviations on each of the nightly temperature profiles varied with the signal-to-noise ratio. Even so, the standard deviations of the mean temperature profiles (derived from traditional Poisson analysis) were almost always

< 5 K and < 3 K at 60 km and 50 km, respectively. Error values were increasingly smaller at lower altitudes. It should be noted that these error bars are generally larger than those reported by other groups due to our conservative analysis technique.

In an effort to account for and represent the overall atmospheric wave activity, we computed root-mean-square (RMS) relative-density perturbations for each data set of sufficient quality. These values indicate the relative atmospheric wave strength throughout the night, with higher values indicating increased wave activity (such values are discussed further in *Senft and Gardner* [1991] and references therein). Details of the calculations involved:

- Computation of the mean background density using *Mitchell et al.* [1990] method II (a third order polynomial fit in natural log domain, with no initial high-pass filtering) from 30 km to 65 km.
- Computation of the relative-density perturbations for each time realization in the 30 km to 50 km regime, where signal-to-noise ratio (SNR) is the highest. The relative-density perturbations were then low-pass filtered at 2 km vertical and 30 min temporal (Kaiser filter) cutoffs. The RMS value over the entire two-dimensional time-altitude range was then computed.

This method resulted in a RMS value for waves with temporal periods between 30 minutes and ~2 times the temporal length and with vertical wavelengths between 2 km and ~40 km (a full two-dimensional power spectrum). As such, the values are mostly dominated by gravity wave activity. We required that the RMS calculation contain at least 2.5 hours worth of data to ensure a relatively stationary estimate of the value. The lower bound and filter cutoffs were based on both the average temporal/vertical periods of commonly seen quasi-monochromatic wave structures as well as SNR considerations. Due to the varying length of the data windows, large period waves could bias the longer data sets compared to the shorter data sets. To verify that this was not the case in this analysis, we estimated the relative stationarity of the wave field using both a sliding four hour data window (length permitting) and a correlation analysis between the length of the data sets and the RMS values. The latter method yielded correlation coefficient values <0.2 (uncorrelated), and the results from the former method are discussed below. Relative statistical errors on the RMS values are < 0.001.

It is important to note that, in addition to gravity wave influence, there could be tidal and planetary wave influence in all values reported here. However, tidal influence is expected to be very small in magnitude at these altitudes (with amplitudes less than ~3 K), as noted from *Forbes* [1982a,b] and as noted from Global Scale Wave Model results [Hagan, M. E., Modeling atmospheric tidal propagation across

the stratopause, this volume]. Planetary wave influence in the RMS values is also expected to be small.

## 3. YEAR-ROUND, MULTI-YEAR TEMPERATURE AND WAVE MEASUREMENTS

In Plate 1, we present the mean monthly temperature profiles from 30 km to 64 km for 1995 through 1998. Also presented in Plate 1 are corresponding typical temperature values from the MSISE-90 model for comparison. It is interesting to note that for any given winter month, the mean temperature profile can vary quite a bit from other similarly related months/years and from the MSISE-90 model. During non-winter months, MSISE-90 values seem to match the observed temperatures fairly well, and the measured temperatures are relatively consistent and smoothly varying on monthly and yearly time-scales.

To better understand the high variability between the monthly and yearly temperature profiles during the winter periods, it is necessary to elucidate the influence that the stratospheric polar vortex has on the Arctic middle atmosphere. Specifically, it has been observed that an instantaneous temperature profile of this region of the atmosphere is related, in part, to the strength of, age of, and location within the polar vortex. As the vortex moves geographically around/over the pole and evolves in time, the observed temperatures over a particular fixed high latitude site will change, corresponding to the inherently different thermal and dynamical structures of the vortex core, jet, and outside the vortex altogether (and also into high pressure cells) [*Duck et al.*, 1998; *Duck*, 1999; Duck, T. J., J. A. Whiteway, and A. I. Carswell, Sudden stratospheric and stratopause warmings: Observations of temperatures in the middle atmosphere above Eureka, this volume, hereafter DWC]. It has also been noted that the polar vortex can move very quickly over Sondrestrom, such that over a period of days Sondrestrom could move between the various vortex structures [Gerrard, A. J., T. J. Kane, J. P. Thayer, T. J. Duck, and J. A. Whiteway, Synoptic-scale study of the Arctic polar vortex's influence on the middle atmosphere, *submitted to J. Geophys. Res.*, 2000, hereafter GKTDW]. This type of activity can further increase the apparent variability at this particular geographic location. Additionally, correlation between gravity wave activity and the various polar vortex structures has been observed, with relatively greater activity in the vortex jet and lower activity in the vortex core or outside the vortex altogether [*Wu and Waters*, 1996a,b; *Whiteway et al.*, 1997].

Applying the above information to the observations at Sondrestrom, the large winter temperature fluctuations from month-to-month, year-to-year, and from MSISE-90 are believed to be due to both the variability in location and

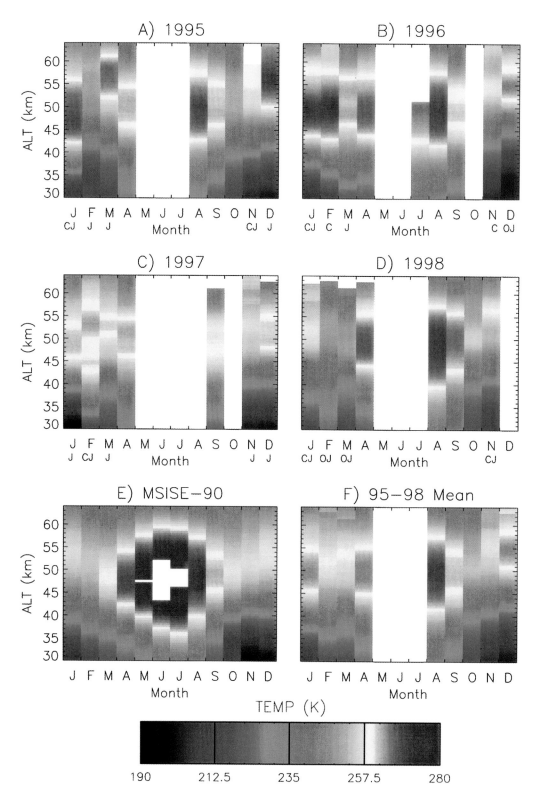

**Plate 1.** Plots A) through D) depict mean monthly temperature profiles from 30 to 64 km at Sondrestrom. Graph E) is the corresponding typical MSISE-90 temperatures for comparison. Graph F) depicts the monthly average of the four years. White on all graphs except E) depicts no data, while on E) it depicts temperature over 280K. For winter periods, the mean location of Sondrestrom relative to the polar vortex for the particular month's observations is listed under the month label. C is for core of the vortex, J is for the jet of the vortex, and O is for outside of the vortex. Two labels indicate a combination of locations/boundary.

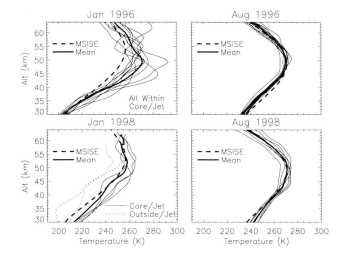

**Figure 2.** Nightly temperature profiles for January and August for 1996 and 1998. Thin solid lines are the mean temperature profiles over each respective data set of Figure 1 in the given month. The thick solid lines are the mean monthly temperature profiles, and the thick dashed lines represent corresponding MSISE-90 temperature profiles for reference (from Plate 1).

strength of the polar vortex structure, as well as increased overall atmospheric wave activity present during winter periods [*Hirota*, 1984; *Lübken and von Zahn*, 1991; *Eckermann et al.*, 1995; *Wu and Waters*, 1996a,b; *Whiteway et al.*, 1997; *Duck et al.*, 1998; *Duck*, 1999; DWC]. As such, it may be a more natural framework to bin winter temperatures by the specific vortex position, but as far as we know no modeled or experimental work has been presented in the literature of any climatologies arranged in such a manner for comparison. In an attempt to remedy the situation, under each winter month label in Plate 1 is the average position of Sondrestrom relative to the various areas of the polar vortex for the measurements taken that month, as determined using geopotential height data from the Free University of Berlin stratospheric analyses (a method of classification related to that introduced by *Whiteway et al.* [1997]).

Conversely, the lower variability in the summer periods can be attributed to both the lack of such variable geophysical structures and the reversal in the stratospheric winds which tends to substantially reduce (but not eliminate) wave activity through critical layering filtering/attenuation. Such relatively constant summer thermal profiles have been observed before by *Lübken and von Zahn* [1991], *Lübken* [1999], etc.

To further demonstrate the overall nature of the nightly, seasonal, and yearly temperature variability, as well as to demonstrate the above concepts, we present the individual nightly temperature profiles for January 1996 and 1998 and August 1996 and 1998 in Figure 2. Also depicted are the computed months' means and the MSISE-90 profiles for reference. It is noted that the variability in the January

months is much greater than that of August months on nightly and yearly scales. For January 1996, all measurements were taken in the vortex core/jet boundary. For January 1998, profiles early in the month were in the vortex core and inner jet, while the last two profiles of the month (denoted by dotted lines) were taken on the vortex jet and jet/outside boundary. We see, in general, warmer temperatures in the upper stratosphere nearer the vortex core, and cooler temperatures when further outside the vortex. Such trends are "expected" considering the overall polar vortex thermal structure noted in references above. The effects of wave activity in each data set are larger in the winter periods than during the summer periods.

We average the four years of temperature data for each month and present these results also in Plate 1. This average would be analogous to the average of numerous "quasi-random" measurements, like those made by numerous short-term rocket campaigns or limited satellite observations combined over a number of years. We note that the MSISE-90 temperatures are now more closely related to these mean, "background" temperatures, as one would expect when the major sources of variability have been "removed" by averaging over the years. The night-to-night variability of the temperature profiles over the four years for each month (i.e., monthly sample standard deviations using the four year mean temperature profile of Plate 1) is presented in Figure 3. As mentioned above, the winter periods have greater overall potential (night-to-night) variability. These trends for the above mean temperatures and

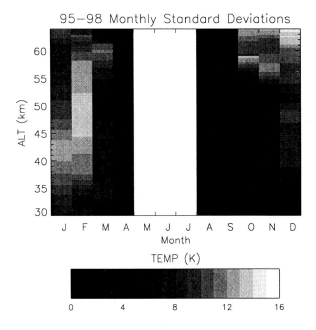

**Figure 3.** Sample standard deviations of the individual temperature profiles for each month over four years. There was insufficient data for MJJ.

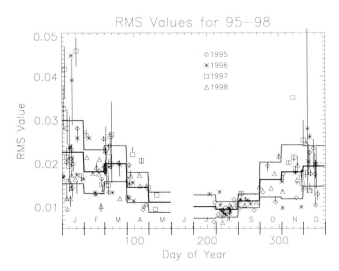

**Figure 4.** Rayleigh Lidar RMS relative-density fluctuations for each day of the year for 1995 (◊), 1996 (∗), 1997 (□), and 1998 (△) are plotted simultaneously with the computed monthly sample mean (over the four years, thick black line) +/- monthly sample standard deviation (over the four years, bounding thick gray lines). Data points give the RMS value over the entire temporal length of the data set. The vertical thin black bars give the range of values from a four hour sliding window indicating the degree of stationarity of the wave field.

standard deviations are consistent with other high latitude temperature measurements [*Lübken and von Zahn*, 1991; *Lübken*, 1999].

The RMS relative-density perturbation values for 1995-1998 recorded by the Rayleigh Lidar are presented in Figure 4. Also plotted are the monthly sample mean values plus/minus the monthly sample standard deviations from the mean. One notes that there is more activity in the winter months and less in the summer months, with winter having greater day-to-day and month-to-month variability than the summer. Spring and fall values indicate transition periods. These year-round measurements are similar to previous observations of a strong annual trend seen in upper stratospheric gravity wave measurements in the Arctic [*Hirota*, 1984; *Eckermann et al.*, 1995]. The RMS values for winter months are more variable due to the effects of the migrating polar vortex (and associated vortex structure) noted above, as well as overall increased gravity and planetary wave activity. The smaller, less variable RMS values in the summer are observed for similar reasons as the relatively constant thermal profiles noted above.

## 4. DISCUSSIONS AND CONCLUSIONS

In the climatological sense, the temperature values in MSISE-90 (which is closely related to the CIRA-86 model below ~80 km due to their common heritage) resemble the presented observations. However, at times there can be dramatic differences between the model values and measurements (e.g., during minor and major stratospheric warmings). Future work will involve investigation of these smaller scale discrepancies. RMS values indicate a relative measure of wave activity similar to that reported in the above references, and future work will involve analysis of gravity wave and planetary wave influences, amplitude growth in altitude, seasonal wave structures, etc.

It is important to note four issues associated with this study. First, the potential for large variability in the winter profiles can have a large impact on the interpretation of short-term measurements on all time-scales. For example, on nightly time-scales significant wave activity (especially gravity wave activity in the vortex jet) can make it difficult to obtain mean background values from one experimental sounding/sample (e.g., from radiosondes, rocket measurements, or satellite observations). On monthly time-scales, the polar vortex can shift in geographical location, thus affecting campaign-based data collection or short-term spaceborne observations. On yearly time-scales, the vortex varies in both strength and variability/stability. As such, short-term coverage (relative to different scales) of the Arctic middle atmosphere should be augmented with longer-term measurements, like that afforded by various permanent, high-frequency sampling, ground-based instruments, in order to account for potential variability.

Second, wave propagation into the upper atmosphere throughout various times of the year will partially depend on lower atmospheric conditions. This can be an especially interesting issue when movement of a large-scale structure, like the polar vortex, occurs over a short period of time [GKTDW]. Though a number of papers have addressed variability in the upper mesosphere associated with polar vortex influence [e.g., *Labitzke et al.*, 1987; *Hauchecorne et al.*, 1987], the issue is still unresolved. Data sets that sample across the middle atmosphere [e.g., *Lübken and von Zahn*, 1991] can be further studied in this manner.

Third, we feel that geophysical variability is not emphasized enough in many models and model/data comparisons of the middle atmosphere. For example, one often sees comparisons of model output to experimental data in the literature. These comparisons would be more meaningful if the potential variability of the model values was also presented. Such values are very important for interpreting data (and even model) products, and as such, should probably be considered as an output parameter in future modeling efforts (e.g., for empirical models).

Finally, these and similar results, taken with one instrument and at one location, cannot give a complete description of the many geophysical processes involved. Using multiple instruments at various locations (e.g., compare Lidar data taken at Poker Flat (65° N, 147° W), ALOMAR

(69° N, 16° E), Eureka (80° N, 86° W), Thule (77° N, 69° W), and Sondrestrom) would allow better characterization of the environment under similar/simultaneous conditions. This would yield a wealth of information on the Arctic middle atmosphere.

*Acknowledgments.* The authors thank members of the Sondrestrom research facility and the stratospheric research group at the Free University of Berlin for stratospheric analyses. We also thank the editors of the monograph, Tom Duck, Keith Soldavin, and Christina Gerrard for their contributions. AJG and TJK were supported in part by NSF Coupling, Energetics and Dynamics of Atmospheric Regions program grant ATM 98-13828 and JPT under NSF Cooperative Agreement ATM 98-13556.

## REFERENCES

Barnett, J. and M. Corney, Temperature data from satellites, in *Handb. MAP 16*, edited by K. Labitzke, J. J. Barnett, and B. Edwards, pp. 3-11 and pp. 47-85, Sci. Comm. For Sol.-Terr. Phys. Secr., Univ. of IL, Urbana, 1985.

Clancy, R. T. and D. W. Rusch, Climatology and trends of mesospheric (58-90 km) temperatures based upon 1982-1986 SME limb scattering profiles, *J. Geophys. Res., 94*, 3377-3393, 1989.

Clancy, R. T., D. W. Rusch, and M. T. Callan, Temperature minima in the average thermal structure of the middle mesosphere (70-80 km) from analysis of 40- to 92-km SME global temperature profiles, *J. Geophys. Res., 99*, 19,001-19,020, 1994.

Duck, T. J., J. A. Whiteway, and A. I. Carswell, Lidar observations of gravity wave activity and Arctic stratospheric vortex core warming, *Geophys. Res. Lett., 25*, 2813-2816, 1998.

Duck, T. J., High Arctic observations of strato-mesospheric temperatures and gravity wave activity, Ph.D. thesis, York University, North York, 1999.

Eckermann, S. D., I. Hirota, and W. K. Hocking, Gravity wave and equatorial wave morphology of the stratosphere derived from long-term rocket soundings, *Q. J. R. Meteorol. Soc., 121*, 149-186, 1995.

Fleming, E. L., S. Chandra, J. J. Barnett, and M. Corney, Zonal mean temperature, pressure, zonal wind, and geopotential height as functions of latitude, in *Adv. Space Res.*, edited by D. Rees, J. J. Barnett, and K. Labitzke, *10*(12), pp. 11-59, 1990.

Forbes, J. M., Atmospheric tides 1, Model description and results for the solar diurnal component, *J. Geophys. Res., 87*, 5222-5240, 1982a.

Forbes, J. M., Atmospheric tides 2, The solar and lunar semidurnal components, *J. Geophys. Res., 87*, 5241-5252, 1982b.

Hauchecorne, A. and M.-L. Chanin, Density and temperature profiles obtained by Lidar between 35 and 70 km, *Geophys. Res. Lett., 7*, 565-568, 1980.

Hauchecorne, A., T. Blix, R. Gerndt, G. A. Kokin, W. Meyer, and N. N. Shefov, Large scale coherence of the mesospheric and upper stratospheric temperature fluctuations, *J. Atmos. Terr. Phys., 49*, 649-654, 1987.

Hedin, A. E., Extension of the MSIS thermosphere model into the middle and lower atmosphere, *J. Geophys. Res., 96*, 1159-1172, 1991.

Hirota, I., Climatology of gravity waves in the middle atmosphere, *J. Atmos. Terr. Phys., 46*, 767-773, 1984.

Labitzke, K., A. H. Mason, H. G. Mueller, Z. Rapoport, and E. R. Williams, Hemispheric synoptic analysis of 95 km winds during the winter of 1983/1984 and comparison with stratospheric parameters, *J. Atmos. Terr. Phys., 49*, 639-648, 1987.

Lübken, F.-J. and U. von Zahn, Thermal structure of the mesopause region at polar latitudes, *J. Geophys. Res., 96*, 20,841-20,857, 1991.

Lübken, F.-J., Thermal structure of the Arctic summer mesosphere, *J. Geophys. Res., 104*, 9135-9149, 1999.

Mitchell, N. J., L. Thomas, and A. K. P. Marsh, Lidar studies of stratospheric gravity waves: A comparison of analysis techniques, *Ann. Geophys., 8*, 705-712, 1990.

Senft, D. C. and C. S. Gardner, Seasonal variability of gravity wave activity and spectra in the mesopause region at Urbana, *J. Geophys. Res., 96*, 17,229-17,264, 1991.

Thayer, J. P., N. B. Nielson, R. E. Warren, C. J. Heinselman, and J. Sohn, Rayleigh Lidar system for middle atmosphere research in the Arctic, *Opt. Eng., 36*, 2045-2061, 1997.

Whiteway, J. A., T. J. Duck, D. P. Donovan, J. C. Bird, S. R. Pal, and A. I. Carswell, Measurements of gravity wave activity within and around the Arctic stratospheric vortex, *Geophys. Res. Lett., 24*, 1387-1390, 1997.

Wu, D. L. and J. W. Waters, Gravity-wave-scale temperature fluctuations seen by the UARS MLS, *Geophys. Res. Lett., 23*, 3289-3292, 1996a.

Wu, D. L. and J. W. Waters, Satellite observations of atmospheric variances: A possible indication of gravity waves, *Geophys. Res. Lett., 23*, 3631-3634, 1996b.

A. J. Gerrard and T. J. Kane, The Pennsylvania State University, 121 Electrical Engineering East, University Park, PA 16802

J. P. Thayer, Radio Science and Engineering Division, SRI International, 333 Ravenswood Ave. G-275, Menlo Park, CA 94025

# Interannual Variability of the Diurnal Tide in the Low-Latitude Mesosphere and Lower Thermosphere During Equinoxes: Mechanistic Model Interpretation of the 1992-96 HRDI Measurements

Valery A. Yudin, Marvin A. Geller, and Ling Wang

*Institute for Terrestrial and Planetary Atmospheres,*
*State University of New York at Stony Brook, Stony Brook, New York, USA*


Stephen D. Eckermann

*Hulburt Center for Space Research, Code 7641.2, Naval Research Laboratory, Washington, DC.*

We present an analysis of the interannual variability of the diurnal tide in wind and temperature observations made by the High Resolution Doppler Imager (HRDI) on the Upper Atmosphere Research Satellite (UARS) in the low-latitude mesosphere and lower thermosphere from 1992-1996. Near the March equinox, the monthly HRDI diurnal tide amplitudes show strong quasi-biennial variability while the phase behavior does not reveal regular interannual variations between 80 and 110 km. Our mechanistic model-data analyses of the mean zonal wind, diurnal tide and gravity wave variances indicate that year-to-year changes in the effective tidal dissipation are a plausible mechanism for producing this strong interannual variability.

## 1. INTRODUCTION

Solar tides play an important role in the dynamics of the mesosphere and lower thermosphere (MLT). Migrating diurnal tides can be so large that they become the dominant component in the meridional wind structure observed by the High Resolution Doppler Imager (HRDI). This remarkable feature has been employed by *Khattatov et al.,* [1997] and *Yudin et al.,* [1997] to derive the "best" estimates of the diurnal tide in the low-latitude MLT region from monthly daytime HRDI winds using the Tuned Mechanistic Tidal Model

Atmospheric Science Across the Stratopause
Geophysical Monograph 123
Copyright by the American Geophysical Union

(TMTM). A review of the TMTM approach is presented by Hagan (this monograph).

Recent studies by *Burrage et al.,* [1995], *Eckermann et al.,* [1997], and *Vincent et al.,* [1998] reveal significant variability in the low-latitude MLT diurnal tide on intraseasonal and interannual timescales. In particular, *Vincent et al.,* [1998] suggested that interannual variability of the diurnal tide appeared to be associated with the quasi-biennial oscillation (QBO) of equatorial mean zonal winds in the stratosphere. Their analysis showed that diurnal amplitudes were larger than the climatological average during the westerly phase of the stratospheric QBO, and weaker during the easterly phase.

Only a few modeling efforts have investigated interannual tidal variability in the low-latitude MLT region. In our previous meeting reports, based on the TMTM simulations of *Yudin et al.,* [1997] and HRDI wind data we recorgnized that the direct effect of mean zonal wind changes in the tidal model could not modify the diurnal tide in a manner that matched the observed differences between March, 1993 and March, 1994. We speculated that year-to-year variations in the effective tidal dissipation above 80 km might explain the observed variability in diurnal amplitudes. We also performed gravity wave (GW) simulations with the Gravity-wave Regional or Global Ray Tracer (GROGRAT) model of *Eckermann and Marks*, [1997] using HRDI mean zonal winds as a background environment. Below 95 km, quite different eddy diffusion structures were produced by saturated GWs during the strong (March, 1993) and weak (March, 1994) HRDI diurnal tide observations. *Hagan et al.*, [1999] also concluded that QBO signatures in HRDI zonal mean winds were unable to make the Global Scale Wave Model tidal simulations in April to match the observed differences in the HRDI diurnal tide amplitudes.

Here we present further analysis of the interannual variability of the diurnal tide in the HRDI wind and temperature observations near the March equinoxes during 1992-96. The HRDI data and revised version of the TMTM are briefly discussed in section 2. Section 3 presents our results for the diurnal tide and eddy dissipation estimated by the TMTM and GROGRAT. Section 4 presents our concluding remarks.

## 2. DATA AND MODELS

The daytime HRDI wind and temperature measurements have been described by *Hays et al.*, [1994] and *Ortland et al.*, [1996], [1998]. Version 11 of the level 3AT HRDI wind and temperature data were used in present analysis. For equinox conditions, we used data during days when the zonal mean wind patterns showed a well-developed westward equatorial jet between 70 and 90 km. According to the 1992-96 HRDI MLT wind data near March, such zonal wind patterns usually persist from the last week of February up to the middle of April.

The numerical model used in this study is a stationary linear model of migrating tides described by *Yudin et al.*, [1997]. Several updates in the specification of the mean background wind and tidal forcing have been made for this study. For ozone tidal forcing calculations we use the monthly averaged Microwave Limb Sounder (MLS) ozone fields and HRDI daytime ozone data in the MLT region for 1993 and 1994. Above 75 km in the subtropics, aliasing of the diurnal tide and the daytime averaged HRDI mean wind was corrected by using the TMTM solutions *Yudin et al.*, [1997]. In

the equatorial region where diurnal wind amplitudes are weaker, the substantial day-to-day variations in the HRDI zonal wind data are suppressed by applying the 5-day sliding window. For the tropospheric water vapor forcing, we used several water vapor datasets from the National Center of Atmospheric Research (NCAR)-National Center for Enviromental Prediction (NCEP) [*Kalnay et al.*, 1996], and European Center for Medium Range Weather Forecasts (ECMWF) [*Trenberth*, 1992] reanalyses. According to our calculations with the 1991-94 NCEP water vapor fields, there was no substantial interannual variations in the strength of the tropospheric diurnal excitation. Model simulations with seasonally invariant tidal dissipation and without effects of the background winds show that tropospheric tidal forcing can initiate some semiannual oscillations of the diurnal amplitudes in the MLT region. However, the amplitude contrasts between the solstice and equinox simulations are significantly less than those observed by HRDI.

Figure 1 illustrates that our model simulations using the ECMWF-93 water vapor fields reproduce reasonably well the seasonal cycle of the diurnal temperature amplitudes across the equatorial stratopause observed by MLS [*Wu et al.*, 1998. It is worth noting that for March, the 1992-94 MLS diurnal temperature ampli-

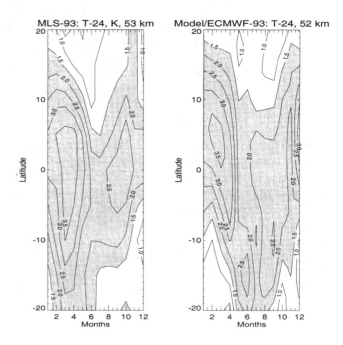

**Figure 1.** Annual cycle of the 1993 diurnal temperature amplitudes observed by MLS at 53 km and simulated by the model at 52 km near the equator. The amplitude values higher than 1.5 K are shaded.

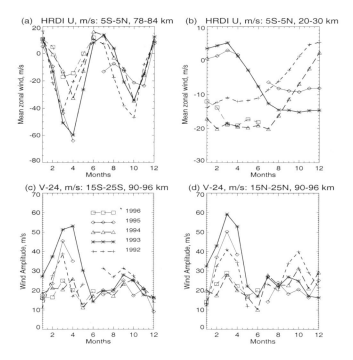

**Figure 2.** Annual cycles of the equatorial $(10°S - 10°N)$ zonal mean winds in the mesosphere between 78 and 84 km (a), and in the stratosphere between 20 and 30 km (b), and the diurnal meridional wind amplitudes at about $20°S$ (c) and $20°N$ (d) between 90 and 96 km as observed by HRDI in 1992-96.

tudes do not reveal substantial interannual variations at 53 km (not shown).

To study GW propagation through the HRDI zonal mean zonal winds, we used the GROGRAT model. A total of 480 waves at 26 latitudes between $60°N$ and $60°S$ were intialized in four azimuths at 18 km. The GW spectrum was specified according to the formulation of *Fritts and VanZandt*, [1993]. Only waves with ground-based periods from 10 min to 4 hours were considered at 18 km. We used a uniform distribution of total GW horizontal wind variance with latitude at the source level with a value of 4 $m^2s^{-2}$. Based on the GROGRAT simulations we calculated the eddy diffusion associated with GW saturation using the relationship between the GW energy dissipation rate and the zonal mean static stability [*Weinstock*, 1990].

## 3. RESULTS

Figure 2 shows calculated annual cycles of equatorial mean zonal winds in the stratosphere (20-30 km) and mesosphere (78-86 km) and diurnal meridional wind amplitudes averaged around the $20°S$ and $20°N$ sub-tropical zones between 88 and 96 km as derived from

the 1992-96 HRDI wind data. Gaps in Figure 2 correspond to insufficient numbers of observations to form a reliable dataset for tidal studies. According to the stratospheric HRDI wind data near March (Figure 2b), the first three years corresponds to different phases of the QBO. A well-developed westward stratospheric jet and weak MLT easterlies were observed by HRDI in March 1994. Conversely, March 1993 corresponds to the eastward phase of the QBO with relatively weak mean winds in the stratosphere and strong westward flow between 70-100 km. March 1992 corresponds to the westward phase, but the strength of stratospheric easterlies is less than in March 1994. With respect to the strength of westward flows in the MLT, March 1992 can be viewed as an intermediate case in comparison with the weak and strong westward winds observed by HRDI in March 1994 and March 1993, respectively. In March 1995 and March 1996, HRDI winds resemble the March 1993 and March 1994 wind patterns, respectively.

The quasi-biennial changes in the 1992-96 HRDI winds during March are accompanied by strong year-to-year variations in the diurnal meridional wind amplitudes (Figures 2c and 2d) and little changes in the phase structures. Figure 3 plots the global structures of the diurnal tide amplitudes and phases for March 1994, and

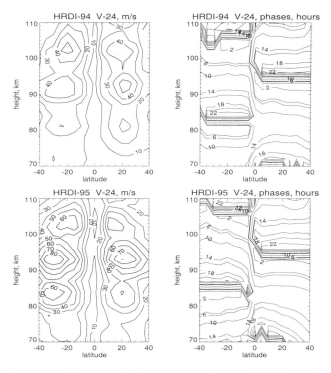

**Figure 3.** The 1994-95 (from top to bottom) HRDI diurnal meridional wind amplitude (left column) and phases (right column). The amplitude values higher than 40 ms$^{-1}$ are shaded.

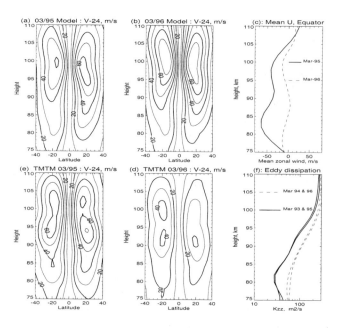

**Figure 4.** The model sensitivity to the equatorial mean zonal wind (top row) and dissipation (bottom). In the top row: meridional wind amplitudes for March 1995 (left plot), March 1996 (middle); vertical profiles of the mean zonal winds at the equator (right). The solid and dashed lines show the 1995 and 1996 mean winds. In the bottom row: meridional wind amplitudes for March 1995 (d), March 1996 (e), diagnosed by the TMTM simulations; the latidudinally averaged vertical profiles of the tidal dissipation (f). The solid and dashed lines present the 1993/95 and 1994/96 eddy dissipation averaged between $30^{o}S$ and $30^{o}N$. The amplitude contours higher than 40 ms$^{-1}$ are shaded.

1995. At 86 km, the correspondance of the interannual changes in the March diurnal wind amplitudes to the QBO phase agree quite well with the biennial variations in diurnal amplitudes observed by *Vincent et al.,* [1998] from 12 years of MF radar data over Adelaide ($35^{o}S$). Above 86 km, the approximate ratios between the diurnal amplitudes observed during the westward and eastward QBO phases vary by a factor of 1.5 and 3 (same order as at $35^{o}S$) but the interannual variations in amplitudes can be up to 30-50 ms$^{-1}$.

In most cases, the application of the least squares fits to the daytime HRDI temperatures provide erroneous results. However, it is possible to look at the daytime sequences of the HRDI temperature ($T$) and stability magnitude $S = d(logT)/dz$ in order to estimate the interannual variations in the diurnal temperature amplitudes. At about 90 km, the differences between early morning and late evening temperatures (not shown) are $\sim20K$ for March 1993 and 1995 and $\sim10K$ for March 1994 and 1996. Downward phase progression with time in vertical profiles of $T$ and $S$ occurs with approximate phase velocities of 1 km hour$^{-1}$. This is comparable

with the vertical wavelength of 24 km for diurnal tide obtained from the wind data (Figure 3).

Keeping in mind the observed correlations between the strength of the westward equatorial jets and the diurnal amplitudes (Figure 2), it is not surprising that the influence of year-to-year changes in the mean wind on the diurnal tide have been tested in previous modeling studies. Figure 4 (top row) illustrates the modeled meridional wind amplitudes for March 1995 (Figure 4a) and March 1996 (Figure 4b). Differences in diurnal amplitudes are only due to year-to-year changes in the HRDI zonal mean winds (Figure 4c). For the strong equatorial westward jets (1993, 1995), modeled diurnal tidal amplitudes above 80 km are about 20 % smaller than those for weak equatorial westward jets (1994 and 1996). These model results can be understood in terms of the equivalent gravity wave mode approximation for tides [*Lindzen,* 1970]. When propagating through strong easterly shear (March 1993 and 1995), westward migrating tidal modes must decrease their amplitudes in the dissipative MLT region due to a Doppler frequency shift of oscillations that shortens their vertical wavelengths. HRDI MLT observations (Figures 2 and 3) show a positive correlation between the diurnal tide amplitude and strength of equatorial westward jets, indicating that the direct effect of the mean zonal wind on the modeled diurnal tide cannot explain the observed interannual changes in tidal amplitudes.

Figure 4 (lower row) shows how the year-to-year variations in tidal dissipation might dominate the effects of the background winds in the TMTM simulations for March 1995 and 1996. These results suggest that, to reproduce the interannual changes in the diurnal amplitudes (Figure 4d and 4e), we need to introduce substantial changes in the tidal dissipation between 75 km and 95 km (Figure 4f). Above 95 km, year-to-year differences in the dissipation are decreased indicating a more uniformly developed dissipative regime for the diurnal tide.

Our TMTM results do not suggest mechanisms for such interannual variability in tidal dissipation. In this study we have performed a next step to interpret the derived interannual changes in the tidal dissipation with an aid of the GROGRAT model.

The 1992-93 GW climatology observed over Kauai ($22^{o}N$) by *Connor and Avery,* [1996] is of particular interest, because the radar is located in the zone of strong interannual variability in the HRDI diurnal tide. Their 1992-93 GW variances as well as the zonal GW variances over Christmas Island ($2^{o}N$) presented by *Eckermann et al.,* [1997] during March 1993 were use in this study to calibrate the modeled GW variances calcu-

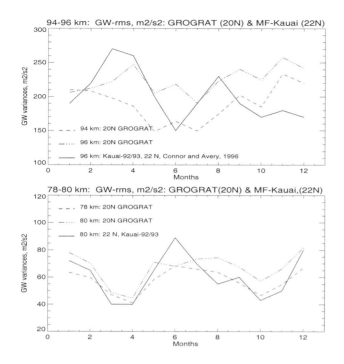

**Figure 5.** The annual cycle of the monthly averaged gravity variances in $m^2s^{-2}$ predicted by the GROGRAT model results and observed by the Kauai MF radar in 1992-93 at about 96 km (top) and 80 km (bottom).

lated by the GROGRAT above 75 km. The GROGRAT results (Figure 5) fit the observed annual cycle of the GW horizontal wind rms over Kauai at 80 km reasonably well. Less agreement is found at 96 km, though the annually averaged value variance of 200 $m^2s^{-2}$ is reproduced . It is worthy to note that the observed semiannual cycle in the GW variances with equinoctial minima at 80 km and equinoctial maxima at 96 km over Kauai is consistent with the semiannual variation of the tidal dissipation derived by the TMTM [*Yudin et al.*, 1997]. During March 1992/93, according to their results, the transition height where equinoctial minima in the tidal dissipation disappear is at 96-98 km with the $K_{zz}$ values $\sim$150-200 $m^2s^{-1}$.

Figure 6 presents model results for the diurnal tide using the eddy dissipation associated with the GW saturation from the GROGRAT simulations for March 1994-95 conditions. Above 80 km, calculated meridional wind and temperature amplitudes show a biennial variability which is comparable with the interannual variations in amplitudes observed by HRDI (Figures 2 and 3). Year-to-year changes in $K_{zz}$ estimated from GROGRAT calculations (Figure 6, right column) are mainly controlled by the equatorial stratospheric and MLT wind profiles. In particular, for March 1995 the filtering effects of the HRDI mean winds produce

the relatively low dissipation (dashed lines ) up to 85-95 km, while for March 1994 more narrow low dissipation areas are located below 85 km. The interannual changes in $K_{zz}$ below 95 km predicted by GROGRAT (Figure 6) qualitatively agree with estimates of tidal dissipation by the TMTM (Figure 4f). It will be important in future studies to validate these mechanistic model indications. In particular further examination of the interannual variability of the GW variances and GW spectra in the low-latitude MLT region are of particular interest.

## 4. CONCLUDING REMARKS

During the March equinox in 1992-96, the HRDI wind and temperature data show strong quasi-biennial variability in diurnal tidal amplitudes in the low-latitude MLT region. At about 86 km, the interannual tidal variability agrees well with long-term Adelaide MF radar observations *Vincent et al.*, [1998]. Both the satellite and MF radar winds do not reveal noticeable interannual changes in diurnal phases.

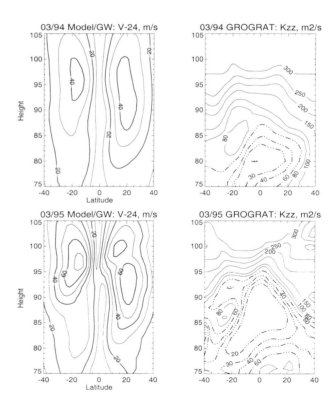

**Figure 6.** The diurnal tide meridional wind amplitudes (right) predicted by model constrained by the eddy dissipation (left) estimated after GROGRAT GW simulations with the 1994 (top), and 1995 (bottom) mean zonal winds. The amplitude contours higher than 40 $ms^{-1}$ are shaded.

In this report, we used a linear tidal model constrained by the 1992-96 HRDI zonal mean wind, 1993-1994 MLS and HRDI ozone data, and the water vapor fields from the 1992-1994 NCEP-NCAR reanalyses to study the response of the diurnal tide to the interannual variations in these input fields and forcings. Our modeling confirmed suggestions from previous model studies that interannual changes in zonal mean background winds, ozone, and water vapor cannot produce the observed biennial variability in the diurnal tidal amplitudes.

Numerical simulations of year-to-year variations in the diurnal tide, tidal dissipation, and dissipation associated with GW breaking in the MLT region were presented in this study using the Tuned Mechanistic Tidal Model (TMTM) and the Gravity-wave Regional or Global Ray Tracer (GROGRAT) for March equinox conditions. Our results indicate that year-to-year changes in the effective tidal dissipation can be considered as one plausible mechanism for the strong interannual variability of the diurnal tide during March.

There are other mechanisms suggested by *Mayr et al.*, [1998] and *Norton and Thuburn*, [1997] for interpretation of the observed semiannual variability in the diurnal tide, (reviewed by *Hagan* in this monograph) that may also be considered possible causes for interannual variability in the diurnal tide. Further theoretical and empirical studies of wave-wave interactions, in particular the relationships between the GW momentum deposition and the strength of the eddy diffusion, are needed to clarify the effects of GWs on the semiannual and interannual variability of the diurnal tide in the low-latitude MLT region.

*Acknowledgments.* We wish to thank Daniel Marsh and Dong Wu for providing access to the HRDI ozone and MLS temperature data.This research was supported by NASA's UARS project.

## REFERENCES

Burrage, M. D. et al., Long-term variability in the solar diurnal tide observed by HRDI and simulated by GSWM, *Geophys. Res. Lett.*, 22, 2641-2644, 1995.

Connor, L. N., and S. K. Avery, A three-year gravity wave climatology of the mesosphere and lower thermosphere over Kauai, *J. Geophys. Res.*, 101, 4065-4077, 1996.

Eckermann, S. D., and C. J. Marks, GROGRAT: a new model of the global propagation and dissipation of atmospheric gravity waves, *Adv. Space Res.*, 20, 1253-1256, 1997.

Eckermann, S. D. et al., Intraseasonal wind variability in the equatorial mesosphere and lower thermosphere: long-term observations from the central Pacific, *J. Atmos. Terr. Phys.*, 59, 603-627, 1997.

Fritts, D. C., and T. E. VanZandt, Spectral estimates of gravity wave energy and momentum fluxes. Part I: Energy dissipation, acceleration, and constraints, *J. Atmos. Sci.*, 50, 3685-3694, 1993.

Hagan, M. E. et al., QBO effects on the diurnal tide in the upper atmosphere, *Earth, Planets, and Space*, 51, 571-577, 1999.

Hagan, M. E., Modeling atmospheric tidal propagation across the stratopause (this monograph).

Hays, P. B., et al., Observations of the diurnal tides from space, *J. Atmos. Sci.*, 51, 3077-3093, 1994.

Kalnay, E. et al., The NCEP-NCAR 40-year reanalysis project, *Bull. Amer. Meteor. Soc.*, 77, 437-471, 1996.

Khattatov, B. V., et al., Diurnal tide as seen by HRDI/UARS Part 2: Monthly mean mean zonal and vertical winds temperature and atmospheric dissipation, *J. Geophys. Res.*, 101, 4423-4435, 1997.

Lindzen, R. S., Internal gravity waves in atmospheres with realistic dissipation and temperature. Part I: Mathematical development and propagation of waves in the thermosphere, *Geophys. Fluid Dyn.*, 1, 303-355, 1970.

Mayr, H. G. et al., Seasonal variations of the diurnal tide induced by gravity wave filtering, *Geophys. Res. Let.*, 25, 943-946, 1998.

Norton W. A., and J. Thuburn, The mesosphere in the Extended UGAMP GCM, in *Gravity Wave Processes and Their Parameterization in Global Climate Models*, 188-197, K. Hamilton (ed), NATO ASI Series, 150, Springer-Verlag, Berlin, 1997.

Ortland, D. A. et al., Measurements of stratospheric winds by the High Resolution Doppler Imager, *J. Geophys. Res*, 101, 10351-10364, 1996.

Ortland, D. A. et al., Remote sensing of mesospheric temperature and $O_2(^1\Sigma)$ volume emission rates with the High Resolution Doppler Imager, *J. Geophys. Res.*, 103, 1821-1835, 1998.

Trenberth, K. E., Global analyses from ECMWF, *NCAR Tech. Note NCAR/TN-373+STR*, 191 pp., 1992.

Vincent, R. A. et al., Long-term MF radar observations of solar tides in the low-latitude mesosphere: Interannual variability and comparisons with the GSWM, *J. Geophys. Res.*, 103, 8667-8683, 1998.

Weinstock, J., Saturated and unsaturated spectra of gravity waves and scale-dependent diffusion, *J. Atmos. Sci.*, 47, 2211-2225, 1990.

Wu, D. L. et al., Equatorial diurnal variations observed in UARS Microwave Limb Sounder temperature during 1991-94 and simulated by the Canadian Middle Atmosphere Model, *J. Geophys. Res.*, , 103, 8909-8917, 1998.

Yudin, V. A. et al., Thermal tides and studies to tune the mechanistic tidal model using UARS observations, *Ann. Geophys.*, 15, 1205-1220, 1997.

M. A. Geller, L. Wang, and V. A. Yudin, Institute for Terrestrial and Planetary Atmospheres, State University of New York at Stony Brook, Stony Brook, NY 11794, USA. (e-mail: mgeller@ccmail.sunysb.edu; lwang@atmsci.msrc.sunysb.edu; valery@dombai.uars.sunysb.edu)

S. D. Eckermann, Hulburt Center for Space Research, Code 7641.2, Naval Research Laboratory. (e-mail:eckerman@ismap4.nrl.navy.mil)

# Free and Forced Interannual Variability of the Circulation in the Extratropical Northern Hemisphere Middle Atmosphere

Kevin Hamilton

*Geophysical Fluid Dynamics Laboratory/NOAA, Princeton, New Jersey*

This paper is an informal review of research relating to the interannual variability of the stratospheric and mesospheric circulation in the northern extratropics. Observations and modelling studies concerning aspects of variability generated internally in the atmosphere are reviewed first, with a particular emphasis on the effects of the tropical quasi-biennial oscillation. Then studies relating to externally forced variability are reviewed, including those that deal with the effects of volcanic aerosols and solar irradiance.

## 1. INTRODUCTION

The interannual variability of the atmospheric circulation tends to increase rapidly from the troposphere into the stratosphere. This is illustrated clearly in Fig. 1 which shows observations of January-mean zonal-mean temperatures at 78N for 20 consecutive years. Away from the surface (and sampling noise associated with vertically-trapped synoptic weather disturbances) the tropospheric temperatures show little year-to-year variation. The variability rises rapidly above the tropopause and the standard deviation peaks at about 10K near 1 hPa. The circulation of the stratosphere is much less constrained by convective and frictional coupling with the surface than that of the troposphere, and this no doubt accounts for part of the increase in variability above the tropopause. The amplitudes of large-scale planetary waves (as well as high-frequency gravity waves) also generally rise with height into the stratosphere, and this may be an additional factor in the enhanced interannual variability of the large-scale circulation in the stratosphere.

The large variability in the extratropical stratosphere has important consequences for modelling and attribu-

tion of stratospheric trends in temperature and ozone concentration. Current predictions are for a substantial stratospheric cooling to occur over the next century as the concentration of well-mixed greenhouse gases (notably $CO_2$) rise [*Austin et al.*, 1992; *Mahfouf et al.*, 1994; *Shindell et al.*, 1998]. This cooling will in turn affect the heterogeneous destruction of ozone in the late winter/spring in the extratropics of both hemispheres. The surface area of sulphate aerosols and polar stratospheric clouds is extremely temperature dependent [*Portmann et al.*, 1996; *Solomon*, 1999], so cooling of even a few K can result in large increases in the rate of chlorine activation and consequent ozone destruction. Modellers are now attempting forecasts of ozone loss for the next century using coupled dynamical-chemical models and scenarios for the anthropogenic release of long-lived trace constituents [*Shindell et al.*, 1998]. There is expected to be a competition between enhanced stratospheric cooling (mostly from $CO_2$ increases) leading to more aerosol surface area and the slow reduction in chlorine loading of the stratosphere in response to the treaty limitations on chlorofluorcarbon release. It is certainly possible that mid and high latitude ozone losses could become more severe in the future. In particular, there is concern that a more severe ozone hole may develop in the NH late winter and spring [*Austin et al.*, 1992; *Shindell et al.*, 1998].

Any useful prediction of the future course of the ozone layer needs to include an adequate treatment of year-

Atmospheric Science Across the Stratopause
Geophysical Monograph 123

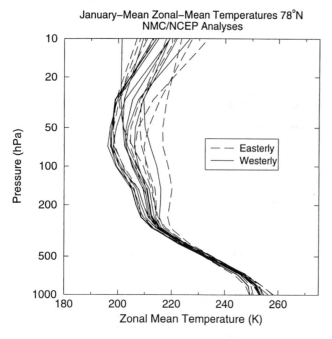

January–Mean Zonal–Mean Temperatures 78°N
NMC/NCEP Analyses

**Figure 1.** January-mean, zonal-mean temperature at 78N for 20 consecutive years (1979-98) computed from the daily analyses of the U.S. National Center for Environmental Prediction. Solid (dashed) lines are for those months in which the mean 40 hPa zonal wind at Singapore (1.3N) was westerly (easterly).

to-year variability, particularly in the extratropical NH winter/spring. The year-to-year variations in heterogeneous ozone destruction caused by interannual meteorological variations will likely be large in the NH, and a prediction needs to accurately characterize the frequency and severity of the deepest ozone loss years.

The present paper is a brief review of recent work relating to internal and externally-forced interannual variability in the extratropical stratosphere and mesosphere. Both observational and modelling studies will be reviewed. The variability to be considered will be divided into *free* and *forced* components, where, following *Lorenz* [1979] the forced variations are defined as those that result from changes in the composition of long-lived trace constituents, solar irradiance, or boundary conditions (notably sea surface temperatures, SSTs). Free variations are then those that are generated solely by processes within the atmosphere. The separation between free and forced variations becomes somewhat ill-defined when the effects of radiatively-active, short-lifetime trace constituents are included. In particular, year-to-year variations in ozone may reflect internal dynamical-chemical processes as well as effects of externally-imposed changes in long-lived trace gases (such as CFCs). The focus here is on variations with pe-

riods of order 1–10 years, rather than on very long term trends (see *Ramaswamy et al.* [2000] for an assessment of observed temperature trends in the stratosphere).

Section 2 gives a general introduction to some aspects of observed and modeled variability in the extratropical stratosphere and mesosphere. Section 3 discusses aspects of free variability, including the interannual variations connected with the tropical quasi-biennial oscillation (QBO). Then Section 4 discusses aspects of forced variability including effects of variations in SSTs, volcanic activity and solar insolation.

## 2.  BASIC ASPECTS OF OBSERVED AND MODELED VARIABILITY

Interannual variability of the large-scale circulation in the extratropical stratosphere is very strongly concentrated in the roughly 6 months spanning late fall through late spring, with only very small year-to-year variations in the remainder of the year. This is true in both hemispheres, although the variability in general is stronger in the NH extratropics. The seasonal dependence of the observed variability points to a strong connection between quasi-stationary planetary wave activity and interannual variability. In the summer, quasi-stationary planetary waves (QSPWs) are almost completely absent from the middle atmospheric circulation and there is remarkably little year-to-year variability. Similarly the interhemispheric asymmetry in QSPWs, which are much stronger in the NH than in the SH, is reflected in the contrast in the observed interannual variability of the zonal-mean circulation in the two hemispheres. Attention in this paper is focussed on the NH, but an up-to-date analysis of observations of the interannual variability of the SH stratospheric circulation is given in *Baldwin and Dunkerton* [1998b].

In the NH, the seasonal evolution normally shows a polar vortex that intensifies and cools through late fall and early winter. In some winters the cooling continues into February, but generally the cooling trend at high latitudes is interrupted by one or more stratospheric warmings. In such warmings the QSPWs in the extratropical stratosphere grow, and the polar vortex is disturbed, often to the point of being largely obliterated. Typically the vortex reestablishes itself after warmings occurring in December-February, although some effects of a mid-winter warming can persist for several weeks. In the period after the beginning of March the interannual variability is most obvious in the timing of the final breakup of the vortex, which can vary by as much as two months [*Waugh et al.,* 1999]. Both the mid-winter and final warmings are often connected with some enhancement of the QSPWs in the troposphere, and interannual anomalies in stratospheric and tropospheric NH winter circulation are observed to have a strong statistical connection (e.g., *Baldwin et al.,* [1994]).

The extent to which the evolution of the stratospheric extratropical flow is directly controlled by the tropospheric QSPWs has been investigated recently by *Kinnersley* [1998]. He performed a 13-year integration of a three-dimensional mechanistic model of the stratospheric circulation forced by the observed large-scale wave fields at 150 hPa. Kinnersley found that in each year the model reproduced quite well the observed development of the NH stratospheric flow in early winter, say through the end of December. The model was somewhat less successful in reproducing the observations in the remainder of each winter season, indicating a somewhat more chaotic nature to the flow evolution later in the winter.

The NH extratropical mesospheric circulation also displays significant interannual variability (e.g. *Lawrence and Randel* [1996]), although this has not been explored as systematically as for the stratosphere. The data available for the mesosphere are considerably more limited than those for the stratosphere, and, in particular, radiometers on operational satellites have not provided useful information on the mesospheric temperature. Satellite-based temperature retrievals above the stratopause are available only from several rather short, non-overlapping records from different instruments on research satellites. An example of the difficulty of using these data is given by *Clancy et al.* [1994], who found that that temperatures obtained from five years of observations from the Solar Mesosphere Explorer instrument disagreed with climatologies based on other satellite observations by ~10K in the 55-65 km range and by as much as 20K above 80 km.

These limitations make it difficult to investigate any long-period variations in the mesosphere, but one aspect relevant to this issue has been studied, namely the relation of the intraseasonal variations in the winter mesospheric circulation to those in the stratosphere [*Labitzke,* 1972, 1981; *Hirota and Barnett,* 1977]. The QSPWs that drive much of the circulation in the winter stratosphere also extend through the mesosphere, although amplitudes typically peak in the upper stratosphere and drop significantly with height in the mesosphere (e.g., *Barnett and Corney* [1985]). The transient warmings of the high-latitude stratosphere associated with amplifying QSPWs have a typical signature in the mesosphere [*Labitzke,* 1972, 1981; *Hirota and Barnett,* 1977]. In particular, some warming in the mesosphere generally precedes the stratospheric warming. However,

once the warming phase phase begins in the stratosphere it is generally accompanied by a high-latitude *cooling* above about 60 km.

The interannual variability of stratospheric circulation has been examined in control runs of a number of comprehensive GCMs [*Rind et al.,* 1988; *Hamilton,* 1995; *Boville,* 1995; *Erlebach et al.,* 1996; *Manzini and Bengtssen,* 1996; *Butchart and Austin,* 1998]. These studies all prescribed a climatological seasonal cycle of sea surface temperature, and assumed no interannual variability in atmospheric composition (except for water vapor) nor in solar irradiance. The general conclusion from all these studies is that GCMs can produce a degree of interannual variance of the extratropical NH stratosphere that is at least roughly of the observed magnitude. An example is shown in Fig. 2, which compares the variance of January-mean zonal-mean temperature in the control simulation using the GFDL SKYHI GCM reported by *Hamilton* [1995] and in several years

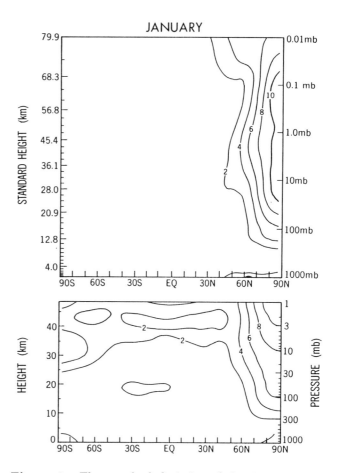

**Figure 2.** The standard deviation of the time series of January-mean zonal-mean temperatures from a control run of the GFDL SKYHI GCM (top) and from NCEP analyses (bottom). The contour interval is 2K.

## Zonal–Mean Temperatures 30 hPa 78°N NMC/NCEP Analyses

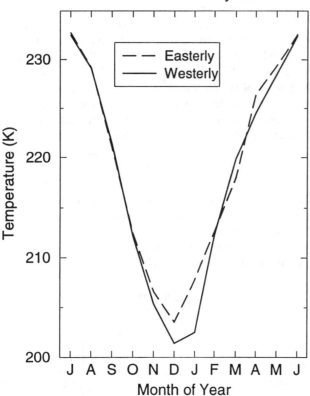

**Figure 3.** The zonal-mean 30 hPa temperature at 78N composited for each calendar month for times when the 40 hPa Singapore winds were westerly (solid) and easterly (dashed). Calculated from 20 years of NCEP analyses.

of NCEP observations. However, there are some interesting differences among the models in their ability to simulate the observed level of variance. For example, *Hamilton* [1995] noted a tendency to overestimate the interannual variance of zonal-mean temperatures in December, while *Manzini and Bengtsson* [1996] found that the ECHAM model produced un unrealistically small variance in the early winter. At this point there has been no systematic attempt to connect the model performance in this regard with aspects of the model formulations.

In the tropical stratosphere, the GCMs discussed in the previous paragraph all have been much less successful in reproducing a realistic degree of variance in the zonal-mean circulation. In this region of the real atmosphere the interannual variability is dominated by the QBO, and this is virtually absent from all the control simulations discussed in these papers. Recently there has been some progress in simulating large interannual

oscillations in the tropical stratosphere in some specially modified GCMs [*Takahashi*, 1996; *Horinouchi and Yoden*, 1998; *Hamilton et al.*, 1999], but there remain some significant deficiencies even in these simulations.

## 3. FREE VARIABILITY

### 3.1. Effects of the Tropical Quasi-biennial Oscillation

The prevailing zonal winds in the equatorial stratosphere are known to undergo a remarkable oscillation from strong westerlies to strong easterlies with a slightly irregular period that averages about 28 months [*Naujokat*, 1986]. The possible connection between this tropical QBO and the circulation in the extratropical stratosphere has been investigated since the early 1960's [*Angell and Korhover*, 1962]. The initial interest was in the possible extratropical forcing of the tropical QBO. However, current thinking is that the QBO is largely forced by the interaction of the mean flow with vertically-propagating waves [*Lindzen and Holton*, 1968; *Holton and Lindzen*, 1972], and that the extratropical QBO variations are secondary effects of this tropically-driven equatorial oscillation.

The QBO influence on the extratropical NH winter stratosphere is evident in Fig. 1, where the results for months when the equatorial prevailing winds at 40 hPa were westerly or easterly are shown by solid and dashed lines, respectively. Even from this sample of 20 years it is apparent that, on average, the high-latitude January stratospheric temperatures are warmer when the 40 hPa equatorial zonal-winds are easterly than when they are westerly. Fig. 3 shows 78N 30 hPa zonal-mean temperature composited for each calendar month for times when the monthly-mean 40 hPa zonal wind at Singapore (1.3N) was westerly vs. times when it was easterly. The tendency for warmer temperatures in the NH winter during easterlies is apparent, but the effect is strongest in January. This general pattern has been confirmed in a number of observational studies. [*Holton and Tan*, 1980, 1982; *Baldwin and Dunkerton*, 1991, 1998a]. Note that *Baldwin and Dunkerton* [1998a] examined the sensitivity of the NH extratropical/tropical QBO correlation when the phase of the QBO was quantified by the equatorial zonal winds at different pressure levels. They found that the strongest correlations with the extratropical NH circulation were obtained for a tropical index based on winds near 40 hPa. *Dunkerton et al.* [1988] report that observations show that about 2/3 of major midwinter warmings occur when the prevailing equatorial wind near 40 hPa is easterly [*Dunkerton et al.* 1988], although these statistics will

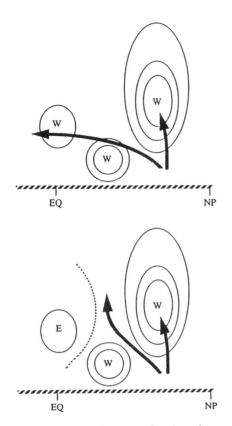

**Figure 4.** A schematic diagram showing the zonal wind structure in the winter hemisphere when the phase of the QBO in the equatorial lower stratosphere is westerly (top) and easterly (bottom). The dashed line in the bottom panel shows the location of the zero mean wind surface. The thick arrows denote the dominant paths of wave activity propagation associated with topographically-forced quasi-stationary planetary waves. From *Hamilton* [1998b].

change somewhat depending on the precise threshold adopted for deciding if a major warming has occurred.

The basic explanation for the observed QBO effect in the extratropics was advanced by *Holton and Tan* [1980, 1982]. Typically, the dominant direction of wave activity propagation for QSPWs forced in the extratropical troposphere is upward and equatorward. If the mean flow in the tropical stratosphere is westerly, QSPWs will be able to penetrate into the tropics and even across the equator without encountering a critical surface (where the zonal-mean zonal wind is zero). By contrast, when the mean flow in the tropical stratosphere is easterly, QSPWs will encounter a critical surface on the winter side of the equator. It is believed that the critical surface should act to partially reflect QSPW wave activity (e.g., *Haynes* [1985]). Thus, when there are easterlies in the tropics, the effective wave guide for QSPW propagation should be narrower and the wave activity at

mid and high latitudes of the winter hemisphere should be stronger. Stronger QSPWs in high latitudes should lead to greater wave-induced drag on the mean flow, and hence a warmer winter stratospheric pole. All this is consistent with the observed Holton-Tan effect and is illustrated schematically in Fig. 4.

Simulating the extratropical QBO effect is an interesting and challenging test for dynamical models of the middle atmosphere. This issue has been studied in an array of mechanistic models of varying complexity [*Holton and Austin*, 1991; *O'Sullivan and Young*, 1992, *O'Sullivan and Dunkerton*, 1994]. Such studies have involved imposing some large-scale wave forcing typical of NH winter at a lower boundary corresponding roughly to the tropopause, and then integrating forward for some period (a few weeks to an entire season). The integrations are performed for initial conditions with tropical easterlies and then repeated with tropical westerlies. Each of these studies has produced simulations at least qualitatively consistent with the observed statistical tendency for a weaker polar vortex when the tropical mean winds are easterly. Some studies, notably *Holton and Austin* [1991], have shown that major midwinter stratospheric warmings are initiated at a lower threshold of QSPW amplification when the equatorial stratospheric winds are easterly than when they are westerly. Meaningful quantitative comparison with observations is not possible from such simplified models, however.

Comprehensive GCMs have also been employed in studies of the QBO effect on the extratropical stratospheric circulation [*Balachandran and Rind*, 1995; *Hamilton*, 1995, 1998a]. These studies involved imposing QBO wind variations in the tropical stratosphere through an arbitrary momentum forcing and then examining the extratropical response. The most complete of these studies, and the one that bears the closest comparison with observations, is that of *Hamilton* [1998a]. *Hamilton* used the GFDL SKYHI model in a 48-year integration with a time-varying tropical momentum forcing that produced a 27-month QBO in the equatorial zonal wind with shear zones of realistic magnitude and descending at a realistic speed. The ability of the model to produce a realistic extratropical QBO response is illustrated in Fig. 5, which shows the same high-latitude temperature composite month-by-month as Fig. 3 but now based on 45 years of the SKYHI imposed-QBO simulation. There is basic agreement in the timing and magnitude of the effect seen in the model results and the observations. The tendency in the easterly QBO phase for a warmer pole is accompanied by a weaker zonal-mean vortex. In fact observations suggest that

## Zonal–Mean Temperatures 28 hPa 78°N SKYHI Simulation

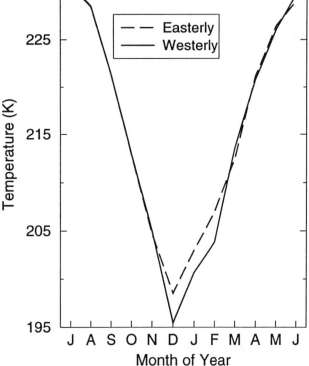

**Figure 5.** The zonal-mean 28 hPa temperature at 78N composited for each calendar month for times when the 40 hPa equatorial zonal-mean winds were westerly (solid) and easterly (dashed). Calculated from 45 years of simulation with the GFDL SKYHI GCM including an imposed QBO momentum source.

the winter-mean vortex near 1 hPa is weaker by ∼10 m-s$^{-1}$ in the easterly QBO phase than in the westerly QBO phase [*Baldwin and Dunkerton*, 1991], and this has been reproduced in the SKYHI experiment of *Hamilton* [1998a]. *Hamilton* [1998a] also notes that roughly 2/3 of the strongest midwinter stratospheric warmings in the model experiment occur when the 40 hPa winds at the equator are anomalously easterly (i.e. easterly relative to the long term mean); this is consistent with the pattern seen in observations [*Dunkerton et al.*, 1988]. *Hamilton* [1998a] also showed that in the model the stationary planetary waves in the NH winter extratropics were affected systematically by the tropical QBO phase and in a manner consistent with that shown in Fig. 4 (i.e. stronger, but more meridionally-confined stationary waves in the NH stratosphere when there are equatorial easterlies at 40 hPa).

A possible complication in the simple Holton-Tan picture of QBO modulation of the extratropical circulation has been introduced recently by *Naito and Hirota* [1997]. They noted that the Holton-Tan correlation looks different if the winters are stratified by the intensity of the solar activity. In particular, the Holton-Tan correlation in late winter empirically has been much stronger near solar minimum than near solar maximum. These results need to be regarded somewhat skeptically, since the total record analyzed by *Naito and Hirota* spanned only 37 years. However, if this empirical result holds up, then it will be a very interesting challenge to explain how the rather modest modulation of solar insolation through the solar cycle could have such a strong influence on tropical-extratropical interactions.

### 3.2. Longer-Period Variability

There have been observational indications of quasi-decadal variability (QDV) in long records of NH stratospheric fields. For example, *van Loon and Labitzke* [1990] found significant positive correlations between measures of solar activity and high-latitude 30 hPa temperatures and geopotential heights. One complication of studying QDV in the real atmosphere is that, given the rather short records available, it may be difficult to distinguish any free QDV from a forced 11-year solar cycle effect. Certainly the 11-year cycle in solar insolation will have at least some modest impact on NH extratropical circulation (see Section 4 below), but the possibility of free QDV in the atmosphere remains. *Baldwin and Dunkerton* [1998a] examined 33-year records of NH potential vorticity at mid-stratospheric levels based on NCEP daily analyses. When they performed time spectral analysis they found clear QBO peaks from the equator to nearly 70N and a biennial peak in mid and high latitudes. They also found a quasi-decadal peak in the extratropics, with at most a weak correlation to measures of solar activity. *Baldwin and Dunkerton* suggested that their quasi-decadal peak could arise from nonlinear interactions between the biennial and QBO signals.

The issue of QDV has also arisen in middle atmospheric models. Control simulations with middle atmospheric GCMs often appear to display an impressive degree of spontaneously-generated quasi-decadal and even longer-period variability in the NH extratropical stratosphere. Fig. 6 shows simulated North Pole DJF stratospheric temperatures from 35 consecutive years of a control simulation using the GFDL SKYHI model. There is an apparent tendency for warm winters and cold winters to group together, and the result is a timeseries

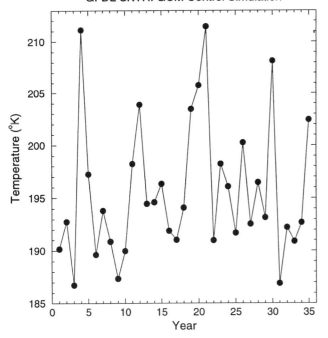

**Figure 6.** Time series of December-February mean North Pole temperatures at 28 hPa from 35 consecutive years of a control simulation with the GFDL SKYHI GCM.

that might be characterized, at least at times, as having an ~8-10 year period of variation. The results in Fig. 6 are from a slightly updated version of the model reported in *Hamilton* [1995], who showed a similar plot for 29 consecutive years of another control simulation. The tendency for clustering of cold and warm winters is apparent in the earlier SKYHI simulation as well. A similar effect seems to be occurring in the control simulation of the Free University of Berlin middle atmospheric GCM reported by *Erlebach et al.* [1996].

An interesting GCM example is shown in Fig. 7, taken from work of *Butchart et al.* [2000]. They integrated a middle atmospheric version of the UKMO Unified Model for 60 years representing 1992-2051, assuming a plausible scenario for the future course of concentration of well-mixed greenhouse gases such as $CO_2$. The detailed space-time evolution of the SSTs were prescribed and taken from a separate greenhouse gas scenario simulation done with a coupled atmosphere-ocean model. Fig. 7 shows the time series of December-February mean temperature at 10 hPa and averaged poleward of 60N. Two realization with the same SSTs and greenhouse gas concentrations, but different atmospheric initial conditions are shown, 11-year running means have been taken, and linear least-squares fits to

the data have been removed. Both runs display some very long-period variability with substantial amplitude. For example in Fig. 7b, from 1995 to 2005 the temperature rises by 3K and then drops back 3K in the next 12 years. The very different time evolution for the two cases suggest that the long-period variations are largely internally-generated rather than being forced by the SSTs or the greenhouse gas concentration changes (which are identical for the two runs). Of course, there must be some residual trend due to the forcing (in particular one anticipates a stratospheric cooling due to the increasing $CO_2$). However, the long-period variability generated spontaneously by the model represent a significant complication in determining the externally-forced trend (note the linear trends removed from Fig. 7a was less than 0.1 K/decade, and in Fig. 7b was

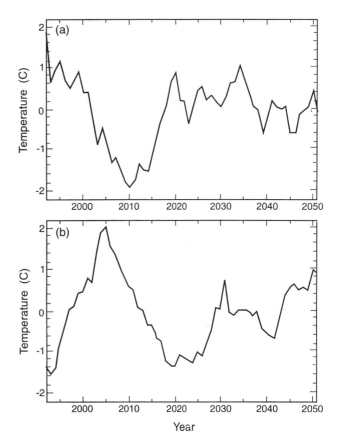

**Figure 7.** Time series of December-February 10 hPa temperature averaged over the area poleward of 60N in two simulations with the UKMO Unified Model GCM. Each run included a realistic scenario for the concentration of greenhouse gases and used a detailed time history of SSTs from a separate coupled ocean-atmosphere GCM run. The two runs shown differed only in their atmospheric initial conditions. In each case a linear least-squares fit has been subtracted from the time series. Redrafted from *Butchart et al.* [2000].

about 0.3 K/decade). In addition, if the real atmosphere behaves like this model, any prediction of future ozone trends will have to take into account the possibility of significantly colder-than-normal periods during which heterogeneous chemical ozone destruction will be greatly enhanced. So, for example, predictions from two-dimensional models with no source of internal dynamical interannual variability may be much too optimistic, since they would not include the possibility of significant internally-generated prolonged cold periods.

The lack of a realistic spontaneous QBO in GCM simulations makes the interpretation of decadal and longer-period variability in model simulations somewhat problematic, as the observed record displays both long-period and quasi-biennial variability. Thus, for example, until the 1990's the observational record does not show a direct analogue of the five consecutive cold winters seen in years 6-10 of the simulation displayed in Fig. 6. However, the real NH polar stratosphere did display a remarkable string of of relatively undisturbed NH winters during the period 1992-1997 [*Pawson and Naujokat*, 1997, 1999]. It is certainly plausible that this unprecedented behavior occurred in response to long-period anthropogenic climate forcing, but the similarity of these observations to the behavior of control GCM simulations (e.g. Fig. 6) raises the possibility that the observations may be explained simply as natural unforced variability.

An interesting issue raised by the GCM simulations is how the circulation in one winter can affect that in succeeding winters, a feature needed to account for more than a random clustering of cold or warm winters. One possibility for this interannual memory in the circulation was suggested by *Scott and Haynes* [1998]. In particular, they ran a mechanistic model of the middle atmosphere forced only by an annually-repeating cycle of near-tropopause large-scale QSPW forcing. They found that, for some range of amplitudes of the wave forcing, the model displayed a strong biennial variability in the polar winter - cold undisturbed winters were always followed by warm winters which were, in turn, followed by cold winters. The summer circulation in the high latitudes was virtually identical from year-to-year, however. The explanation offered by *Scott and Haynes* relied on the latitude-dependence of the effective timescale for zonal-mean radiative relaxation to act to restore the stratospheric zonal-mean flow. At high latitudes the effective timescale is less than a season allowing little scope for interannual variations if the wave forcing is the same each year. At low latitudes the timescale can become of order years, and this, of

course, helps account for the existence of the QBO near the equator. Scott and Haynes found that in the subtropics the timescale was long enough to allow anomalies to persist from one winter into the next. So during a cold undisturbed winter at high latitudes the mean flow in the subtropics tends to develop an easterly anomaly. In the next winter this easterly mean flow anomaly acts to slightly reduce the width of the effective waveguide for QSPWs (Fig. 4) thus leading to the evolution of a more disturbed polar winter, which also produces a westerly subtropical mean flow anomaly. So the interannual memory in their model is effectively stored in the mean flow momentum in the subtropics (an example of the *flywheel effect* described initially by *Sneider and Fels*, [1988]).

In the very simple model considered by *Scott and Haynes* the mean flow memory produced a strong biennial signal. In a more complete model (or in the real atmosphere) this mechanism for interannual memory might be expected to lead to a broader spectrum of interannual variations. This idea has received some support in preliminary analysis of results of a very recent SKYHI model experiment. This essentially repeated the control run of Fig. 6, but included an extra damping on the zonal-mean momentum in the tropics and subtropics that constrained the flow to be close to the long-term mean climatology from the control run. The results (not shown) suggest that this constraint does indeed significantly suppress high-latitude NH variability on interannual timescales.

The GCMs and simpler models discussed thus far have all employed prescribed ozone distributions. In more complete models or in the real atmosphere, the coupling of ozone chemistry with radiation and dynamics allows the possibility of another source of interannual memory for the circulation. There has been some speculation that cold winters will lead to high-latitude ozone losses that may not be fully restored chemically by the next fall and winter. With less high-latitude ozone heating the vortex may become colder and more stable. This is a positive feedback that has been suggested as a possible amplification of anthropogenic ozone trends. However, such a feedback could also contribute to long-period natural variability as well. This effect has not been studied extensively, but one modeling study has shown that it could possibly be significant. *Mahlman et al.* [1994] ran a GCM with simple treatment of heterogeneous ozone destruction and that included ozone feedbacks on the radiative heating. Their focus was on the development of the Antarctic ozone hole. They showed that the interannual memory associated with the chemical/radiative/dynamical feedback played a role in deepening the ozone hole as their integration progressed.

## 4. FORCED VARIABILITY

### 4.1. Solar Insolation Variations

The output from the sun is observed to undergo a roughly 11-year cycle. Variations in total output are rather small (~ .2% peak-to-peak), but the variability rises at the short wavelength end of the spectrum. Thus at the UV wavelengths where ozone heating is significant and, more particularly, at the extreme UV wavelengths responsible for odd oxygen production, the solar cycle variations are larger. *Lean et al.* [1997] find that the solar output averaged over the 200-300 nm range varies by 1.1% over the solar cycle.

*Angell* [1991] used radiosonde and rocketsonde data to examine tropical and NH solar cycle variations of temperature. In the upper stratosphere he found variations of order 1K with perhaps half this in the lower stratosphere. This is consistent as well with a later analysis of rocketsonde data by *Dunkerton et al.* [1998]. Analysis of satellite radiometer data reported in *Ramaswamy et al.* [2000] also suggest a statistically significant solar cycle temperature variation of about 1K in the tropical mid and upper stratosphere, but results were not significant in the extratropics. Indeed, the limited stratospheric data record makes it difficult to reliably isolate the solar cycle contribution to temperature variability in the extratropics, particularly if the atmosphere can generate significant unforced extratropical QDV.

Model calculations of the effect of solar cycle variations on the middle atmosphere have been attempted by a number of investigators using one-dimensional and two-dimensional coupled dynamical-chemical-radiative models [*Garcia et al.*, 1984; *Brasseur*, 1993]. The experiments are typically quite simple, running the model with solar input spectra typical of solar minimum and then repeating for solar maximum conditions. The results indicate the solar cycle tropical stratopause variations of ~1-2K peak-to-peak, with much smaller effects in the tropical lower stratosphere. The model solar cycle temperature variations in the extratropics are comparable to those in the tropics during the summer season, but are much smaller in winter.

### 4.2. Effects of Volcanic Aerosols

Explosive volcanic eruptions can produce significant abrupt changes in stratospheric composition and radiative balance. There have been three major eruptions in the period since the advent of routine stratospheric observations: Mt. Agung (1963), El Chichon (1982), and Mt. Pinatubo (1991), all at fairly low latitudes. In each case the observations show that the region between about 30S and 30N and in the 100-50 hPa layer warmed by ~2-3K, with the warming slowly diminishing over a ~2 year period as the volcanic aerosol decayed [*Angell*, 1997]. The tropical/subtropical warming is strongest in the lower stratosphere but can be appreciable even up to 10 hPa. The tropical warming seems to be accompanied by a cooling of comparable magnitude in the NH polar stratosphere (at least for the first half-year after the eruptions), although no statistically significant conclusions can be drawn for the extratropical effects.

### 4.3. Sea Surface Temperature Variations

The influence of SSTs on the interannual variability of tropospheric circulation has received a great deal of attention. Observations show that monthly or seasonal-mean tropospheric circulation anomalies in the tropics are strongly correlated with tropical SST anomalies. Extratropical tropospheric circulation anomalies are also clearly, but more weakly, correlated with tropical SSTs [*van Loon and Madden*, 1981; *Pan and Oort*, 1983]. In particular, the extratropical tropospheric stationary wave pattern in NH winter appears to be significantly correlated with the SST in the equatorial Pacific (which itself undergoes a strong Southern Oscillation on timescales from ~2-7 years).

The causal connections between SST anomalies and the seasonal-mean tropospheric circulation have been studied extensively in GCM simulations. *Lau* [1985] compared the standard deviation of monthly-mean geopotential heights in the troposphere for a GCM run with annually-repeating climatological SSTs and for a run with the same model, but with observed SSTs for a 30 year period. In the tropics the run with climatological SSTs underestimates the interannual variability of tropospheric geopotential by a large factor (typically ~5) at almost all locations. This is consistent with the tropical prevailing circulation being under the strong control of the SST distribution. The inclusion of SST variability does not seem to significantly raise the total geopotential variance in the extratropical troposphere, but the circulation in the extratropics in the model does display a statistical correlation with the tropical SSTs similar to that observed. The significance and extent of extratropical SST influence on tropospheric circulation is still unresolved [*Lau and Nath*, 1994].

The SST effects on stratospheric circulation have been investigated less extensively. Observational stud-

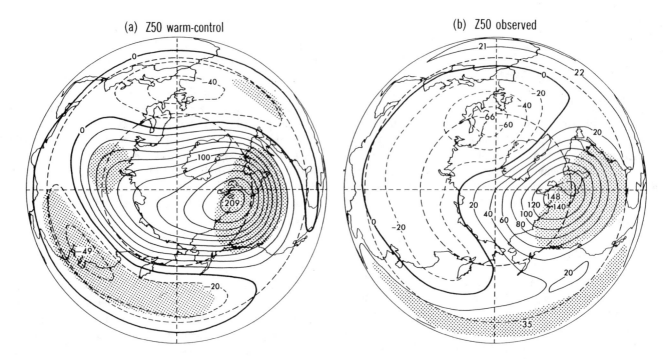

**Figure 8.** The December-February 50 hPa height anomalies associated with warm conditions in the tropical Pacific Ocean. (a) GFDL SKYHI model results averaged over 10 single year integrations with an imposed warm anomaly in the tropical Pacific SST minus a 25-year mean from a control run with climatological SSTs. (b) Difference between 50 hPa heights in the Free University of Berlin observational analyses for an average of 9 winters corresponding to mature phases of El Nino events and 18 winters in which the Southern Oscillation was in neither its El Nino or La Nina phase. In both panels the contour interval is 20 m, dashed contours denote negative values (i.e. lower heights for warm tropical SSTs), and shading denotes regions that are judged significantly different from zero using a 95% criterion in a two-tailed t-test. From *Hamilton* [1995].

ies have focussed on the possible influence of the warm and cold Southern Oscillation (SO) extremes on the NH extratropical winter stratospheric circulation. Panel b of Fig. 8, reproduced from *Hamilton* [1993], shows the 50 hPa geopotential anomaly composited over 9 NH winters corresponding to the warm extreme (i.e. mature El Nino) of the SO. The effects are not negligible, with the positive anomaly composite as large as 148 *m* over northern Canada, but they are still too small to be clearly seen in a single year. The shading shows the region judged to be significantly different from zero at the 95% level. Even with 9 cases for the composite the area of statistical significance is very small. The general pattern seen in this observational study is consistent with that of *Baldwin and O'Sullivan* [1995].

Most GCM simulations of the middle atmosphere have employed climatological SST, and there seem to be no published results of extensive stratospheric GCM experiments using detailed observed SST timeseries, i.e experiments analogous to those that have been done for the troposphere [*Lau*, 1985; *Lau and Nath*, 1994].

Perhaps the most relevant GCM experiment is that of *Hamilton* [1995] who ran the SKYHI model for 29 years in control mode, then ran 10 September-March simulations with an imposed SST anomaly field in the tropical Pacific that corresponds roughly to that observed in the warmest phase of an extremely strong El Nino. Fig. 8a shows the 50 hPa geopotential averaged over the El Nino NH winters minus that averaged over the winters of the control run. The overall similarity to the observed composite in Fig. 8b is quite apparent. Even the somewhat larger peak geopotential anomalies in the model are understandable, since the imposed SST anomalies were stronger than in average observed El Nino events. However, once again the effect is small enough that even a 10-year composite has statistical significance only over a very small area.

The relatively limited influence of SST variability on the NH extratropical stratosphere was also noted by *Butchart et al.* [2000]. In particular the two 60-year integrations of their GCM (Fig. 7) imposed identical detailed time series of the SST anomalies. *Butchart et al.* found little correlation between the time series of stratospheric variables in the two integrations, again suggesting the dominance of unforced variability.

### 4.4. Summary

There is reasonably strong observational and/or modeling evidence that each of the forcing mechanisms discussed here has a significant effect on NH extratropical stratospheric circulation. However, all these effects appear to be dominated, at least in winter, by the very substantial unforced natural variability in the extratropical middle atmosphere. Thus a detailed understanding of the interannual forcings may be somewhat useful in interpreting the observed record, but much of the typical variability of the NH extratropical stratospheric circulation should be reproduced in traditional control experiments with GCMs using annually-repeating atmospheric composition, solar insolation and ocean boundary conditions.

## 5. CONCLUSION

This paper has reviewed literature relating to the interannual variability in the extratropical stratosphere. This subject involves a rather broad range of aspects of atmospheric science, from purely dynamical considerations of forced *versus* chaotic flow evolution to the physical/chemical processes involved in solar-terrestrial relations. Some aspects, such as the effects of the tropical QBO on extratropical circulation, have been widely investigated and may now be basically understood. Other aspects, such as longer-term spontaneous variability, are much less well characterized and understood.

Modelling extratropical stratospheric variability and predicting how it might change in the future is now regarded as a very important aspect of climate prediction, particularly for forecasts of the response of the ozone layer to expected anthropogenic effects. The present review has highlighted some encouraging aspects of current modelling capabilities, but there are still some very significant challenges remaining. Certainly the credibility of stratospheric climate predictions by GCMs would be enhanced if such models would spontaneously simulate a realistic QBO in the tropical atmosphere. While there has been some recent progress in this regard, the current state of GCMs is still inadequate. It is also very desirable to have an understanding of the nature of the quasi-decadal variability that has been observed in the extratropical stratosphere and to relate the observed quasi-decadal variability to long-term variations seen in some GCM simulations. Other aspects that need to be investigated further include the possible roles of ocean-atmosphere interaction and chemical-radiative-dynamical interaction in generating interannual variability in the stratosphere.

*Acknowledgments.* The author thanks John Lanzante, Steve Garner, Jerry Mahlman, Mark Baldwin, Bill Randel and the official reviewers for comments on the manuscript and Neal Butchart and his colleagues at the UKMO for permission to use their unpublished results.

## REFERENCES

Angell, J.K. Stratospheric temperature change as a function of height and sunspot number during 1972-89 based on rocketsonde and radiosonde data. *J. Climate, 4*, 1170-1180, 1991.

Angell, J.K. Stratospheric warming due to Agung, El Chichon and Pinatubo taking into account the quasi-biennial oscillation. *J. Geophys. Res., 102*, 9479-9485, 1997.

Angell, J.K. and J. Korshover, The biennial wind and temperature oscillations of the equatorial stratosphere and their possible extension to higher latitudes. *Mon. Wea. Rev., 90*, 127-132, 1962.

Austin, J., N. Butchart and K.P. Shine, Probability of an Arctic ozone hole in a doubled $CO_2$ climate. *Nature, 360*, 221-225, 1992.

Balachandran, N.K., and D. Rind, Modeling the effects of solar variability and the QBO on the troposphere/stratosphere system. Part 1: The middle atmosphere. *J. Climate, 8*, 2058-2079, 1995.

Baldwin, M.P., X. Cheng and T.J. Dunkerton, Observed correlation between winter-mean tropospheric and stratospheric anomalies, *Geophys. Res. Lett., 98*, 23079-23090, 1993.

Baldwin, M.P., and T.J. Dunkerton, Quasi-biennial oscillation above 10 mb. *Geophys. Res. Lett., 18*, 1205-1208, 1991.

Baldwin, M.P., and T.J. Dunkerton, Biennial, quasi-biennial and decadal oscillations of potential vorticity in the northern stratosphere. *J. Geophys. Res., 103*, 3919-3928, 1998a.

Baldwin, M.P., and T.J. Dunkerton, Quasi-biennial modulation of the Southern Hemisphere polar vortex, *Geophys. Res. Lett., 25*, 3343-3346, 1998b.

Baldwin, M.P., and D. O'Sullivan, Stratospheric effects of ENSO-related tropospheric circulation anomalies, *J. Climate, 4*, 649-667, 1995.

Barnett, J.J. and M. Corney, Planetary waves, *Middle Atmosphere Porgram Handbook, 16*, 86-137, 1985.

Boville, B.A., Middle atmosphere version of the CCM2 (MACCM2): annual cycle and interannual variability. *J. Geophys. Res., 100*, 9017-9039, 1995.

Brasseur, G., The response of the middle atmosphere to long-term and short-term solar variability: A two-dimensional model. *J. Geophys. Res., 98*, 23079-23090, 1993.

Butchart, N., and J. Austin, Middle atmosphere climatologies from the troposphere-stratosphere configuration of the UKMO unified model. *J. Atmos. Sci., 55*, 2782-2809 1998.

Butchart,N., J. Austin, J. Knight, A. Scaife and M. Gallani, The response of the stratospheric climate to projected changes in the concentrations of well-mixed greenhouse gases from 1992 to 2051, *J. Climate*, in press, 2000.

Clancy, R.T., D.W. Rusch and M.T. Callan, Temperature minima in the average thermal structure of the middle mesosphere (70-80 km) from analyses of 40-to-92 km SME global temperature profiles, *J. Geophys. Res., 99*, 19001-19020, 1994.

Dunkerton, T.J., D.P. Delisi and M.P. Baldwin, Distribution of major stratospheric warmings in relation to the quasi-biennial oscillation, *Geophys. Res. Lett., 15*, 136-139, 1988.

Dunkerton, T.J., D.P. Delisi and M.P. Baldwin, Middle atmosphere cooling trend in historical rocketsonde data, *Geophys. Res. Lett., 25*, 3371-3374, 1998.

Erlebach, P., U. Langematz and S. Pawson, Simulations of sudden stratospheric warmings in the Berlin troposphere-stratosphere-mesosphere GCM. *Ann. Geophys., 14*, 443-463, 1996.

Garcia, R.R., S. Solomon, R.G. Roble and D.W. Rusch, A numerical study of the response of the middle atmosphere to the 11-year solar cycle. *Planet. Space Sci., 32*, 411-423, 1984.

Haigh, J.D., The impact of solar variability on climate. *Science, 272*, 981-984, 1996.

Hamilton, K., An examination of observed Southern Oscillation effects in the Northern Hemisphere stratosphere. *J. Atmos. Sci., 50*, 3468-3473, 1993.

Hamilton, K., Interannual variability in the Northern Hemisphere winter middle atmosphere in control and perturbed experiments with the SKYHI general circulation model, *J. Atmos. Sci., 52*, 44-66, 1995.

Hamilton, K., Effects of an imposed quasi-biennial oscillation in a comprehensive troposphere-stratosphere-mesosphere general circulation model. *J. Atmos. Sci., 55*, 2393-2418, 1998a.

Hamilton, K., Dynamics of the tropical middle atmosphere: A tutorial review. *Atmos.-Ocean, 36*, 319-354, 1998b.

Hamilton, K., R.J. Wilson and R. Hemler. Middle atmosphere simulated with high vertical and horizontal resolution versions of a GCM: Improvement in the cold pole bias and generation of a QBO-like oscillation in the tropics. *J. Atmos. Sci., 56*, 3829-3846, 1999.

Haynes, P.H., Nonlinear instability of a Rossby wave critical layer. *J. Fluid Mech., 161*, 493-511, 1985.

Hirota, I. and J.J. Barnett, Planetary waves in the winter mesosphere: preliminary analysis of the Nimbus 6 PMR results. *Quart. J. Roy. Meterolo. Soc., 103*, 487-498, 1977.

Holton, J.R. and J. Austin, The influence of the QBO on sudden stratospheric warmings. *J. Atmos. Sci., 48*, 607-618, 1991.

Holton, J.R. and R.S. Lindzen, An updated model for the quasi-biennial cycle of the tropical stratosphere. *J. Atmos. Sci., 29*, 1076-1080, 1972.

Holton, J.R. and H.-C. Tan, Influence of the equatorial quasi-biennial oscillation on the global circulation at 50 mb *J. Atmos. Sci., 37*, 2200-2208, 1980.

Holton, J.R. and H.-C. Tan, Quasi-biennial oscillation in the Northern Hemisphere lower stratosphere *J. Meteor. Soc. Japan, 60*, 140-148, 1982.

Horinouchi, T. and S. Yoden, Wave-mean flow interaction

associated with a QBO-like oscillation simulated in a simplified GCM. *J. Atmos. Sci., 55*, 502-526, 1998.

Kinnersley, J.S., Interannual variability of stratospheric zonal wind forced by the northern lower-stratospheric large-scale waves. *J. Atmos. Sci., 55*, 2270-2283, 1998.

Labitzke, K., Temperature changes in the mesosphere and stratosphere connected with circulation changes in winter. *J. Atmos. Sci., 29*, 756-766, 1972.

Labitzke, K., Stratospheric-mesospheric midwinter disturbances: a summary of observed characteristics. *J. Geophys. Res., 86*, 9665-9678, 1981.

Labitzke, K. and M.P. McCormick, Stratospheric temperature increases due to Pinatubo aerosols. *Geophys. Res. Lett., 19*, 207-210, 1992.

Lau, N.-C., Modeling the seasonal dependence of the atmospheric response to observed El Ninos in 1962-76. *Mon. Wea. Rev., 113*, 1970-1996, 1985.

Lau, N.-C. and M.J. Nath, A modelling study of the relative roles of tropical and extratropical SST anomalies in the variability of the global atmosphere-ocean system. *J. Clim., 7*, 1184-1207, 1994.

Lawrence, B.N. and W.J. Randel, Variability in the mesosphere observed by the Nimbus 6 pressure modulated radiometer. *J. Geophys. Res., 101*, 23475-23489, 1996.

Lean, J.L., G.J. Rottman, H.L. Kyle, T.N. Woods, J.R. Hickey, L.C. Puga, Detection and parameterization of variations in solar mid- and near-ultraviolet radiation (200-400 nm) *J. Geophys. Res., 102*, 29939-29956, 1997.

Lindzen, R.S. and J.R. Holton, A theory of the quasi-biennial oscillation. *J. Atmos. Sci., 25*, 1095-1107, 1968.

Lorenz, E.N., Forced and free variations of weather and climate, *J. Atmos. Sci., 36*, 1367-1376, 1979.

Mahfouf, J.F., D. Cariolle, J.-F. Geleyn and B. Timbal, Response of the Meteo-France climate model to changes in $CO_2$ and sea surface temperature. *Climate Dynamics, 9*, 345-362, 1994.

Mahlman, J.D., J.P. Pinto and L.J. Umscheid, Transport, radiative, and dynamical effects of the Antarctic ozone hole: A GFDL "SKYHI" model experiment. *J. Atmos. Sci., 51*, 489-508, 1994.

Manzini, E. and L. Bengtsson, Stratospheric climate variability from a general circulation model and observations. *Climate Dyn., 12*, 615-639, 1996.

Naito, Y. and I. Hirota, Interannual variability of the Northern winter stratospheric circulation related to the QBO and solar cycle. *J. Meteor. Soc. Japan, 75*, 925-937, 1997.

Naujokat, B., An update of the observed quasi-biennial oscillation of the stratospheric winds over the tropics. *J. Atmos. Sci., 43*, 1873-1877, 1986.

O'Sullivan, D. and R.E. Young, Modelling the quasi-biennial oscillation's effect on the winter stratospheric circulation. *J. Atmos. Sci, 49*, 2437-2448, 1992.

O'Sullivan, D. and T.J. Dunkerton, Seasonal development of the extratropical QBO in a model of the middle atmosphere. *J. Atmos. Sci, 51*, 3706-3721, 1994.

Pan, Y.-H. and A.H. Oort, Global climate variations associated with sea surface temperature anomalies in the eastern equatorial Pacific Ocean for the 1958-73 period. *Mon. Wea. Rev., 111*, 1244-1258, 1983.

Pawson, S. and B. Naujokat, Trends in daily wintertime

temperatures in the Northern stratosphere. *Geophys. Res. Lett., 24*, 575-578, 1997.

Pawson, S. and B. Naujokat, The cold winters of the middle 1990's revisited. *J. Geophys. Res., 104*, 14209-14222, 1999.

Portmann, R.W., S. Solomon, R.R. Garcia, L.W. Thomason, L.R. Poole and M.P. McCormick, Role of aerosol variations in anthropogenic ozone depletion in the polar regions. *J. Geophys. Res., 101*, 22991-23006, 1996.

Ramaswamy, V., M.-L. Chanin, J. Angell, J. Barnett, D. Gaffen, M. Gelman, P. Keckhut, Y. Koshelkov, K. Labitzke, J.-J.R. Lin, A. O'Neill, J. Nash, W. Randel, R. Rood, K. Shine, M. Shiotani and R. Swinbank, Stratospheric temperature trends: observations and model simulations. *Rev. Geophys.*, in press, 2000.

Randel, W.J., F. Wu, J.M. Russell III, J.W. Waters and L. Froidevaux, Ozone and temperature changes in the stratosphere following the eruption of Mt. Pinatubo. *J. Geophys. Res., 100*, 16753-16764, 1995.

Rind, D., R. Suozzo and N.K. Balachandran, GISS global climate-middle atmosphere model, Pt. 2, Model variability due to interactions between planetary waves, the mean circulation, and gravity wave drag, *J. Atmos. Sci, 40*, 371-386, 1988.

Scott, R.K. and P.H. Haynes, Internal interannual variability of the extratropical circulation: the low-latitude flywheel. *Quart. J. Roy. Meteor. Soc., 124*, 2149-2173, 1998.

Shindell, D.T., D. Rind and P. Lonergan, Increased polar stratospheric ozone losses and delayed eventual recovery owing to increasing greenhouse gas concentrations. *Nature, 392*, 589-592, 1998.

Solomon, S., Stratospheric ozone depletion: a review of concepts and history, *Rev. Geophys., 37*, 275-316, 1999.

Sneider, R.K. and S. Fels, The flywheel effect in the middle atmosphere. *J. Atmos. Sci., 45*, 3996-4004, 1988.

Takahashi, M., Simulation of the stratospheric quasi-biennial oscillation using a general circulation model. *Geophys. Res. Lett., 23*, 661-664, 1996.

van Loon, H., and R. Madden, The Southern Oscillation Part I: Global associations with pressure and temperature in northern winter. *Mon. Wea. Rev., 109*, 1150-2262, 1981.

van Loon, H., and K. Labitzke, Association between the 11-year solar cycle and the atmosphere. Part IV: The stratosphere, not grouped by the phase of the QBO. *J. Clim., 3*, 827-837, 1990.

Waugh, D.W., W.J. Randel, S. Pawson, P.A. Newman and E.R. Nash, Persistence of the lower stratospheric polar vortices. *J. Geophys. Res., 104*, 27191-27201, 1999.

K. Hamilton Geophysical Fluid Dynamics Laboratory/NOAA Princeton University P.O. Box 308 Princeton, NJ 08542

# Future Changes in Upper Stratospheric Ozone

## K. W. Jucks

*Harvard-Smithsonian Center for Astrophysics, Cambridge, Massachusetts*

## R. J. Salawitch

*Jet Propulsion Laboratory, Pasadena, California*

It is commonly thought that as the level of stratospheric chlorine declines due to the ban on the use of chlorofluorocarbons, the concentration of stratospheric ozone will rise. The recovery of stratospheric ozone may first become apparent in the upper stratosphere because, for this region of the atmosphere, there is a direct correspondence between the concentration of $O_3$ and the abundance of ClO. We use a photochemical model to examine changes in $[O_3]$ due to a 15% reduction in the level of stratospheric chlorine, which should occur around the year 2010. This forcing results in a maximum increase in $[O_3]$ of about 4% near 42 km altitude. The abundance of upper stratospheric $O_3$ is relatively insensitive to changes in the direct transport of $O_3$ owing to its short photochemical lifetime. However, $[O_3]$ in this region is sensitive to changes in the abundance of radical precursors, which are controlled by circulation, thermodynamics, and tropospheric source concentrations. Of particular concern is the possibility that increased concentrations of greenhouse gases might alter stratospheric circulation and/or temperature in the tropopause region. We show that plausible changes in $[CH_4]$, $[H_2O]$, $[NO_y]$, and temperature may either mitigate or enhance the expected recovery of upper stratospheric $[O_3]$ during the next decade. Consequently, a concerted effort to monitor the temporal evolution of $[CH_4]$, $[H_2O]$, $[NO_y]$, and temperature is required to quantify the expected recovery of upper stratospheric $[O_3]$ in the next decade.

## 1. INTRODUCTION

It has been suggested that the recovery of $O_3$ due to the decreasing burden of stratospheric chlorine may first become apparent in the upper stratosphere [e.g., *WMO, 1999*, Chapters 6 and 12]. The photochemical lifetime of ozone in this region of the atmosphere is relatively short compared to the time scale for redistribution of $O_3$ by stratospheric circulation. As a result, the concentration of $O_3$ in the upper stratosphere responds in an indirect manner to changes in the transport field (e.g., direct transport of $O_3$ is essentially unimportant for determining the concentration of $O_3$) and is decoupled from chemical effects of sulfate aerosols due to the unimportance of heterogeneous chemistry. The concentration of upper stratospheric ozone is thought to respond directly to changes in the concentrations of molecules that

Atmospheric Science Across the Stratopause
Geophysical Monograph 123
Copyright 2000 by the American Geophysical Union

limit the loss of $O_3$, namely the reactive radicals ClO, $HO_2$, and $NO_2$. Catalytic cycles involving each of these radicals contribute significantly to photochemical loss of upper stratospheric ozone.

It is expected that the total inorganic chlorine loading ($Cl_y$) of the stratosphere will decline over the next several decades due to legislation that has essentially banned the use of chlorofluorocarbons (CFCs) and other organic chlorinated source gases [*WMO*, 1999, Chapter 1; *Montzka et al.*, 1999]. A chlorine emission scenario based on this legislation predicts about a 15% decline in $Cl_y$, relative to contemporary levels, by the year 2010 [e.g. *WMO*, 1999, Chapter 1]. Photochemical models indicate that, all else being equal, this decrease in $Cl_y$ will lead to a decrease in [ClO] ([X] is used to denote the mixing ratio of species X) and an increase in [$O_3$]. The apparent insensitivity of upper stratospheric $O_3$ to transport and aerosol loading may result in recovery of $O_3$ first being documented for this region of the atmosphere [WMO, 1999, Chapters 6 and 12].

It is likely, however, that variations in the abundance of upper stratospheric $O_3$ during the next several decades will be affected by factors besides just [ClO]. The concentration of $HO_2$ is controlled primarily by the abundance of $H_2O$ and $CH_4$. The photochemical lifetimes of these source molecules are considerably longer than the lifetime of $O_3$, and as a result the abundances of these gases in the upper stratosphere are influenced by transport processes within the stratosphere as well as by concentrations in the tropical source region of fresh stratospheric air. Recent observations have indicated increases in the concentration of upper stratospheric $H_2O$ [*Evans et al.*, 1998; *Nedoluha et al.*, 1998a] and decreases in the abundance of $CH_4$ [*Nedoluha et al.*, 1998b] that are of sufficient magnitude to affect $O_3$ during the next several decades, should these changes persist. The concentration of [$NO_2$] in the upper stratosphere is proportional to abundance of total odd nitrogen, $NO_y$. The level of $NO_y$ is controlled by the abundance of $N_2O$ as well as by transport and kinetic processes [e.g., *Rosenfield and Douglass*, 1998]. Although the measured rise in the tropospheric burden of $N_2O$, about 0.2 to 0.3 %/year, is too small to contribute significantly to changes in upper stratospheric $O_3$ over the next several decades, variations in upper stratospheric [$NO_y$] induced by changes in transport have the potential to affect [$O_3$]. Finally, future decreases in the temperature of the upper stratosphere caused by the radiative effects of increased concentrations of greenhouse gases may also have important consequences for [$O_3$].

The purpose of the present study is to quantify the effects on [$O_3$] of the expected decrease in [$Cl_y$] over the next decade taking into account plausible changes in [$H_2O$], [$CH_4$], [$NO_y$], and temperature. We use a photochemical steady state model, for conditions at midlatitudes during equinox, to calculate the change in the profile of [$O_3$] for present conditions (e.g., the time of peak chlorine loading in the stratosphere) compared to conditions for the year 2010. We first illustrate how [$O_3$] would be affected by a 15% reduction in [$Cl_y$]. We then illustrate the effects on [$O_3$] of the persistence of observed recent changes in [$H_2O$], [$CH_4$], and temperature (we lack an observational basis to define observed changes in [$NO_y$], although the potential exists using the HALOE data sets for [NO] and [$NO_2$] in the upper stratosphere). Several open questions regarding the kinetics of stratospheric photochemistry are considered in the context of the change in [$O_3$] over the next decade.

At present we lack a theoretical basis to predict in a meaningful manner how upper stratospheric [$H_2O$], [$CH_4$], [$NO_y$], and temperature will vary in the future. The abundances of these quantities in the upper stratosphere are controlled by the large scale transport field that is driven by breaking gravity and planetary waves [e.g., *Holton et al.*, 1995]. Stratospheric $H_2O$ is also affected by thermodynamic processes near the tropopause [e.g., *Newell and Gould-Stewart*, 1981]. It is not yet possible to account quantitatively for the observed year-to-year variations of [$H_2O$] and [$CH_4$] with existing models [e.g., *Nedoluha et al.*, 1998a]. There is considerable uncertainty concerning the effect of aerosol heating on stratospheric circulation [e.g., *Nedoluha et al.*, 1998b] as well as the possibility that long-term changes in the large scale transport field may be caused in part by the rising burden of greenhouse gases [e.g., *Hood et al.*, 1997; *Steinbrecht et al.*, 1998; *Fusco and Salby*, 1999]. Since we lack a basis to predict future variations of [$H_2O$], [$CH_4$], [$NO_y$], and temperature, our final set of calculations focuses instead on defining the change in each of these quantities that would be required to offset the expected increase in [$O_3$] due to declining [$Cl_y$]. We show that upper stratospheric [$O_3$] is likely to be affected by factors other than just the decline of [$Cl_y$] during the next decade. Our study highlights the need to define from observation, future changes in [$H_2O$], [$CH_4$], [$NO_y$], and temperature in order to fully understand the effects on [$O_3$] of declining levels of [$Cl_y$].

## 2. MODEL CALCULATIONS

The calculations performed here use a photochemical model that has been used for many applications to investigate observations made from satellites [e.g. *Michelsen et al.*, 1996], aircraft [e.g., *McElroy and Salawitch*, 1989; *Salawitch et al.*, 1994], and balloons [e.g., *Chance et al.*, 1996; *Jucks et al.*, 1998]. The model is constrained by observations of long-lived radical precursors such as [$H_2O$] and [$CH_4$], the total abundance of inorganic chlorine ([$Cl_y$]), bromine ([$Br_y$]), and odd nitrogen ([$NO_y$]), as well as temperature and aerosol loading. For most past applications of the model, [$O_3$] also has been constrained by observation because these studies have typically focused on our understanding of measured abundances of species such as $Cl_y$, $NO_y$, or $HO_x$ ($OH+HO_2$) that are sensitive to the concentration of $O_3$. Since the goal of this study is to assess the sensitivity of [$O_3$] resulting from changes in source gases and temperature, [$O_3$] is relaxed to photochemical steady state in a manner similar to that described by *McElroy and Salawitch* [1989], who compared calculated [$O_3$] to measurements from Atmospheric Trace Molecule Spectroscopy Experiment (ATMOS). This is valid for the upper stratosphere (altitudes above 30 km), where the photochemical lifetime of $O_3$ is short, but is not valid for the lower stratosphere (altitudes below 25 km), where transport will have a significant effect on the concentration of ozone [e.g., *Ko et al.*, 1989]. Our analysis focuses on the upper stratosphere.

We chose as a base atmosphere constraints from northern mid-latitude balloon measurements made by the FIRS-2 spectrometer in September 1992 at 35°N [*Johnson et al.*, 1995a]. The altitude distribution of the long-lived radical precursors observed on this flight are representative of conditions seen on a number of other mid-latitude flights. Simulations for different times of year will yield similar results for the upper stratosphere since the relative contributions to the photochemical loss of ozone do not vary significantly between seasons. The relative contribution to loss of upper stratospheric $O_3$ by the ClO cycle maximizes at higher latitudes due mainly to lower abundances of $CH_4$, which leads to an increase in the ClO/HCl ratio. While suitable constraints on the latitudinal distribution of precursors exist to fully explore the latitudinal dependence of the results presented here, such an exercise is beyond the scope of the present study. We set the aerosol concentrations to background levels based on the SAGE II zonal monthly mean climatology [*Thomason et al.*, 1997]. Aerosol concentrations vary considerably in the lower stratosphere, but not in the upper stratosphere, which is the region being investigated here.

The concentration profile of $O_3$ is calculated at each altitude (2 km intervals are used) assuming photochemical steady state (e.g, a balance of production and loss,

integrated over 24 hours) for each chemical species in the model. An initial profile for [$O_3$] based on observation is used for the first iteration of the model. The profile of [$O_3$] is then adjusted, and photolysis rates are recalculated, until steady state for [$O_3$] is achieved at each altitude. In this manner, we fully account for the so-called "self healing effect" (e.g., increases in [$O_3$] at a specific altitude mitigate increases in [$O_3$] at lower altitudes due to decreased transmission of UV radiation, leading to decreased production of $O_3$ from photolysis of $O_2$). Our model results are insensitive to the initial profile of [$O_3$]. All calculations, unless otherwise indicated, use reaction rates and absorption cross sections from the latest JPL compendium [*DeMore et al.*, 1997] plus the product yields of the ClO+OH reaction reported by *Lipson et al.* [1997].

The advantage of using a constrained photochemical model for these simulations is that this model has been shown to represent the measured abundances of [ClO], [$NO_2$], and [$HO_2$] to within ±20% in the upper stratosphere [e.g., *Chance et al.*, 1996; *Jucks et al.*, 1996; 1998; *Michelsen et al.*, 1996; *Osterman et al.*, 1997; *Sen et al.*, 1998]. We illustrate the measurement/model agreement in Figure 1. These panels show the comparisons between FIRS-2 observations of $HO_x$, $NO_y$, and $Cl_y$ species with calculated values for part of the flight on April 29, 1997 near 68°N. The calculations use the reaction rate constants described above. Full discussion of the top two panels can be found in *Jucks et al.* [1998] and *Jucks et al.* [1999].

Because of this reasonable agreement, the removal rates for $O_3$ by the various catalytic families for the contemporary atmosphere are represented in a realistic manner by this model provided the rate coefficients of the few key reactions that actually limit loss of $O_3$ are well known [e.g., *Jucks et al.*, 1996; *Osterman et al.*, 1997]. While the chemical mechanism within most two-dimensional (2D) models compares favorably to the chemical mechanism within our model, there is considerable variability in the mixing ratio of [$NO_y$] found by 2D models in the 30 to 50 km region [*Park et al.*, 1999]. Consequently, the results of similar perturbation calculations using a 2D model will rely somewhat on the initial distribution of [$NO_y$]. We believe it is instructive to pursue the perturbation calculations using distributions for [$NO_y$] and other radical precursors that are based on observation.

*2.1. Scenarios*

Three scenarios are chosen for this study in addition to the aforementioned 15% reduction in [$Cl_y$], which are summarized in table 1. The first of these sce-

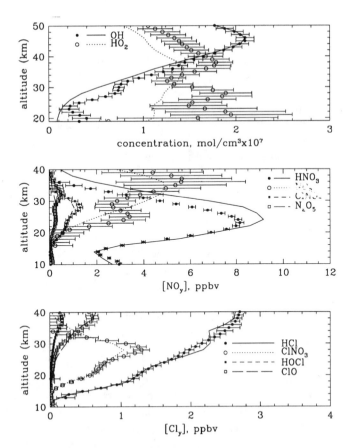

**Figure 1.** An example of the comparisons between stratospheric measurements and calculations of reactive radicals and reservoirs that are measured by the FIRS-2 thermal emission spectrometer of the Smithsonian Astrophysical Observatory. These observations occured near 68°N at 9:30 AM local solar time. A more complete discussion and analysis of the comparisons shown in top two panels are contained in *Jucks et al.* [1998] and *Jucks et al.* [1999].

narios examines the consequences of a 0.1 ppmv increase in [CH$_4$], which is based on extrapolation of the historical mean growth rate of tropospheric [CH$_4$] ( 10 ppbv/yr) over the next decade [e.g., Figure 2-14 of *WMO*, 1999 and *Etheridge et al.*, 1996]. Tropospheric growth rates of CH$_4$ exhibit considerable year-to-year variability, ranging from 0 to 15 ppbv/yr, for reasons that are not completely understood [*WMO*, 1999, Chapter 1]. Recent HALogen Occultation Experiment (HALOE) observations of upper stratospheric [CH$_4$] show a significant decline (10 to 20 ppbv/yr) between 1991 and 1997 [*Evans et al.*, 1998; *Nedoluha et al.*, 1998b; *Randel et al.*, 1999] with important consequences for [ClO] [*Froidevaux et al.*, 1999] and [O$_3$] [*Siskind et al*, 1998]. *Nedoluha et al.* [1998b] hypothesized this change is due to decreased upwelling following the eruption of Mt. Pinatubo. It is important to note

that more recent HALOE observations of upper stratospheric [CH$_4$] indicate a reversal of this trend [*W. Randel et al.*, this volume]. The HALOE observations suggest that changes in circulation can exert a dominant control in the year-to-year variations of upper stratospheric [CH$_4$]. We have chosen the long-term trend for [CH$_4$] driven by the tropospheric source for our base CH$_4$ scenario in Section 3. The variations in [CH$_4$] discussed later in Section 4 are representative of the magnitude of change that may be induced by variations in stratospheric circulation.

The second additional scenario considered is a 1 ppmv increase in [H$_2$O] over the entire mid-latitude profile. This perturbation is based on extrapolation over the next decade of the rise in upper stratospheric H$_2$O observed by HALOE, which ranges from 55 to 150 ppbv/yr [*Evans et al.*, 1998; *Nedoluha et al.*, 1998a; *Randel at al.*, 1999]. It is believed that this change is due primarily to an increase in the abundance of [H$_2$O] for air entering the stratosphere. Upper stratospheric [H$_2$O] exhibits less sensitivity to changes in circulation than does upper stratospheric [CH$_4$] because of differences in the shape of the profiles vs altitude of each species. The stratospheric abundance of total hydrogen [2xCH$_4$+H$_2$O], which can not strongly be affected by transport (slight variations may arise due to H$_2$ [e.g., *Abbas et al.*, 1996b]), shows a significant increase over the same time period. An increase in the abundance of [H$_2$O] for air entering the stratosphere has also been suggested based on the earlier observations of *Oltmans and Hofmann* [1995] and *Abbas et al.* [1996a]. The observed trends in [H$_2$O] could be explained by a rise in temperature of only a few tenths of a degree for air entering the tropical stratosphere, which would be difficult to detect directly [*Nedoluha et al.*, 1998a]. It is important to note that the observed time series of [H$_2$O] exhibits nonlinear variations due to numerous forcings, such as the quasi-biennial and seasonal variations in stratospheric winds [*Evans et al.*, 1998], and appears to have leveled off during 1997 [*Randel et al.*, 1999]. Consequently, our scenario of a 1 ppmv rise in [H$_2$O] over the next decade must be considered at best a conservative upper limit as to the future course of [H$_2$O].

The final additional scenario considered below extrapolates the observed decrease in stratospheric temperature over the next decade. The reduction in stratospheric temperature is based on observations collected during the past 15 years, shown in Figure 5A in *WMO* [1999], based on radiosonde and satellite data sets, which range from 2 to 4 K from the lower to upper stratosphere. However, trends in upper stratospheric temperature from different observations platforms show

Table 1. Scenarios used for possible changes in upper stratospheric Ozone

| Scenario | $\Delta$ Cl$_y$ | $\Delta$ H$_2$O | $\Delta$ CH$_4$ | $\Delta$ Temp. |
|---|---|---|---|---|
| 1 | -15% | 0 ppmv | 0 ppmv | 0 K |
| 2 | -15% | 1 ppmv | 0 ppmv | 0 K |
| 3 | -15% | 0 ppmv | .1 ppmv | 0 K |
| 4 | -15% | 0 ppmv | 0 ppmv | -2 to -4 K |

considerable differences [*WMO*, 1999, Chapter 5]. Also, the cause of the observed cooling of the upper stratosphere is not well understood. While the observed rise in greenhouse gases and the observed depletion of O$_3$ are both predicted to cool the upper stratosphere, the observed cooling appears to be larger than calculated [*WMO*, 1999, Chapter 5]. Some of this difference may be due to changes in [H$_2$O], whose trend in the 1980s is not well defined [*WMO*, 1999, Chapter 5]. We consider the effects on [O$_3$] of an extrapolation of the observed stratospheric cooling because it appears that the majority of the change in temperature is not due to chemical depletion of [O$_3$][*WMO*, 1999, Chapter 5] .

## 3. RESULTS

Figure 2a illustrates the relative contributions to the loss of ozone of the different catalytic cycles, calculated assuming steady state at 35°N, equinox. At 42 km, the calculated contributions from HO$_x$, NO$_x$, and Cl$_x$ cycles are nearly equal, with a significant contribution from the O+O$_3$ reaction. Below and above 42 km, NO$_x$ and HO$_x$ cycles dominate, respectively. The relative contribution to loss of ozone from cycles involving Cl$_x$ has two peaks, near 20 and 42 km. These altitudes correspond roughly to the same regions that exhibit maximum observed declines in the concentration of O$_3$ over the past several decades [*WMO*, 1999, Chapter 4]. Since the altitude range between 40 and 45 km is where the relative contribution of the Cl$_x$ cycles is strongest, it's the ideal place to study the effects of reduced Cl$_y$ and changes in other source gases on future ozone concentrations.

Figure 3 shows the change in calculated O$_3$ for the year 2010 compared to the contemporary atmosphere. The first scenario considered is the anticipated 15% decline in Cl$_y$. As expected, the 15% decrease in [Cl$_y$] results in an increase in ozone at almost all altitudes. The effects of the decrease in [Cl$_y$] are most prominent at 42 km, which is the altitude with the highest relative contribution of Cl$_x$ cycles to photochemical loss of O$_3$ for the contemporary atmosphere (Figure 2a). The calculated increase in ozone at 42 km is roughly 3.5%.

The secondary maximum in the recovery of O$_3$ near 22 km also corresponds to a maximum in the contribution of the Cl$_x$ cycles in the lower stratosphere. Between 24 and 30 km, the changes in [O$_3$] are small because photochemical loss of ozone loss is dominated by NO$_x$ cycles whose precursor, N$_2$O, is not changed in any of these scenarios.

Our theoretical framework, as noted above, is not suited for accurate representation of [O$_3$] trends below about 30 km. The photochemical lifetime of O$_3$ for the simulation of the contemporary atmosphere is shown in Figure 2b. Below about 30 km, the lifetime of O$_3$ exceeds 1 month, and direct transport of O$_3$ becomes important [e.g., *Ko et al.*, 1989]. However, we have chosen to show the results of our O$_3$ recovery calculations for the entire stratosphere because the sensitivity of [O$_3$] at various altitudes to perturbations in H$_2$O, CH$_4$, and

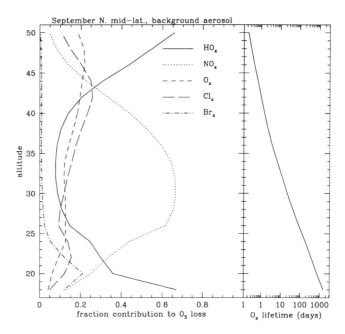

**Figure 2.** The relative contributions to the photochemical loss of ozone from the HO$_x$, NO$_x$, Cl$_x$, Br$_x$, and O+O$_3$ cycles as a function of altitude in stratosphere at mid latitudes in the fall (left panel); the photochemical lifetime of odd oxygen (O+O$_3$) as a function of altitude (right panel).

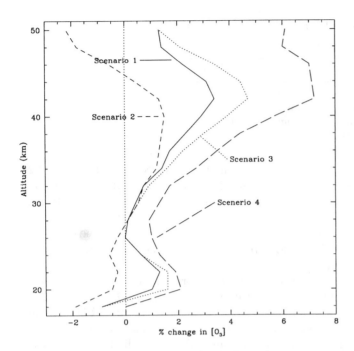

**Figure 3.** The change in photochemical steady state ozone concentrations as a function of altitude for 4 scenarios, one where the constrained $Cl_y$ is reduced by 15%, one where $Cl_y$ is reduced by 15% and $H_2O$ is increased by 1.0 ppmv, one where $Cl_y$ is reduced by 15% and $CH_4$ is increased by 0.2 ppmv, one where $Cl_y$ is reduced by 15% and the temperature changes over the past 15 years is applied to the constrained temperatures.

temperature illustrated in Figure 3 (e.g., lowest sensitivity near 30 km to changes in $H_2O$ and $CH_4$) should be broadly representative of the response found by multidimensional models subject to the same forcings. Our approach ignores any change in ozone in the lower stratosphere from changes in processes that transport ozone or change in aerosol loading. A more complete discussion of the lower stratosphere can be found in Chapter 7 of *WMO* [1999].

The smaller degree of ozone recovery calculated for the 18 to 22 km region compared to the 40 to 44 km region is caused by several factors. The ozone production rate calculated in the lower stratosphere for the year 2010 is lower than the production rate calculated for the contemporary atmosphere because the recovery of $[O_3]$ at higher altitudes in the 2010 simulation leads to less transmission of radiation (due to absorption by $O_3$) in the long wavelength tail of the $O_2$ bands that are responsible for the production of odd oxygen. Also, since we have chosen a profile for sulfate aerosols representative of background conditions, the relative contribution of the $Cl_x$ cycles to loss of $O_3$ is larger near 42 km than at 20 km. Had we used instead a profile for sulfate

aerosol loading based on volcanically perturbed conditions observed following the eruption of Mt. Pinatubo, the contribution to $O_3$ loss from the $Cl_x$ cycles would be larger at 20 km than at 42 km. The sensitivity of the recovery of lower stratospheric $O_3$ to sulfate aerosol loading is not surprising, given the well documented dependence on aerosol loading of the depletion of total column $O_3$ at northern mid-latitudes during the past several decades [e.g., *Solomon et al.*, 1998].

The direct correspondence between the altitudes with the largest increase in ozone and the greatest relative contribution of $Cl_x$ cycles shows that changes in chlorine compounds have little effect on the partitioning of the $HO_x$ or $NO_y$ families. This is illustrated in the top panel of Figure 4, which shows the relative change in the contributions to ozone loss from $Cl_x$, $HO_x$, and $NO_x$ cycles for a 15% reduction in $Cl_y$. The $Cl_x$ cycles are reduced by between 12 and 15% while $O_3$ loss from the $HO_x$ and $NO_x$ cycles increases by less than 3%. These

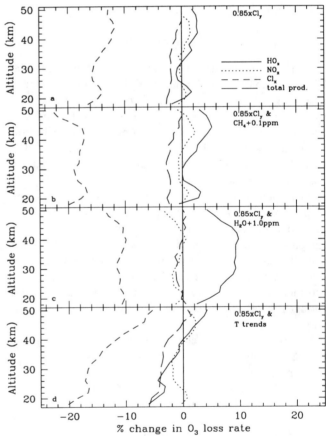

**Figure 4.** The relative rate of change in the $Cl_y$, $NO_y$, and $HO_x$ catalyzed ozone loss cycles for the 4 scenarios in figure 2. Also shown is the change in the ozone production rate.

small increases are caused by an increase in $HO_x$ from the decreased $Cl_y$ (lower $HO_x$ loss from the OH+HCl reaction) and by a small increase in the $NO_x/NO_y$ ratio caused by a decrease in [ClO] (due to the ClO+NO reaction). The reduction in the production rate of $O_3$, caused by the increase in [$O_3$] at higher altitudes, is also illustrated in Figure 4.

We now turn our attention to investigating the effects on [$O_3$] for the year 2010 of the continuation of recently observed trends in [$H_2O$], [$CH_4$], and temperature, in addition to the 15% decrease in [$Cl_y$]. An increase in [$CH_4$] by 0.1 ppmv adds to the increase in ozone caused by reduced [$Cl_y$], as illustrated in Figure 3. This is due primarily to the change in $Cl_y$ partitioning caused by the increase in [$CH_4$]. The reaction $Cl+CH_4$ is the primary pathway for production of HCl. An increase in [$CH_4$] will shift the $Cl_y$ partitioning away from reactive chlorine, Cl and ClO, and toward the reservoir species HCl. This is illustrated in Figure 4b which shows that the decrease in $O_3$ loss by the $Cl_x$ cycles is larger than for the baseline simulation with only changing [$Cl_y$]. Such effects have been described recently to explain the response of [ClO] observations by the Microwave Limb Sounder (MLS) instrument on board the Upper Atmosphere Research Satellite (UARS) to changes in [$CH_4$] observed by HALOE, also on UARS [Siskind et al., 1998; Froidevaux et al., 1999]. In this simulation, the effects on upper stratospheric $O_3$ of the decrease in [ClO] resulting from increased [$CH_4$] more than offset the effects of increased [$HO_x$]. Increased [$HO_x$] and decreased [ClO] for this simulation also have the subtle effect of raising the $NO_x/NO_y$ ratio, which leads to greater $O_3$ loss by the $NO_x$ cycle in the middle and upper stratosphere. These effects are illustrated in Figure 4b, where the changes in ozone loss by the $HO_x$ and $NO_x$ cycles increase slightly relative to the changes found for the [$Cl_y$] only perturbation (Figure 4a).

An increase in [$H_2O$] of 1 ppmv, together with the 15% decrease in [$Cl_y$], reduces the increase in [$O_3$] at 40 km by about 50% and completely negates the increases in [$O_3$] above 45 km. The changes in [$O_3$] are a direct effect of the increase in concentrations of $HO_x$ resulting from elevated [$H_2O$], which is the primary source of $HO_x$ in the upper stratosphere. This is illustrated in Figure 4c, which shows a nearly 10% increase in the $HO_x$ driven $O_3$ loss cycles compared to the baseline simulation. Similar decreases in the recovery of ozone are seen in the lower stratosphere, where $HO_x$ cycles are the main contributors to photochemical loss of ozone. Increased [$HO_x$] also has indirect effects on [$O_3$] because OH plays an important role in the partitioning between HCl and free chlorine and the partitioning be-

tween $NO_x$ and $NO_y$. These effects are also shown in Figure 4c. The reaction OH+HCl is the primary pathway for conversion of HCl back to free chlorine. Increases in [OH] will raise the calculated concentration of ClO and hence increase the contributions of the $Cl_x$ cycles to ozone loss. The $NO_x/NO_y$ ratio is also higher in this simulation compared to the baseline case, due also to higher abundances of OH.

The measured trends in temperature over the past 15 years show a significant decrease in the upper stratosphere, on the order of 2-4 K/decade [WMO, 1999, Chapter 5]. The temperature trends for the scenario considered here were taken from Figure 5A of the Scientific Assessment of Ozone Depletion report [WMO, 1999], and projected over the next decade. The reduction in stratospheric temperature results in an increase in [$O_3$], as illustrated in Figure 3. The main effect of the decreased temperature in our simulation is the change in gas phase reaction rate constants. We do not take into account the effects of changes in transport, which is heavily inter-dependent on temperature (temperature has a minor effect on transmission and photolysis rates, which is included in the model). For all altitudes, the perturbed temperature profile is still too warm for heterogeneous chemical reactions to be important. Of course, greenhouse induced stratospheric cooling could dramatically increase the depletion of polar ozone in the lower stratosphere due to heterogeneous chemistry [e.g, Shindell et al., 1998].

In the upper stratosphere, all of the rate limiting steps for the catalytic cycles that destroy ozone involve the reaction of a radical with atomic oxygen, which is in photochemical equilibrium with ozone. One reaction, $O+O_3$, has a large positive temperature dependence and slows considerably as temperatures decrease, increasing concentrations of ozone. All of the other rate limiting reactions have a small negative temperature dependence; thus, one might expect these reactions to become faster as temperature decreases, causing a decrease in calculated steady-state ozone. However, the reaction that converts atomic oxygen back to ozone, $O+O_2+M \rightarrow O_3+M$ is very temperature dependent and speeds up considerably as temperature decreases [DeMore et al., 1997]. Speeding up this reaction shifts the equilibrium between O and $O_3$ toward $O_3$, lowering the calculated concentration of atomic oxygen, and thus slowing down the rate of every limiting reaction for destruction of upper stratospheric odd oxygen. Other reactions controlling the concentrations of the reactive radicals involved in the rate limiting steps for loss of ozone are also temperature sensitive. The rate of OH+$HO_2$, an important $HO_x$ sink, becomes faster as temperature decreases which leads to slower ozone loss due to $HO_x$. Similarly, the reaction $OH + NO_2$ speeds up, lowering the $NO_x$ concentrations and the $NO_x$ catalyzed ozone loss reaction rates.

Another factor that contributes to variations in ozone in the upper stratosphere is the level of solar UV irradiance, which has been observed to fluctuate as a result of the 27-day period of solar rotation and the 11-year variation in the frequency of solar sun spots and faculae (commonly referred to as the 11-year solar cycle). We have not varied solar irradiance in this study. Two previous studies that have evaluated the effects of variations in solar irradiance during the declining phase of the 11 year solar cycle reported a decrease in $[O_3]$ on the order of 0.5%/year [*Chandra et al.*, 1996; *Siskind et al.*, 1998]. The effects of the 27-day variation in solar irradiance on stratospheric $O_3$ are small, on the order of 0.2 to 0.3% [*Hood and Cantrell*, 1988]. To properly quantify the cause of changes in ozone over any specific time period the variation of solar irradiance must of course be considered. However, we have not included these variations in our study since they are believed to be cyclical, and thus do not drive either long-term depletion or recovery of $[O_3]$ over decadal time periods [e.g., *Jackman et al.*, 1996; *Lean et al.*, 1997].

## 4. EFFECTS OF TRANSPORT

All of the above changes in source gas concentrations and temperature are based on past changes in stratospheric or tropospheric values. However, the actual changes that will occur over the next 15 to 20 years in $[CH_4]$, $[H_2O]$, and temperature may depend strongly on changes in stratospheric transport processes. Stratospheric transport could be affected by climate change due to increased concentrations of greenhouse gases [*Hood et al.*, 1997; *Steinbrecht et al.*, 1998; *Fusco and Salby*, 1999]. The effect of tropospheric climate change on stratospheric circulation is not well understood in a quantitative manner [*WMO*, 1999, Chapter 5]. While changes in transport will not directly affect ozone levels in the upper stratosphere, these changes could have a strong indirect effect by changing the distribution of source gases that ultimately lead to the photochemical loss of ozone. The mixing ratio of $H_2O$ that enters the stratosphere is strongly regulated by temperatures in the tropical upper troposphere [e.g. *Holton et al.*, 1995]. Temperatures in this region are controlled by the magnitude of convective events and the balance between heating and cooling rates [*Holton et al.*, 1995].

The abundance of upper stratospheric $CH_4$ is not only dependent on its tropospheric source mixing ratio, but are also controlled by transport processes that affect the mean age of the stratospheric air at any altitude and latitude [*Nedoluha et al.*, 1998b]. The magnitude of this transport is driven by the so-called "wave driven pump" at mid latitudes, where the breaking of gravity and planetary waves drives both uplift in the tropics and horizontal mixing at mid latitudes [*Holton et al.*, 1995]. A weakening of wave activity will result in slower transport and an increase in the age of air in the upper stratosphere. This will result in lower concentrations of $CH_4$ because it will have more time to be oxidized to $H_2O$. Because $CH_4$ is converted to $H_2O$, decreases in wave activity will also be reflected by an increase in $[H_2O]$. Increases in wave activity will have the effect of increasing concentrations of $CH_4$ and decreasing concentrations of $H_2O$. Various general circulation model simulations suggest a substantially different manner of changes in the characteristics of the planetary wave activity due to rising concentrations of greenhouse gases [*WMO*, 1999, Chapter 5] (most of these calculations are for a "doubled $CO_2$" environment). It is not possible, therefore, to discern how upper stratospheric $[CH_4]$ will respond to tropospheric climate forcings.

Temperatures in the upper stratosphere are driven by both heating from UV absorption by ozone, long wave cooling by $CO_2$, $H_2O$, and $O_3$, by the solar input, and by transport [*Forster and Shine*, 1999; *Garcia and Boville*, 1994]. Thus changes in temperature are coupled to both future variations in ozone and changes in stratospheric transport. Changes in reaction rates from a decrease in temperature will tend to enhance the increases in ozone concentrations from a decrease in $Cl_y$ (see Figure 3). More ozone will result in higher heating rates from UV absorption. This increase in temperature will speed up the rates of the photochemical loss cycles, thus somewhat mitigating the effects of reduced reactive chlorine concentrations. However, as mentioned above, the observed decline in upper stratospheric temperature over the past several decades is believed to be considerably larger than the amount that can be attributed to just the radiative effects of ozone. If upper stratospheric cooling is indeed related to tropospheric climate change, this forcing will tend to enhance the recovery of $[O_3]$.

The future change to stratospheric circulation can not currently be estimated in a reliable manner [*WMO*, 1999, Chapter 5]. The role of increased concentrations of greenhouse gases on stratospheric transport is a subject of vigorous current scientific research. It would be desirable to treat future variations in the long-lived precursors in a self consistent manner within a chemical-transport model. For example, speeding up the stratospheric circulation should lead to a well-defined rela-

Table 2. Changes needed to negate/double a 15% reduction in $[Cl_y]$ on $[O_3]$ at 40 km

| Parameter | Negate | Double |
|-----------|--------|--------|
| $H_2O$ | +2.2 ppmv | -1.7 ppmv |
| $CH_4$ | -0.23 ppmv | +0.21 ppmv |
| Temperature | +2.0 K | -1.8 K |
| $NO_y$ | +1.8 ppbv | -1.7 ppbv |

tionship between changes in $[H_2O]$, $[CH_4]$, $[NO_y]$, and temperature in the upper stratosphere ($[Cl_y]$ near 40 km is unlikely to be strongly affected by stratospheric circulation, since the organic source molecules have nearly completely decomposed at this altitude [e.g., *Zander et al.*, 1996] and $Cl_y$ does not have a high altitude sink). It is questionable, however, whether the results of such a calculation would be meaningful given the facts that: a) stratospheric $[H_2O]$ is controlled not only by the large scale transport field but also by small-scale thermodynamics at the tropopause that are certainly subject to possible climatic influence and difficult to simulate in global models; b) upper stratospheric $[NO_y]$ is quite sensitive to variations in mesospheric temperature [e.g., *Rosenfield and Douglass*, 1999]; c) future variations in stratospheric $[CH_4]$ may be affected by changes in its tropospheric source that are difficult to predict.

We will therefore proceed by determining the amount of change needed in $[H_2O]$, $[CH_4]$, $[NO_y]$, and temperature to both double and offset the increase in upper stratospheric ozone caused by the expected future 15% decrease in $[Cl_y]$ by the year 2010. The results of these calculations are shown in Table 2. We stress that these calculations are meant to serve as a guide to the magnitude of the change for each quantity that "is important" using the effect on $O_3$ of the 15% decrease in $[Cl_y]$ as a metric. An approximately 2.0 ppmv increase in $[H_2O]$ is needed to negate the recovery in $[O_3]$ at 40 km caused by a 15% reduction in $[Cl_y]$. This is a roughly 30% change in the current mixing ratio of $H_2O$. The change in $[H_2O]$ observed by both HALOE and ground based observations at 40 km over the time period 1992 to 1997 is 0.7 ppmv [*Nedoluha et al.*, 1998a]. While a 2.0 ppmv increase in stratospheric $[H_2O]$ seems large, it corresponds to only a few degrees change in the saturation temperature at the locations of troposphere to stratosphere exchange. A 0.2 ppmv decrease in $[CH_4]$ will negate the recovery in $[O_3]$ at 40 km. In the upper stratosphere this corresponds to a roughly 20-30% decrease in $[CH_4]$. While this may seem like a large change, it is of similar magnitude to the changes observed in upper stratospheric $[CH_4]$ over the 1990s by HALOE [*Froidevaux et al.*, 1999]. A temperature increase of 2 K is

necessary to negate the recovery of upper stratospheric $[O_3]$. There is no indication that upper stratospheric temperature is likely to rise by this much over the next decade; indeed, a temperature decrease of about 2 K is plausible based on an extrapolation of present trends. In this case, the temperature forcing could strongly enhance the recovery of ozone due to declining levels of $Cl_y$.

The magnitude of the changes in $[H_2O]$, $[CH_4]$, and temperature given in Table 2 are all within the range of plausible variations that may occur over the next several decades. We lack an observational basis to evaluate the context of the variations in $[NO_y]$ given in Table 2, but we note that this change is on the order of that calculated in a doubled $CO_2$ atmosphere [*Rosenfield and Douglass*, 1998]. These calculations are meant to underscore the critical need for continued, accurate monitoring of changes in $[H_2O]$, $[CH_4]$, $[NO_y]$, and temperature over the next several decades to quantify the relation between upper stratospheric $[O_3]$, $[Cl_y]$, and $[ClO]$.

## 5. KINETICS UNCERTAINTIES

The rate coefficients of many of the reactions that govern loss of ozone in the upper stratosphere have significant uncertainties. These uncertainties must be resolved to better understand ozone photochemistry in the upper stratosphere. These uncertainties lie in the reactions that control the partitioning between reactive (ClO and Cl) and reservoir (HCl) chlorine species, reactions that control the concentrations of $HO_x$, and reactions that affect the $NO_x/NO_y$ ratio and ozone loss by the $NO_x$ cycle.

A number of recent observations of $Cl_y$ species in the upper stratosphere have been shown to be consistent with models that include a roughly 5-7% branching of the ClO+OH reaction to form HCl [e.g., *Chance et al.*, 1996; *Michelson et al.*, 1996]. Recent laboratory kinetics data also show a branching on the order of 5-7% [*Lipson et al.*, 1997]. Such an HCl production channel, which is included in all the model calculations shown above, acts to reduce the contribution to ozone loss by the $Cl_x$ cycles compared to calculations based on the currently recommended set of reaction rates. However, a new laboratory kinetics study [*Kegley-Owens et al,* 1999] obtains a much faster overall rate constant for the OH+ClO reaction than that found by *Lipson et al* [1997]. If the 5-7% branching ratio can be applied to this new rate, then all the model calculations shown here will require significant revision. Under this scenario, models would significantly underestimate the observed ClO/HCl ratio in the upper stratosphere and we would be searching for either missing sinks of HCl in our current model or the reason for an overestimation of the source of HCl in our model (e.g., possible misrepresentation of HCl production by Cl+CH$_4$ reaction, either due to uncertainties in the rate coefficient or inaccuracies in calculated [Cl]). These two laboratory measurements of the rate of ClO+OH must be reconciled to better understand the partitioning of Cl$_y$ in the upper stratosphere, the relative contribution of Cl$_x$ cycles to photochemical loss of ozone, and the expected future recovery of upper stratospheric [O$_3$].

The concentrations of HO$_x$ species in the upper stratosphere significantly affect the photochemical loss rates of ozone in a number of fashions. First, HO$_x$ cycles are a significant contributor to loss of ozone, especially above 42 km. Also, OH is the main oxidant for HCl, and thus plays a significant role in determining the ClO/HCl ratio. Third, the reaction OH+NO$_2$ is the primary source of HNO$_3$ in the stratosphere. As a result, OH plays a significant role in determining the partitioning of NO$_x$/NO$_y$. A number of observations of abundances of OH in the upper stratosphere/mesosphere give conflicting comparisons with calculated concentrations in photochemical models [*Summers et al.,* 1997; *Jucks et al.,* 1998; *Burnett and Minschwaner,* 1998]. Observations of [OH] above 50 km from the MAHRSI satellite instrument show values that are roughly 25% lower than calculated [*Summers et al.,* 1997]. The MAHRSI data have been interpreted in one of two ways; either there is a nearly 50% discrepancy for the rates of the reactions that affect the partitioning between OH and HO$_2$ (the reaction O+HO$_2$ was chosen) or there is a significant underestimation in the loss rate of HO$_x$ (OH+HO$_2$). However, further investigation of this data set shows that the OH concentrations between 40 and 50 km are slightly larger than the model calculations with a transition near 50 km [*Summers and Conway,* this mongraph]. Ground based observations of column OH, which are dominated by contributions from the upper stratosphere and lower mesosphere, show columns that are roughly 25% higher than calculated, especially at low solar zenith angles near noontime [*Burnett and Minschwaner,* 1998]. Interpreting these data sets to determine the cause of the discrepancies is difficult because they either lack sufficient altitude resolution or coverage, they lack simultaneous observations of HO$_2$, and there are fundamental differences in the sense of the discrepancy (e.g., "too much" [OH] in the model of *Summers et al.;* "too little" [OH] in the model of *Burnett and Minschwaner*).

Observations of [OH], [HO$_2$], [H$_2$O], and [O$_3$] have been obtained between 20 and 50 km from balloon platforms by the FIRS-2 spectrometer [*Chance et al.,* 1996; *Jucks et al.,* 1998]. In the upper stratosphere, between 30 and 50 km, the observed concentrations of OH are found to be in statistical agreement with calculated values while observations of [HO$_2$] are roughly 25% higher than calculated. The FIRS-2 observations are consistent with both a 25% discrepancy in one or a combination of the reactions affecting HO$_x$ partitioning above 35 km (either O+OH or O+HO$_2$) and a 25% discrepancy in one or a combination of the reactions that affect production and loss of HO$_x$ (either OH+HO$_2$ or O($^1$D)+H$_2$O, the only significant loss and production reactions above 35 km) [*Jucks et al.,* 1998]. However, these observations were insufficient to pinpoint which of the above reactions was most likely the cause of the discrepancy. Furthermore, the FIRS-2 observations are not sensitive to [OH] above 50 km and cannot currently be directly compared to the MAHRSI observations of [OH] above 50 km, the altitudes where the observations are significantly lower than the calculations. The FIRS-2 observations are consistent with the MAHRSI observations between 40 and 50 km. Changing any individual rate or combination of reaction rates cannot account for the comparisons between measurements and models over the altitude range from the middle stratosphere to the mesosphere. There appears to be missing chemistry, possibly in the mesosphere [*Summers and Conway,* this mongraph]. More HO$_x$ observations such as these from different latitudes and a full range of altitudes are needed to further resolve these ambiguities. Reduced uncertainties for the rates of the HO$_x$ reactions mentioned above based on additional laboratory studies are also necessary. Even so, the scenarios needed to explain the FIRS-2 observations of HO$_x$ do not adequately represent other observations of [OH] in the stratosphere.

Recent observations of [NO$_2$] and [HNO$_3$] in the middle stratosphere suggest large uncertainties in our quantification of NO$_x$/NO$_y$ partitioning [*Jucks et al.,* 1999; *Osterman et al.,* 1999]. These observations show higher [NO$_2$] and lower [HNO$_3$] than calculated by the models. For the FIRS-2 observations, this discrepancy even holds with models that include new laboratory rates for the reactions OH+NO$_2$ [*Dransfield et al.*, 1999; *Brown et al.*, 1999a] and OH+HNO$_3$ [*Brown et al.*, 1999b], both of which will increase the modeled NO$_x$/NO$_y$ [*Jucks et al.*, 1999]. A similar discrepancy in our understanding of NO$_x$/NO$_y$ partitioning has also been noted for in situ observations in the lower stratosphere, although in this case use of the new laboratory rates for the above reactions resolves the discrepancy to within 10% [*Gao et al.*, 1999]. The cause of the remaining discrepancy suggested by the FIRS-2 observations is not evident, but could possibly be due to an error in the calculated photolysis rate of HNO$_3$. This discrepancy suggests that the effects of decreased [Cl$_y$] on [O$_3$] may be lower than expected in the middle stratosphere. This chemistry will have little effect above 35 km, because at these altitudes most of NO$_y$ is in the form of NO$_x$. A new laboratory observation of the rate constant for the reaction of NO$_2$+O, the rate limiting reaction of the NO$_x$ catalyzed loss of ozone, has also been measured to be 5 to 10% faster than the currently recommended value [*Gierczak et al.*, 1999]. Including this reaction rate will increase the relative contribution of the NO$_x$ cycles to loss of [O$_3$] and lower the calculated recover of [O$_3$] from decreased [Cl$_y$].

The different kinetic scenarios needed to explain each of the above observations of reactive radicals will have varying effects on the calculated recovery of upper stratospheric ozone. These effects are quantified in Figure 5, where the calculated change in ozone for a 15% decrease in [Cl$_y$] is determined from models using different measured or proposed reaction rates that affect the radical concentrations. The 4 different models use the following sets of rates: the JPL97 reaction rates [*DeMore et al.*, 1997], the JPL97 rates with the *Lipson et al.* [1997] yields for the two product channels of the ClO+OH reaction, the above model plus the new laboratory rates for the OH+NO$_2$ [*Dransfield et al.*, 1999], OH+HNO$_3$ [*Brown et al.*, 1999b], and O+NO$_2$ *Gierczak et al.*, 1999 reactions, and finally a model with all the above changes and a 50% reduction in the rate of O+HO$_2$ that was one of the scenarios suggested by *Summers et al.* [1997] to account for the MAHRSI observations of mesospheric [OH]. In the upper stratosphere, only two reaction rate changes significantly affect the calculated recovery of ozone compared to the JPL97 calculation: the branching of the ClO+OH reaction and the change in the rate of O+HO$_2$. These are the only two reactions considered that significantly change the relative contribution of ozone loss attributed to Cl$_y$ as shown in figure 2. Even these effects are sub-

tle (changes in the recovery of [O$_3$] of less than 20%), less than the changes calculated from plausible future variations in the abundances of [CH$_4$], [H$_2$O], and temperature.

We note that one of the reactions chosen, the 50% reduction in the rate of O+HO$_2$, was chosen arbitrarily as it was one scheme to explain the MAHRSI OH above 50 km. However, the balloon [*Jucks et al.*, 1998; and MAHRSI data [*Summers and Conway*, this monograph] between 40 and 50 km are not consistent with such a model. There appears to be missing HO$_x$ chemistry in the mesosphere [*Summers and Conway*, this monograph] which could yield a different curve than that shown with the reduced O+HO$_2$ rate. It most likely will have its largest effects above 50 km and not at 40 km where the effects of Chlorine mediated loss of ozone has it largest relative impact. This could potentially explain the difficulties in simulating the HALOE ozone trends above 45 km from trends in Cl$_y$, H$_2$O, and CH$_4$ during the 1990s by *Siskind et al.* [1998]. While it is important to resolve the kinetic uncertainties that affect the abundances of [ClO], [OH], [HO$_2$], and [NO$_2$], the model calculations shown in Figure 5 suggest the largest uncertainties in quantifying the future recovery of upper stratospheric [O$_3$] as a result of decreasing chlorine may still lie in the future course of the concentration of radical precursors.

One further complication to the recovery of ozone in the upper stratosphere is the issue of whether the calculated production and loss rates of ozone are in balance above 40 km. *Jucks et al.* [1998] and *Osterman et al.* [1997], both of which used the same photochemical model described here, reported a significant imbalance between production and loss of ozone (loss exceeded production) that increased with rising altitude above 40 km. These studies analyzed balloon-borne observations obtained by the FIRS-2 and MkIV spectrometers, respectively. However, *Crutzen et al.* [1997] and *Groos et al.* [1998] reported close balance between production and loss of [O$_3$] at all altitudes between 35 and 55 km. These studies also used the same photochemical model, which of course is a different model than the one used here, and analyzed data obtained by instruments aboard UARS. It is important that this discrepancy be resolved: it may be due to differences in the treatment of photolysis of O$_2$, some other unaccounted for model difference, or a fundamental difference between the balloon and satellite observations. If the calculated imbalance between production and loss of ozone is correct, this discrepancy indicates that either the production rate of O$_3$ is underestimated or the loss rate is

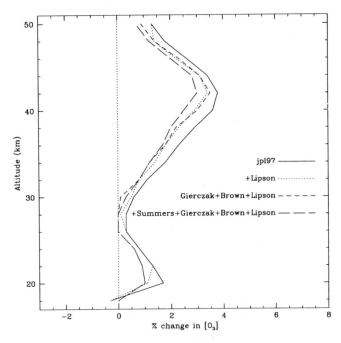

**Figure 5.** The changes in the photochemical steady ozone concentrations for the scenario with a 15% reduction in $Cl_y$ using models with varying reaction rates for key reactions. The base JPL97 calculation uses no production of HCl from ClO+OH. The next calculation used the *Lipson et al.* [1997] formulation for HCl production from that reaction, the next calculation has further changes in the new rates for OH+NO$_2$ and OH+HNO$_3$ [*Brown et al.*, 1999a; 1999b] and the measured rate for O+NO$_2$ [*Gierzak et al.*, 1999], and the final calculation includes the proposed 50% reduction of O+HO$_2$ from *Summers et al.* [1997].

overestimated. Barring "missing chemistry", any uncertainty in the production rate of ozone would mostly result from errors in the radiative transfer near 200 nm, which should have only a minor effect on the relative contributions of the different cycles contributing to loss of ozone. Above 45 km, where the discrepancy is most apparent, HO$_x$ cycles are the dominant contributor to the photochemical loss rates of ozone. If the errors are in the rate constants that affect either concentrations of HO$_2$ or the rate limiting reaction O+HO$_2$, they may affect the relative contribution between Cl$_x$ and HO$_x$ cycles near 42 km.

The other main uncertainty in the upper stratosphere is the actual concentration of atomic oxygen in both the ground, O($^3$P), and excited, O($^1$D), electronic states. The concentrations of these species control the concentrations of HO$_x$, the loss rates of ozone, and the atmospheric lifetimes of long lived species. Neither concentration has ever been measured to sufficient precision or accuracy in the stratosphere to test photochemical

model calculations. Processes that affect these concentrations are the photolysis rates of O$_2$ and O$_3$, the collisional deactivation of O($^1$D), and the O+O$_2$ recombination reaction. Future observations of the stratospheric abundances of O($^3$P) and O($^1$D), either directly or by proxy, are required to establish whether model representation of ozone loss rates and HO$_x$ production rates are truly accurate.

## 6. CONCLUSIONS

We have shown that plausible future variations in the abundances of [CH$_4$], [H$_2$O], and temperature can have comparable effects on the level of upper stratospheric ozone as the forcing due to the expected 15% decline in [Cl$_y$] over the next decade. Upper stratospheric ozone is also sensitive to future variations in [NO$_y$], although we lack an observational basis to define how [NO$_y$] might vary. For example, the "recovery" of [O$_3$] at 42 km due to a 15% decline in [Cl$_y$] would be negated by either a 2.2 ppmv rise in [H$_2$O], a 0.23 ppmv drop in [CH$_4$], a 2 K rise in temperature, or a 1.8 ppbv rise in [NO$_y$]. The recovery of upper stratospheric [O$_3$] due to declining [Cl$_y$] would be enhanced by either a decrease in [H$_2$O], a rise in [CH$_4$], a drop in temperature, or a decline in [NO$_y$]. The calculations presented in our study demonstrate that, if the changes in ozone over the next 10 to 20 years are to be explained in context with the expected decline in chlorine loading, then future changes in the abundances of CH$_4$, H$_2$O, NO$_y$, and temperature must be also quantified.

It is important to note that the large (e.g., 7%/decade) depletion of upper stratospheric [O$_3$] observed over the past several decades was almost certainly caused by the build-up of [Cl$_y$], due to industrial activity, over that time period. Although past variations in [CH$_4$], [H$_2$O], [NO$_y$], and temperature would have played a role in determining past changes in upper stratospheric [O$_3$], these changes were almost certainly smaller than the changes in [O$_3$] induced by the build-up of [Cl$_y$]. Observations of [CH$_4$] obtained by ATMOS at northern midlatitudes in 1985 and 1994 reveal approximately similar levels (e.g., Figure 5 of *Gunson et al.* [1990] and Figure 2 of *Abbas et al.* [1996b]). The increase in [H$_2$O] reported by *Oltmans and Hofmann* [1995] over the time period 1981 and 1995 at 24 to 26 km, the highest altitude bin of their study, is only 0.34 ± 0.34 ppmv. A 0.34 ppmv rise in [H$_2$O] is too small, by at least an order of magnitude, to have caused the observed drop in upper stratospheric [O$_3$] between 1981 and 1995. The observed cooling of the upper stratosphere over the same time period is expected, on theoretical grounds, to have led to a slight rise in $[O_3]$ [e.g., Figure 6-21 of *WMO*, 1999]. We lack direct observational evidence of changes in $[NO_y]$ over this time period. However, the rise in the tropospheric abundance of $[N_2O]$ is too small to for changes in $[NO_y]$ due to variations in its source to have had an appreciable effect on $[O_3]$ [e.g., *Jackman et al.*, 1996; *Nevison and Holland*, 1997]. Final evidence for the causal relationship between the depletion of upper stratospheric ozone and the build-up of anthropogenic chlorine is provided by the close agreement between the shape of the observed and calculated profiles of the trend in $[O_3]$. Both curves peak at 40 km [e.g., Figures 6-29, 6-20, and 6-21 of *WMO*, 1999], the altitude at which the contribution of chlorine to the loss of ozone maximizes. Had variations in $[NO_y]$ been responsible for significant changes in $[O_3]$ over the past several decades, we would have expected the profiles of $[O_3]$ depletion to peak near 30 km, which is not observed.

Since 1992, $[CH_4]$ and $[H_2O]$ have been measured by HALOE on board UARS. Future observations of these two quantities will be made by the High-Resolution Dynamics Limb Sounder (HIRDLS) instrument on the Earth Observing Satellite (EOS) Chem platform, scheduled to be launched in 2002, the Mid Infrared Passive Atmospheric Sounder (MIPAS) instrument on board the European ENVISAT satellite, and the ILAS-II instrument on board the Japanese ADEOS-II satellite. However, there is no guarantee of overlap between HALOE observations and measurements from future instruments. If the HALOE instrument is not operational at the time these future instruments first obtain measurements then validation of the retrievals of $[H_2O]$ and $[CH_4]$ from the future satellite instruments with observations from the instruments used to validate HALOE retrievals is absolutely required to combine the data sets for analysis of long term changes. Such measurements would also help to resolve the up to 20% discrepancies in observed $[H_2O]$ between different satellite and in situ observations [e.g. *Mote et al.*, 1996; *Rosenlof et al.*, 1997]. These biases must be understood in order to quantify future changes in $[H_2O]$.

Past observations of $[NO_y]$ in the upper stratosphere have incomplete time coverage. Most measurements of $[NO_y]$ come from ATMOS [e.g., *Rinsland et al.*, 1996] and balloon [e.g., *Osterman et al.*, 1999; *Jucks et al.*, 1999] remote sensing observations of the individual constituents and in situ balloon observations of total $NO_y$ [*Kondo et al.*, 1996]. HIRDLS has the capability to measure all of the individual constituents of $NO_y$. These observations will need to be validated

against the instruments that measured the past $[NO_y]$ profiles in order to begin to understand the begin to understand the temporal evolution of $[NO_y]$.

Measurements of the long term changes of $[O_3]$ in the upper stratosphere come mostly from HALOE (since 1992) [*Siskind et al.*, 1998], SBUV data sets, SAGE I/II, and Umkehr data [*WMO*, 1998]. Future observations of the vertical profile of $[O_3]$ are expected from the EOS Chem, ENVISAT and ADEOS-II satellites. ADEOS-II will only be making observations near high latitudes and will not get global coverage. The accuracy of future observations of $[O_3]$ will need to be understood to within a few percent or better in order to combine these data with past measurements and better quantify long term changes in ozone. Most of the changes in $[O_3]$ calculated in the present study are quite small, about half the changes in ozone near 40 km in the past 10 years in the upper stratosphere. It will be a particular challenge to define such small changes in upper stratospheric $[O_3]$ if it is necessary to combine data from different instruments, as will be the case in the unfortunate event that HALOE and/or SAGE II are not operational when data from these future satellite instruments are first obtained. Again, a concerted effort to "validate" measurements of $[O_3]$ by HALOE and SAGE II at the end of the life cycle of these instruments, using balloon-borne instruments that could also be used for comparison to the future satellite observations, would provide a quantitative basis for "tying together" data from the various platforms.

Many of the instruments used to validate the UARS and SAGE II observations have undergone significant improvements in the accuracy and precision of the measurements they can obtain [e.g. *Jucks et al.*, 1998; *Osterman et al.*, 1999] since the time of those validation campaigns [e.g. *Gille et al.*, 1996]. Further balloon campaigns designed both to validate the UARS and SAGE II instruments near the end of their lifetimes and to further address some of the above mentioned unresolved photochemical issues in the middle and upper stratosphere would be extremely important for quantifying future changes in upper stratospheric $[O_3]$ as well as pinpointing the factors responsible for variations in $[O_3]$. These balloon instruments would provide a critical baseline for connecting UARS and SAGE II observations to those of EOS Chem and ENVISAT if these same instruments are used to then validate the future satellite observations. Understanding the future recovery of upper stratospheric $[O_3]$ is dependent on the design and execution of a plan to assure that the future observations of $[O_3]$, $[CH_4]$, $[H_2O]$, and $[NO_y]$ can be combined with contemporary measurements in a manner that is meaningful for discerning trends.

*Acknowledgments.* We thank Dave Siskind for some useful comments to this manuscript. The work at SAO was supported by NASA grant NSG–5175. Research at the Jet Propulsion Laboratory, California Institute of Technology, was performed under contract with NASA.

## REFERENCES

Abbas, M. M., et al., Seasonal variations of water vapor in the lower stratosphere inferred from ATMOS/ATLAS-3 measurements of $H_2O$ and $CH_4$, *Geophys. Res. Lett., 23*, 2401-2404, 1996a.

Abbas, M. M., et al., The hydrogen budget of the stratosphere inferred from ATMOS measurements of $H_2O$ and $CH_4$, *Geophys. Res. Lett., 23*, 2405-2408, 1996b.

Brown, S. S., R. K. Talukdar, and A. R. Ravishankara, Rate constants for the reaction $OH+NO_2+M \rightarrow HNO_3+M$ under atmospheric conditions, *Chem. Phys. Lett., 299*, 277-284, 1999a.

Brown, S. S., R. K. Talukdar, and A. R. Ravishankara, Reconsideration of the rate constant for the reaction of hydroxyl radicals with nitric acid, *J. Phys. Chem. A, 103*, 3031-3037, 1999b.

Burnett, C. R., and K. Minschwaner, continuing development in the regime of decreased atmospheric OH at Fritz Peak, Colorado, *Geophys. Res. Lett., 25*, 1313-1316, 1998.

Chance, K. V., et al., Simultaneous measurements of stratospheric $HO_x$, $NO_x$, and $Cl_x$: Comparison with a photochemical model, *J. Geophys. Res., 101*, 9031, 1996.

Chandra, S., et al., $O_3$ variability in the upper stratosphere during the declining phase of the solar cycle 22, *Geophys. Res. Lett., 23*, 2935-2938, 1996.

Crutzen, P. J., Mesospheric mysteries, *Science, 277*, 1951-1952, 1997.

DeMore, W. B., et al., Chemical kinetics and photochemical data for use in stratospheric modeling: Evaluation number 12, *JPL Pub. 97-4*, 1997.

Dransfield, T. J., K. K. Perkins, N. M. Donahue, J. G. Anderson, M. M. Sprengnether, and K. L. Demerjian, Temperature and pressure dependent kinetics of the gas-phase reaction of the hydroxyl radical with nitrogen dioxide, *Geophys. Res. Lett., 26*, 687-690, 1999.

Etheridge, D. M., P. P. Steele, R. L. Langenfelds, R. J. Francey, and J.-M. Barnola, Natural and anthropogenic changes in atmospheric $CO_2$ over the last 1000 years from air in Antarctic ice and firn, *J. Geophys. Res., 101*, 4115-4128, 1996.

Evans, S. J., R. Toumi, J. E. Harries, M. P. Chipperfield, and J. M. Russell III, Trends in stratospheric humidity and the sensitivity of ozone to these trends, *J. Geophys. Res., 103*, 8715-8725, 1998.

Forster, P. M. de f., and K. P. Shine, Stratospheric water vapour changes as a possible contributor to observed stratospheric cooling, *Geophys. Res. Lett., 26*, 3309-3312, 1999.

Froidevaux, L., J. W. Waters, W. G. Read, P. S. Connell, D. E. Kinnison, and J. M. Russell III, Variations in the free chlorine content of the stratosphere (1991-1997): An thropogenic, volcanic, and methane influences, *J. Geophys. Res., ,* in press, 1999.

Fusco, A. C., and M. L. Salby, Interannual variations of total ozone and their relationship to variations of planetary wave ativity, *J. Clim., 12*, 1619-1629, 1999.

Garcia, R. R., and B. A. Boville, "Downward control" of the mean meridional circulation and temperature distribution of the polar winter stratosphere, *J. Atmos. Sci., 51*, 2238-2245, 1994.

Gierczak T., J. B. Burkholder, and A. R. Ravishankara, Temperature dependent rate coefficient for the reaction $O(^3P)+NO_2 \rightarrow NO+O_2$, *J. Phys. Chem. A, 103*, 877-883, 1999.

Gille, J. C., S. T. Massie, and W. G. Mankin, Preface for Evaluation of UARS data, *J. Geophys. Res., 101*, 9539-9539, 1996.

Grooss, J.-U., R. Muller, G. Becker, D. S. McKenna, and P. J. Crutzen, The upper stratospheric ozone budget: An update of calculations based on HALOE data, *J. Atmos. Chem.,* in press, 1999.

Gunson, M. R., et al., Measurements of $CH_4$, $N_2O$, CO, $H_2O$, and $O_3$ in the middle atmosphere by the Atmospheric Trace Molecule Spectroscopy experiment on Spacelab-3, *J. Geophys. Res., 95*, 13867-13882, 1990.

Holton, J. R., P. H. Haynes, M. E. McIntyre, A. R. Douglass, R. B. Rood, and L. Pfister, Stratospheric-tropospheric exchange, *Rev. Geophys., 33*, 403-439, 1995.

Hood, L. L., J.P. McCormack, and K. Labitzke, An investigation of dynamical contributions to mid latitude ozone trends in winter, *J. Geophys. Res., 102*, 13079-13093, 1997.

Hood, L. L. and S. Cantrell, Stratospheric ozone and temperature responses to short term solar ultraviolet variations: Reproducibility of low-latitude response measurements, *Ann. Geophys., 33*, 403-439, 1995.

Jackman, C. H., E. L. Fleming, S. Chandra, D. B. Considine, and J. E. Rosenfield, Past, present, and future modeled ozone trends with comparisons to observed trends, *J. Geophys. Res., 101*, 28753-28767, 1996.

Johnson, D. G., K. W. Jucks, W. A. Traub, and K. V. Chance, Smithsonian stratospheric far-infrared spectrometer and data reduction system, *J. Geophys. Res., 100*, 3091, 1995a.

Johnson, D. G., W. A. Traub, K. V. Chance, and K. W. Jucks, Estimating the abundance of ClO from simultaneous remote sensing measurements of $HO_2$, OH, and HOCl, *Geophys. Res. Lett., 22*, 1869, 1995b.

Jucks, K. J., D. G. Johnson, K. V. Chance, W. A. Traub, R. J. Salawitch, and R. A. Stachnik, Ozone production and loss rate measurements in the middle stratosphere, *J. Geophys. Res., 101*, 28785-28792, 1996.

Jucks, K. J., D. G. Johnson, K. V. Chance, W. A. Traub, J. J. Margitan, G. B. Osterman, R. J. Salawitch, and Y. Sasano, Observations of OH, $HO_2$, $H_2O$, and $O_3$ in the upper stratosphere; Implications for $HO_x$ photochemistry, *Geophys. Res. Lett., 25*, 3935–3938, 1998.

Jucks, K. J., D. G. Johnson, K. V. Chance, W. A. Traub, and R. J. Salawitch, Nitric acid in the middle stratosphere as a function of altitude and aerosol loading, *J. Geophys. Res., 104*, 26715-26724, 1999.

Ko, M. K. W., N. D. Sze, and D. K. Weisenstein, The roles of dynamical and chemical processes in determining the stratospheric concentration of ozone in one-dimensional and two-dimensional models, *J. Geophys. Res.*, *94*, 9889-9896, 1989.

Kondo, Y., et al., $NO_y$ correlation with $N_2O$ and $CH_4$ in the midlatitude stratosphere, *Geophys. Res. Lett.*, *23*, 2369-2372, 1996.

Lean, J. L., G. J. Rottman, H. L. Kyle, T. N. Woods, J. R. Hickey, and L. C. Puga, Detection and parameterization of variations in solar mid- and near-ultraviolet radiation (200 - 400 nm), *J. Geophys. Res.*, *102*, 29939-29956, 1997.

Lipson, J. B., et al., Temperature dependence of the rate constant and branching ratio for the OH+ClO reaction, *J. Chem. Soc., Faraday Trans.*, *93*, 2665, 1997.

McElroy, M. B. and R. J. Salawitch, Stratospheric ozone: impact of human activity, *Planet. Space Sci.*, *37*, 1653-1672, 1989.

Michelsen, H. A., et al., Stratospheric chlorine partitioning: Constraints from shuttle-borne measurements of [HCl], [ClNO3], and [ClO], *Geophys. Res. Lett.*, *23*, 2361, 1996.

Montzka, S. A., J. H. Butler, J. W. Elkins, T. M. Thompson, A. D. Clarke, and L. T. Lock, Present and future trends in the atmospheric burden of ozone-depleting halogens, *Nature*, 398, 690-694, 1999.

Mote, P. W., et al., An atmospheric tape recorder: The imprint of tropical tropopause temperatures on stratospheric water vapor, *J. Geophys. Res.*, *101*, 3989-4006, 1996.

Nedoluha, G. E., et al., Increases in middle atmospheric water vapor as observed by the Halogen Occultation Experiment and the ground-based Water Vapor Millimeter-wave Spectrometer from 1991 to 1997, *J. Geophys. Res.*, *103*, 3531-3543, 1998a.

Nedoluha, G. E., et al., Changes in upper stratospheric $CH_4$ and $NO_2$ as measured by HALOE and implications for changes in transport, *Geophys. Res. Lett.*, *25*, 987-990, 1998b.

Nevison, C. D. and E. A. Holland, A reexamination of the impact of anthropogenically fixed nitrogen on atmospheric $N_2O$ and the stratospheric $O_3$ layer, *J. Geophys. Res.*, *102*, 25519-25536, 1997.

Newell, R. E. and S. Gould-Stewart, A stratospheric fountain?, *J. Atmos. Sci.*, 38, 2789-2796, 1981.

Oltmans, S. J., and D. J. Hofmann, Increase in lower-stratospheric water vapour at a mid-latitude Northern Hemisphere site from 1981 to 1994, *Nature, 374*, 146-149, 1995.

Osterman, G. B., R. J. Salawitch, B. Sen, G. C. Toon, R. A. Stachnik, H. M. Pickett, J. J. Margitan, and D. B. Peterson, Balloon-borne measurements of stratospheric radicals and their precursors: implications for the production and loss of ozone, *Geophys. Res. Lett.*, *24*, 1107-1110, 1997.

Osterman, G. B., B. Sen, G. C. Toon, R. J. Salawitch, J. J. Margitan, and J.-F. Blavier, The partitioning of reactive nitrogen species in the summer Arctic stratosphere, *Geophys. Res. Lett.*, *26*, 1157-1160, 1999.

Park, J., M. K. W. Ko, R. A. Plumb, and C. Jackman (eds), Report of the 1998 Models and Measurements Workshop II, NASA Reference Publication, National Aeronautics and Space Administration, Washington, DC, 1999.

Randel, W. J., F. Wu, J. M. Russell, J. W. Waters, and M. Santee, Space-time patterns of trends in stratospheric

constituents derived from UARS measurements, *J. Geophys. Res.*, accepted, 1999.

Randel, W., F. Wu, J. M. Russell III, and J. Nash, Interannal Changes in Stratospheric Constituents and Global Circulation Derived from Satellite Data, this volume, 2000.

Rinsland, C. P., et al., ATMOS/ATLAS-3 measurements of stratospheric chlorine and reactive nitrogen partitioning inside and outside the November 1995 Antarctic vortex, *Geophys. Res. Lett.*, *23* 2365-2368, 1996.

Rosenfield, J. E., and A. R. Douglass, Doubled $CO_2$ effects on $NO_y$ in a coupled 2D model, *Geophys. Res. Lett.*, *25* 4381-4384, 1998.

Rosenlof, K. H., et al., Hemispheric asymmetries in water vapor and inferences about transport in the lower stratosphere, *J. Geophys. Res.*, *102*, 13213-13234, 1997.

Salawitch, R. J., et al., The distribution of hydrogen, nitrogen, and chlorine radicals in the lower stratosphere: Implications for changes in $O_3$ due to emission of $NO_y$ from supersonic aircraft, *Geophys. Res. Lett.*, *21*, 2547, 1994.

Sen B., et al., Measurements of reactive nitrogen in the stratosphere, *J. Geophys. Res.*, *103*, 3571, 1998.

Siskind, D. E., L. Froidevaux, J. M. Russell, and J. Lean, Implications of upper stratospheric trace constituent changes observed by HALOE for $O_3$ and ClO from 1992 to 1995, *Geophys. Res. Lett.*, *25*, 3513-3516, 1998.

Solomon, S., et al., Ozone depletion at mid-latitudes: Coupling of volcanic aerosols and temperature variability to anthropogenic chlorine, *Geophys. Res. Lett.*, *25*, 1871-1874, 1998.

Steinbrecht, W., H. Claude, U. Kohler and K. P. Hoinka, Correlations between tropopause height and total ozone: Implications for long-term changes, *J. Geophys. Res.*, *103*, 19183-19192, 1998.

Summers, M. E., et al., Implications of satellite OH observations for middle atmospheric $H_2O$ and ozone, *Science*, *277*, 1967-1970, 1997.

Summers, M. E., and R. R. Conway, Insights into middle atmospheric hydrogen chemistry from analysis of MAHRSI observations of hydroxyl (OH), this volume, 2000.

Thomason, L. W., L. R. Poole, and T. Deshler, A global climatology of stratospheric aerosol surface area density deduced from Stratospheric Aerosol and Gas Experiment II measurements: 1984-1994, *J. Geophys. Res.*, *102*, 8967, 1997.

Woodbridge, E. L., et al., Estimates of total organic and inorganic chlorine in the lower stratosphere from *in situ* measurements during AASE II, *J. Geophys. Res.*, *100*, 3057, 1995.

World Meteorological Organization (WMO); Scientific Assessment of Ozone Depletion, 1999.

World Meteorological Organization (WMO); SPARC/IOC/GAW Assessment of trends in the vertical distribution of ozone, report No. 43, Geneva, 1998.

K. W. Jucks, Harvard-Smithsonian Center for Astrophysics, 60 Garden Street, Cambridge, MA 02138. (email: jucks@cfa.harvard.edu)

R. J. Salawitch, Jet Propulsion Laboratory, Pasadena, CA 91109.

# Ground-based Microwave Observations of Middle Atmospheric Water Vapor in the 1990s

Gerald E. Nedoluha, Richard M. Bevilacqua, R. Michael Gomez, and Brian C. Hicks

*Naval Research Laboratory, Washington, D. C.*

James M. Russell III

*Hampton University, Center for Atmospheric Sciences, Hampton, VA*

Brian J. Connor

*National Institute of Water and Atmospheric Research, Lauder, New Zealand*

Ground-based microwave measurements of middle atmospheric water vapor measurements from three sites of the Network for the Detection of Stratospheric Change (NDSC) are presented. The measurements are made by the Water Vapor Millimeter-wave Spectrometer (WVMS) instruments at Table Mountain, California, Mauna Loa, Hawaii, and Lauder, New Zealand. We include an extended description of the measurement and retrieval methods, and a long-term comparison with measurements from HALOE. The agreement with HALOE is generally very good, however there are some differences in the shape of the water vapor profile in the upper mesosphere. These differences may have important implications for water vapor chemistry in this region. The long-term data record obtained by the WVMS instruments, which have been operating since 1992, shows a large increase in water vapor during the first half of the 1990s, in agreement with the measurements from HALOE. The cause of this increase is still not completely understood, however it may be related to the 1991 eruption of Mount Pinatubo.

## 1. INTRODUCTION

In the two decades since the water vapor measurements of *Radford et al.* [1977], a number of groups have used ground-based microwave measurements to obtain estimates of water vapor in the middle atmosphere. Measurements have been made at 22.2 GHz, the frequency of the $6_{16}$-$5_{23}$ rotational transition of water vapor, by *Bevilacqua et al.* [1983, 1985, 1987, 1990], *Olivero et al.* [1986], *Tsou et al.* [1988], *Nedoluha et al.* [1995, 1996, 1997, 1998a, 1999], *Jarchow* [1999], and *Seele and Hartogh* [1999]. The earliest of these measurements

consisted of short campaigns in which the goal was to obtain a reasonably quantitative estimate of a typical water vapor mixing ratio profile. With the exception of the observations by *Radford et al.* [1977], all of the measurements found peak water vapor mixing ratios of ~4-8 ppmv. The first study of monthly variations in the water vapor mixing ratio was made by *Bevilacqua et al.* [1985].

Long-term ground-based measurements of middle atmospheric water vapor began with the formation of the Network for the Detection of Stratospheric Change (NDSC) in 1990. This network currently consists of 20 primary and 29 complementary stations distributed across the globe from which numerous important atmospheric species, including ozone, ClO, $NO_2$, and others are monitored. Instruments deployed at these sites include lidars, ultraviolet/visible spectrographs, Fourier transform infrared spectrometers, and microwave radiometers. As part of this network extensive datasets of water vapor

Atmospheric Science Across the Stratopause
Geophysical Monograph 123
Copyright 2000 by the American Geophysical Union

**Table 1.** Deployment History of the WVMS Instruments

| Measurement Site | Instrument | Measurement Period |
|---|---|---|
| Table Mountain, California (34.4°N, 242.3°E) | WVMS1 | Jan. 23 1992 to Oct. 13 1992* |
| Lauder, New Zealand (45.0°S, 169.7°E) | WVMS1 | Nov. 3, 1992 to Apr. 21, 1993 |
| Table Mountain, California (34.4°N, 242.3°E) | WVMS1 | May 17, 1993 to Nov. 9, 1993 |
| Lauder, New Zealand (45.0°S, 169.7°E) | WVMS1 | Jan. 14, 1994 to present |
| Table Mountain, California (34.4°N, 242.3°E) | WVMS2 | Aug. 19, 1993 to Nov. 18, 1997** |
| Table Mountain, California (34.4°N, 242.3°E) | WVMS3 | Sep. 20, 1995 to Feb. 3, 1996 |
| Mauna Loa, Hawaii (19.5°N, 204.4°E) | WVMS3 | Mar. 4, 1996 to present |

\* - Because of large uncertainties in the instrumental pointing during this period these measurements have not been used in long-term studies.

\*\* - Narrow filters with a bandwidth of 50 kHz were added on Jan. 7, 1994.

measurements using 22.2 GHz ground-based microwave radiometers have been collected. The Naval Research Laboratory has positioned three Water Vapor Millimeter-wave Spectrometers (WVMS) at NDSC sites. The deployment history is given in Table 1. Measurements of water vapor at 22.2 GHz are also currently being made by the Max-Planck-Institut für Aeronomie. This dataset consists of 19 months of data from Lindau, Germany (51°N, 10°E), and also high latitude data from ALOMAR, Norway (69°N, 16°E) for the period October 1995 to October 1996 and July 1997 to the present [*Jarchow, 1999; Seele and Hartogh, 1999*].

In this paper we present a review of measurement and retrieval techniques used to obtain water vapor profiles from the WVMS measurements, and we review some of the scientific results obtained from this data. We also update the published record of WVMS measurements and present some results obtained from this new data, including the detection of the mesospheric QBO at midlatitudes and the latest estimate of trends in middle atmospheric water vapor.

## 2. MEASUREMENT AND RETRIEVAL METHODS

The calculation of water vapor mixing ratio profiles with microwave techniques relies upon the sensitivity of the lineshape to pressure. Given a relationship between pressure and altitude it is then possible to calculate the water vapor emission that is observed over a particular pressure range, and thereby to calculate a mixing ratio for a particular range. At 22.2 GHz this technique allows retrievals up to ~80-90 km, at which point the Doppler broadening becomes greater than the pressure broadening.

The WVMS radiometer is based on a high electron mobility transistor (HEMT) amplifier which is cooled to 20 K, and provides 20 dB gain over a 500-MHz bandwidth. Each WVMS instrument measures the spectrum using several filterbanks, with the filters evenly spaced at frequencies symmetric about the 22.2 GHz

resonance frequency. The WVMS1 spectrometer contains twenty 200-kHz filters, twenty 2-MHz filters, and ten 40-MHz filters. The WVMS2 and WVMS3 spectrometers contain ten 50-kHz filters, twenty 200-kHz filters, thirty 2-MHz filters, and thirty 14-MHz filters. The system is internally calibrated with a noise diode, which is itself calibrated periodically by alternately placing a liquid nitrogen load and an ambient temperature load in the beam. Radiation enters the system through a scalar feed horn, which has a FWHM beam size of ~8°. The atmospheric radiation is reflected from an aluminum plate, which is inclined at 45° to the axis of the motor and the horn and can be rotated to vary the measurement angle.

The data taking procedure consists first of a tipping measurement, in which the instrument is pointed at a series of elevation angles. Using the assumption of a spherical atmosphere it is then possible to determine the contribution of the atmosphere and of the instrumental receiver temperature to the total observed temperature. The optical depth of the atmosphere at 22.2 GHz is then solved for by combining the atmospheric contribution to the observed brightness temperature with an independent estimate of the atmospheric temperature. This optical depth is dominated by the tropospheric contribution, and is of interest primarily because it attenuates the signal from water vapor in the middle atmosphere.

While the total optical depth can be obtained from a single channel by using the tipping measurement, retrieving a profile requires the use of multiple filters. Because the observed signal is dominated by the tropospheric emission, any unknown difference in the gain of these filters can introduce a large error in the spectral variation with frequency. We therefore employ a Dicke switching technique, in which we alternate between a measurement near zenith (a "reference" measurement), and one at ~60°-70° from zenith (a "signal" measurement). In order to noise balance these two measurements, a spectrally flat absorber is placed so as to be in the beam when the mirror is in the reference position. Further

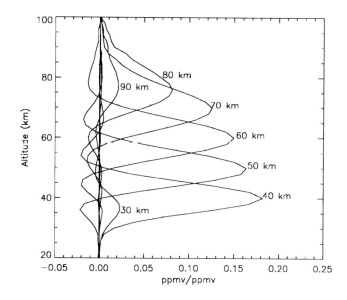

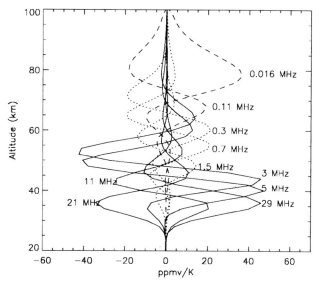

**Figure 1.** Top: The averaging kernels at selected altitudes for the WVMS2 instrument at Table Mountain for the period January 16-22, 1996. Each line represents the sensitivity of the retrieval at an indicated altitude to perturbations over a range of 2 km altitude bins. Bottom: The contribution functions for selected filters for the same period. Each line indicates the sensitivity of the retrieved mixing ratio as a function of altitude to a 1K perturbation in brightness temperature at the indicated frequency (measured from line center). The line styles correspond to different filter widths: 2 MHz (solid), 200 kHz (dotted), and 50 kHz (dashed).

details on the measurement technique are given in *Nedoluha et al.* [1995].

One of the most important sources of error in ground-based microwave measurements of water vapor is the presence of undulations in the spectral baseline that have

periods comparable to, or larger than, the width of the measured spectrum. In order to reduce this source of error, we move the entire feedhorn assembly by λ/4 between scans. By averaging together separately the scans taken in the two positions, we can significantly reduce the effects of any reflections that occur from the environment in front of the feedhorn.

In order to retrieve a mixing ratio from the measured spectrum we require a forward model which contains all of the physics required to calculate the spectrum as a function of the mixing ratio profile. For our calculation we have used a pressure-broadened half-width from *Liebe* [1993], which is given by

$$\Delta v_{1/2} = (2.811 \text{ GHz})[(P/1000 \text{ mb})(300 \text{ K}/T)^{0.69} + 4.8 (P_{H2O}/1000 \text{ mb})(300 \text{ K}/T)].$$

The lower state energy = $8.86987 \times 10^{-14}$ ergs, and line strength (at 300 K) = $0.13146 \times 10^{-19}$ cm² MHz, are taken from the JPL line catalog [*Pickett et al., 1998*].

The above linestrength includes the ~5% increase incorporated in the JPL catalog in the early 1990s and in the tables given by *Liebe* [1993] following ground-level emission measurements by *Westwater* [1993]. This linestrength has been applied to all previous publications of WVMS data. The precise linestrength and linewidth of the 22.2 GHz transition remain the subject of spectroscopic studies. Recent measurements by *Pol et al.* [1998] are ~1% higher than the linestrength given above. The *Pol et al.* [1998] results also suggest that the current estimate of the linewidth is ~7% too small. In order to maintain consistency with previous results, we have not incorporated the *Pol et al.* [1998] results into the WVMS retrieval scheme at this point. The adoption of a 7% increase in linewidth in the forward model would result in an increase in the retrieved mixing ratio of ~5% at 40 km, and ~7% in the upper mesosphere. We note that if the forward model were adjusted to reflect the results of *Pol et al.* [1998], then the retrieved middle atmospheric mixing ratios would be within ~2% of the values obtained with the pre-1990 parameters.

The mixing ratio profile is retrieved using the optimal estimation technique [*Rodgers, 1976; Rodgers, 1990*]. Use of this technique requires an estimate of uncertainties in both the measurement and in the a priori knowledge of the atmosphere. The primary sources of measurement error that are incorporated into the retrieval scheme are random noise, which limits the accuracy of the retrievals primarily in the upper mesosphere, and baseline error, which is the most important source of error in the lower stratosphere [*Nedoluha et al., 1995*]. The measurements are combined with an a priori estimate of the water vapor profile in such a way as to produce a statistically optimal solution given the relative uncertainties in all of these terms.

In order to estimate the sensitivity and resolution of the WVMS measurements we follow *Rodgers* [1990] in defining an averaging kernel which describes the change in the derived mixing ratio at a given altitude, given a perturbation at any altitude. The resulting kernels represent a linear approximation of the sensitivity of the model to mixing ratio changes at different altitudes. In Figure 1 we show typical averaging kernels for a 1 week integration period for a WVMS instrument. One feature which is apparent in Figure 1 is that the averaging kernels show a significant decrease in amplitude below ~40 km. While the sensitivity of a microwave measurement to variations in the lower stratosphere is fundamentally limited by the spectral width of the instrument, the rapid decrease in sensitivity between 40 and 30 km in these retrievals is not the result of this limitation, but is instead primarily caused by uncertainties in the instrumental baseline (mimicking a weak, broad emission or absorption line contribution) which we have included in our estimate of the error. In the remainder of this paper we will therefore show mixing ratio profiles only down to 40 km. The sensitivity of the measurements begins to drop off above ~70 km due to decreasing signal-to-noise in the emission from these altitudes. Note that while the retrievals still exhibit some sensitivity to atmospheric perturbations at altitudes as high as 100 km, the retrieval at 90 km is much more sensitive to perturbations at ~76-78 km than to those at 90 km. Hence, even if variations in the retrieved mixing ratios are seen above 80 km, this cannot necessarily be interpreted as a measurement of variability above 80 km. The upper altitude of mixing ratio plots for WVMS measurements has therefore been set at 80 km.

The sensitivity of the retrieved profile to changes in measured brightness temperature at different frequencies is also shown in Figure 1. This figure shows that channels closest to line center are the ones with the most sensitivity for retrieving water vapor in the upper mesosphere. As is clear from the figure, an increase in observed emission in any particular channel may actually result in a decrease in the water vapor mixing ratio at many altitudes. An increase in brightness temperature in channels far from line center, especially those that are used in establishing the linear baseline (such as the 21 MHz channel) may actually cause a net decrease in the measured column water vapor in the middle atmosphere.

There are two major sources of measurement error that have not been included in the retrieval algorithm because they vary on timescales longer than the weekly timescales at which the measurements are made. The most important source of uncertainty that varies on timescales ~1 year is the accuracy of the pointing. Because the water vapor retrieval relies upon an accurate estimate of the air mass, an error in the pointing of the instrument will result in an error in the retrieved mixing ratio. While the sensitivity of the near zenith measurement to pointing errors is small, a pointing error of 1° when the instrument is pointing at a typical "signal" angle of 70° from zenith will result in an error of ~5% in the retrieved mixing ratio. The pointing is currently checked using a laser alignment system, and the instrument stand is stable at the precision to which we are able to measure the angles (~0.2°). Thus we estimate that our current pointing is accurate to within ~0.2°. However, for data obtained prior to 1995 we estimate an error of ~1°.

The WVMS systems are internally calibrated with a noise diode which is itself calibrated by measuring the difference between the temperatures of a liquid nitrogen load and an ambient temperature load, both of which function as black body sources. Variations in the attenuation of the signal from the solid-state noise diode introduced errors of ~5-10%. Removal of the attenuators placed directly after the noise diode in the WVMS3 instrument (a change subsequently implemented in December 1995 on WVMS2 at Table Mountain, and in October 1996 at Lauder) significantly reduced the calibration uncertainty. This improvement, combined with improvements in the techniques used to calibrate the noise diode itself, now make it possible to consistently calibrate the system at the 1% level.

Forward model errors can also contribute to uncertainties in the data. The most important variable source of forward model errors is probably that resulting from uncertainties in the atmospheric temperature. Data from the National Centers for Environmental Prediction (NCEP) are used to determine temperature and pressure up to 1 mb. Above this altitude the temperature profile as a function of altitude is smoothly merged with the temperature from the MSISE90 model [*Hedin*, 1991]. The effect of a perturbation in middle atmospheric temperatures of 5K (approximately the average difference between the MSISE90 and NMC models in the stratosphere) smoothed with a correlation scale height of 8 km on the retrieved mixing ratio has been estimated to be ~0.1 ppmv at all retrieved altitudes [*Nedoluha et al.*, 1995]. Another source of forward model error is the uncertainty in the spectral parameters. Given the differences in measurements of linewidth and linestrength made during the past decade [*Pol et al.*, 1998], we estimate that the errors in the spectral parameters will result in nearly altitude independent uncertainties of ~5% in the retrieved water vapor mixing ratio.

## 3. WVMS MEASUREMENTS AND COMPARISONS

A typical spectrum for observations at Table Mountain is shown in Figure 2. This spectrum covers the same period used for the calculations in Figure 1. The measurement consists of a 500-scan integration, which corresponds to ~1 week of observation time. This

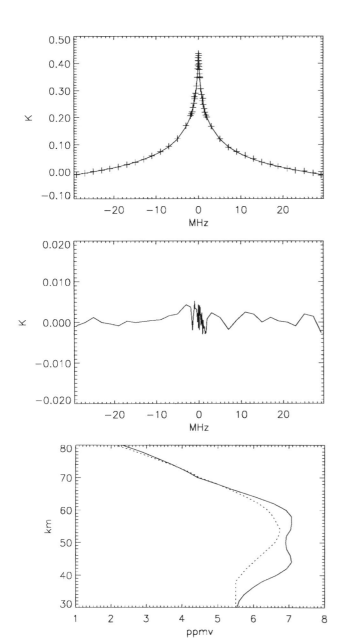

**Figure 2.** Top: The spectrum for WVMS observations at Table Mountain for the period January 16-22, 1996. Crosses mark the observed mean intensity and center frequency at each filterbank spectrometer channel, while the continuous curve is the lineshape synthesized from the retrieved vertical mixing ratio profile. Middle: The difference between the measured spectrum and the model fit. Bottom: The retrieved water vapor mixing ratio (solid) and the a priori profile used in all retrievals (dashed).

integration period has been chosen in order to provide adequate signal-to-noise so that retrievals show significant sensitivity at altitudes up to ~80 km (see Figure 1). Shorter periods of integration would be more appropriate for studies of short term variations at lower altitudes. Also

shown in Figure 2 is the water vapor mixing ratio retrieved from the measured spectrum, and the residual (measured – model) brightness temperatures.

Plate 1 shows the complete dataset of water vapor mixing ratios measured at each of the three sites where WVMS instruments have been deployed. The time of each 500-scan retrieved profile is indicated by tick marks at the top of the graph. These profiles are then smoothed using a Gaussian with a (1/e) width of 25 days.

Each instrument has been tested and validated over an extended period against a working instrument at Table Mountain. The results of the WVMS1-WVMS2 comparison are given by *Nedoluha et al.* [1996], and that of the WVMS2-WVMS3 comparison is shown by *Nedoluha et al.* [1999]. Comparisons have also been conducted with five satellite instruments [*Nedoluha et al.,* 1997]. Comparisons where made with the Halogen Occultation Experiment (HALOE) [*Russell et al.,* 1993], the Microwave Limb Sounder (MLS) [*Barath et al.,* 1993], and the Improved Stratospheric and Mesospheric Sounder (ISAMS) [*Taylor et al.,* 1993] aboard the UARS satellite. The comparisons with HALOE (version 17) and MLS (version 3) retrievals provided an opportunity to validate observed temporal variations. Measurements made by the Millimeter-wave Atmospheric Sounder (MAS), and the Atmospheric Trace Molecule Spectroscopy Experiment (ATMOS) provided the opportunity for coincident measurements with up to four space-based instruments. In *Nedoluha et al.* [1997] it was found that the retrieved mixing ratios from most of these instruments differed from the average of all of the instruments by <1 ppmv at most altitudes from 40 km to 80 km.

While the failure of the 183 GHz receiver on the MLS instrument in April 1993 severely limits the period over which comparisons can be made with the WVMS instruments, comparisons between the WVMS and HALOE instruments are available over several years. In Figure 3 we show the average of all of the coincident measurements (within ±7 days and ±5° latitude) made between HALOE and the WVMS instruments at Table Mountain, Lauder, and Mauna Loa. In the plots for each site we show the average of the WVMS retrievals, the average of the HALOE (version 19) retrievals, and the average of the HALOE retrievals after convolving these retrievals with the appropriate averaging kernel for each WVMS instrument. The average convolved profiles agree to within better than 1 ppmv everywhere. We note that if the *Pol et al.* [1998] linewidths were adopted the mixing ratios retrieved by the WVMS instruments would be 5-7% higher, thus increasing the difference in the HALOE and WVMS measurements throughout much of the stratosphere and lower mesosphere.

The average HALOE profiles near the Table Mountain and Mauna Loa latitudes shown in Figure 3 exhibit a local

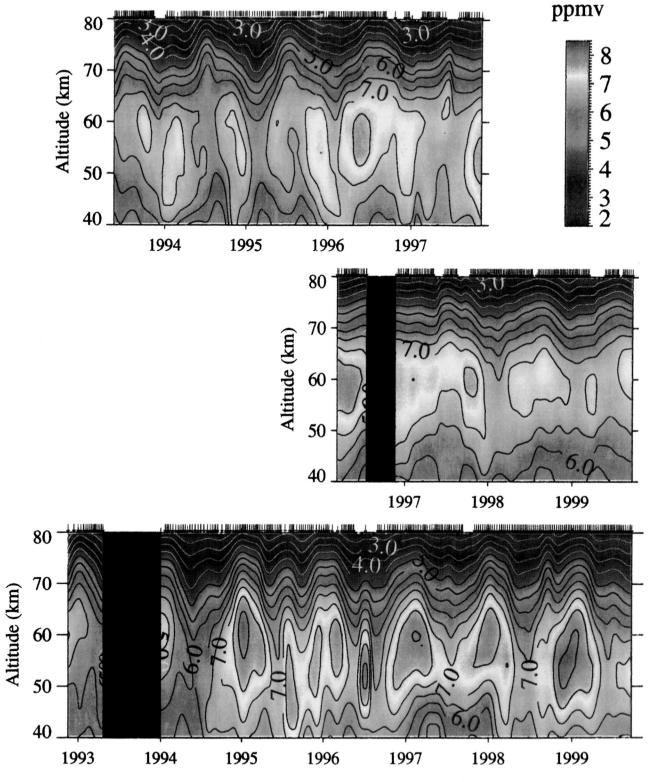

**Plate 1.** The water vapor mixing ratios measured by WVMS instruments at Table Mountain (34.4°N, 242.3°E) (top), **Mauna Loa** (19.5°N, 204.4°E) (middle), and Lauder (45.0°S, 169.7°E) (bottom). Small tick marks on the top of each plot indicate times of individual retrievals. The mixing ratios shown are calculated using retrievals obtained from spectral integrations taken over ~1 week (500 scans). These profiles are then smoothed using a Gaussian with a (1/e) width of 25 days.

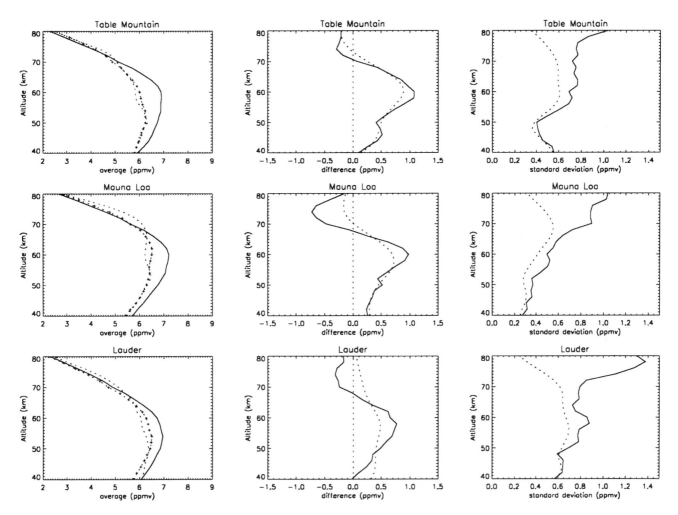

**Figure 3.** A comparison of all coincident measurements (within ±7 days and ±5° latitude) between the WVMS instruments and HALOE. The data has been binned separated according to site. Row 1: comparisons at Table Mountain (34.4°N, 242.3°E). Row 2: comparisons at Mauna Loa (19.5°N, 204.4°E). Row 3: comparisons at Lauder (45.0°S, 169.7°E). Column 1: average of WVMS retrievals (solid), average of HALOE retrievals (dotted), average of convolved HALOE retrievals (dotted with crosses). Column 2: Difference of WVMS and HALOE retrievals (solid), and after convolution of the HALOE retrievals (dotted). Column 3: Standard deviation of the WVMS-HALOE difference (solid), and after convolution of the HALOE retrievals (dotted).

minimum in water vapor near 60 km. This feature, which is especially pronounced near the equator and in the summer hemisphere, has been previously noted by several authors [*Nedoluha et al.,* 1998a; *Siskind and Summers,* 1998; *Nedoluha et al.,* 1999]. *Summers and Siskind* [1999] and *Summers and Conway* [2000] have proposed that a narrow layer of increased water vapor near 70 km could result from surface recombination of O and $H_2$ on meteoric dust, but even this model does not reproduce the local minimum near 60 km.

Because of the coarser resolution of the WVMS retrievals relative to those of the HALOE measurement, such a local minimum is not directly observable in the WVMS retrievals. Nevertheless, such a feature will impact the retrieved WVMS profiles, and should show up as a broad peak in the WVMS-HALOE difference profiles with a maximum near 60 km. Figure 3 shows that the difference between the average WVMS and convolved HALOE profiles at all three sites increases with increasing altitude up to ~60 km. The average difference between the HALOE and WVMS measurements is therefore largest in the altitude range where the HALOE profiles show the unexpected local minimum. Above 60 km the mixing ratios measured by the three WVMS instruments decrease relative to HALOE. This steeper gradient in the upper mesospheric mixing ratio of the WVMS measurements relative to HALOE covers the altitude range where the HALOE v19 retrievals show the 70 km layer of increased water vapor.

While the WVMS measurements may cast some doubt on the validity of the 60 km minimum and the 70 km layer in the HALOE profile, we do note that measurements of OH made by the Middle Atmosphere High Resolution Spectrograph Investigation (MAHRSI) [*Conway et al.,* 1996] provide some support for the HALOE observations. Given the short chemical lifetime of OH, it is expected that OH and $H_2O$ would have similar vertical distributions [*Brasseur and Solomon,* 1984]. *Summers et al.* [1997a] and *Summers and Conway* [2000], show that the increase in OH observed by MAHRSI between 65 and 70 km is consistent with the layer of $H_2O$ observed by HALOE at this altitude.

*Pumphrey* [1999] shows that some of the MLS data indicates a "double peak" structure, but notes that when differences in resolution between HALOE and MLS are taken into account the mesospheric peak in the v18 HALOE data rises ~1.0 ppmv above the 60 km minimum, while the MLS measurements show only an increase of only ~0.2 ppmv. This difference would be somewhat smaller if the MLS data were compared to the v19 HALOE data, since the double peak amplitude in the v19 data is ~0.5 ppmv smaller. *Pumphrey* [1999] also shows evidence of a small local minimum (~0.2 ppmv) at ~56 km in the ATMOS data from Atlas 1. Further detailed comparisons of mesospheric water vapor measurements will be required in order to establish the veracity of the 60 km minimum and the 70 km layer observed by HALOE.

In addition to showing average differences between the WVMS and HALOE measurements, Figure 3 also shows the standard deviation of these differences for each of the three sites. The standard deviation of the convolved mixing ratios differs by 0.25-0.7 ppmv, with the largest difference occurring at Lauder and the smallest at Mauna Loa. The smaller differences at Mauna Loa probably reflect both the smaller tropospheric optical depth for observations from that site, and the improvements made to WVMS instrument stability in the years since measurements began at Lauder. The large difference between convolved and unconvolved comparisons in the upper mesosphere results from the decreasing sensitivity of the WVMS observations above ~70 km. This difference is largest at Lauder, primarily because of the absence of 50 kHz filters in this instrument (cf. *Nedoluha et al.,* [1996], Figure 2), but also in part because of the larger tropospheric optical depth at the Lauder site.

### 3.1 Variations in the Upper Stratosphere and Lower Mesosphere

The length of the WVMS data set has proven to be extremely valuable in validating the large increase in water

vapor observed by HALOE during the period 1991-1997 [*Nedoluha et al.,* 1998a]. In Figure 4 we show the water vapor mixing ratio at 50 km as measured by the WVMS instruments, and the coincident HALOE measurements (within ±5° latitude). The curved lines show the fit of the data (calculated using singular value decomposition [*Press et al.,* 1992]) to the function

$$f = a_0 + a_1\sin(2\pi t) + a_2\cos(2\pi t) + a_3\sin(4\pi t) + a_4\cos(4\pi t) + a_5\sin((12/27)2\pi t) + a_6\cos((12/27)2\pi t) + a_7 t \qquad (1)$$

The straight lines show the linear component of this fit ($a_0 + a_7 t$). Here $t$ is given in years, so the first two pairs of sinusoidal terms represent the annual and semi-annual contributions, while the third pair of sinusoidal terms represents a 27-month QBO. In a previous analysis [*Nedoluha et al.,* 1998a] the QBO term was not included in the fit to the Table Mountain measurements; however, as we shall show, the WVMS and HALOE data show a clear mesospheric QBO signature at this mid-latitude site. In this analysis we include a QBO term in all fits for the Mauna Loa comparisons and in the 70 km comparison at Table Mountain.

Figure 4 shows that from the beginning of the HALOE measurement period until November 1997 (the end of the WVMS Table Mountain data record) there was a clear increase in water vapor at 50 km. The increase ranges from 0.07 ppmv/yr (~1.1%/yr) for the HALOE measurements near Table Mountain, to 0.23 ppmv/year (~3.3%/yr) for the WVMS measurements at Lauder. The reason for the larger seasonal cycles in the WVMS data is not well understood, but this does suggest that seasonal variations in local conditions at the surface may, through a mechanism that is as yet not understood, affect the WVMS measurements.

The trend fitting algorithm used in this study does provide a formal error for the trend, but this formal error does not necessarily provide a good approximation for the uncertainty in the calculated trends, at least for the WVMS instrument. The largest uncertainty in the trend results from uncertainties in the instrumental pointing, and the pointing does not vary randomly from one event to the next, but instead generally remains stable over extended periods and can change suddenly when repairs or improvements are made to the system. As was discussed in Section 2, the estimated uncertainty in the pointing before 1995 is ~±1°. To obtain a conservative estimate of the trend error, we assume that the pointing was perturbed randomly with an uncertainty of ±1° every ~6 months. The resultant uncertainty in the linear trend for the period through November 1997 shown in Figure 4 is then ~1%/yr (~0.06 ppmv/yr). The magnitude of the mixing ratio error introduced by calibration errors was initially comparable to that introduced by the pointing errors (see Section 2), but

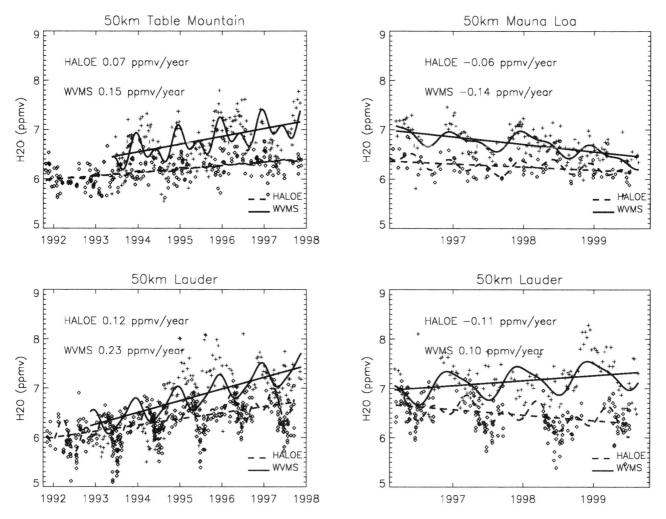

**Figure 4.** Retrieved water vapor mixing ratios at 50 km for WVMS instruments (crosses) at Table Mountain (top left), Mauna Loa (top right), and Lauder (both bottom panels). HALOE data coincident within ±5° latitude is also shown (diamonds). The HALOE data has been convolved with an appropriate set of averaging kernels for each site. Also plotted are the fits derived from (1), and the linear components of the fits ($a_0 + a_7 t$). A QBO component has been included only in the fit to the Mauna Loa data.

these errors varied on shorter (~monthly) timescales. The effect of these errors on the linear trend estimate should therefore be much smaller. Above ~1 mb, where the MSISE90 model is used to estimate temperatures for the retrievals, it is possible that unaccounted for trends in atmospheric temperature could introduce spurious trends in the water vapor retrieval. Estimates of decadal temperature changes are ~1K [WMO, 1999], which would introduce an error of ~0.5%/decade in the water vapor mixing ratios retrieved in the mesosphere, where the temperature is taken from the MSISE90 model. All other errors of comparable magnitude vary over shorter timescales, and therefore have a much smaller impact on multi-year trends than the pointing errors.

The estimates of the errors in the trends given above are, necessarily, very uncertain. The trends in the WVMS

data for the 1991-1997 period are larger than those measured by HALOE by more than the 0.06 ppmv/year error estimate, although at Table Mountain a small error (~0.02 ppmv/year) in the HALOE trend would result in an overlap of the error bars for the derived trends. We also note that the trend in water vapor measured by HALOE in the lower mesosphere at latitudes near Table Mountain is smaller than that measured at other latitudes over the same time period [Nedoluha et al., 1998a]. The WVMS results at Table Mountain are therefore in better agreement with the global trend measured by HALOE than with those measurements coincident with the Table Mountain site. Larger apparent errors in the trend of the WVMS instrument at Lauder may reflect the fact that the pointing in this instrument was perturbed somewhat less often than at Table Mountain.

Despite the differences in the trends, we note that for the initial period shown in Figure 4 all of the instruments measure increases that are much larger than the water vapor increase of ~0.015 to ~0.032 ppmv/yr measured in the lower stratosphere by *Oltmans and Hofmann* [1995] over the period 1981 to 1994. Studies by *Evans et al.* [1998] and *Randel et al.* [1999] of the stratospheric trend in the HALOE water vapor measurements show a similarly large increase.

*Randel et al.* [1999] used an empirical orthogonal function time projection to conclude that both the global $H_2O$ and $H_2O+2CH_4$ mixing ratios peaked near the beginning of 1996, and decreased very slightly in the following years. Figure 4 indicates that the large rate of increase that occurred in the first half of the 1990s is no longer taking place, in agreement with the analysis of HALOE data by *Randel et al.* [1999]. Figure 4 also shows that since March 1996, when the WVMS3 instrument was deployed at Mauna Loa, the WVMS measurements and the coincident HALOE measurements indicate a decrease in water vapor mixing ratio at this site. At Lauder the HALOE measurements also show a decrease, but the WVMS measurements continue to show an increase, albeit at a much smaller rate than for the earlier period. Since these measurements only cover a period of ~3 years it is too early to draw any definitive conclusions about a trend during this period, especially given the importance of the QBO to the mixing ratios measured near Mauna Loa. Several additional years of WVMS and HALOE measurements will be required to conclusively confirm whether the global stratospheric water vapor mixing ratio is now decreasing.

### 3.2 Upper Mesospheric Variations

In the upper mesosphere the water vapor profile decreases sharply with increasing altitude as water is photodissociated. Despite the seasonal increase in the photodissociation rate near the summer solstice, the water vapor mixing ratio generally reaches a maximum in the summer because the timescale for vertical advection is shorter than that for photodissociation at these altitudes. Plate 1 shows this summertime peak in mesospheric mixing ratio, with increasing amplitude of the seasonal variation correlated with increasing latitude of a site. The measurements from the Lauder site show the largest annual variation despite the fact that this instrument has a somewhat reduced sensitivity to upper mesospheric variations relative to the other instruments (see Section 2). In addition to this summertime maximum in water vapor, the WVMS measurements, especially at Lauder, often show a small peak near the winter solstice. This increase is probably the result of increased diffusion near the solstice resulting from variations in gravity wave breaking [*Thomas et al.*, 1984; *Garcia and Solomon*, 1985;

*Bevilacqua et al.*, 1990; *Nedoluha et al.*, 1996]. The larger wintertime peak at Lauder relative to that at the other sites is probably due, at least in part, to the higher latitude of this site, but we also note that it is consistent with studies that show that there is more transmission of gravity waves through the austral stratosphere than in the boreal stratosphere [*Shine*, 1989; see also *Siskind*, 2000]

In Figure 5 we show the entire record of the HALOE data at 70 km within ±5° of the latitudes of the Mauna Loa and Table Mountain sites, and the WVMS measurements from these sites. As in all other plots, the HALOE data has been convolved with appropriate averaging kernels for each site. Since most of the measurements at Table Mountain were made with an instrument that included 50 kHz filters, we have used a typical averaging kernel from measurements that include these filters. These filters were, however, not added until January 7, 1994, and for this reason the WVMS observations before this period show less seasonal variability. The cause of the lower variability at in the WVMS measurements at Table Mountain towards the of 1997 is not understood, since the 50 kHz filters were used in these measurements.

In addition to the annual variation at both sites shown in Figure 5, a gradual increase from 1991 to ~1996 is apparent in the data. There is also an apparent variation with a timescale of somewhat longer than 2 years which produces deep wintertime minima in the winters of 1992-1993, 1994-1995, and 1997-1998, and more shallow minima in the other years. This interannual variation in the depth of the wintertime minima is also seen in the coincident WVMS measurements at Table Mountain and at Mauna Loa. The deepest wintertime minimum in the WVMS data from Table Mountain occurred in the winter of 1994-1995, in agreement with the HALOE data for the four winters in which WVMS measurements are available. Similarly, the deepest wintertime minimum over three winters of WVMS Mauna Loa measurements occurred in 1997-1998. The timescale of this variation suggests that this interannual variation is related to the QBO, and that the mesospheric water vapor even at a mid-latitude site such as Table Mountain may be affected by the QBO.

The mechanism by which the QBO is affecting water vapor in this region is not clearly understood. In the stratosphere many $H_2O$ anomalies are mirrored by anomalies in methane, suggesting a link with upwelling velocity [*Randel et al.*, 1998], but the amount of $CH_4$ at 70 km is too small to have any significant affect on water vapor. *Randel et al.* [1998] did, however, also find regions with anomalous amounts of $H_2O$ in the stratosphere that did not mirror $CH_4$ anomalies, and there is some suggestion in the HALOE data that such regions are propagating into the mesosphere. Alternatively, changes in stratospheric winds associated with the QBO may be affecting gravity wave propagation into the mesosphere, and thereby altering the diffusion in this region.

## 70km Table Mountain

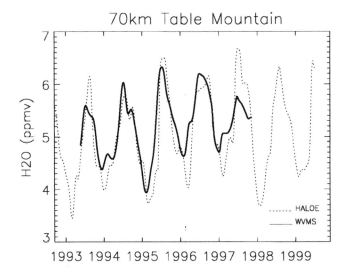

## 70km Mauna Loa

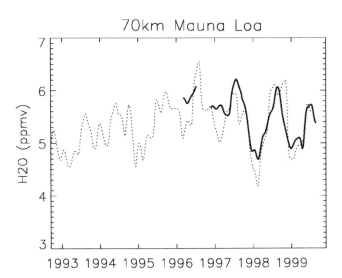

**Figure 5.** Retrieved water vapor mixing ratios at 70 km, smoothed as in Figure 3, for WVMS instruments (solid) at Table Mountain (top) and Mauna Loa (bottom). Also shown is convolved coincident HALOE data (dashed).

In Figure 6 we show the WVMS and HALOE measurements at 70 km, and superimpose on this data a fit calculated using (1), and the linear component of this fit. Since photodissociation plays an important part in determining the water vapor mixing ratio at this altitude, interannual variations in Lyman-$\alpha$ will produce interannual variations in the water vapor mixing ratio [*Huang and Brasseur*, 1993]. The solar cycle index decreased from 1991 to ~1996, and began to increase again in 1997 [*Randel et al.*, 1999]. We would therefore expect some increase in water vapor at this altitude for the 1991 to 1997 period shown in Figure 6 even in the absence of any stratospheric increase in water vapor. The trends calculated for each instrument and period in Figure 6 have

the same sign as those shown in Figure 4. There is an increase during the period covered by the Table Mountain observations, and (with the exception of the WVMS observations from Lauder, which show almost no trend at all) a decrease during the period covered by the Mauna Loa observations. The increase in mesospheric water vapor data from HALOE was first documented by *Chandra et al.* [1997] for the period 1991-1995. *Chandra et al.* [1997] also found that the increase during this period was larger than that which was expected from the variation in Lyman-$\alpha$. Given the increase that was occurring in the stratosphere during this period, it is likely that the observed increase in the mesosphere was caused both by the decrease in Lyman-$\alpha$ and by an increase in water entering the mesosphere from the stratosphere.

## 4. DISCUSSION

We have shown that there is, in general, good agreement in the water vapor profiles retrieved by HALOE and by the WVMS instruments. Both show similar seasonal cycles in the upper mesosphere, and there is good agreement in measurements of the mesospheric QBO from Table Mountain and Mauna Loa. There remains some disagreement in the precise shape of the profile in the mesosphere, with the profiles measured by HALOE suggesting that the presently understood water vapor photochemistry in this region may need some modification.

The most surprising result of the $H_2O$ measurements in the first half of the 1990s was the large increase that occurred during this period. While there was no reason to distrust the validity of the increase in $H_2O$ observed by HALOE during the first half of the 1990s, the availability of several years of WVMS measurements from two NDSC sites certainly helped to better establish the validity of the HALOE results. For other species measured by HALOE (such as $CH_4$, $NO$, and $NO_2$) where no independent confirmation of the unusually large multi-year changes existed for the first half of the 1990s, the confirmation of a large water vapor increase helps to establish that this was a period of significant change in the middle atmosphere, and therefore supported the plausibility that the other species also experienced significant changes.

The observed increase in water vapor was coincident with a significant decrease in stratospheric $CH_4$. Since the quantity $2CH_4+H_2O$ is nearly conserved in the stratosphere, some of the observed water vapor increase can be attributed to an increase in oxidation of the available methane to water vapor [*Nedoluha et al.*, 1998a]. Even if no decrease had been observed in stratospheric $CH_4$, some increase in water vapor would be expected to result from the oxidation of an increasing amount of $CH_4$ entering the stratosphere. Measurements by *Dlugokencky et al.* [1998], however, show that the global trend in $CH_4$

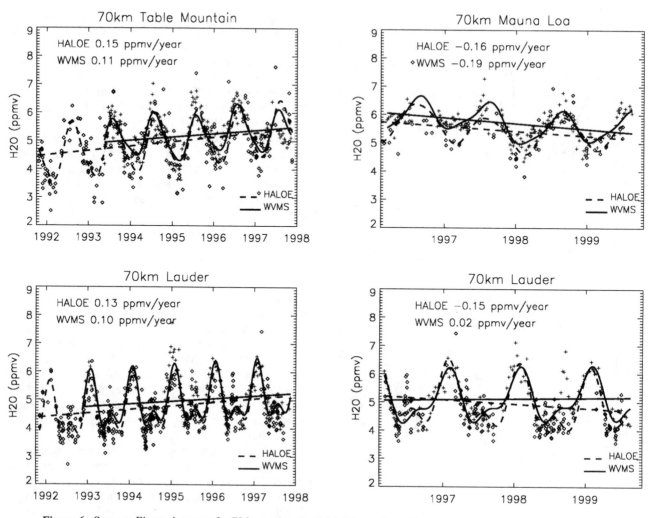

**Figure 6.** Same as Figure 4, except for 70 km. Also, the Table Mountain and Mauna Loa fits both include QBO terms.

entering the stratosphere is <0.01 ppmv/year for the period 1984-1996, and this rate appears to be decreasing. The combined effect of these changes in methane contribute at most ~1/2 of the increase in water vapor. The remaining fraction can most plausibly be attributed to an increase in water vapor entering the stratosphere from the troposphere.

A prolonged increase in water vapor mixing ratio at the rate observed in the early 1990s would have a significant impact on ozone recovery in the upper stratosphere, where ozone recovery may first become apparent. *Jucks and Salawitch* [2000] find that a 2.0 ppmv increase in water vapor will negate the recovery of ozone caused by a 15% reduction of $Cl_y$. *Siskind et al.* [1998], however, suggest that for the standard $HO_x$ chemistry, and for the some previously suggested perturbations thereof, the model $O_3$ may be too sensitive to changes in $H_2O$. In addition to affecting ozone, *Forster and Shine* [1999] have shown that a prolonged increase in stratospheric water vapor would also cause significant cooling of the stratosphere.

The apparent termination of the large upward trend in stratospheric and mesospheric water vapor that was seen in the first half of the 1990s still leaves open the question of its cause. Certainly the solar cycle does play some role in the upper mesosphere, and this region clearly does show a decrease in water vapor mixing ratio since ~1996. There is, however, no well understood mechanism that relates the solar cycle to changes in lower mesospheric or stratospheric water vapor. In addition, the increased oxidation of methane that played an important part in the water vapor increase does not appear to be related to variations in the solar cycle [*Nedoluha et al.,* 1998b]. The only major perturbation of the stratosphere that occurred in the early 1990s was the eruption of Mount Pinatubo in June, 1991. This eruption may have allowed more water to enter the stratosphere, possibly by warming the tropical tropopause in regions where water enters the stratosphere, but we note that several recent studies indicate a cooling of the tropical tropopause since 1979, and none show a warming associated with the Mount Pinatubo eruption [*Simmons et al.*, 1999; *Randel et al.*, 2000; *Zhou et al., submitted manuscript*, 2000]. There are, however, indications that the eruption may have altered stratospheric transport [*Rosenfield et al.*, 1997; *Eluszkiewicz et al.*, 1997; *Nedoluha et al.*, 1998b]. If the changes in water vapor in the first half of the 1990s were caused by the eruption of Mount Pinatubo, then (in the absence of any other major perturbations to the stratosphere and to stratosphere-troposphere exchange) we might expect a slow decrease in water vapor in the stratosphere and lower mesosphere in the next few years, even after the solar cycle reaches its maximum in ~2000. Alternatively, if the solar cycle is playing some yet to be understood role in changing water vapor in the stratosphere, as is suggested by the fact that both stratospheric and upper mesospheric water vapor mixing ratios seem to peak near the minimum of the solar cycle, then we might expect the water vapor mixing ratio to begin increasing again after ~2000. Continued monitoring of middle atmospheric water vapor by satellite and ground-based measurements should help us to determine the importance of these external perturbations on the composition of the Earth's atmosphere.

*Acknowledgments.* The WVMS project has benefited greatly from the assistance of S. McDermid, D. Walsh, and R. Cageao who have provided on-site support for the WVMS instruments at Table Mountain and Mauna Loa for many years. We thank the HALOE team for making their data readily available, P. Newman's group at NASA Goddard and the Climate Prediction Center (CPC) of the National Centers for Environmental Prediction (NCEP) for providing daily temperature and pressure data for all three sites, and D. Siskind for making his NRL 2D model runs available. This project has been funded for many years by NASA under the Upper Atmospheric Research Program, and has also received funding from the Department of Defense Strategic Environmental Research and Development Program (SERDP).

## REFERENCES

Barath, F. T., et al., The upper atmosphere research satellite microwave limb sounder instrument, *J. Geophys., Res.*, 98, 10751-10762, 1993.

Bevilacqua, R. M., et al., An observational study of water vapor in the midlatitude mesosphere using ground-based microwave techniques, *J. Geophys. Res.*, 88, 8523-8534, 1983.

Bevilacqua, R. M., et al., Possible seasonal variability of mesospheric water vapor, *Geophys. Res. Lett.*, 12, 397-400, 1985.

Bevilacqua, R. M., W. J. Wilson, and P. R. Schwartz, Measurements of mesospheric water vapor in 1984 and 1985: results and implications for middle atmospheric transport, *J. Geophys. Res.*, 92, 6679-6690, 1987.

Bevilacqua, R. M., et al., The seasonal variation of water vapor and ozone in the upper mesosphere: Implications for vertical transport and ozone photochemistry, *J. Geophys. Res.*, 95, 883-893, 1990.

Brasseur, G., and S. Solomon, Aeronomy of the middle atmosphere, D. Reidel, 2nd edition, 1986.

Chandra S., C. H. Jackman, E. L. Fleming, and J. M. Russell, The seasonal and long term changes in mesospheric water vapor, *Geophys. Res. Lett.*, 24, 639-642, 1997.

Conway, R. R., et al., Satellite measurements of hydroxyl in the mesosphere, *Geophys. Res. Lett.*, 23, 2093-2096, 1996.

Dlugokencky, E. J., et al., Continuing decline in the growth rate of the atmospheric methane burden, *Nature 393/4*, 447-450, 1998.

Eluszkiewicz, J., et al., Sensitivity of the residual circulation diagnosed from the UARS data to the uncertainties in the input fields and to the inclusion of aerosols, *J. Atmos. Sci.*, 54, 1739-1757, 1997.

Evans, S. J., et al., Trends in stratospheric humidity and the sensitivity of ozone to these trends, *J. Geophys. Res.*, 103, 8715-8725, 1998.

Finger, F. G., et al., Evaluation of NMC upper-stratospheric temperature analyses using rocketsonde and lidar data, *Bull. Am. Meteorol. Soc.*, 74, 789-799, 1993.

Forster, P. M. F., and K. P. Shine, Stratospheric water vapour changes as a possible contributor to observed stratospheric cooling, *Geophys. Res. Let.*, 26, 3309, 1999.

Garcia, R. R., and S. Solomon, The effect of breaking gravity waves on the dynamics and chemical composition of the mesosphere and lower thermosphere, *J. Geophys. Res.*, 90, 3850-3868, 1985.

Hedin, A. E., Extension of the MSIS thermosphere model into the middle and lower atmosphere, *J. Geophys. Res.*, 96, 1159-1172, 1991.

Huang, T. Y. W., and G. P. Brasseur, Effect of long-term solar variability in a two-dimensional interactive model of the middle atmosphere, *J. Geophys. Res.*, 98, 20413-20427, 1993.

Liebe, H. J., G. A. Hufford, and M. G. Cotton, Atmospheric propagation effects through natural and man-made obscurants, *AGARD conference proceedings 542*, 1993.

Jarchow, C., Bestimmung atmosphaerischer Wasserdampf- und Ozonprofile mittels bodengebundener Millimeterwellen-Fernerkundung, Ph.D. thesis, Max-Planck-Institut fuer Aeronomie, 1999.

Jucks, K. W., and R. J. Salawitch, Future changes in upper stratospheric ozone, *this volume*, 2000.

Nedoluha, G. E., et al., Ground-based measurements of water vapor in the middle atmosphere, *J. Geophys. Res.*, 100, 2927-2939, 1995.

Nedoluha, G. E., et al., Measurements of water vapor in the middle atmosphere and implications for mesospheric transport, *J. Geophys. Res.*, 101, 21,183-21,193, 1996.

Nedoluha, G. E., et al., A comparative study of mesospheric water vapor measurements from the ground-based Water Vapor Millimeter-wave Spectrometer and space-based instruments, *J. Geophys. Res.*, 102, 16,647-16,661, 1997.

Nedoluha, G. E., et al., Increases in middle atmospheric water vapor as observed by the Halogen Occultation Experiment and the ground-based Water Vapor Millimeter-wave Spectrometer from 1991 to 1997, *J. Geophys. Res.*, 103, 3531-3543. 1998a.

Nedoluha, G. E., et al., Changes in upper stratospheric $CH_4$ and $NO_2$ as measured by HALOE and implications for changes in transport, *Geophys. Res. Let.*, *25*, 987-990, 1998b.

Nedoluha, G. E., et al., Measurements of middle atmospheric water vapor from low latitudes and midlatitudes in the Northern Hemisphere, 1995-1998, *J. Geophys. Res.*, *106*, 19257-19266, 1999.

Olivero, J. J., et al., Solar absorption microwave measurement of upper atmospheric water vapor, *Geophys., Res. Let.*, *13*, 197-200, 1986.

Oltmans, S. J., and D. J. Hofmann, Increase in lower-stratospheric water vapour at a mid-latitude Northern Hemisphere site from 1981 to 1994, *Nature*, *374*, 146-149, 1995.

Pol, S. L. C., C. S. Ruf, and S. J. Keihm, Improved 20- to 32-GHz atmospheric absorption model, *Radio Science*, *33*, 5, 1319-1333, 1998.

Pumphrey, H. C., Validation of a new prototype water vapor retrieval for the UARS microwave limb sounder, *J. Geophys. Res.*, *104*, 9399-9412, 1999.

Pickett, H. M., et al., Submillimeter, millimeter, and microwave spectral line catalog, *JQRST*, *60*, 883-890, 1998.

Radford, H. E., et al., Mesospheric water vapor measured from ground-based microwave observations, *J. Geophys. Res.*, *92*, 472-278, 1977.

Randel, W. J., et al., Seasonal cycles and QBO variations in the stratospheric $CH_4$ and $H_2O$ observed in UARS HALOE data, *Jour. Atmos. Sci.*, *55*, 163-185, 1998.

Randel, W. J., et al., Space-time patterns of trends in stratospheric constituents derived from UARS measurements, *J. Geophys. Res.*, *104*, 3711-3727, 1999.

Randel, W.J., F. Wu and D. Gaffen, Interannual variability of the tropical tropopause derived from radiosonde data and NCEP reanalyses. *J. Geophys. Res, in press*, 2000.

Rodgers, C. D., Retrieval of atmospheric temperature and composition from remote measurements of thermal radiation, *Rev. Geophys.*, *14*, 609-624, 1976.

Rodgers, C. D., Characterization and error analysis of profiles retrieved from remote sounding measurements, *J. Geophys. Res.*, *95*, 5587-5595, 1990.

Rosenfield, J. E., et al., Stratospheric effects of Mount Pinatubo aerosol studied with a coupled two-dimensional model, *J. Geophys. Res.*, *102*, 3649-3670, 1997.

Russell, J. M. III, et al., The halogen occultation experiment, *J. Geophys. Res.*, *98*, 10777-10797, 1993.

Seele, C., and P. Hartogh, Water vapor of the polar middle atmosphere: Annual variation and summer mesosphere conditions as observed by ground-based microwave spectroscopy, *Geophys. Res. Let.*, *26*, 1517-1520, 1999.

Shine, K., Sources and sinks of zonal momentum in the middle atmosphere diagnosed using the diabatic circulation, *Quart. J. Roy. Met. Soc.*, *115*, 265-292, 1989.

Simmons, A. J., et al., Stratospheric water vapor and tropical tropopause temperatures in ECMWF analyses and multi-year simulations, *Quart. J. Roy. Meteor. Soc.*, *125*, 353-386, 1999.

Siskind, D. E., et al., Implications of upper stratospheric trace constituent changes observed by HALOE for $O_3$ and ClO from 1992 to 1995, *Geophys. Res. Lett.*, *25*, 3513-3516, 1998.

Siskind, D. E. and M. E. Summers, Implications of enhanced mesospheric water vapor observed by HALOE, *Geophys. Res. Let.*, *25*, 2133-2136, 1998.

Siskind D. E., On the coupling between middle and upper atmospheric odd nitrogen, *this volume*, 2000.

Summers, M. E., and R. R. Conway, Insights into middle atmospheric hydrogen chemistry from analysis of MAHRSI observations of hydroxyl (OH), *this volume*, 2000.

Summers, M. E., et al., Implications of satellite OH observations for middle atmospheric $H_2O$ and ozone, *Science*, *277*, 1967-1970, 1997a.

Summers, M. R., et al., Seasonal variation of middle atmospheric $CH_4$ and $H_2O$ with a new chemical-dynamical model, *J. Geophys. Res.*, *102*, 3502-3526, 1997b.

Summers, M. E. and D. E. Siskind, Surface recombination of O and $H_2$ on meteoric dust as a source of mesospheric water vapor, *Geophys. Res. Let.*, *26*, 1837-1840, 1999.

Taylor, F. W., et al., Remote sensing of atmospheric structure and composition by pressure modulator radiometry from space: The ISAMS experiment on UARS, *J. Geophys. Res.*, *98*, 10799-10814, 1993.

Thomas, R. J., C. A. Barth, and S. Solomon, Seasonal variations of ozone in the upper mesosphere and gravity waves, *Geophys. Res. Lett.*, *11*, 673-676, 1984.

Tsou, J.-J., J. J. Olivero, and C. L. Croskey, Study of variability of mesospheric $H_2O$ during spring 1984 by ground-based microwave radiometric observations, *J. Geophys. Res.*, *93*, 5255-5266, 1988.

Westwater, E. R., Groundbased Microwave Radiometry, in Atmospheric Remote Sensing by Microwave Radiometry, Janssen, M. A., ed., J. Wiley & Sons, 145-213, 1993.

WMO, Scientific assessment of ozone depletion, 1998. World Meteorological Organization, global ozone research and monitoring project, *44*, Geneva, Switzerland, 1999.

R. M. Bevilacqua, R. M. Gomez, B. C. Hicks, and G. E. Nedoluha, Naval Research Laboratory, Code 7227, Washington, DC 20375. (nedoluha@nrl.navy.mil).

B. J. Connor, National Institute of Water and Atmospheric Research, Private Bag 50061, Omakau, Central Otago, New Zealand (connor@kea.lauder.cri.nz).

J. M. Russell III, Dept. of Physics, Hampton University, Hampton, VA 23668. (JAMES.RUSSELL@hamptonu.edu).

# Interannual Changes in Stratospheric Constituents and Global Circulation Derived From Satellite Data

William J. Randel and Fei Wu

*Atmospheric Chemistry Division, National Center for Atmospheric Research, Boulder, CO*


J. M. Russell III

*Atmospheric Sciences, Hampton University, Hampton, VA*


J. M. Zawodny

*Atmospheric Chemistry Division, NASA Langley Research Center, Hampton, VA*


John Nash

*UK Meteorological Office, Bracknell, Berkshire, UK*

The characteristics of low-frequency interannual changes in stratospheric constituents are examined using measurements from the HALOE [1991-1999] and SAGE II [1984-1998] satellite instruments. Both the increasing $H_2O$ and decreasing $CH_4$ observations from the first several years of HALOE measurements are found to be absent (or partially reversed) in the data since ~ 1996. Ozone and $NO_x$ ($NO + NO_2$) observations from HALOE also show 'trends' over ~ 1991-1995 that become flat in the latter record. Comparisons with the longer-term SAGE II measurements of ozone and $NO_2$ demonstrate that changes over the short period 1991-1995 are not representative of decadal-scale trends, but rather are episodic in nature. One possibility is that these changes reflect a prolonged response to the Mt. Pinatubo volcanic eruption in 1991. Stratospheric temperatures and derived circulation statistics are examined for the period 1979-1999. While Pinatubo produced enhanced tropical upwelling in the stratosphere for 1-2 years, other sources of stratospheric variability probably also contribute to low-frequency constituent 'trends.'

## 1. INTRODUCTION

Decadal-scale changes in temperature and circulation of the stratosphere and mesosphere are anticipated due to increases in greenhouse gas concentrations, and decreases in stratospheric ozone [e.g., *WMO*, 1999]. Similarly,

concentrations of constituents in the middle atmosphere are expected to change due to trends in tropospheric source gases (e.g., $CO_2$, $N_2O$, and $CH_4$), trends in chlorine affecting ozone chemistry, and changes in dynamics and transport. Indeed, observed variations in long-lived constituents can be a sensitive measure of changes in (integrated) large-scale transport, which may not otherwise be observed in dynamical quantities directly. In addition to externally forced changes, the middle atmosphere can exhibit significant interannual variability due solely to internal dynamics [e.g., *Scott and Haynes*, 1998; *Hamilton*,

Atmospheric Science Across the Stratopause
Geophysical Monograph 123

this volume]. Although constituent observations alone cannot identify 'forced' versus 'natural' changes, it is instructive to quantify the interannual variability in the long time series of data now available from satellites.

The long-term record (1991-present) of constituent observations from the Halogen Occultation Experiment (HALOE) [*Russell et al.*, 1993] has provided an unprecedented view of interannual variations in stratospheric constituents. Some of the most intriguing and unanticipated results from the first several years of HALOE data were observations of increasing water vapor throughout the stratosphere, together with decreasing $CH_4$ above 35 km [*Nedoluha et al.*, 1998a, b; *Evans et al.*, 1998; *Randel et al.*, 1999]. The causes of these trends are still unknown, and one key question is do these 'trends' continue in the longer record of HALOE measurements. HALOE has also observed trend-like changes in other constituents. *Russell et al.* [1996] used HALOE observations of increasing HF and HCl near the stratopause to show consistency with tropospheric emission rates, and effectively close the stratospheric budgets of chlorine and fluorine. The HF observations were updated by *Considine et al.* [1997] to show a slowdown in the rate of increase. *Anderson et al.* [2000] analyzed the HCl and HF time series through 1999, and tied the stratospheric changes to tropospheric source variations linked to the Montreal Protocol. *Siskind et al.* [1998] have studied the relationships between chlorine, methane and ozone changes observed by HALOE during 1992-1995, and *Froidevaux et al.* [2000] has quantified long-term changes in stratospheric chlorine compounds during the HALOE observing period.

The objective here is to update the HALOE constituent observations through 1999 to quantify the spatial and temporal characteristics of observed interannual changes. The $H_2O$ and $CH_4$ variations are shown to be coherent throughout the stratosphere and lower mesosphere, demonstrating interannual coupling between these regions. In order to provide a longer-term perspective on interannual variability, we include SAGE II observations of ozone and $NO_2$ for the period 1984-1998. These are the longest record of stratospheric constituents measured by a single satellite instrument, and it is instructive to compare the SAGE II and HALOE records of these same species. Finally, we briefly examine some diagnostics of changes in stratospheric circulation over the period 1979-1999. Although extensive diagnostics of constituent transport circulations are beyond the scope of this paper, we present some diagnostics of interannual changes in stratospheric temperature and planetary wave variability which give qualitative measures of large-scale circulation changes.

## 2. DATA AND ANALYSES

### 2.1. HALOE

The HALOE instrument uses solar occultation measurements to provide high quality observations of stratospheric $H_2O$, $CH_4$, HF, HCl, ozone, NO, and $NO_2$ [*Russell et al.*, 1993]. The HALOE data here are from the version 19 (v19) retrieval algorithm, spanning the time period October 1991 – December 1999. We use HALOE level 3a data, with vertical sampling on standard UARS pressure levels (6 per decade of pressure; vertical resolution of ~ 2.5 km). The $H_2O$ and $CH_4$ measurements extend over ~ 15-80 km, while the other species are analyzed here for ~ 15-55 km. Sunrise and sunset HALOE measurements are combined for all constituents except NO and $NO_2$. We bin the data into monthly samples on a 4° latitude grid prior to the statistical analyses discussed below. Because HALOE does not observe polar regions during much of the year, we focus results on the latitude band 60°N-S.

The HALOE measurement approach is ideal for examining long-term changes in constituents because it uses a ratio method. Retrievals are performed on limb transmission versus height, obtained by dividing the signal observed while viewing the sun through the atmospheric limb to that measured when viewing the unattenuated solar signal (prior to or after an occultation event). This method is virtually self-calibrating in terms of systematic errors such as detector responsivity or optics throughput changes, i.e., these errors are removed by the ratio. Other instrument checks routinely performed (e.g., spectral, gas cell, pointer tracker) show no significant long-term changes. In summary, all indications from the HALOE instrument and experiment support the validity of the observed long-term changes in constituents.

### 2.2. SAGE II

Ozone and $NO_2$ observations are also available from SAGE II measurements spanning November 1984- December 1998 [*McCormick et al.*, 1989; *Cunnold et al.*, 1991]. We use here an experimental version of the v6.0 retrieval (termed v6.0b), which incorporates improved separation of constituent and aerosol effects together with improved calibration and altitude registration characteristics. Although these new data products are still undergoing scientific validation analyses, we note that the results of interest here are also evident in prior versions of SAGE II data (i.e., retrieval v5.96). Since SAGE II is a solar occultation instrument, the space-time sampling (and our data analysis) is similar to that for HALOE.

## 2.3. Stratospheric Circulation Diagnostics

We include analyses of the long record of stratospheric temperature measurements derived from the Stratospheric Sounding Unit (SSU) on board the series of NOAA operational satellites. The time series here are monthly averages spanning the time period November 1978-May 1998. These data have been adjusted using temporal overlap between individual satellite instruments to provide homogeneous time series of stratospheric radiance measurements; they provide more temporal continuity in the middle and upper stratosphere than available using operational NMC/CPC stratospheric temperatures [e.g., *Finger et al.*, 1993]. These measurements are from the 3 original SSU channels, plus 5 additional 'synthetic' channels derived from differences between nadir and off-nadir SSU measurements [see *Nash*, 1988]. The radiance measurements correspond to layer mean temperatures over ~ 10-15 km thick levels; the weighting functions for these temperature measurements can be found in Chapter 5 of *WMO* [1999]. We furthermore include some diagnostics of stratospheric eddy forcing derived from the NCEP [*Kalnay et al.*, 1996] and ECMWF [*Gibson et al.*, 1997] reanalyses, and from the NCEP/CPC stratospheric analyses [*Finger et al.*, 1993]. The estimates of interannual changes in these derived quantities are subject to some uncertainties, and we include results from these separate analyses to provide one estimate of such uncertainties.

## 2.4. Data Analyses

Our objective here is to quantify the low-frequency trend-like variability in the constituent data sets. In general these interannual changes are relatively small compared to oscillatory variations associated with the seasonal cycle and the quasi-biennial oscillation (QBO), although this statement depends on the species in question and the specific location. An example where the seasonal and QBO components are large is for $H_2O$ in the low latitude middle stratosphere, as shown in the time series of Figure 1. For the occultation measurements we fit the seasonal cycle at each latitude and pressure using harmonic regression analysis (Figure 1a). The deseasonalized interannual anomalies (Figure 1b) show large variations with an approximate 2-year period, which are coherent with the QBO in equatorial zonal winds. A regression onto the QBO winds over 70-10 mb (using two orthogonal basis

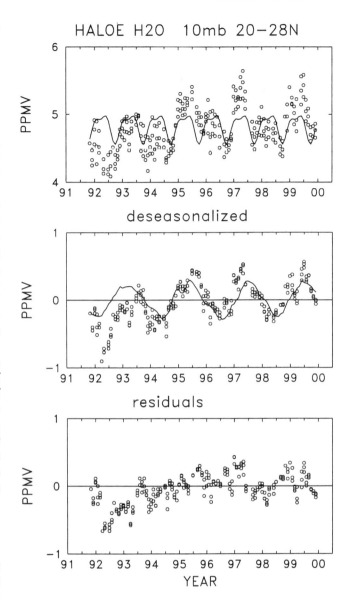

**Figure 1.** (top) Time series of HALOE $H_2O$ measurements at 10 mb over 20-28°N. The solid line shows the (repeating) seasonal cycle fit. (middle) Corresponding deseasonalized $H_2O$ anomalies, together with the QBO fit derived from regression. (bottom) Residual anomalies (deseasonalized anomalies minus the QBO fit).

functions) provides a dynamically-based accurate empirical fit of the QBO variability, as shown in Figure 1b [the details of the seasonal and QBO regression fits are discussed in *Randel and Wu*, 1996 and *Randel et al.*, 1998]. The residual of the QBO-fit (Figure 1c) shows the low-frequency changes of most interest for the current work:

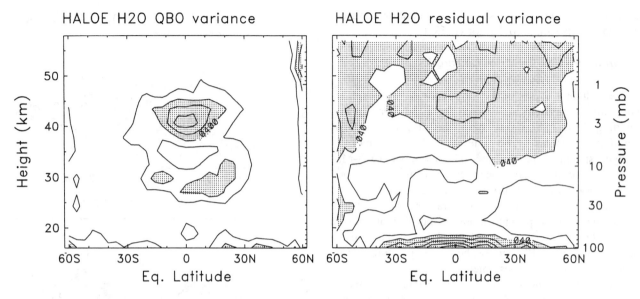

**Figure 2.** Meridional cross sections of interannual variance in HALOE $H_2O$ data associated with the QBO (left) and the residual (right) for data over 1991-1999. The contour interval is 0.02 ppmv$^2$.

the $H_2O$ variations in Figure 1c show a slight increase during 1991-1995, and relatively constant values thereafter. Note that these changes are evident in the 'full' time series (Figure 1a), but are highlighted by removing the seasonal and QBO signals.

An important point is that the QBO is often a large component of variability in constituent data, and one which complicates interpretation of "other" low-frequency changes in relatively short time records. Figure 2 shows the temporal variance in HALOE $H_2O$ data during 1991-1999 associated with the QBO and with the residual interannual anomalies (calculated as in Figure 1). The distinctive equatorially-centered spatial structure of the QBO is evident in Figure 2, contrasting with the global character of the residual. Figure 2 illustrates that the relative importance of the QBO depends on the altitude and latitude region of interest; for the HALOE and SAGE II constituent data the QBO is generally large between 20-45 km and 40°N-S.

Our philosophy here is to isolate the low-frequency interannual changes in constituents using the least amount of filtering necessary. Time series of area-averaged deseasonalized anomalies illustrate the variations most simply; we remove the QBO signal for ozone and $NO_x$ variations in the middle stratosphere in some of the diagnostics below, using the regression fits described above. Anomalies averaged over each hemisphere are shown to illustrate the global character of the 'trends.' Smoothed versions of the global anomalies are derived by use of a moving Gaussian weighting, with a half-width of

15 months (roughly equivalent to 3-year running means), intended to minimize QBO effects. Although the overall variability is not linear for most species, we calculate linear trends over 5-year samples at the beginning and end of the record in order to quantify spatial structure of changes during these times. The trends are calculated by standard regression analyses, and standard errors of the statistical fits are derived using bootstrap resampling of the original anomalies [*Efron and Tibshirani*, 1993].

## 3. CONSTITUENT OBSERVATIONS

### 3.1. $CH_4$ and $H_2O$

The overall structure and variability of $CH_4$ and $H_2O$ are tightly coupled in the stratosphere, because $CH_4$ oxidation is a principal source of stratospheric $H_2O$ [e.g., *Remsberg et al.*, 1984]. The oxidation reactions produce approximately two molecules of $H_2O$ for each one of $CH_4$; this photochemical conversion occurs as air ages in the middle and upper stratosphere, and hence anticorrelated changes in $CH_4$ and $H_2O$ signify variations in the age of air induced by transport. Empirical studies have shown the quantity $D = H_2O + 2CH_4$ is an approximately conserved parameter throughout much of the stratosphere [*Dessler et al.*, 1994; Remsberg et al., 1996], and we thus examine the variability of $CH_4$, $H_2O$, and $H_2O + 2CH_4$.

Figure 3 shows time series of hemispherically-averaged deseasonzlied anomalies in $CH_4$, $H_2O$, and $H_2O + 2CH_4$ near the stratopause. Included on these plots are smooth

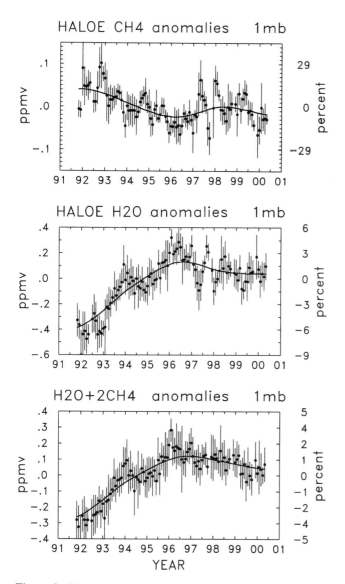

**Figure 3.** Time series of deseasonalized anomalies in $CH_4$ (top), $H_2O$ (middle) and $H_2O + 2CH_4$ (bottom) at 1 mb (~ 48 km) derived from HALOE data. Circles denote the monthly global means, and the vertical lines show plus and minus one standard deviation of all observations about the mean. The smooth curves show a moving average of the global means. Left axes are mixing ratio units, and right axes show relative percentage changes.

curves showing the low frequency interannual variations of the corresponding global means. The $CH_4$ anomalies show a general decrease during 1991-1995, and a slight increase since 1996. Water vapor shows just the opposite low frequency changes: increases during 1991-1995, and a leveling off or slight decrease thereafter. These 1991-1995 changes correspond to the trends reported by *Nedoluha et al.* [1998a, b], *Evans et al.* [1998] and *Randel et al.* [1999], but they have clearly not continued beyond 1996. The

variable part of total hydrogen ($H_2O + 2CH_4$) shows similar behavior as $H_2O$ alone, demonstrating that variations in $CH_4$ oxidation produce only a fraction (approximately 20%) of the $H_2O$ change during 1991-1995 (i.e., there are sources of $H_2O$ variability separate from the $CH_4$ source). Note the compensating $CH_4$-$H_2O$ deviations in 1997-1998 in Figure 3, suggestive of global transport variability on relatively fast time scales.

Time series of deseasonalized $H_2O + 2CH_4$ anomalies at 31, 10 and 3 mb (near 24, 32 and 40 km, respectively), are shown in Figure 4. Each of these time series shows similar low-frequency changes as the ~ 50 km data in Figure 3, i.e., a clear increase during the first half of the HALOE record, and near-constant or slightly decreasing values thereafter. The time variations for $H_2O$ alone are similar (not shown). The global nature of the low-frequency changes, and the dramatic differences between the first half and latter half of the HALOE record, are illustrated in Figure 5, which show linear trends in $H_2O$ for the 5-year periods 1992-1996 and 1995-1999. The positive $H_2O$ trends (of ~ 1-2% per year) seen throughout the stratosphere and lower mesosphere for the 1992-1996 period are absent (or reversed) for the period since 1995. Similar calculations for $CH_4$ (Figure 6) also reveal dramatic differences: the intriguing 1992-1996 decrease in $CH_4$ over 35-65 km is reversed or absent in the recent record. These calculations demonstrate that the interannual variations in the short HALOE record are not representative of monotonic decadal-scale trends, but rather are suggestive of episodic changes. One exception to this is the $CH_4$ increase in the lower stratosphere (~ 15-30 km), which is positive (of order ~ 0.5-1.0% per year) over the entire HALOE record (Figure 6). The slight decrease in trend magnitude in this region for the latter half of the record is consistent with the slowing of tropospheric $CH_4$ increases reported by *Dlugokencky et al.*, 1998.

The largest $H_2O$ trends seen in Figure 5 occur in the upper mesosphere, over ~ 70-80 km, and correspond to 5-year changes of order ± 20-30%. *Chandra et al.* [1997] have analyzed $H_2O$ variability in HALOE data and a 2-D model, and demonstrated that large changes in the upper mesosphere are primarily due to changes in solar output (variations in solar Lyman $\alpha$ associated with the 11-year solar cycle, cause out-of-phase changes in $H_2O$). Figure 7 shows time series of global-mean HALOE $H_2O$ anomalies at 0.01 mb (~ 80 km), together with time series of the Lyman-$\alpha$ irradiance measured by the UARS SOLSTICE instrument [*London et al.*, 1993], showing this temporal anti-correlation. It is interesting that the overall low-frequency variations in $H_2O$ at 80 km are similar to changes

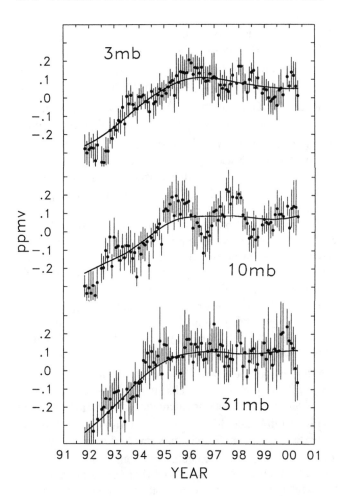

The HF measurements show a monotonic increase throughout the HALOE record, with a slowing of the trend rate after approximately 1997. The HCl time series show a clear maximum in ~ 1996-1997, with decreasing values for 1998-1999. The differing behaviors of HF and HCl are related to variations in the respective tropospheric fluorine and chlorine emissions, as demonstrated in *Anderson et al.* [2000]. The clear turnover in stratospheric HCl is associated with a similar change in tropospheric chlorine loading, which had an observed maximum in ~ 1993 [*Montzka et al.*, 1999]. The linear HCl and HF increases during 1992-1996 fit the tropospheric emissions with an approximate 5-year time lag [*Russell et al.*, 1996; *Anderson et al.*, 2000], roughly consistent with the inferred age of air in the upper stratosphere [*Hall et al.*, 1999]. However, the time lag between peak chlorine amounts in the troposphere (~ 1993) and upper stratosphere (~1996-1997) is only ~ 3-4 years, and the reason for this somewhat shorter time lag (i.e., earlier turnover of HCl than expected) is not understood at present.

### 3.3. Ozone and $NO_x$

Ozone and $NO_x$ ($NO + NO_2$) are photochemically coupled in the middle atmosphere, because $NO_x$ is the main mechanism for catalytic ozone destruction over ~ 25-40 km [e.g., *Osterman et al.*, 1997]. HALOE provides sunrise and sunset measurements of both NO and $NO_2$. Both constituents exhibit strong diurnal cycles, and sunset $NO_x$ is somewhat larger than sunrise $NO_x$ due to daytime photolysis of $N_2O_5$ [which builds up during the night; e.g., *Nevison et al.*, 1996]. Here we analyze the sunset observations of $NO_x$ together with ozone to demonstrate coupled variability.

Time series of ozone and $NO_x$ anomalies at 10 mb are shown in Figure 9, and here we have removed the QBO signal together with the seasonal cycle, because the QBO is relatively large for these constituents in the middle stratosphere [*Randel et al.*, 1996]. The month-to-month variability in these constituents appears significantly larger than that observed in the longer lived species H2O, CH4, HF and HCl (Figures 3, 4 and 8). In terms of low frequency changes, ozone shows a gradual decline over ~ 1991-1996, coincident with increasing $NO_x$, while both quantities are relatively constant since ~ 1996. The spatial structure of the anti-correlated ozone-$NO_x$ changes over 1992-1996 are illustrated by the linear trends shown in Figure 10. Ozone trends show a decrease in the tropical middle stratosphere over ~ 30-40 km, while $NO_x$ shows an increase everywhere below ~ 40 km and decreases over 40-50 km (note the spatially coincident maxima seen in O3 and

**Figure 4.** Time series of global anomalies in H2O + 2CH4 at 31, 10 and 3 mb (near 24, 32 and 40 km, respectively). Details are the same as in Figure 3.

at stratospheric levels (Figures 3-4) although *Chandra et al.* [1997] suggest that direct (photochemical) solar effects are mostly important above ~ 70 km (consistent with the vertical structure of largest trends seen in Figure 5).

### 3.2. HF and HCl

HF and HCl in the middle atmosphere are similar in that they both originate primarily as the photochemical breakdown products of tropospheric chlorofluorocarbons (CFC's). HF has no significant natural sources, while about 80% of stratospheric HCl originates from CFC's. Time series of HF and HCl near the stratopause are shown in Figure 8, extending the HALOE time series shown in *Russell et al.* [1996], *Considine et al.* [1997], and *WMO* [1999] and covering the same time period analyzed by *Anderson et al.* [2000]. In Figure 8 we show the full HF and HCl time series rather than deseasonalized data, in order to illustrate the magnitude of the absolute changes.

$NO_X$ trends near 35 km). Ozone and $NO_X$ trends for the period 1995-1999 (not shown) exhibit almost no significant values (e.g., Figure 9 ), so that the changes in the early HALOE record were not characteristic of longer-term (decadal) trends. This is also consistent with the fact that the 1992-1996 ozone trends (Figure 10) are completely different from the ~ 15-20 year ozone trends derived from SAGE or SBUV satellite data [see *WMO*, 1999], which exhibit no significant trends in the tropical middle stratosphere (as shown below). Two other aspects of interest in Figure 10 are (1) increases in tropical ozone near ~ 25 km, and (2) near-global $NO_X$ decreases over ~ 40-50 km. The latter is due almost entirely to decreases in NO, as $NO_2$ contributes little to $NO_X$ above 40 km.

A longer-term perspective of recent ozone and $NO_2$ variability is provided by SAGE II observations spanning 1984-present. SAGE II does not measure NO, so we focus on variability of $NO_2$ in comparisons with HALOE. Figure 11 show time series of anomalies in tropical ozone at 10 mb (~ 32 km), and $NO_2$ at 15 mb (~ 30 km) for both SAGE II and HALOE measurements. Here we have removed the respective QBO variations, which are relatively large for ozone and $NO_2$ in this region [*Randel and Wu*, 1996]. This comparison shows reasonable overall agreement between the HALOE and SAGE II data during 1992-1998, and in particular the low-frequency changes seen in HALOE data over ~ 1991-1995 (decreases in ozone and increases in $NO_2$) are also evident in the SAGE II results. However, the longer-term perspective of SAGE II observations suggests that the changes during this time period were anomalous and not representative of decadal-scale trends. Rather, the close association with the Pinatubo eruption suggests volcanic effects as a likely causal mechanism, although some substantial low-frequency changes also occur in the SAGE II data prior to Pinatubo (note the anti-correlated ozone-$NO_2$ variations during 1988-1990). Overall the HALOE-SAGE II agreement and the coherent changes in ozone and $NO_2$ are strongly suggestive of true atmospheric variability (as opposed to instrumental artifacts).

To further emphasize the variability of short-term 'trends,' Figure 12 compares SAGE II ozone trends for the short period 1992-1996 versus those calculated for 1984-1998. The 1992-1996 trends show largest changes in the tropics, with decreases over ~ 30-45 km and increases near 25 km; the spatial patterns and magnitudes are very similar to the HALOE results in Figure 10. In contrast, the 14-year SAGE II trends in Figure 12 are completely different, with significant negative trends only in the extratropical upper stratosphere, and small changes in the tropics. This clearly demonstrates that observed changes for the several years following Mt. Pinatubo do not reflect decadal-scale variations in the stratosphere.

## 4. STRATOSPHERIC TEMPERATURE AND CIRCULATION CHANGES

The observed low-frequency variations seen in the constituent data sets (particularly the long-lived $CH_4$ and $H_2O$ data) are suggestive of possible changes in stratospheric circulation: higher $CH_4$ near the stratopause can be a signature of enhanced tropical upwelling (bringing up photochemically 'younger' air with relatively more $CH_4$). Such variations in stratospheric circulation could result from the Pinatubo volcanic eruption in 1991 [*Nedoluha et al.*, 1998b], or from changes due to solar forcing, tropospheric wave driving of the stratosphere, or internal dynamic variability. The relative magnitudes of interannual variations in mean or eddy transport circulations associated with these different processes is poorly known at present, because such circulations are difficult to diagnose from observed or assimilated circulation statistics, particularly for small interannual variations. A more simplified and qualitative approach is to examine circulation statistics which are related to the strength of the mean stratospheric Brewer-Dobson circulation. Here we examine stratospheric temperature anomalies, which are indicative of circulation changes on the global scale. A second diagnostic of interannual changes in the mean circulation is provided by estimates of the tropospheric planetary wave forcing, as quantified in meteorological analyses (discussed below).

Time series of global mean temperature anomalies spanning the lower to upper stratosphere are shown in Figure 13. These are an update of the stratospheric temperature time series shown in *WMO* [1999]. Apparent in these data are (1) episodic warming in the lower-middle stratosphere for the El Chichon (1982) and Pinatubo (1991) volcanic eruptions, (2) significant long-term cooling, particularly in the upper stratosphere, and (3) an 11-year solar signal in the middle and upper stratosphere (which together with the negative trend produces a 'stair-step' variation near 42 km in Figure 13). The spatial structure of trends and solar cycle variations in the SSU temperature data are illustrated in Figure 14, with the solar component calculated by regression onto a smoothed F10.7-cm solar radio flux time series. These results are similar to those shown in Chapter 5 of *WMO* [1999], but updated using data through 1979-1998. Significant negative trends are observed over much of the stratosphere, with largest values (approximately -2K/decade) near the stratopause. The solar variations in the data are largest in the tropical middle and upper stratosphere, with approximately 1 K variations between solar minimum and maximum. The mean Brewer-Dobson circulation changes associated with the trend and solar cycle variations in the stratosphere have yet to be quantified, but are an important topic for future study.

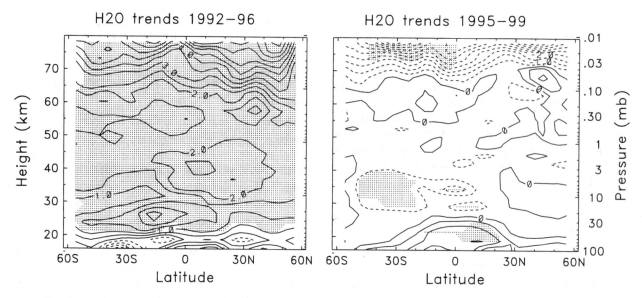

**Figure 5.** Meridional cross section of linear trends in HALOE $H_2O$ measurements for the period 1992-1996 (left) and 1995-1999 (right). Contours interval is 0.5% per year in both panels; shaded areas indicate statistically significant trends.

The volcanic eruptions of El Chichon and Pinatubo produce strong global mean warmings that last for 1-2 years (Figure 13). The spatial structure of the temperature anomalies in SSU data following the Pinatubo eruption are shown in Figure 15; these are calculated as average anomalies for one year following the eruption (July 1991-June 1992), with prior statistical removal of the trend, solar and QBO temperature signals. Shaded regions in Figure 15 show where the temperature anomalies are statistically significant, in the sense of being larger than twice the local standard deviation of annual mean anomalies over 1979-1997 (a measure of 'natural' variability of annual means). These temperature anomaly patterns show a warming of the lower stratosphere below ~ 27 km over much of the globe, with maxima in the tropics of ~ 1.5 K. The maximum lower stratospheric warming was probably in excess of 3 K [*Labitzke and McCormick*, 1992], but this signal is less in the thick layer SSU temperatures. In the middle and upper stratosphere, significant warm anomalies are observed in extratropics of each hemisphere, with relative cooling in the tropics. There is a good deal of global symmetry in the patterns in Figure 15, and although the magnitudes of the anomalies are not large, their persistence for an entire year is remarkable. Although not shown here, overall similar anomalies are found for the year following the El Chichon eruption in 1982.

Included in the Pinatubo results in Figure 15 are arrows indicating (qualitatively) the mean meridional circulation anomalies which would be in balance with the upper stratospheric temperature patterns [see *Andrews et al.*, 1987]. These show sinking motion in high latitudes of both

hemispheres and, by continuity, anomalous upwelling in the tropics. Calculated model results [e.g., *Brasseur and Granier*, 1992; *Rosenfield et al.*, 1997] and diagnostic analyses [*Eluszkiewicz et al.*, 1997] of the Pinatubo tropical circulation following the Pinatubo eruption show upward circulation anomalies also in the tropical lower stratosphere; the direct aerosol radiative heating (between ~ 15-30 km) results in local warming (as in Figure 15) plus induced upward motion. Hence the inferred global circulation anomalies for the year following the Pinatubo eruption are upward in the tropics throughout the entire stratosphere, with compensating downward motion over both NH and SH extratropics. This enhanced tropical upwelling is qualitatively consistent with the relatively high $CH_4$ anomalies during 1992-1993 seen in Figure 3.

A fundamental driver of the stratospheric circulation is the vertical flux of planetary wave activity from the troposphere into the winter stratosphere [*Andrews et al.*, 1987; *Haynes et al.*, 1991; *Holton et al.*, 1995]. Interannual variations in wave forcing should therefore be reflected in changes in mean circulation; analyses of wave forcing diagnostics are thus complementary to studies of global temperature variability. A concise diagnostic for the amount of planetary wave forcing from the troposphere into the stratosphere is given by the vertical component of the Eliassen-Palm (EP) flux in the lower stratosphere, which is proportional to the zonal mean eddy heat flux $\overline{v'T'}$ [*Andrews et al.*, 1987]. Figure 16 shows time series of winter-mean (November-March) eddy heat flux in the NH (averaged over 40-70°N at 100 mb), for the years 1979-1999 (1979=November 1978 to March 1979). Three

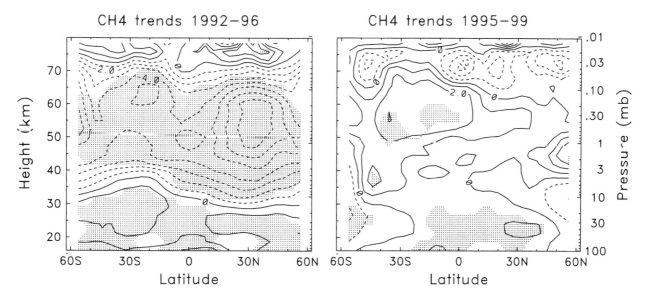

## CH4 trends 1992-96   CH4 trends 1995-99

**Figure 6.** Linear trends in HALOE CH4 measurements for the periods 1992-1996 (left) and 1995-1999 (right). Contour interval is 1% per year.

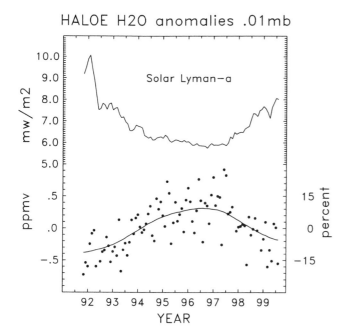

## HALOE H2O anomalies .01mb

Solar Lyman-α

**Figure 7.** Time series of global anomalies in HALOE H₂O at 0.01 mb (~ 80 km), together with variability of the solar Lyman-α flux measured by the SOLSTICE instrument on UARS.

separate estimates of $\overline{v'T'}$ are shown in Figure 16, derived from NCEP and ECMWF reanalyses and Climate Prediction Center (CPC) stratospheric analyses; these give slightly different results which provide one estimate of the uncertainty in this derived quantity (see Newman and Nash, 2000, for detailed discussion). The time series in Figure 16 show interannual variations in wave forcing of up to ~ ±

15% for the winter mean. Of particular note is the relative minimum in wave forcing for the years 1993-1997, compared to the pre-1990 mean; this minimum is somewhat accentuated in the CPC results. This reduced wave driving can result is a weakened Brewer-Dobson circulation (reduced tropical upwelling), which could be reflected in the relatively low CH4 anomalies during ~ 1994-1997 seen in Figure 3. Furthermore, the increased wave forcing during 1998-1999 seen in Figure 16 could by this mechanism account for the increased CH4 observed during this period in Figure 3. However, these arguments are qualitative, and detailed modeling studies are probably required to isolate tropospheric forcing effects from volcanic, solar or other mechanisms.

## 5. SYNTHESIS AND DISCUSSION

Beyond the seasonal cycle and QBO variations, the most obvious characteristic of the HALOE constituent data are a change in low-frequency interannual 'trends.' The remarkable increases in H₂O throughout the stratosphere, and decreases in CH4 in the upper stratosphere-lower mesosphere, seen in the early HALOE record have not continued beyond ~ 1996 (and partially reverse in 1998-1999). In fact, there are virtually no significant trends in the H₂O and CH4 data over the period 1995-1999 (except for increases in CH4 in the lower stratosphere). Similar temporal variability is observed in HALOE measurements of ozone and NOₓ, i.e., a flattening of 'trends' apparent in the early record. The longer term perspective of SAGE II

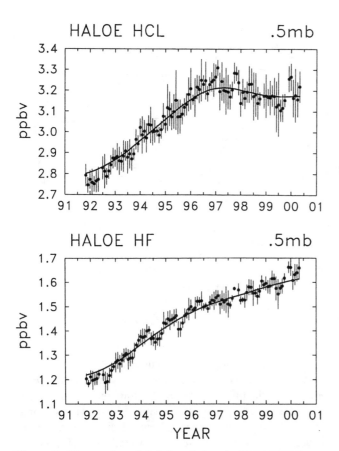

HALOE HCL                .5mb

HALOE HF                 .5mb

YEAR

**Figure 8.** Time series of global variations in HALOE HCl (top) and HF (bottom) at 0.5 mb (~ 53 km). Note these data are not deseasonalized. Details are the same as Figure 3.

data demonstrates that the ozone and $NO_2$ changes over ~ 1992-1996 are not characteristic of decadal-scale variations, but are more episodic in character. One possibility is that these constituent variations reflect a prolonged recovery from the Mt. Pinatubo eruption in 1991, since more recent data exhibit values similar to pre-Pinatubo averages. HF is the only stratospheric constituent measured by HALOE that continues to increase after 1997. HCl decreases after this time, which is evidence for the atmospheric response to the Montreal Protocol regulations [*Anderson et al.*, 2000].

It is reasonable to question the reality of the interannual constituent changes found in HALOE data, particularly for $CH_4$ and $H_2O$ where other global measurements are unavailable. There are several aspects of these data which suggest that the changes discussed here are real. First, careful analyses of instrument stability over time show no obvious problems that would translate into spurious constituent trends. Second, there is substantial coherence in both space and time for variations in the separate HALOE constituents (e.g., the $CH_4$, $H_2O$, ozone and $NO_x$ all show

similar low frequency temporal changes). Third, there is overall good agreement between the HALOE and SAGE II ozone and $NO_2$ variations during 1991-1998. Also, the increases in HALOE $H_2O$ over the early record (1991-1996) were confirmed in ground-based remote sensing measurements in both hemispheres [*Nedoluha et al.*, 1998a]. Thus there are good reasons to believe the overall patterns of variability seen in the HALOE measurements. However, an important point is that 'trends' derived from short time records, with arbitrary starting and ending points, are not representative of decadal-scale change. Thus interpretation of trend-like changes in the HALOE data is a continuing question at this point.

The upper stratospheric $CH_4$ and $NO_x$ changes suggest interannual variations in the global stratospheric circulation. Pinatubo is one mechanism for such variability in the UARS time frame, together with interannual changes in tropospheric wave forcing or solar cycle effects. An increase in tropical upwelling (say, during 1991-1992) will produce 'younger' air in the upper stratosphere, with correspondingly more $CH_4$ and less $H_2O$, qualitatively

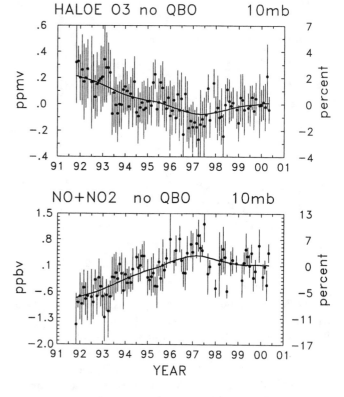

HALOE O3 no QBO         10mb

NO+NO2   no QBO         10mb

YEAR

**Figure 9.** Time series of global anomalies in HALOE ozone (top) and sunset $NO_x$ (NO + $NO_2$) (bottom) at 10 mb (~ 32 km). These time series have the seasonal and QBO variations removed. Details are the same as in Figure 3.

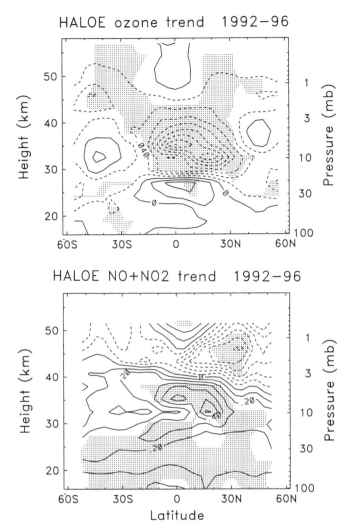

**Figure 10.** Cross sections of linear trends over 1992-1996 for HALOE ozone (top) and sunset $NO_x$ (NO + $NO_2$) (bottom). Contour intervals are 0.02 ppmv/year and 0.10 ppbv/year, respectively.

consistent with Figure 3. It is also consistent with the vertical and temporal structure of $NO_x$ changes in Figure 10-11, i.e., younger air has less $NO_x$ below ~ 40 km and more above this altitude, so that an equilibration to "background" conditions can produce the $NO_x$ 'trends' in Figure 10. However, aerosol-induced chemical changes also act to decrease stratospheric $NO_x$ [see for example the model calculations of *Rosenfield et al.*, 1997], so that the decrease below 40 km immediately after Pinatubo (see Figure 11) may be a mixed dynamical-chemical response. The presence of oppositely-signed $NO_x$ changes above 40 km (where chemical aerosol effects are small) argue for at least some component of a dynamical change. Presumably,

ozone in the middle stratosphere (~30-40 km) responds to the $NO_x$ changes (as evidenced by the high spatial anti-correlation of trends in Figure 10).

Long records of stratospheric temperature measurements show an impulsive response to Pinatubo with equilibration in ~ 2 years (to an apparently different background, i.e., Figure 13). The Pinatubo temperature anomalies are warm in the lower stratosphere over most of the globe, but only over the extratropics in the upper stratosphere (with slight cooling of the tropical upper stratosphere). The associated mean meridional circulation anomalies are suggested to be upward throughout the depth of the stratosphere in the tropics, with compensating

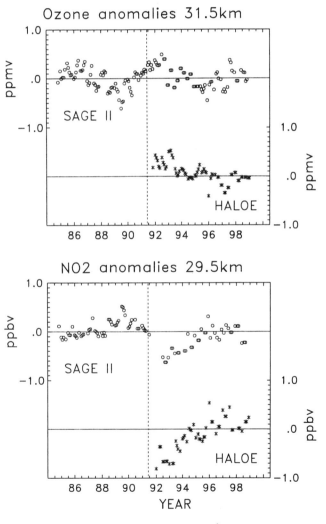

**Figure 11.** Time series of SAGE II and HALOE anomalies in ozone at 10 mb (top) and $NO_2$ at 15 mb (bottom). These data are averages over 20°N-20°N. The dashed lines indicate the eruption of Mt. Pinatubo (June 1991).

SAGEII ozone trend 1992-96    SAGEII ozone trend 1984-98

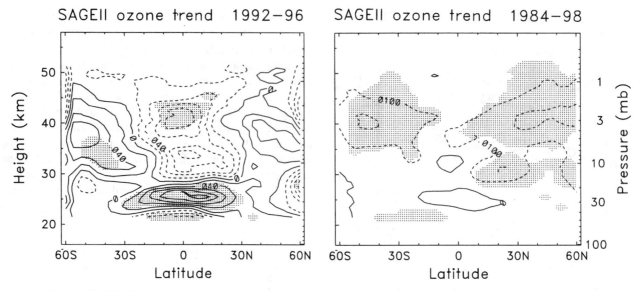

**Figure 12.** Trends in ozone density derived from SAGE II data for 1992-1996 (left) and 1984-1998 (right). The contour interval for 1984-1998 is half that for 1992-1996.

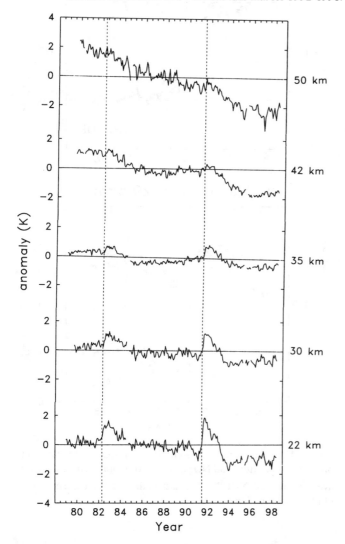

downward flow over extratropics (Figure 15), and this represents a strengthening of the time-mean Brewer-Dobson circulation. We note that the radiative effects of enhanced tropical ozone in the middle and upper stratosphere (seen in Figure 11) could furthermore strengthen the tropical upwelling after Pinatubo. An analysis of NH winter tropospheric wave forcing of the stratosphere (Figure 16) shows substantial year-to-year variability, but the middle 1990's are remarkable for a series of years with relatively weak forcing, which could translate into a weakened Brewer-Dobson circulation. This would give the same sense of circulation changes as that inferred from the Pinatubo temperature signal alone, i.e., a weakened circulation after 1993. We have not attempted to directly estimate anomalies in the stratospheric Brewer-Dobson circulation, but note here that modest changes in tropical upwelling in idealized models can significantly impact $CH_4$ near the stratopause [*D. Waugh*, personal communication, 1999].

One further factor for low-frequency variability of $CH_4$ near the stratopause is the photochemical loss associated with chlorine, which accounts for ~ 1/3 of the calculated total loss at these altitudes [*C. Granier*, personal communication 1997; *Lary and Toumi*, 1997]. The observed changes in HALOE HCl near the stratopause

**Figure 13.** Time series of global temperature anomalies derived from several SSU channels (from bottom to top, channels 15x, 26x, 26, 27 and 47x, respectively). Each time series provides an estimate of the temperature in a ~ 10-15 km layer, centered near the altitudes indicated at right. Dashed lines indicate the eruptions of El Chichon and Pinatubo.

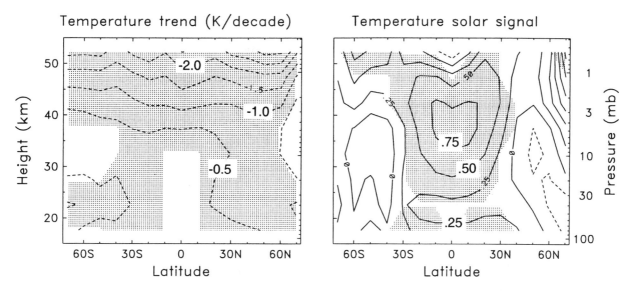

**Figure 14.** Spatial patterns of the trend (left, contours of 0.5 K/decade) and solar cycle variation (right, contours of 0.25 K/100 units of F10.7 solar radio flux) derived from SSU data for 1979-1998. Solar maximum-solar minimum variations of F10.7 are of order 130 units, so the solar cycle temperature changes are of order 1 K. Shading in each figure denotes a statistically significant signal.

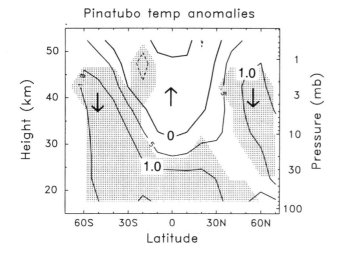

**Figure 15.** Cross sections of SSU temperature anomalies for the year following the Mt. Pinatubo eruption (July 1991-June 1992). Shaded regions denote where the anomalies are statistically significant as compared to 'natural' variability (as discussed in the text). Arrows in the upper stratosphere denote the sense of the associated anomalies in meridional circulation.

(Figure 8) are indicative of chlorine variations which influence $CH_4$. The observed HCl maximum in 1996-1997 is nearly coincident with the $CH_4$ minimum, and the slight increase in $CH_4$ during 1998-1999 (Figure 3) occurs when HCl is decreasing. The influence of these chlorine changes may contribute to circulation variability in explaining the $CH_4$ 'trends'.

An outstanding problem not addressed by this change in stratospheric circulation is the source of the $H_2O$ variability not accounted for by $CH_4$ changes (i.e., the $H_2O + 2CH_4$ changes seen in Figures 3-4). In light of the episodic changes in other constituents observed here, one possibility is that stratospheric $H_2O + 2CH_4$ was anomalously low for the period 1991-1993, and the early HALOE trends represent a return to equilibrium. The first place to look for the cause of such anomalies is near the source region for stratospheric $H_2O$, i.e., near the tropical tropopause [as suggested by *Nedoluha et al.*, 1998a]. Figure 17 shows time series of tropical tropopause temperature anomalies over 1979-1997 derived from global radiosonde measurements [from *Randel et al.*, 2000]. These data show near constant or decreasing temperatures during the period of increasing HALOE $H_2O + 2CH_4$ (~ 1991-1995), and overall similar results are derived from NCEP or ECMWF reanalysis tropopause statistics [*Randel et al.*, 2000]. This demonstrates that the source of the early HALOE 'trends' in $H_2O$ (or anomalously low values in ~ 1991-1993) was not the large-scale mean tropopause temperature. This forces the search for causes of $H_2O + 2CH_4$ variability to more subtle changes in transport across the tropopause, or to sources which are internal to the stratosphere.

The similarity of $H_2O$ variability throughout the stratosphere with the (much larger) changes near 80 km (compare Figures 3, 4 and 7) is possibly suggestive of a solar cycle influence below the mesosphere. However, the detailed calculations of *Chandra et al.* [1997] suggest that

## Stratospheric wave forcing

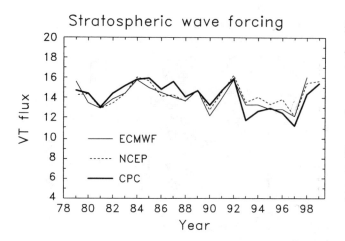

**Figure 16.** Time series of winter mean (November-March) zonal mean eddy heat flux over 40-70°N at 100 mb. This quantity is proportional to the vertical wave activity flux from the troposphere to the stratosphere. Results are shown from calculations based on NCEP and ECMWF reanalyses and CPC stratospheric analyses.

## Tropopause temp 20N−20S

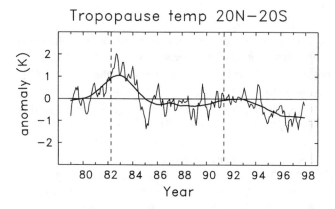

**Figure 17.** Time series of tropical tropopause temperature anomalies over 1979-1999, derived from an ensemble of tropical radiosonde measurements. Dashed lines indicate the El Chichon and Pinatubo eruptions. From *Randel et al.* [2000].

direct (photochemical) solar effects on $H_2O$ are relatively small below ~ 60 km. The fact that the upper stratospheric $CH_4$ variations (Figure 3) show temporal changes qualitatively similar to the solar cycle (Figure 7) is also intriguing. One possibility is that solar variability could cause small changes in stratospheric circulation which contribute to transport effects. This could be of importance as well for other species which are chemically linked to $CH_4$ [*Siskind et al.*, 1998]. It may take a long time series of measurements (at least through the current solar maximum) to empirically separate the direct effects of solar variability.

The continuing observations of stratospheric constituents by HALOE and SAGE II provide an ever changing perspective of interannual variability in the stratosphere. We can anticipate further understanding as observations extend through the current solar cycle, and into the next decade.

*Acknowledgments.* We acknowledge support from NASA under the Atmospheric Chemistry Modeling and Analysis Program and the UARS Guest Investigator Program. We gratefully acknowledge insightful discussions and comments from Rolando Garcia, Jim Holton, Doug Kinnison, David Considine, Arun Gopalan and two anonymous reviewers. Marilena Stone expertly prepared the manuscript. NCAR is sponsored by the National Science Foundation.

## REFERENCES

Anderson, J., J. M. Russell, S. Solomon, and L. Deaver, HALOE confirmation of stratospheric chlorine decreases in accordance with the Montreal Protocol., *J. Geophys. Res.*, in press, 2000.

Andrews, D. G., J. R. Holton, and C. B. Leovy, *Middle Atmosphere Dynamics.* Academic Press, 489 pp., 1987.

Brasseur, G., and C. Granier, Mount Pinatubo aerosols, chlorofluorocarbons, and ozone depletion, *Science, 257,* 1239-1242, 1992.

Chandra, S., C. H. Jackman, E. L. Fleming, and J. M. Russell III, The seasonal and long term changes in mesospheric water vapor, *Geophys. Res. Lett., 24,* 639-642, 1997.

Considine, G. D., L. Deaver, E. E. Remsberg, and J. M. Russell III, HALOE observations of a slowdown in the rate of increase of HF in the lower mesosphere, *Geophys. Res. Lett., 24,* 3217-3220, 1997.

Cunnold, D., H. J. Wang, L. Thomason, J. Zawodny, and J. Logan, SAGE (v5.96) ozone trends in the lower stratosphere, *J. Geophys. Res.*, in press, 1999.

Cunnold, D. M., et al., Validation of SAGE II $NO_2$ measurements, *J. Geophys. Res., 96,* 12,913-12,925, 1991.

Dessler, A. E., E. M. Weinstock, E. J. Hintsa, J. G. Anderson, C. R. Webster, R. D. May, J. W. Elkins, and G. S. Dutton, An examination of the total hydrogen budget of the lower stratosphere, *Geophys. Res. Lett., 21,* 2563-2566, 1994.

Dlugokencky, E. J., K. A. Masarie, P. M. Lang, and P. P. Tans, Continuing decline in the growth rate of atmospheric methane burden, *Nature, 393,* 447-450, 1998.

Efron, B., and R. J. Tibshirani, *An Introduction to the Bootstrap,* vol. 57, *Monographs on Statistics and Applied Probability,* 436 pp., Chapman and Hall, New York, 1993.

Eluszkiewicz, J., D. Crisp, R. G. Grainger, A. Lambert, A. Roche, J. Kumer, and J. Mergenthaler, Sensitivity of the residual circulation diagnosed from UARS data to the uncertainties in the input fields and to the inclusion of aerosols, *J. Atmos. Sci., 54,* 1739-1757, 1997.

Evans, S. J., R. Toumi, J. E. Harries, M. P. Chipperfield, and J. M. Russell III, Trends in stratospheric humidity and the sensitivity of ozone to these trends, *J. Geophys. Res., 103,* 8715-8725, 1998.

Finger, F. G., M. E. Gelman, J. D. Wild, M. L. Chanin, A. Hauchecorne, and A. J. Miller, Evaluation of NMC upper-stratospheric temperature analyses using rocketsonde and lidar data, *Bull. Am. Meteorol. Soc., 24,* 789-799, 1993.

Froidevaux, L., et al., Variations in the free chlorine content of the stratosphere (1991-1997): Anthropogenic volcanic and methane influences, *J. Geophys. Res.,* in press, 2000.

Gibson, J. K., P. Kallberg, S. Uppala, A. Hernandez, A. Nomura, and E. Serrano, *ERA Description. ECMWF re-analysis project Report No. 1,* Reading, UK, 1997.

Hall, T. M., D. W. Waugh, K. A. Boering, and R. A. Plumb, Evaluation of transport in stratospheric models, *J. Geophys. Res., 104,* 18,815-18,839, 1999.

Hamilton, K., Free and forced interannual variability of the circulation in the extratropical northern hemisphere middle atmosphere, 2000, this volume.

Haynes, P. H., C. J. Marks, M. E. McIntyre, T. G. Shepher, and K. P. Shine, On the 'downward control' of extratropical diabatic circulation by eddy-induced zonal mean forces, *J. Atmos. Sci., 48,* 651-678, 1991.

Holton, J. R., P. H. Haynes, M. E. McIntyre, A. R. Douglass, R. B. Rood, and L. Pfister, Stratosphere-troposphere exchange, *Rev. Geophys., 33,* 403-439, 1995.

Kalnay, E., et al., The NCEP/NCAR 40-year reanalysis project, *Bull. Am. Met. Soc., 77,* 437-471, 1996.

Labitzke, K., and M. P. McCormick, Stratospheric temperature increases due to Pinatubo aerosols, *Geophys. Res. Lett., 19,* 207-210, 1992.

Lary, D. J., and R. Toumi, Halogen catalyzed methane oxidation, *J. Geophys. Res., 102,* 23,421-23,428, need a year.

London, J., G. Rottman, T. Woods, and F. Wu, Time variations of solar UV irradiance as measured by the SOLSTICE (UARS) instrument, *Geophys. Res. Lett., 20,* 1315-1318, 1993.

McCormick, M. P., J. M. Zawodny, R. E. Viega, J. C. Larsen, and P. H. Wang, An overview of SAGE I and II ozone measurements, *Planet. Space Sci., 37,* 1567-1586, 1989.

Montzka, S., et al., Present and future trends in the atmospheric burden of ozone-depleting halogens, *Nature, 398,* 690-694, 1999.

Nash, J., Extension of explicit radiance observations by the stratospheric sounding unit into the lower stratosphere and lower mesosphere, *Quart. J. Roy. Meteor. Soc., 114,* 1153-1171, 1988.

Nedoluha, G. E., R. M. Bevilacqua, R. M. Gomoz, D. E. Siskind, and B. C. Hicks, Increases in middle atmospheric water vapor as observed by the Halogen Occultation Experiment and the ground-based Water Vapor Millimeter-Wave Spectrometer from 1991 to 1997, *J. Geophys. Res., 103,* 3531-3543, 1998a.

Nedoluha, G. E., D. E. Siskind, J. T. Bacmeister, R. M. Bevilacqua, and J. M. Russell III, Changes in upper stratospheric $CH_4$ and $NO_2$ as measured by HALOE and implications for changes in transport, *Geophys. Res. Lett., 25,* 987-990, 1998b.

Nevison, C. D., S. Solomon, and J. M. Russell III, Nighttime formation of $N_2O_5$ inferred from Halogen Occultation Experiment sunrise/sunset $NO_x$ ratios, *J. Geophys. Res., 101,* 6741-6748, 1996.

Newman, P. A., and E. R. Nash, Quantifying the wave driving of the stratosphere, *J. Geophys. Res.,* submitted, 2000.

Osterman, G. B., R. J. Salawitch, B. Sen, G. C. Toon, R. A. Stachnik, H. M. Pickett, J. J. Margitan, J. F. Blavier, and D. B.

Peterson, Balloon-borne measurements of stratospheric radicals and their precursors: Implications for the production and loss of ozone, *Geophys. Res. Lett., 24,* 1107-1110, 1997.

Randel, W. J., and F. Wu, Isolation of the ozone QBO in SAGE II data by singular value decomposition, *J. Atmos. Sci., 53,* 2546-2559, 1996.

Randel, W. J., F. Wu, J. M. Russell III, A. Roche, and J. W. Waters, Seasonal cycles and QBO variations in stratospheric $CH_4$ and $H_2O$ observed in UARS HALOE data, *J. Atmos. Sci., 55,* 163-185, 1998.

Randel, W. J., F. Wu, J. M. Russell III, and J. Waters, Space-time patterns of trends in stratospheric constituents derived from UARS measurements, *J. Geophys. Res., 104,* 3711-3727, 1999.

Randel, W. J., F. Wu, and D. Gaffen, Interannual variability of the tropical tropopause derived from radiosonde data and NCEP reanalyses, *J. Geophys. Res.,* accepted, 2000.

Remsberg, E. E., J. M. Russell III, L. L. Gordley, J. C. Gille, and P. L. Bailey, Implications of stratospheric water vapor distributions as determined from the Nimbus 7 LIMS Experiment, *J. Atmos. Sci., 41,* 2934-2945, 1984.

Remsberg, E. E., P. P. Bhatt, and J. M. Russell III, Estimates of the water vapor budget of the stratosphere from UARS HALOE data, *J. Geophys. Res., 101,* 6749-6766, 1996.

Rosenfield, J. E., D. B. Considine, P. E. Meade, J. T. Bacmeister, C. H. Jackman, and M. R. Schoeberl, Stratospheric effects of Mount Pinatubo aerosol studied with a coupled two-dimensional model, *J. Geophys. Res., 102,* 3649-3670, 1997.

Russell III, J. M., et al., The halogen occultation experiment, *J. Geophys. Res., 98,* 10,777-10,797, 1993.

Russell III, J. M., M. Luo, R. J. Cicerone, and L. E. Deaver, Satellite confirmation of the dominance of chlorofluorocarbons in the global stratospheric chlorine budget, *Nature, 379,* 526-529, 1996a.

Scott, R. K., and P. H. Haynes, Internal interannual variability of the extratropical circulation: the low-latitude flywheel, *Quart. J. Roy. Meteor. Soc., 124,* 2149-2173, 1998.

Siskind, D. E., L. Froidevaux, J. M. Russell III, and J. Lean, Implications of upper stratospheric trace constituent changes observed by HALOE for $O_3$ and $ClO$ from 1992 to 1995, *Geophys. Res. Lett., 25,* 3513-3516, 1998.

Tie, X., G. P. Brasseur, B. Briegleb, and C. Granier, Two-dimensional simulation of Pinatubo aerosol and its effects on stratospheric ozone, *J. Geophys. Res., 99,* 20,545-20,562, 1994.

World Meteorological Organization (WMO), Scientific Assessment of Ozone Depletion: 1998, *WMO Report 44,* Geneva, Switzerland, 1999

John Nash, UK Meteorological Office, Bracknell, Berkshire, UK

William J. Randel, Atmospheric Chemistry Division, National Center for Atmospheric Research, Boulder, CO 80307

J. M. Russell III, Center for Atmospheric Sciences, 23 Tyler Street, Hampton University, Hampton, VA 23668

Fei Wu, Atmospheric Chemistry Division, National Center for Atmospheric Research, Boulder, CO 80307

J. M. Zawodny, Atmospheric Chemistry Division, NASA Langley Research Center, Hampton, VA

# Recent Improvements in Middle Atmosphere Remote Sounding Techniques: The CRISTA-SPAS Experiment

Klaus U. Grossmann

*Physics Department, University of Wuppertal, 42097 Wuppertal, Germany*

In November 1994 and in August 1997 the CRyogenic Infrared Spectrometers and Telescopes for the Atmosphere (CRISTA) experiment was flown aboard the ASTRO-SPAS (ASTROnomical Shuttle PAllet Satellite) freeflying platform launched by the Space Shuttle. The main aim of CRISTA is to detect small dynamical features in trace constituents of the middle atmosphere. The instrument is therefore equipped with three telescopes which collect the infrared radiation from three different air volumes at the earth limb. High recording speed and high radiometric sensitivity is obtained from cryogenic cooling of the optics and of the detectors. Radiation is analysed by grating spectrometers in the wavelength range from 4 μm to 71 μm and in the altitude regime from the upper troposphere to the middle thermosphere. The spatial resolution as well as the latitudinal coverage of the measurement net is further improved by making use of the satellite pointing and manoevering capabilities. The improvements in the performance of middle atmosphere limb soundings obtained from the combination of the CRISTA instrument concept and the ASTRO-SPAS pointing capabilities are illustrated by a few examples from the two missions. These results include narrow streamers of ozone rich air observed in the subtropics at 30 km, sharp transport barriers in $O_3$, and small scale temperature fluctuations over the Andes which were identified as gravity waves. A special thermospheric observation mode provided the first global measurement of the fine structure emission of atomic oxygen at 63 μm.

## 1. INTRODUCTION

Our understanding of the chemical and physical processes which govern the earth's middle atmosphere has tremendously increased over the past few decades. Significant improvements have been achieved in sensor technology as well as in the modeling capabilities. Experimental data

from satellite based remote sensing instruments which yield global coverage as well as sufficient temporal and spatial resolution are of extra-ordinary importance for the development and improvement of middle atmosphere models. It is now widely accepted that dynamical processes play an important role for the distribution of trace gases in the middle atmosphere. Early satellite experiments documented large scale planetary waves in the stratosphere in a number of parameters like temperature or ozone mixing ratios [*Drummond et al.*, 1980; *Krueger et al.*, 1980; *Gille and Russell*, 1984]. On the other hand in-situ or quasi local experiments flown on aircrafts, on balloons or on rockets frequently recorded spatial or temporal structures in their

Atmospheric Science Across the Stratopause
Geophysical Monograph 123

measurements on very small scales. Examples can be found or are cited in *Offermann* [1987], *Pfister et al.* [1993], *Murphy et al.* [1993], and *Thrane et al.* [1994]. Unfortunately the in-situ techniques are confined to occasional soundings over selected places on the globe only. The frequent detection of localized small scale structures suggests, however, that they are indeed of global significance. There is therefore a need for global data on spatially resolved structures of large to small scales in the middle atmosphere and thus a need for improved experiments. The spatial resolution or the density of the measurement grid obtainable from space based instruments is limited in one way or the other. Nadir sounders can be built with excellent horizontal resolution as all visible imaging instruments show. Their vertical resolution is poor, however, and often only of the order of a scaleheight or more. Limb sounding instruments are much better in the vertical direction offering resolutions of 1 km or slightly above. In the horizontal plane their sampling volume is along the line of sight (LOS) of the order of about 200 km and may be a few km perpendicular to it. The measurement grid obtained from a limbsounder is usually not very dense. For typical low earth orbit satellites the atmospheric volumes sensed by the optical instrument from successive orbits are separated by up to 2500 km (near the equator) and less at higher latitudes. Due to the precession of the orbital plane with time the measurement grid gets denser as time progresses.

Remote sensing of middle atmosphere temperatures by satellite instruments started as early as 1963 with the MRIR (Medium Resolution Infrared Radiometer) on board of the TIROS-7 satellite [*Kennedy and Nordberg*, 1967]. MRIR was a one channel radiometer for the 15 μm band of $CO_2$. Data were taken from around the nadir direction. The temperatures derived from MRIR essentially cover one broad altitude layer in the lower stratosphere (near 18 km). The MRIR technique was subsequently improved towards more channels with higher spectral resolution in order to obtain simultaneous information from different altitude layers. The improved versions were flown on several other satellites of the TIROS series as well as on the NOAA and NIMBUS satellites. High spectral resolution was later obtained using gas absorption cells to determine the optical passband (SCR - Selective Chopper Radiometer on Nimbus -4 and -5, and with passband scanning as PMR – Pressure Modulator Radiometer on Nimbus-7). A detailed survey about these temperature sensing experiments was given by *Barnett* [1980]. The altitude resolution of these nadir looking experiments is, however, restricted to the finite width of the weighting functions in the different optical passbands and generally is of the order of a scale height only. The first orbiting infrared limb sounder offering good

vertical resolution was the Limb Radiance Inversion Monitor (LRIM) on Nimbus-6 in 1975. This 4 channel radiometer with solid methane cooled detectors yielded global maps of temperature, ozone, and water vapor [*Gille et al.*, 1980]. The extension of this technique to 6 channels in the Limb Infrared Monitor of the Stratosphere (LIMS) on Nimbus-7 allowed the first global scale simultaneous observations of $HNO_3$, $H_2O$, and $NO_2$ in the middle atmosphere in addition to its temperature and ozone measurements [*Gille and Russell*, 1984]. LIMS ozone and temperature data extended into the lower mesosphere ($\geq 0.1$ hPa). Together with LIMS another trace gas monitoring experiment was flown on the same satellite. The SAMS instrument (Stratospheric and Mesospheric Sounder, *Drummond et al.* [1980]) used the PMR technique and allowed measurements of the emissions from $CO_2$, CO, $CH_4$, NO, $N_2O$, and $H_2O$. A further enhancement of the number of simultaneously recorded species was achieved with the set of experiments on the Upper Atmosphere Research Satellite (UARS, *Reber et al.* [1993]) launched in September 1991. Enhanced spectral resolution and higher sensitivities compared to LIMS were reached by the Cryogenic Limb Array Etalon Spectrometer (CLAES). The first global maps of $N_2O_5$, $ClONO_2$, CFC-11, and CFC-12 were determined by this solid neon cooled infrared spectrometer [*Roche et al.*, 1993]. CLAES took data with 2.5 km vertical resolution on a 500 km along track grid between 80°S and 80°N latitude. As in the case of LRIR and LIMS the measurement period of CLAES was limited by the amount of cryogen on board (19 months). The PMR measurement principle was also used on UARS in the Improved Stratospheric and Mesospheric Sounder (ISAMS, *Taylor et al.* [1993]). A closed cycle mechanical cooler is used in ISAMS to cool the radiation detectors. Much higher sensitivities than with the uncooled or only radiatively cooled SAMS detectors were reached. The technical requirement to use cryogenically cooled detectors in order to reach sufficient sensitivity for infrared emission measurements has prevented these instruments from yielding longterm data series so far. Solar backscatter or solar absorption spectroscopy has higher intensity levels available for analysis than the IR emission technique but is restricted to daylight conditions or to orbital sunrise/sunset. A one channel photometer (SAM II: Stratospheric Aerosol Measurement II) measuring the absorption of solar near infrared radiation during orbital sunrise/sunset was flown on NIMBUS-7 and yielded global data about the stratospheric aerosol mass loading [*McCormick et al.*, 1981]. Later experiments added further channels to include trace gas measuring capability like ozone in the case of SAGE (Stratospheric Aerosol and Gas Experiment, *Kent and McCormick* [1984]; *McCormick et al.* [1984]). NO$_2$ and H$_2$O channels were added to the SAGE II instrument launched in October 1984 [*Mauldin et al.*, 1985]. The Halogen Occultation Experiment (HALOE) on UARS employes solar occultation radiometry in the mid infrared [*Russell et al.*, 1993; *Russell and Pierce*, this volume). Species measured by HALOE include O$_3$, H$_2$O, HCl, HF, NO, NO$_2$, CH$_4$, and aerosols in addition to temperatures. First global cross sections of HF [*Luo et al.*, 1994] and of HCl [*Russell et al.*, 1996] were derived from this experiment. The lack of low temperature cooling requirements together with its self calibrating absorption technique makes HALOE very well suited for longterm trace gas monitoring. However, the number of vertical profiles obtainable per day is limited to 32 only (16 sunrise and 16 sunset soundings) resulting in a poor spatial density of the measurement grid. A denser horizontal grid of a large number of simultaneously measured species is obtained from the Global Ozone Monitoring Experiment (GOME) on the ERS-2 platform (Earth Remote Sensing Satellite 2) launched April 25, 1995. This nadir scanning spectrometer analyses solar backscatter radiation in the UV, VIS, and near-IR during daytime [*Burrows et al.*, 1999].

The tremendous improvement in instrument performance over the past years focused mainly on higher spectral resolution and higher sensitivities in order to measure more species at the same time and with higher precision. The spatial resolution and coverage has essentially remained unchanged. High spatial and temporal data density can be achieved through the combination of various measures. Multi-azimuth limb scanning or parallel operation of several limb pointing telescopes can be combined with the high measuring speed of a cryogenically cooled thermal infrared sensor allowing day as well as night measurements. High performance attitude control in combination with spacecraft manoevrability and operational flexibility can furthermore enhance the data density. The parallel use of three telescopes pointing to the earth's limb at different horizontal angles has been used by the CRyogenic Infrared Spectrometers and Telescopes for the Atmosphere (CRISTA). Integrated in the freeflying ASTRO-SPAS (ASTROnomical Shuttle PAllet Satellite) platform, CRISTA was flown twice on Space Shuttle missions in 1994 and 1997. CRISTA is the first instrument which combined high sensitivity, sufficiently high spectral resolution, and a high spatial density of measurement points in all three dimensions. The CRISTA experiment is described in detail below to demonstrate the capabilities of such systems and to encourage further and similar experimental setups.

Better spatial resolution is also planned for the next generation of middle atmosphere sounders. Infrared limb emission radiometry combined with multi azimuth scanning is used by the High Resolution Dynamics Limb Sounder (HIRDLS) to be launched on EOS-CHEM [*Dials et al.*, 1998]. The HIRDLS line of sight can be commanded to various azimuths along the horizon. As a standard horizontal measurement grid size 400x500 km is chosen with 1 km vertical resolution. This can, however, be modified during the mission. As the HIRDLS detectors will be cooled by a mechanical refrigerator a global monitoring of O$_3$, H$_2$O, N$_2$O, NO$_2$, HNO$_3$, N$_2$O$_5$, CH$_4$, CFC-11, CFC-12, aerosols, and of temperature is expected at the above given resolution and over the course of several years. Solar occultation and backscatter measurements are planned from ENVISAT-1 (ENVIronmental SATellite) by means of the SCIAMACHY instrument (Scanning Imaging Absorption Spectrometer for Atmospheric CHartography, *Bovensmann et al.*, [1999]). SCIAMACHY is based on the GOME design but includes nadir and limb backscatter soundings as well as solar and lunar occultation. The line of sight of this experiment can be scanned across track (in the nadir viewing mode) or along the limb in the limb viewing geometry. High spatial resolution is obtainable here as well, however, these measurements are restricted to daytime. An infrared limb sounding Michelson interferometer (MIPAS - MIchelson Passive Atmospheric Sounder) on ENVISAT-1 will derive trace gas densities during day and night [*Fischer and Oelhaf*, 1996]. Also this instrument will have some azimuth scanning capabilities in order to enhance the horizontal resolution.

## 2. CRISTA INSTRUMENT OVERVIEW

As outlined above the key issue for the CRISTA instrument is to perform trace gas measurements in the middle atmosphere on a grid which is very dense in all three dimensions.

This is achieved by using three independent telescopes which simultaneously sense three spatially separated atmospheric volumes and by cooling of the instrument to cryogenic temperatures in order to reach a high measuring speed. The viewing direction of one telescope (named the center telescope) is in the main symmetry axis of CRISTA. The other two (lateral) telescopes have their optical axes oriented at +/- 18° relative to the center telescope. A sketch of the observation geometry is given in Figure 1. For a 300 km orbit, an 18° angular separation leads to about 600 km separation at the stratospheric tangent points. When the orientation of the CRISTA instrument is such that the

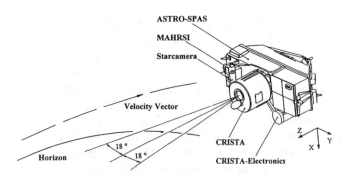

**Figure 1.** CRISTA measurement geometry. (For details see text.)

center telescope looks at about 180° relative to the flight vector the measured volumes form three tracks spaced by 600 km. This distance compares to the distance of two adjacent orbits of about 2500 km near the equator in the case of a single LOS sensing instrument. The space between the subsatellite ground tracks is thus filled more or less evenly (see Plate 1). In flight direction the spacing of the measurement points is determined by the time which is needed to perform one altitude scan and varies between about 250 km and 500 km depending upon the type of vertical scan mode executed. CRISTA measures in the mid- and far-infrared spectral regime which requires cooling of the optics and of the radiation detectors to cryogenic temperatures in order to reach the desired measurement performance. The cooling is provided by a bath cryostat in a combination of supercritical and subcooled helium. CRISTA was built for exploratory missions with a typical mission duration of about one week on a freeflying satellite launched by the NASA Space Shuttle. In this approach instrument mass and electrical power consumption are of less concern as relatively large instruments can be accomodated on the Shuttle. In Figure 2 an overview cross section of CRISTA is given.

*2.1 Spectrometers*

The optical subsystem of CRISTA consists of three Herschel type off-axis telescopes each followed by an Ebert-Fastie grating spectrometer [*Ebert*, 1889; *Fastie*, 1953] sensitive in the mid-infrared spectral regime between about 4 μm and 13.9 μm. These spectrometers are labelled SR-, SCS-, and SL- spectrometers for the right, center, and left telescope, respectively. The center telescope in addition feeds an Ebert-Fastie grating spectrometer operating in the mid- to far-infrared wavelength range (9.3 μm to 71 μm, labelled SCL- spectrometer). The throughput of the spectrometers is $1.72 \times 10^{-4}$ $cm^2 sr$ in the mid infrared and $4.7 \times 10^{-4}$ $cm^2 sr$ in the far infrared.

The spectral resolution of all spectrometers is $\lambda/\Delta\lambda \approx 500$. This resolution is sufficient to securely identify the signatures of the gas emissions listed in Table 1. The wavelength separation is done by reflection gratings blazed to 10.8 μm for the mid-infrared spectrometers and to 60.4 μm for the far-infrared spectrometer. The spectral signals are recorded by up to 8 detectors in each spectrometer. Si:Ga and Ge:Ga bulk photoconductors are used for the mid- and far-infrared wavelength range. Selected channels of the center telescope mid-infrared spectrometer are equipped with Si:As BIB detectors. The noise equivalent spectral radiance (NESR) of the various detectors is between $1.6 \times 10^{-10}$ and $1.8 \times 10^{-9}$ $W/cm^2$ sr $cm^{-1}$ in case of the mid-IR Si:Ga detectors. The far-infrared germanium detectors are between $5.0 \times 10^{-10}$ and $7.6 \times 10^{-9}$ $W/cm^2$ sr $cm^{-1}$. The channels using Si:As BIB detectors exhibit NESR values between $1.9 \times 10^{-11}$ and $3.4 \times 10^{-9}$ $W/cm^2$ sr $cm^{-1}$. This comparatively large range of detector performance is in part due to the scatter in the detector quality and in part due to the operating conditions.

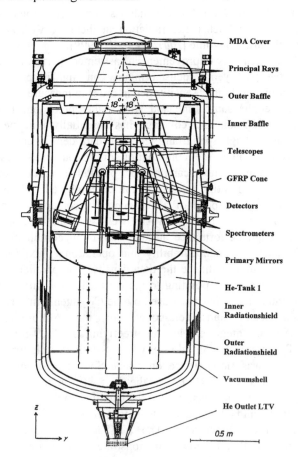

**Figure 2.** Cross section of the CRISTA instrument in the y-z plane. During flight the z-axis points towards the earth horizon, the x-axis points downwards.

The detector mounts in each spectrometer including detectors, light guides, filters, and preamplifiers are thermally isolated from the spectrometer structure. Their temperature is actively controlled to 13.0 K in case of the lateral telescope spectrometers (SR and SL). The respective detector units in the central telescope spectrometers SCS and SCL are not temperature stabilized. Their operating temperature varies between 3.25 and 3.60 K during the course of the mission. The temperatures and their variations constitute a compromise between optimum detector performance of the three detector types used here (Si:Ga, Ge:Ga, and Si:As BIB) and performance constraints of the cryostat.

The combination of moderate spectral resolving power with up to 8 parallel measuring high sensitivity detectors allows to take a complete spectrum in 1.13 seconds. Depending upon the type of measurement mode the time required for one altitude scan is between 0.5 and 1 minute which translates into an along track resolution of about 250 to 500 km.

## 2.2 Telescopes

The off-axis Herschel telescopes have a focal length of 1000 mm. The primary mirrors are spherical with a diameter of 120 mm. Two spherical mirrors in Czerny-Turner mount form the secondary. Straylight suppression is an important issue for all infrared limbsounding instruments. Telescope straylight is minimized in several ways: Superpolished low scatter surfaces for all telescope mirrors which are illuminated by radiation from the earth, implementation of a Lyot stop [Lyot, 1930] situated between the two secondary mirrors to remove diffracted radiation from the rim of the primary mirror, and the careful design of the two stage entrance baffle. The telescope primary mirrors are tiltable in order to scan the telescope line of sight through the atmosphere. They are mounted on the axes of brushless DC motors which operate at cryogenic temperatures. The motors can be positioned to about 1 arcsec. The same type of motor is used to turn the gratings in the spectrometers.

The grating spectrometers require narrow and long entrance slits to achieve the desired spectral resolution and to gain enough throughput. The image of the entrance slits in the atmosphere has to be parallel to the horizon for good vertical resolution. Adjusting the entrance slit of one telescope (as for example the center telescope of CRISTA) to the horizon is simply a matter of the spacecraft attitude and can be done for all orbit altitudes. In the case of the CRISTA instrument the line of sights of two lateral telescopes form angles of +18° or -18° with the center telescope. In this case the curvature of the horizon has to be

**Table 1.** Trace gas emissions measured by CRISTA

| Trace gas | Wavelength | Altitude range |
|-----------|------------|----------------|
| $CO_2$ | 4.3μm | 11 - 120 km |
| CO | 4.7μm | * |
| NO | 5.3μm | 100 - 180 km |
| $NO_2$ | 6.2μm | 11 - 40 km |
| $H_2O$ | 6.3μm | 11 - 70 km |
| $CH_4$ | 7.7μm | 11 - 70 km |
| $N_2O$ | 7.8μm | 11 - 40 km |
| $CF_4$ | 7.8μm | 11 - 70 km |
| $N_2O_5$ | 8.0μm | 20 - 40 km |
| $O_3$ | 9.6μm | 11 - 95 km |
| $CF_2Cl_2$ | 10.8μm | 11 - 30 km |
| $HNO_3$ | 11.3μm | 11 - 35 km |
| $CFCl_3$ | 11.8μm | 11 - 25 km |
| Aerosol | 12.0μm | 11 - 35 km |
| $HO_2NO_2$ | 12.5μm | * |
| $CCl_4$ | 12.6μm | 20 km |
| $CO_2$ (Temp) | 12.6μm | 11 - 70 km |
| $O_3$ | 12.7μm | 20 - 55 km |
| $ClONO_2$ | 12.8μm | 20 - 40 km |
| Center telescope only: | | |
| $CO_2$(Temp) | 15.0μm | 40 - 150 km |
| $N_2O$ | 17.0μm | 40 - 45 km |
| $H_2O$ | 58μm | 40 - 80 km |
| HF | 61μm | 40 - 65 km |
| $O(^3P)$ | 63μm | 80 - 180 km |
| HCl | 69μm | * |

* *under study*

taken into account. At the angular position of the lateral telescopes the horizon and thus the spectrometer entrance slit is tilted by +/- 5.6° relative to the horizon of the center telescope in the case of a 300 km orbit altitude. This tilt is built into the CRISTA optics so that all three entrance slits can be kept parallel to the earth horizon at the same time. An orbit considerably different from 300 km altitude would require hardware modifications. Optimum use of the three telescope design also restricts the orbits to near circular ones.

## 2.3 Cryogenics

High instrument sensitivities in the mid- and far-infrared can only be obtained with cryogenically cooled detectors and with sufficient suppression of the thermal emission of the instrument itself. Due to the use of three telescopes and the resulting large entrance opening the influx of thermal radiation from the troposphere and the hard earth as well as from direct and reflected sunlight is quite large. The average heat input into the outer part of the entrance baffle (see Figure 2) is 5 - 6 W. The cooling of the CRISTA optics therefore requires a comparatively large amount of cryogen which is provided by a 725 ltr capacity helium tank. The helium in this He-tank-1 is kept in the supercritical state thus facilitating the cryogen management under zero-g conditions in space. Supercritical helium is also used as cryogen by the CIRRIS-1A experiment [*Bartschi et al.*, 1992] and by the Infrared Background Signature Survey (IBSS) experiment [*Seidel and Passvogel*, 1994]. The operating pressure is controlled to about 3 bar. The temperature of the cryogen increases from about 5.3 K at the begin of the mission to up to 12 K at the end. These temperatures are low enough to suppress all thermal radiation from the optic components to negligible values. The large heat input into the entrance opening of CRISTA is removed by cooling of the entrance baffles by means of the exhaust  He gas from the tank. Two radiation shields are cooled as well by the exhaust gas. The exhaust helium finally leaves the instrument through a gas diffuser at the lower end of the cryostat vacuum shell (Low Thrust Vent: LTV in Figure 2) in order not to produce any thrust or torque which could disturb the satellite attitude control. As the mid-infrared bulk detectors are temperature stabilized to 13.0 K the He-tank-1 temperatures are sufficiently low to cool these detectors. The Si:As BIB detectors and the far-infrared Ge:Ga detectors require much lower operating temperatures. The cryogenic system of CRISTA  therefore comprises a second and much smaller tank (He-tank-2, 55 ltr capacity) containing subcooled helium. The helium in this tank (not shown in Figure 2) is pumped down  prior to launch to a vapour pressure of 100 mbar corresponding to a temperature of 2.5 K. He-tank-2 is then hermetically closed and remains so throughout the mission whithout degassing. Its temperature rises to about 3.6 K at the end of the mission. The cooled parts of the cryostat including the optics are mounted to the experiment vacuum shell by means of a glass fiber reinforced plastic (gfrp) cone which provides sufficient thermal insulation while taking all the mechanical loads occuring during ground handling and flight of the instrument. A Motorized Door Assembly (MDA) from the Goddard Space Flight Center Get Away Special program is used as vacuum tight cover for the instrument. The MDA is closed during all ground handling, ascent, descent, and during all ASTRO-SPAS manoevers which could potentially lead to an influx of ambient gas into the cold CRISTA optics.

### 2.4 Calibrations

Large efforts were undertaken to thoroughly calibrate CRISTA. The calibrations were done as late as possible before launch under consideration of the satellite and of the Shuttle integration schedule. The measurements included wavelength calibrations using molecular absorption features by placing gas cells in front of the CRISTA telescopes and absolute radiometric sensitivity determinations with a cryogenic blackbody source. The blackbody diameter is larger than the telescope beam and therefore fills the complete field of view of the CRISTA telescopes. A separate calibration was performed with each telescope. The temperature of the blackbody can be varied between 10 K and 260 K. This large temperature range is necessary to obtain calibration data for all intensities seen during the atmospheric soundings and for all wavelengths from the mid- to the far-infrared. The LOS of the three telescopes was measured relative to the axis of the star camera which serves as the primary attitude reference for the ASTRO-SPAS carrier. All calibrations were done in the period of 4 - 6 month before launch and were repeated immediately after the return of the instrument from the mission. The instrument sensitivities determined from the blackbody calibration before and after the mission agree within ±2.5% with most channels being below ±1 %. An exception are the BIB detectors used at short (4 – 5 μm) wavelengths which showed deviations of up to ±4.3 %.

Part of the inside surface of the instrument cover (MDA) is blackened and serves as a relative radiometric calibration source during flight. Measurements of the signals generated by the MDA calibration source were made immediately after the absolute radiometric calibration before flight, several times during the mission whenever the cover was closed, and just prior to the post-flight calibration sequence. The sensitivity variations seen during the mission are for most channels <1% and < 4.5 % at maximum.

Of major concern are relaxation effects in the photo-conducting dertectors. The detector signals are not alone a function of the instantaneous radiation intensity but also depend upon the radiation history over some time period before. The deviations from an ideal behaviour are less than a few % during calibrations for all detectors. During flight the relaxation effects are mostly below 5 - 10 % but can be as large as 25 % under extreme conditions (very large gradients within one spectrum). To quantify this behaviour the detectors were subjected to infrared radiation pulses of varying length, intensity and and frequency as a function of the operating conditions. From these pulse response data correction functions for the flight data are derived [*Ern*, 2000]. This correction procedure eliminates most of the observed relaxation effects. The remaining errors due to detector relaxations are a few %. As the operating temperatures of the SCS and SCL detectors varied during the flight (cf. chapter 2.3) all calibrations of the center telescope were done at several discrete detector temperature levels.

## 3. MISSION

For its flight CRISTA is integrated in the freeflying ASTRO-SPAS platform . The ASTRO-SPAS is a multi-use platform designed for short duration astronomy and earth atmosphere missions. The ASTRO-SPAS with its scientific instruments is launched by the NASA Space Shuttle. On orbit the satellite is taken out of the Shuttle's cargo bay and subsequently released for an autonomous flight of about one week. During this time the ASTRO-SPAS trails the Shuttle by up to 100 km. After completion of the measuring sequences the ASTRO-SPAS is retrieved and carried back to ground for refurbishment. The ASTRO-SPAS was launched 4 times. The first and the third missions carried the ORFEUS (Orbiting and Retrievable Far and Extreme Ultraviolet Spectrometers) as the prime instrument which analysed the far-UV radiation from a large number of astronomical objects [*Appenzeller et al.*, 1995]. The two astronomy missions were launched in September 1993 and in November 1996. The two atmospheric missions carried CRISTA as the main instrument. In this configuration the mission is called CRISTA-SPAS. In addition to the CRISTA instrument the CRISTA-SPAS is equipped with the Middle Atmosphere High Resolution Spectrograph Investigation (MAHRSI) experiment of the Naval Research Laboratories [*Conway et al.*, 1999]. MAHRSI is bore-sighted with the CRISTA center telescope. Measurements of the two instruments are therefore taken in about the same volume of air. Technical and programmatic aspects of the ASTRO-SPAS system are given by *Wattenbach and Moritz* [1997].

The ASTRO-SPAS system offers a number of outstanding features which make it particularly suited for high spatial density measurements in the earth's middle atmosphere as the full benefit from instruments like CRISTA is only obtained in combination with the appropriate carrier and its operational capabilities. Based on the primary requirements for observations of astronomical point sources the ASTRO-SPAS is equipped with a precision cold gas pointing control system which receives its primary attitude information from a star camera. The alignment between the CRISTA optical axes and the star camera is verified during the mission by pointing CRISTA to Mars (CRISTA-1) or to Jupiter (CRISTA-2). The real time pointing accuracy at the tangent altitude is ±2 km. In the post-flight analysis the accuracy is ±200 m. Altitude scans are performed by tilting the CRISTA telescopes primary mirrors. CRISTA takes mesurements from the upper troposphere to the middle thermosphere (Table 1). This large altitude range is covered by several altitude scan modes which focus either on the troposphere/stratosphere or on the mesosphere/thermosphere regime.

The ASTRO-SPAS attitude control system in its latest version (CRISTA-2) allowed to point the CRISTA telescope to any desired direction along the limb and to perform slewing manoevers around the local vertical axis. This capability enabled several observation geometries which will be adressed below. Of special importance was the fact that the observation mode could be replanned and modified within a few hours during the mission. This flexibility made it possible to shorten the required test and to introduce additional observation geometries which enhanced the scientific return from this mission.

The two CRISTA-SPAS missions took place in November 1994 and in August 1997. In both cases the orbital inclination was 57° and the orbit altitude was 300 km circular. CRISTA as well as the ASTRO-SPAS performed flawlessly throughout the missions. CRISTA recorded about 50000 altitude profiles of the gases given in Table 1 during its first mission and another 44000 profiles during CRISTA-2. The data are reduced following the method described by *Riese et al.* [1999a]. For the second flight the data evaluation is still at an early stage, however.

Parallel to the missions extended validation campaigns for CRISTA and MAHRSI were undertaken [*Lehmacher and Offermann*, 1997] which comprised airplane flights, rocket and ballon launches, and a large number of parallel operating ground based instruments all over the world. Additional data on parameters not measurable by the satellite instruments were thus obtained together with some information on the dynamical state of the atmosphere just prior to and after the CRISTA-SPAS mission periods.

## 4. RESULTS

### 4.1 Streamers

CRISTA-2 ozone mixing ratios obtained during the standard stratospheric observation mode are presented in Plate 1 for 30 km altitude. The data shown here were recorded during one day (day 222, August 10, 1997) and demonstrate the density of the measurement grid. Each symbol represents one complete altitude scan of all the gases listed in Table 1. The three traces of data points from the three telescopes are clearly seen. Plate 1 is a representation of the standard stratospheric observation mode during CRISTA-2. In this mode CRISTA-SPAS slews with constant angular velocity backwards and forwards around the local vertical axis. These slews are synchronised with the orbital motion of the carrier such that CRISTA looks northwards (at +72° from the orbit plane) at the northernmost point of each orbit, looks backwards (i.e. within the orbital plane but at 180° to the velocity vector) while above the equator, and looks southwards (at -72° from the orbital plane) at the southernmost point of the orbit. Along with the limb viewing geometry this mode (called "ping-pong" mode) extends the latitudinal coverage of the CRISTA measurements to about ±74°. In Figure 3 the observational geometry is sketched for a few orbits. As there were also some special observations made during this day the distribution of the data in Plate 1 is somewhat uneven. Specially between 80°E and 140°E there are some regions not covered by CRISTA on this day because of the execution of the high spatial resolution "hawkeye" mode above Indonesia (see below). Figure 4 shows the zonal mean mixing ratios between 26 km and 44 km averaged over 2° in latitude. In the northern summer hemisphere ozone concentrations decrease gradually from the equator to the pole and Plate 1 reveals no pronounced longitudinal structures. In the winter hemisphere the ozone zonal mean values show a steep decline towards southern mid-latitudes (in the altitude interval between about 26 km and 36 km) and a flat mixing ratio gradient between 20°S and 40°S. The low gradients are probably the result of large scale horizontal mixing [*Plumb*, 1996] since the photochemical lifetime of mid-stratosphere odd oxygen is weeks [*Brasseur and Solomon*, 1984]. Streamers are responsible for a significant part of the large scale transport of trace gases from high latitudes to the tropics and vice versa [*Chen et al.*, 1994]. In Plate 1 several features are recognized. A well developed streamer brings ozone rich air from the central Pacific subtropics to mid-latitudes near the southern tip of South America and then leads back to the subtropics west of South Africa. The horizontal width of this streamer is about 800 km and its vertical extention is near 10 km. A second streamer developed east of Africa on August 10 and became more pronounced later on. The two streamer pattern moved eastward around the pole during the time period of the CRISTA-2 mission. This is consistent with the development of the polar vortex. The vortex in the lower stratosphere (18 km) on day 222 was centered at the pole and was only slightly elongated towards the Central Pacific and towards Africa. Near 30 km, however, the vortex was well streched towards Africa and towards the Central Pacific. Plate 2 shows the potential vorticity as given by the DAO analysis for August 10 [*Schubert et al.*, 1993]. Tropical air is drawn to mid and high latitudes near the mid latitude edges of the vortex. Over and west of South America there are remnants of high pv air around which the tropical air is guided south and then back to low latitudes.

In the region over the southern Indian ocean a secondary vortex formed causing the development of the second streamer. A similar situation at slightly higher altitudes and for $H_2O$ and $N_2O$ was reported by *Randel et al.* [1993].

Also during CRISTA-1 several such streamers were found in a number of trace gases [*Offermann et al.*, 1999]. *Riese et al.* [1999b] modeled the distributions of CFC-11 and of $N_2O$ analysed by CRISTA-1. The National Center for Atmospheric Research on Ozone in the Stratosphere and its Evolution (NCAR-ROSE) model was used in a version in which the dynamics are driven by assimilated winds from the U.K. Meteorological Office. The authors generally found good agreement between modeled and measured mixing ratios including many dynamically induced features. A pronounced streamer extending over North America, the North Atlantic, Europe, and Central Asia is well reproduced by the model at the begin of the CRISTA-1 period but 2 days later some discrepancies show up. Other differences are encountered in the tropics, subtropics, and in the southern hemisphere. Differences of this kind are only detectable with an instrument providing a high density measurement net in both space and time. A spatial resolution less than that of CRISTA would smear out such structures. As atmospheric 3-D models go to smaller and smaller scales there will be a need to study detail structures over a variety of geographical locations. Such structures are highly variable in time and space. The two streamer pattern found by CRISTA-2 moved about 1500 km in 3 days. A more detailed examination of these and other features could be done by pointing the instrument to the (moving) structures thus requiring real time data evaluation combined with short term forecasting models and near real time spacecraft pointing control. Several successful attempts in this direction have been made in the past few years. In April 1996 the MSX satellite was launched carrying a set of cryogenically cooled mid-infrared sensors as well as visible and UV instrumentation for earth atmosphere and astronomical observations [*Mill et al.*, 1994]. The MSX satellite similar to the ASTRO-SPAS can be pointed to any allowed direction in the earth atmosphere and is thus capable to specifically study local features. An earlier cryogenically cooled infrared sensor was the Shuttle borne CIRRIS 1A experiment launched in April 1991 as part of the Shuttle mission STS-39 [*Bartschi et al.*, 1992]. CIRRIS 1A was mounted on a steerable platform in the Shuttle cargo bay. Besides pointing the instrument to the earth limb several objects including aurora features were specifically examined. In this way very high spatial resolution soundings of streamers appear feasible by looking onto the same structure from several

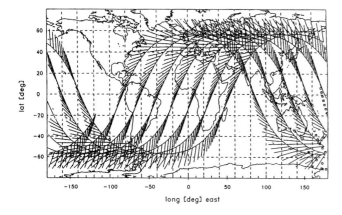

**Figure 3.** Sketch of the standard observation mode geometry. Only 7 orbits are plotted as example. The horizontal bars denote the ray path from the observed atmospheric volume to CRISTA. Only the center telescope is shown for clarity in one minute increments.

consecutive orbits. Such an observational geometry was tested during CRISTA-2 by looking for small scale structures over Indonesia.

*4.2 Southern subtropical transport barrier*

The steep gradient in the 30 km ozone mixing ratios near 10°S latitude in Plate 1 and in Figure 4 is an indication for the (southern) subtropical transport barrier which impedes meridional mass exchange in the middle stratosphere [*Trepte and Hitchman*, 1992; *Plumb*, 1996]. From Plate 1 it is evident that the transport barrier during the CRISTA-2 period extends all around the globe interrupted only in the regions of the two streamers. Over Indonesia a special very high spatial resolution measurement (called "hawkeye" – mode) was carried out on this and on the following three days. This geographical region was chosen because of the very localised and frequent deep convection cells found here [*Chen and Houze*, 1997]. The convection causes the air to penetrate into the lower stratosphere and does therefore contribute to the troposphere-stratosphere exchange. Small scale variations in atmospheric parameters are expected over this area in the lower stratosphere. During three consecutive orbits the same geographical region is observed within slightly more than 3 hours increasing the spatial resolution considerably. If ascending and descending orbits are combined the density of measurements is further increased. Such a sequence takes about 13 hours. In Plate 3 a detail map of the "hawkeye region" from Plate 1 is given showing the resulting net of ozone concentrations at 30 km. As can be seen from this plate, data from 3 ascending and from 2 descending orbits

are included. The average distance between the data points within the area between the equator and 10°S and between 110°E and 120°E is about 165 km. The decrease of ozone mixing ratios across the transport barrier is quite steep and there is not much zonal variation in the latitude of the transport barrier as is also seen in Plate 1. Analysing data from a differential absorption lidar on board a NASA DC-8 aircraft during the 1992 AASE-II campaign (Airborne Arctic Stratospheric Expedition II) *Grant et al.* [1994] found similar steep gradients in the ozone concentrations between 20 and 25 km. Ozone at 25 km dropped from 5.5 ppm to below 4 ppm from 25°N to 17°N along the flight path of the aircraft at the end of January 1992. The ozone depletion in the tropics is attributed by the authors to high aerosol concentrations from the Pinatubo eruption. The steep gradients in the latitudinal distribution of both ozone and aerosols could be an indication of the northern (winter hemisphere) transport barrier. Comparing individual ozone profiles of CRISTA-2 in the vicinity of the transport barrier mixing ratio decreases close to 0.7 ppm between 30 and 32 km are found over horizontal distances as low as 100 km and with a time separation of about 90 min. This is a much steeper gradient than seen in the aircraft data. The transport

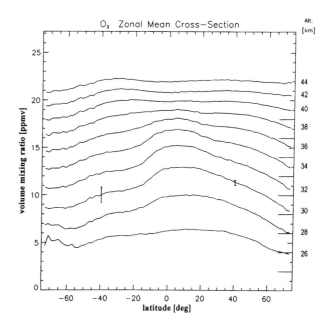

**Figure 4.** Zonal mean ozone mixing ratios for August 10, 1997, averaged over 2° in latitude. The variability (as standard deviation) given for 30 km altitude and at 40°N is typical for all northern hemisphere zonal mean values. In the southern hemisphere the variability increases about linearly from the transport barrier near 10°S to the high latitudes accessible by CRISTA.

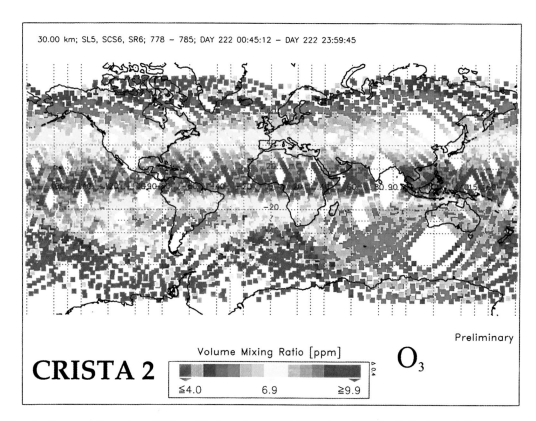

**Plate 1.** Ozone mixing ratios at 30 km altitude on August 10, 1997 from CRISTA-2. Each symbol represents a complete altitude scan of all gases measurable by CRISTA.

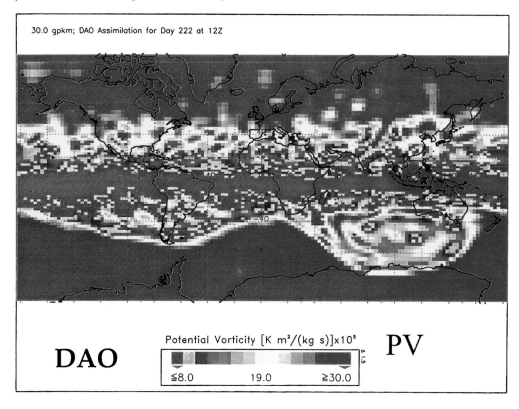

**Plate 2.** Potential vorticity map at 30 km for August 10, 1997, from the GSFC DAO analysis.

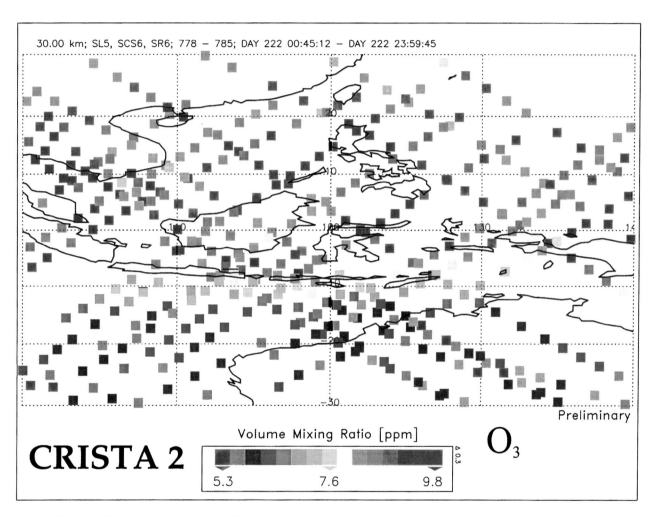

**Plate 3.** Map of ozone soundings at 30 km during the high spatial resolution ("hawkeye") mode over Indonesia on August 10, 1997

barrier during CRISTA-2 thus appears to be quite sharp and its shape in longitudinal direction is very smooth. The modelling of several CRISTA-1 trace gas distributions by *Riese et al.* [1999b] revealed discrepancies between the model and the experimental data particularly in the tropics. The data from the CRISTA-2 "hawkeye" mode demonstrate that variations in trace gas distributions are present over very short temporal and spatial scales even in the mid stratosphere. These small scale structures will be a good test for the dynamics of 3-dimensional models.

*4.3 Small scale variability*

Joint data analyses of different experiments which are not measuring in the same volume of air or not at the same time are often difficult to interpret because the atmospheric conditions are different for the two soundings. Atmospheric small scale variability can be quite high even over short distances. This is illustrated by looking at the high tropical ozone mixing ratios of Plate 3 in the area between the equator and 20°N and in the longitude interval 100°E to 140°E. In this region no significant latitudinal or longitudinal gradients or apparent structures are seen. Nevertheless, the differences between any two profiles measured within 10 min and located in the above given region are found to go up to 1.4 ppm. The observed differences from 406 profile pairs exhibit an rms value of 0.59 ppm for a maximum time difference of 10 min and distances between 200 km and 2500 km. The rms variability in the ozone concentrations is higher by a factor of 4.5 compared to the statistical error of the CRISTA data (0.13 ppm standard deviation). The variability is therefore believed to be of atmospheric origin due to local dynamics like small scale waves or turbulence. There is some indication that the differences between data points decrease for distances below 500 km. Above 500 km and up to 2500 km there is no significant variation with distance. In regions of increased dynamical activity such as streamers the small scale variability can of course be considerably higher. Comparisons with other experiments even at horizontal scales of a few 100 km may therefore be hampered if todays high precision instruments are used.

*Bacmeister et al.* [1999] compared ozone, CFC-11, and $NO_y$ concentrations obtained from the CRISTA-1 flight with those measured during the ASHOE/MAESA campaign (Airborne Southern Hemisphere Ozone Experiment and Measurements to Assess the Effects of Stratospheric Aircraft). Experimental data obtained on board the ER-2 aircraft on November 2 and 4, 1994, are compared to CRISTA measurements. The CRISTA measurements began on November 4 only and the experiments were not coordinated. *Bacmeister et al.* [1999] therefore used trajectory mapping techniques to advect all measurements to the same time (12.00 UT on November 5, 1994, in this study). The data points of the two experiments are then all at the same point in time but not at the same location. A comparison is then made by selecting closest pairs or by interpolating the data points to a common location. This technique thus takes atmospheric dynamics into account as far as contained in the GSFC DAO analysis which was used here. Generally, a good agreement between the advected aircraft and CRISTA data was found when the satellite data were compared to the aircraft measurements taken November 4. The November 4 ER-2 flight took its course roughly along the US-Canadian west coast in a region of high CRISTA data density. This is in contrast to the situation of the November 2 aircraft flight which took place over regions where the later CRISTA data density was lower by about a factor of 2. Here the agreement between CRISTA and the ER-2 data is poor. A high measurement density apparently is crucial for a good intercomparison. Even in case of the good overall agreement with the November 4 aircraft results there is considerable scatter in individual data comparisons. Local small scale or short lived disturbances like gravity waves or others in one or both data sets remain uncorrected because they are not considered in the model dynamics. The comparison of 72 ozone mixing ratio pairs at a potential temperature of 480 K yielded statistical differences of up to 0.9 ppmv which is of the same magnitude as the profile to profile variations seen by CRISTA-2 near the equator at 30 km.

*4.4 Intercomparisons*

In order to improve the comparison of data sets in the presence of small scale variations, joint measurements of CRISTA and meteorological rockets launched from Wallops Island, VA, and of CRISTA and ozone sonde balloons launched from Hohenpeissenberg, Germany, were done at the same time and in the same volume of air during CRISTA-2. These measurements were performed essentially in order to validate the CRISTA measurements and its data reduction procedure. 33 rockets were launched from Wallops Island to determine mesospheric and stratospheric temperatures, densities, and winds over a 4 week period centered around the CRISTA-2 mission [*Lehmacher and Offermann*, 1997]. During 6 selected orbits the CRISTA field of view was pointed at such an angle to the orbit plane that the measurement volume at tangent altitude passed east of the Wallops Flight Facility. Rockets carrying inflatable falling sphere payloads were launched such that they were at measuring altitude at the same time when CRISTA was sensing this region. In 4 occasions two rockets were launched close together in order to have one measurement in the mesosphere and one measurement in the stratosphere jointly with CRISTA. The rocket measurements are of course very local values whereas the CRISTA analysis volume is about 20 x 200 x 2 km$^3$. The average minimum distance between 10 falling spheres and the CRISTA air volume at zero time difference was 33 km [*Lehmacher et al.*, 1999]. This can be considered as zero miss distance in relation to the CRISTA measurement volume. The agreement between the CRISTA and the rocket temperatures under these conditions was much better than during a similar comparison done for CRISTA-1. The absolute differences are about 30% lower for CRISTA-2 compared to CRISTA-1 where the average miss distance of the 10 closest falling sphere experiments was 200 km within a time interval of ±2 hours. The absolute agreement between the two techniques although within the combined error bars is not really convincing. Falling sphere temperatures are higher than CRISTA at 60 km and lower at 70 km by 5K and 8K, resp. There is, however, some evidence that these differences or part of it are due to the falling sphere technique [*Luebken et al.*, 1994].

A special effort was undertaken to make correlative observations of OH in the middle atmosphere between the MAHRSI experiment on board CRISTA-SPAS and the airborne THOMAS (Tera Hertz OH Measurement Airborne Sounder) experiment. MAHRSI analyses the daytime uv spectrum in the limb mode to derive OH and NO concentrations [*Conway et al.*, 1999]. THOMAS measures the far infrared OH rotational transition at 83.87 cm$^{-1}$ by heterodyne techniques looking sideways out of the DLR Falcon aircraft at an angle between 5° and 20° above the horizon [*Englert et al.*, this volume]. During the CRISTA-2 mission the pointing capabilities of the ASTRO-SPAS and the flight pattern of the DLR Falcon were used to orient MAHRSI and THOMAS to the same volume of air at the same time. A total of 5 aircraft flights over central Europe were carried out and yielded good agreement between the two different OH techniques.

*4.5 Gravity wave signatures*

Small scale structures in the middle atmosphere include gravity waves propagating upwards from the troposphere into the stratosphere and mesosphere where they eventually break. Wave breaking modifies the overall circulation which then influences energetics and composition. The global distribution of gravity waves, their characteristics and their temporal development is therefore an important pre-requisite for middle atmosphere modelling. Not much

is known about the spatial and temporal distribution of these waves globally. *Eckermann and Preusse* [1999] derived gravity wave signatures from CRISTA-1 temperatures in the middle stratosphere. They found gravity waves over the Andes in the southern part of South America as well as over other mountain areas (central Asia, Europe, Alaska). These waves were generated by wind interaction with mountain ridges (mountain waves). The waves over South America typically had vertical wavelengths between $\lambda_z$ = 5 km and $\lambda_z$ = 15 km and were recorded at altitudes between 15 and 30 km. Wave breaking was seen above 30 km in agreement with theory. In a particular case a wave with $\lambda_z$ = 6.5 km is analysed. The temperature perturbations due to this wave are shown in Plate 4. The profiles were measured during three consecutive CRISTA scans separated by about 200 km. The exponential increase of the wave amplitude with height and the wave breaking above 30 – 35 km is obvious. From this wave also the horizontal wavelength was derived being either 130 km or 400 km. There is an ambiguity due to the discrete sampling by CRISTA but there is also some evidence that the shorter value is more probable. The derived gravity wave fields were compared to the model prediction by the Naval Research Laboratory Mountain Wave Forecast Model (MWFM). A generally satisfactory agreement was obtained. In areas where the model predicted high wave activity the CRISTA data also show these waves. A numerical simulation of these waves is discussed by *Tan and Eckermann* [this volume]. The measurement yielded additional wave activity fields not associated with mountains. These wave are possibly generated by time varying non stationary sources and are therefore not predictable by the MWFM. Gravity waves originating from non mountain sources have been reported by ground based stations analysing upper mesosphere airglow emissions [*Taylor and Hill*, 1991] or ionospheric disturbances recorded by sounding rockets [*Kelley*, 1997]. A satellite based measurement of a stratospheric gravity wave generated by a thunderstorm was obtained by the infrared experiment on the MSX satellite [*Dewan et al.*, 1998]. A first inspection of the CRISTA-2 temperature fluctuations in the stratosphere also reveals high gravity wave activity not caused by mountain waves.

*4.6 Thermospheric atomic oxygen*

Atomic oxygen is one of the main constituents in the lower thermosphere. It is produced by photodissociation of molecular oxygen mainly in the Schumann-Runge bands and continuum. The dissociation energy remains as stored chemical energy with the oxygen atoms. At 90 km their

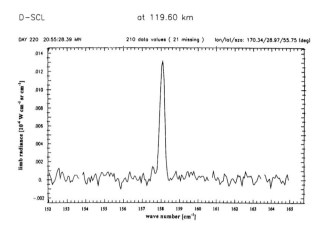

D-SCL                    at 119.60 km

DAY 220  20:55:28.39 MN          210 data values ( 21 missing )   lon/lat/sza: 170.34/28.97/55.75 (deg)

**Figure 5.** Single spectrum of a far infrared channel taken at about 120 km altitude. The spectral feature is the atomic oxygen fine structure emission at 63 μm. The line width is determined by the spectrometer resolution.

lifetime is several days increasing rapidly with height [*Brasseur and Solomon*, 1984]. A transport of oxygen can thus move the stored energy to quite distant regions where it can be released in exothermic recombination reactions [*Riese et al.*, 1994]. Radiative cooling of the thermosphere is mainly due to atomic oxygen through essentially three processes: (1) Collisional excitation of the $CO_2$ - $v_2$ - band with subsequent emission at 15 μm [*Sharma and Wintersteiner*, 1990; *Vollmann and Grossmann*, 1997], (2) collisional excitation of the NO v=1 levels and subsequent emission in the NO 5.3 μm band [*Kockarts*, 1980; *Gordiets et al.*, 1982], and (3) the fine structure emission of atomic oxygen at 63 μm [*Gordiets et al.*, 1982].

The knowledge of the atomic oxygen densities is therefore extremely important for all thermospheric models. In the past satellite measurements have been performed using in-situ techniques as well as indirect remote sounding instruments. Most of these data were used as basis for the MSIS model [*Hedin*, 1991]. Local soundings have resulted in a wide scatter of values which was attributed in part to experimental problems and in part to the natural variability [*Gumbel*, 1997]. A new approach to global measurements of atomic oxygen in the lower thermosphere was done with CRISTA. The atomic oxygen in its electronic ground state $^3P_J$ (J=2,1,0) is thermally excited in the lower thermosphere and radiates in two magnetic dipole transitions $^3P_1 \rightarrow ^3P_2$ at a wavelength of 63 μm and $^3P_0 \rightarrow ^3P_1$ at 147 μm. For the altitudes observed by CRISTA (below 185 km), the emission remains in LTE [*Grossmann and Vollmann*, 1997]. The first global measurement of the 63 μm line was performed with CRISTA-1 [*Grossmann et al.*, 1997]. For the second mission the instrument sensitivity was improved by about a factor of 10 and a larger portion of the mission

was devoted to thermospheric measurements. A spectrum recorded at an altitude of about 120 km during the CRISTA-2 mission is shown in Figure 5. The atomic oxygen emission line is the only spectral feature observed at these altitudes and wavelengths. The signal to noise ratio is near 10 at this altitude. The limb emission as observed by a satellite instrument was modeled by *Sharma et al.* [1988] using atomic oxygen density distributions close to those given in the CIRA reference atmosphere. The calculations show that at around 130 km the 63 μm line becomes optically thick under these conditions. The observed line radiances reach a maximum at this altitude. Below 130 km the radiances decrease because of reabsorption along the line of sight and the decrease of temperature with decreasing altitude. In Figure 6 two altitude profiles of this emission from the two CRISTA missions are shown. The profiles are averages over 84 profiles between 30°S and 30°N in case of CRISTA-1 and 144 profiles in the same latitude band for CRISTA-2. As can be seen from the figure both profiles have their maxima considerably lower than 130 km. This can only be explained by atomic oxygen densities that are much lower then those used for the radiance model. It should be noted that the atmospheric temperature has only a minor effect on the radiance profile due to the very low excitation energy of the O ($^3P_1$) level. Atomic oxygen concentrations were retrieved from the measured CRISTA-2 radiances as described in *Grossmann et al.* [2000]. The zonal mean atomic oxygen densities from the mesosphere/thermosphere mode on August 15/16, 1997, are plotted in Plate 5(a). The densities are averaged over 10° in latitude. Highest concentrations are found at southern mid latitudes and low concentrations over polar regions in general agreement with the large scale

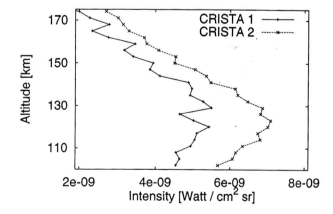

**Figure 6.** Altitude profiles of the atomic oxygen fine structure emission at 63 μm from CRISTA-1 and CRISTA-2. Data are from the latitude band of 30° around the equator. The instrumental noise in these profiles is about ±15% and ±1% for CRISTA-1 and -2, resp.

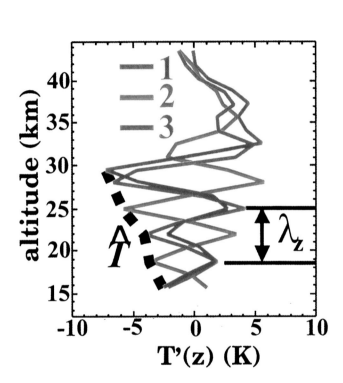

**Plate 4.** Temperature perturbations T'(z) from three successive CRISTA-1 scans over South America (labeled 1, 2, and 3). The data were taken on 6 November 1994. The wave amplitude increases exponentially with altitude as illustrated by the minima of the temperature fluctuations $\check{T}$.

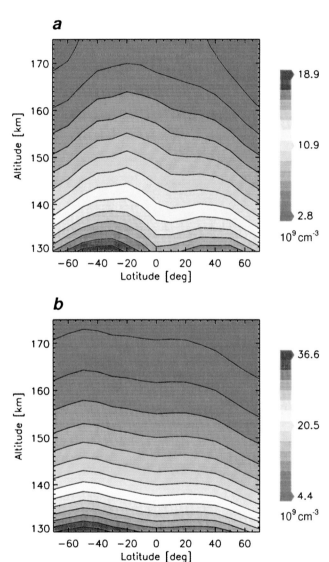

**Plate 5.** Zonal mean atomic oxygen concentrations on August 15/16, 1997. Upper panel (a) shows CRISTA measurements, the lower panel (b) gives MSIS-90 model predictions for the same time period.

thermosphere dynamics [*Fuller-Rowell*, 1995]. In Plate 5(b) zonal mean oxygen densities as predicted by the MSIS model [*Hedin*, 1991] are shown for comparison. The model densities are sampled and averaged in the same way as the measured values. The relative distributions are similar but detail differences remain to be examined. The main difference, however, is the absolute scale. The measured densities are on the average about 40 % below the model predictions. This large discrepancy is well above the absolute CRISTA measurement and retrieval error in this channel of +15%/-25% for zonal mean values. The two CRISTA missions took place under solar minimum conditions. Low atomic oxygen during low solar activity but at F-region altitudes were also found by *Schoendorf and Oliver* [1998] from EISCAT incoherent scatter data over northern Scandinavia. Their findings agree well with the CRISTA derived low oxygen densities.

## 5. SUMMARY AND OUTLOOK

The CRISTA experiments on the ASTRO-SPAS satellite have shown that considerable improvements in the spatial density of remote trace gas measurements in the middle atmosphere can be achieved by means of an appropriate instrument concept in combination with good satellite pointing and manoevering capabilities. CRISTA recorded trace gas emissions in a latitude band from 74°S to 74°N during day and night. The size of the measurement grid is better than 500 x 600 km globally. Several special observation modes were carried out which brought the spatial density of the data points down to 100 km over a limited area and allowed zero miss-distance/-time data comparisons with other experiments. The two CRISTA experiments detected streamers in several stratopheric trace gases. The streamers are variable in time and space. The mission time of CRISTA allows, however, the observation over a roughly one week period only. Small scale structures like a streamer or the shape and sharpness of the subtropical transport barrier are spatially resolved within a short measurement time period of a few hours. The combination of high spatial and good temporal data density considerably improves studies of the overall small scale atmospheric variability.

The features seen by CRISTA-1 are reasonably well modelled by todays assimilated wind driven 3-D models. In detail there are still significant discrepancies which require improvements specially in the small scale dynamics of the models. Gravity wave signatures isolated from the CRISTA-1 temperature measurements have been a valuable test of the model forecast of mountain waves. Other gravity waves are found in the CRISTA data of both flights and are currently being analyzed.

The high sensitivity of CRISTA enabled measurements up into the middle thermosphere. The first global measurement of the atomic oxygen fine structure emission was done by CRISTA. The oxygen densities derived from this emission are considerably lower than the prediction of the MSIS model throughout the lower thermosphere above 130 km. The radiance data alone indicate that the low oxygen concentrations extend down to 120 km. The low atomic oxygen content of the lower thermosphere may be linked to the low solar activity which prevailed during the CRISTA periods.

Trace gas measurements in the middle atmosphere and with global coverage have so far somewhat neglected the density of the measurement net. The data obtained from the two CRISTA experiments demonstrate the importance of high spatial resolution measurements combined with high detectability of large set of trace gases. They are snapshots of one week duration only. Longterm measurements with similar or better spatial resolution are clearly the next step.

*Acknowledgement.* The extraordinary support by the NASA teams at Headquarters, KSC, JSC, and MSFC, and by the satellite team of DASA is gratefully acknowledged. The CRISTA experiments are funded by Deutsche Agentur für Raumfahrtangelegenheiten, Bonn, and Deutsches Zentrum für Luft- und Raumfahrt, Bonn, through grants 50 OE 8503 and 50 QV 9501.

Plate 4 is excerpted with permission from Eckermann and Preusse, Science, Vol 286, pp 1534-1537, Copyright 1999, American Association for the Advancement of Science.

## REFERENCES

Appenzeller, I., et al., Medium resolution FUV spectroscopy of PKS 2155-304., *Astrophys. J., 439*, L33-L37, 1995

Bacmeister, J. T., et al., Intercomparison of satellite and aircraft observations of ozone, CFC-11, and $NO_y$ using trajectory mapping, *J. Geophys. Res., 104*, 16379-16390, 1999

Barnett, J. J., Satellite measurements of middle atmosphere temperature structure, *Phil. Trans. R. Soc. Lond., A 296*, 41-57, 1980

Bartschi, B., et al., Cryogenic Infrared Radiance Instrumentation for Shuttle (CIRRIS 1A): Instrumentation and flight performance, *Proc. SPIE Int. Soc. Opt. Eng., 1765*, 64-74, 1992

Bovensmann, H., et al., SCIAMACHY: Mission objectives and measurement modes, *J. Atmos. Sci., 56*, 127-150, 1999

Brasseur, G., and S. Solomon, *Aeronomy of the middle atmosphere*, D.Reidel, Dordrecht, 1984

Burrows, J. P., et al., The Global Ozone Monitoring Experiment (GOME): Mission concept and first scientific results, *J. Atmos. Sci., 56*, 151-175, 1999

Chen, P., et al., Isentropic mass exchange between the tropics and the extratropics in the stratosphere, *J. Atmos. Sci., 51*, 3006-3018, 1994

Chen, S. S., and R. A. Houze, Jr., Interannual variability of deep convection over the tropical warm pool, *J. Geophys. Res., 102*, 25783-25796, 1997

Conway, R. R., et al., Middle Atmosphere High Resolution Spectrograph Investigation, *J. Geophys. Res., 104*, 16327-16348, 1999

Dewan, E. M., et al., MSX satellite observations of thunderstorm-generated gravity waves in mid-wave infrared images of the upper stratosphere, *Geophys.Res. Lett., 25*, 939-942, 1998

Dials, M. A., et al., A description of the High Resolution Dynamics Limb Sounder (HIRDLS) instrument, *Proc. SPIE Int. Soc. Opt. Eng., 3437*, 84-91, 1998

Drummond, J. R., et al., The stratospheric and mesospheric sounder on Nimbus 7, *Phil. Trans. R. Soc. Lond., A296*, 219-241, 1980

Ebert, H., Zwei Formen von Spektrographen, *Wiedemanns Annln., 38*, 489-493, 1889

Eckermann, S. D., and P. Preusse, Global measurements of stratospheric mountain waves from Space, *Science, 286*, 1534-1537, 1999

Englert, C. R., et al., THOMAS 2.5 THz measurements of middle atmospheric OH: Comparison with MAHRSI observations and model results, *this volume*

Ern, M., *Relaxationseffekte der CRISTA-Infrarotdetektoren und ihre Korrektur*, Dissertation, University of Wuppertal, 2000

Fastie, W. G., A small plane grating monochromator, *J. Opt. Soc. Am., 42*, 641-647, 1952

Fischer, H., and H. Oelhaf, Remote sensing of vertical profiles of atmospheric trace constituents with MIPAS limb emission spectrometers, *Applied Optics, 35*, 2787-2796, 1996

Fuller-Rowell, T. J., The dynamics of the lower thermosphere, in: The upper mesosphere and lower thermosphere: A review of experiment and theory, R. M. Johnson and T. L. Killeen, eds., AGU, *Geophysical Monograph 87*, 23-36, 1995

Gille, J. C., P. L. Bailey, and J. M. Russell III, Temperature and composition measurements from the l.r.i.r. and l.i.m.s. experiments on Nimbus 6 and 7, *Phil. Trans. R. Soc. Lond., A 296*, 205-218, 1980

Gille, J. C., and J. M. Russell III, The limb infrared monitor of the stratosphere: Experiment description, performance, and results, *J. Geophys. Res.,89*, 5125-5140, 1984

Gordiets, B. F., et al., Numerical modeling of the thermospheric heat budget, *J. Geophys. Res., 87*, 4504-4514, 1982

Grant, W. B., et al., Aerosol-associated changes in tropical stratospheric ozone following the eruption of Mount Pinatubo, *J. Geophys. Res., 99*, 8197-8211, 1994

Grossmann, K. U., and K. Vollmann, Thermal infrared measurements in the middle and upper atmosphere, *Adv. Space Res., 19*, 631-638, 1997

Grossmann, K. U., M. Kaufmann, and K. Vollmann, The fine structure emission of thermospheric atomic oxygen, *Adv. Space Res., 19*, 595-598, 1997

Grossmann, K. U., M. Kaufmann, and E. Gerstner, A global measurement of lower thermosphere atomic oxygen densities, *submitted to Geophys. Res. Lett.*, 2000

Gumbel, J., *Rocket-borne optical measurements of minor constituents in the middle atmosphere*, thesis, Department of Meteorology, Stockholm University, 1997

Hedin, A. E., Extension of the MSIS thermosphere model into the middle and lower atmosphere, *J. Geophys. Res., 96*, 1159-1172, 1991

Kelley, M. C., In situ ionospheric observations of severe weather-related gravity waves and associated small-scale plasma structure, *J. Geophys. Res., 102*, 329-335, 1997

Kennedy, J. S., and W. Nordberg, Circulation features of the stratosphere derived from radiometric temperature measurements with the TIROS VII satellite, *J. Atmos. Sci., 24*, 711-719, 1967

Kent, G. S., and M. P. McCormick, SAGE and SAM II measurements of global stratospheric aerosol optical depth and mass loading, *J. Geophys. Res., 89*, 5303-5314, 1984

Kockarts, G., Nitric oxide cooling in the terrestrial atmosphere, *Geophys. Res. Lett., 7*, 137-140, 1980

Krueger, A. J., et al., Satellite ozone measurements, *Phil. Trans. R. Soc. Lond., A296*, 191-204, 1980

Lehmacher, G. A., and D. Offermann, *CRISTA/MAHRSI Campaign 2 Handbook*, University of Wuppertal, 1997

Lehmacher, G. A., et al., Zero miss time and zero miss distance experiments for validation of CRISTA 2 temperatures, *Adv. Space Res.*, in press, 1999

Luebken, F.-J., et al., Intercomparison of density and temperature profiles obtained by lidar, ionization gauges, falling spheres, datasondes and radiosondes during the DYANA campaign, *J. Atmos. Terr. Phys., 56*, 1969-1984, 1994

Luo, M., et al., Observations of stratospheric hydrogen fluoride by Halogen Occultation Experiment (HALOE), *J. Geophys. Res., 99*, 16691-16706, 1994

Lyot, B., La couronne solaire étudiée en dehors des éclipses, *C. R. Acad. Sci., 191*, 834-837, 1930

Mauldin, L. E., et al., Stratospheric Aerosol and Gas Experiment II instrument: A functional description, *Opt. Eng., 24*, 307-312, 1985

McCormick, M. P., et al., High-latitude stratospheric aerosols measured by the SAM II satellite system in 1978 and 1979, *Science, 214*, 328-331, 1981

McCormick, M. P., et al., Satellite and correlative measurements of stratospheric ozone: Comparison of measurements made by SAGE, ECC balloons, chemiluminescent, and optical rocket-sondes, *J. Geophys. Res., 89*, 5315-5320, 1984

Mill, J. D., et al., Midcourse Space Experiment: Introduction to the spacecraft, instruments, and scientific objectives, *J. Spacecraft Rockets, 31*, 900-907, 1994

Murphy, D. M., et al., Reactive nitrogen and its correlation with ozone in the lower stratosphere and the upper troposphere, *J. Geophys. Res., 98*, 8751-8773, 1993

Offermann, D., The MAP/GLOBUS campaign 1983: Introduction, *Planet. Space Sci., 35*, 515-524, 1987

Offermann, D., et al., Cryogenic Infrared Spectrometers and Telescopes for the Atmosphere (CRISTA) experiment and middle atmosphere variability, *J. Geophys. Res., 104*, 16311-16325, 1999

Pfister, L., et al., Gravity waves generated by a tropical cyclone during the STEP tropical field program: A case study, *J. Geophys. Res., 98*, 8611-8638, 1993

Plumb, R. A., A "tropical pipe" model of stratospheric transport, *J. Geophys. Res., 101*, 3957-3972, 1996

Randel, W. J., et al., Stratospheric transport from the tropics to middle latitudes by planetary-wave mixing, *Nature, 365*, 533-535, 1993

Reber, C. A., et al., The Upper Atmosphere Research Satellite (UARS) mission, *J. Geophys. Res., 98*, 10643-10647, 1993

Riese, M., D. Offermann, and G. Brasseur, Energy released by recombination of atomic oxygen and related species at mesopause heights, *J. Geophys. Res., 99*, 14585-14593, 1994

Riese, M., et al., Cryogenic Infrared Spectrometers and Telescopes for the Atmosphere (CRISTA) data processing and atmospheric temperature and trace gas retrieval, *J. Geophys. Res., 104*, 16349-16367, 1999a

Riese, M., et al., Three-dimensional simulation of stratospheric trace gas distributions measured by CRISTA, *J. Geophys. Res., 104*, 16419-16435, 1999b

Roche, A. E., et al., The cryogenic limb array etalon spectrometer (CLAES) on UARS: Experiment description and performance, *J. Geophys. Res., 98*, 10763-10775, 1993

Russell, J. M., III, and R. B. Pierce, Observations of polar descent and coupling in the thermosphere, mesosphere, and stratosphere provided by HALOE, *this volume*

Russell, J. M., III, et al., The Halogen Occultation Experiment, *J. Geophys. Res., 98*, 10777-10797, 1993

Russell, J. M., III, et al., Validation of hydrogen chloride measurements made by the Halogen Occultation Experiment from the UARS platform, *J. Geophys. Res., 101*, 10151-10162, 1996

Schoendorf, J., and W. L. Oliver, A comparison of thermospheric [O] derived at EISCAT with [O] predicted by MSIS, *Geophys. Res. Lett., 25*, 2119-2122, 1998

Schubert, S. D., R. B. Rodd, and J. Pfaendtner, An assimilated dataset for earth science application, *Bull. Am. Meteorol. Soc., 74*, 2331-2342, 1993

Seidel, A., and T. Passvogel, Cryosystems for the infrared missions German Infrared Lab. (GIRL), Infrared Background Signature (IBSS), and Infrared Space Observatory (ISO), *Proc. SPIE Int. Soc. Opt. Eng., 2268*, 318-330, 1994

Sharma, R. B., H. B. Harlow, and J. P. Riehl, Determination of atomic oxygen density and temperature of the thermosphere by remote sensing, *Planet. Space Sci., 36*, 531-538, 1988

Sharma, R. B., and P. P. Wintersteiner, Role of carbon dioxide in cooling planetary atmospheres, *Geophys. Res. Lett., 17*, 2201-2204, 1990

Tan, K. A., and S. D. Eckermann, Numerical simulations of mountain waves in the middle atmosphere over the southern Andes, *this volume*

Taylor, M. J., and M. J. Hill, Near infrared imaging of hydroxyl wave structure over an ocean site at low latitudes, *Geophys. Res. Lett., 18*, 1333-1336, 1991

Taylor, F. W., et al., Remote sensing of atmospheric structure and composition by pressure modulator radiometry from space: The ISAMS experiment on UARS, *J. Geophys. Res., 98*, 10799-10814, 1993

Thrane, E. V., et al., A study of small-scale waves and turbulence in the mesosphere using simultaneous in situ observations of neutral gas and plasma fluctuations, *J. Atmos. Terr. Phys., 56*, 1797-1808, 1994

Trepte, C. R., and M. H. Hitchman, Tropical stratospheric circulation deduced from satellite aerosol data, *Nature, 355*, 626-628, 1992

Vollmann, K., and K. U. Grossmann, Excitation of 4.3 $\mu$m $CO_2$ emissions by $O(^1D)$ during twilight, *Adv. Space Res., 20*, 1185-1189, 1997

Wattenbach, R., and K. Moritz, Astronomical Shuttle Pallet Satellite (ASTRO-SPAS), *Acta Astronautica, 40*, 723-732, 1997

K. U. Grossmann, University of Wuppertal, Dept. of Physics, 42097 Wuppertal, Germany. (e-mail: gross@wpos2.physik.uni-wupertal.de)

# THOMAS 2.5 THz Measurements of Middle Atmospheric OH: Comparison With MAHRSI Observations and Model Results

Christoph R. Englert[1],

Birger A. Schimpf[2], Manfred Birk[2], Franz Schreier[2]

*Institut für Optoelektronik, Deutsches Zentrum für Luft- und Raumfahrt e.V., Oberpfaffenhofen, Germany*

Robert R. Conway, Michael H. Stevens, Michael E. Summers

*E. O. Hulburt Center for Space Research, Naval Research Laboratory, Washington DC*

In this work OH observations of the middle atmosphere performed by the improved 2.5 THz heterodyne spectrometer THOMAS (Tera Hertz OH Measurement Airborne Sounder) during the 1997 THOMAS/MAHRSI (Middle Atmosphere High Resolution Spectrograph Investigation) campaign are presented. Results of the THOMAS OH measurements are compared to simultaneous MAHRSI OH observations and to photochemical model calculations using both standard $HO_x$ chemistry and a recently proposed change in $HO_x$ chemistry. The 40 km – 90 km OH column densities measured by THOMAS and MAHRSI show excellent agreement, the 50 km – 90 km OH column densities agree well within the measurement uncertainties throughout the diurnal cycle. The comparison of the observed and modeled OH column densities in the upper stratosphere and mesosphere suggests that neither model is capable of simultaneously reproducing the measurements in both regions.

## 1. INTRODUCTION

Chemical reactions involving the $HO_x$ family (H, OH, and $HO_2$) play a significant role in middle atmos-

pheric ozone chemistry. However, recent measurements of $HO_x$ in the upper stratosphere and mesosphere present an inconsistent picture, indicating either measurement errors or deficiencies in our understanding of atmospheric odd hydrogen. Up to now, measurements of middle atmospheric OH could not be compared directly because they were carried out at different locations, at different times, or covered altitude regions that do not overlap. For instance, retrievals from the 1994 space borne observations of mesospheric OH by MAHRSI [*Conway et al.*, 1996, *Summers et al.*, 1997, *Conway et al.*, 1999] extend down to 50 km, whereas the measurement of stratospheric OH profiles by a balloon borne instrument (FIRS–2) covers altitudes below 50 km [*Jucks et al.*, 1998]. For the balloon borne meas-

[1] Now at Upper Atmospheric Physics Branch, Space Science Division, Code 7641, E. O. Hulburt Center for Space Research, Naval Research Laboratory, Washington, D.C., 20375–5352.

[2] Now at Remote Sensing Technology Institute, Deutsches Zentrum für Luft- und Raumfahrt e.V., Oberpfaffenhofen, D–82230 Weßling, Germany.

Atmospheric Science Across the Stratopause
Geophysical Monograph 123

urement, the reported OH concentrations at 50 km are in agreement with the standard chemistry prediction, while the 1994 MAHRSI results were found to be 30 % – 40 % lower than the calculated model values at 50 km [*WMO*, 1998]. To resolve these problems, coordinated atmospheric measurements with different measurement techniques and overlapping altitude regions are desirable in order to rule out potential unknown systematic errors.

OH measurements covering the middle and upper stratosphere and the mesosphere were performed in August 1997 with the improved airborne 2.5 THz heterodyne spectrometer THOMAS. These measurements were coordinated in space and time with measurements of mesospheric and upper stratospheric OH from the second MAHRSI flight. In the following, we give a brief overview of the THOMAS instrument, the inversion technique and the method applied to compare THOMAS observations to MAHRSI results. Finally, the results of the THOMAS OH measurements are presented and compared to MAHRSI measurements as well as to photochemical model calculations.

## 2. EXPERIMENT

### 2.1. Instrument

The THOMAS instrument is a further development of a 2.5 THz heterodyne spectrometer for extraterrestrial observations that was built at the Max–Planck–Institute for Radio Astronomy in Bonn, Germany [*Röser*, 1991]. At DLR (German Aerospace Center, Oberpfaffenhofen), the instrument design was modified to allow the observation of OH thermal emission lines in earth's atmosphere from the DLR research aircraft FALCON, using an up–looking geometry. The operation of THOMAS above the tropopause is desirable because of the significant signal absorption due to water vapor in the far–infrared at lower altitudes. The first measurement flights were performed in 1994/95, proving the airworthiness of the instrument [*Titz et al.*, 1995a; *Titz et al.*, 1995b]. Between these flights and the 1997 THOMAS measurements, the spectrometer was significantly improved and a new aircraft window was built [*Englert et al.*, 1999a; *Englert et al.*, 1999b; *Englert*, 1999].

The instrumental modifications increased the signal to noise ratio by more than a factor of five, while improving the spectral resolution by about a factor of two. Figure 1 shows a typical spectrum measured by THOMAS after the improvements. A strong water vapor line and the OH triplet (two strong and one weak line near 83.869 cm$^{-1}$) can clearly be recognized.

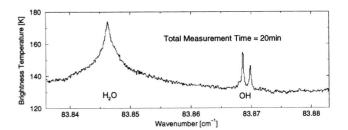

**Figure 1.** Calibrated spectrum measured during the THOMAS/MAHRSI campaign in August 1997.

The photometric accuracy of THOMAS was investigated using laboratory measurements of methanol emission lines and was included in the error budget.

### 2.2. Measurement Geometry and Time Coverage

The measurement locations and timing were selected to achieve nearly zero miss distance and zero miss time with respect to the MAHRSI observations for the upper stratosphere and mesosphere. Due to the different viewing geometries of the instruments (THOMAS: up–looking, 70° – 85° from zenith, MAHRSI: limb viewing, tangent heights between about 35 km and 90 km) a perfect match of the probed air volume is not possible. However, since the upper stratospheric and mesospheric OH column amount is dominated by OH between 40 km and 60 km, the FALCON flight route was planned in a way that the air volume observed by THOMAS at 50 km coincided with the tangent point track of MAHRSI, whenever possible. Plate 1 shows the measurement geometry for August 15, 1997. The two lines indicating the layer entrance and exit points of a virtual MAHRSI 50 km tangent height line of sight (layer thickness: 5 km) were added to the plate to illustrate the extent of the air volume measured by MAHRSI. Minimal miss distance and time was achieved for ten MAHRSI overflights on 4 days between August 10, and August 15, 1997. The coordinated measurements range over a local solar time (LST) interval between about 7 am and 5 pm. An additional sunset flight was performed with THOMAS to extend the time coverage to about 8 pm LST.

## 3. DATA ANALYSIS

### 3.1. Inversion

A best fit of a radiative transfer calculation to the measured spectrum is used to determine OH concentration profiles from the up–looking thermal emission spectra. The information about the vertical distribution of OH is contained in the pressure dependent line width (OH line parameters were adopted from Park et al. [1996] (pressure broadening and temperature exponent) and the HITRAN data base (line position and strength)). The inversion of concentration profiles from atmospheric emission data is generally known to be an ill–posed problem. Regularization, i.e. the introduction of additional constraints, is necessary to get a numerically stable and physically meaningful solution.

The THOMAS spectra have been processed using Phillips–Tikhonov regularization (PTR) as developed by *Hansen* [1992] and first applied to atmospheric retrieval by *Schimpf and Schreier* [1997]. In contrast to optimal estimation [*Rodgers*, 1976] PTR does not require specification of an a priori profile and its covariance matrix. Instead, PTR stabilizes the solution of the least squares problem by adding a smoothness constraint, and Hansen's L–curve criterion is used to automatically determine the optimal smoothing, allowing the extraction of as much information as possible from the measurement [*Schimpf*, 1999, *Englert et al. manuscript in preparation*].

In addition to the OH concentration profile, the inversion algorithm also calculates the averaging kernel matrix. In theory, the inverted spectrum is the product of the averaging kernel matrix and the true concentration profile, which means that the averaging kernel matrix rows determine the sensitivity of the retrieved OH concentration at a particular altitude to the OH concentration at other altitudes. Due to the smoothing of the inverted profile by the averaging kernel, it is commonly used to characterize the altitude resolution of the retrieved profile. The middle panel of Plate 2 shows typical THOMAS averaging kernel rows, illustrating how the resulting profile is smoothed on the altitude grid of the inversion. In the middle and upper stratosphere, the averaging kernel rows are strongly peaked and the typical full width at half maximum of the rows is around 25 km. In the mesosphere however, the averaging kernel rows are not strongly peaked, showing that in this region no height information is available from the spectrum. This behavior is a consequence of the fact that in the mesosphere the line shape of OH thermal emission is no longer dominated by pressure broadening, but by temperature broadening, which results in a nearly altitude independent line shape in the mesosphere.

### 3.2. Determination of Weighted Column Densities

Due to the substantial smoothing of the OH profiles inverted from THOMAS data, a direct comparison to high vertical resolution MAHRSI OH profiles or photochemical model results is not possible. Multiplication of the high vertical resolution profiles with a THOMAS averaging kernel can be used to match the vertical resolution of the profiles to be compared. The comparisons presented here utilize *weighted column densities*, defined as the atmospheric OH profile smoothed by a THOMAS averaging kernel and integrated over altitude as illustrated in Plate 2. The left panel of Plate 2 shows two photochemical model OH profiles representing the standard odd hydrogen chemistry [*DeMore et al.*, 1997] (henceforth referred to as Model A) and revised odd hydrogen chemistry (henceforth referred to as Model B) after *Summers et al.* [1997]. In the right panel the same profiles are shown after smoothing with the averaging kernel which is depicted in the middle panel. In addition to the model profiles, a THOMAS measured profile is included in the right panel of Plate 2. The shaded area illustrates the 40 km – 90 km weighted OH column density. The smoothing of the OH profiles prior to the integration makes the weighted column densities dependent on OH below the lower limit of integration. For the 50 km to 90 km weighted OH column, the sensitivity to OH at lower altitudes is about 20 % whereas it is about 15 % for the 40 km to 90 km weighted OH column.

The different altitude coverage of THOMAS and MAHRSI further complicates the comparison of the results of the two experiments. OH profiles inverted from individual MAHRSI limb scans, that are used for the minimum miss time and distance comparisons, cover the altitude region between about 50 km and 90 km. In order to smooth the single scan MAHRSI profiles using a THOMAS averaging kernel, the MAHRSI profiles have to be extended into the stratosphere. The extension into the upper stratosphere was performed based on averaged MAHRSI limb scan data [*Conway et al.*, 2000] that allowed a reliable inversion of OH concentrations down to about 38 km. Below 38 km the extension was made using standard photochemistry results (Model A) that show reasonable agreement with MAHRSI results at 38 km [*Conway et al.*, 2000]. Using this technique, 50 km to 90 km weighted OH column densities have been derived from single MAHRSI limb scans. 40 km to 90 km weighted column densities have been derived from averaged MAHRSI limb scans extended by Model A results below 38 km.

### 3.3. Error Assessment

The assessment of possible errors in the THOMAS weighted OH column amounts was performed using a typical noise contaminated synthetic spectrum. The synthetic spectrum has been inverted including perturbations due to assumed error sources, in order to estimate the effects on the weighted OH column density.

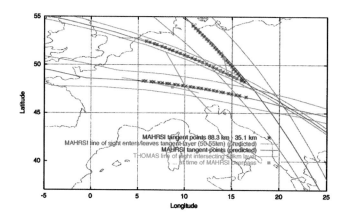

**Plate 1.** Measurement geometry of August 15, 1997. The 'predicted' data is the result from initial calculations.

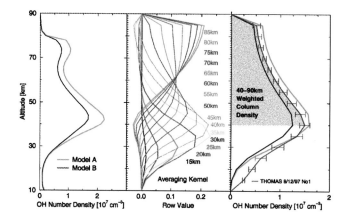

**Plate 2.** Left panel: Model OH profiles using the standard (Model A) and a revised chemistry (Model B). Center panel: typical THOMAS averaging kernel representing the mapping of the high resolution model profiles to the low resolution domain of THOMAS retrieval results. The rows correspond to fifteen altitude layers of equal spacing between 15 km and 85 km. Right panel: Both model profiles, smoothed with the averaging kernel so that they can be compared directly to the corresponding profile retrieved from a THOMAS measurement (LST ≈ 13:00). (Error bars illustrate the statistical error only.) The shaded area represents the weighted column density between 40 km and 90 km.

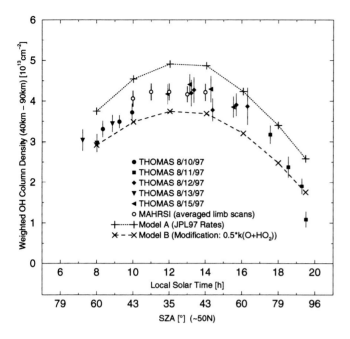

**Figure 2.** Comparison of 40 km – 90 km weighted OH column densities derived from THOMAS observations, MAHRSI results from averaged limb scans, and two photochemical model calculations. The MAHRSI profiles from averaged limb scans cover the 38 km – 90 km altitude region. Model A uses standard odd hydrogen chemistry [*DeMore et al.*, 1997], model B uses a 50 % reduction in the rate coefficient for $O+HO_2 \rightarrow OH+O_2$ as discussed in *Summers et al.*, [1997].

For the 1997 THOMAS measurements, eighteen systematic error sources are investigated this way and the total systematic error for the 40 km – 90 km as well as the 50 km – 90 km weighted OH column density is found to be on the order of ±15 %. The error sources regarded here include instrumental errors, uncertainties in atmospheric pressure and temperature, variations in interfering gas concentrations and uncertainties in spectral line and continuum data. The statistical error of the inversion is typically in the range of ±7 %.

## 4. RESULTS

Figure 2 shows 40 km – 90 km weighted column densities as a function of LST that were derived from THOMAS data, averaged MAHRSI limb scans and from photochemical model results. Model A uses standard odd hydrogen chemistry whereas Model B incorporates a 50 % decrease in the key $OH/HO_2$ partitioning reaction ($O+HO_2 \rightarrow OH + O_2$) as discussed by *Summers et al.*, [1997]. The comparison of the 40 km to 90 km weighted OH column densities shows excellent agreement of THOMAS and MAHRSI observations. The upper stratospheric and mesospheric OH column abun-

dances observed by both instruments are about 15 % lower than the model prediction using the standard odd hydrogen chemistry (Model A). The model results using the revised chemistry are about 10 % lower than the experimental values.

Figure 3 shows 50 km – 90 km weighted OH column densities derived from THOMAS observations, MAHRSI single scan observations coinciding in space and time with the THOMAS observations and photochemical model results as a function of LST. The 50 km – 90 km weighted OH column densities measured by THOMAS are about 15 % lower than Model A results and about 15 % greater than predicted by Model B. The values derived from coincident MAHRSI single limb scans show agreement with THOMAS to better than about 10 %, but are generally lower than the THOMAS observations.

The overall shape of the diurnal variation of the OH weighted column densities depicted in Figures 2 and 3 is in agreement with model predictions. This shows, that neither local variations in OH, nor changes between the days of observation have been significant regarding the statistical error of the presented experimental data.

The error bars depicted in Figures 2 and 3 represent the statistical errors due to measurement noise only. As

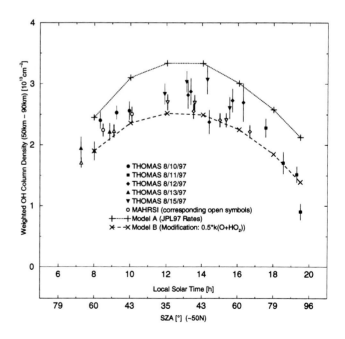

**Figure 3.** Comparison of 50 km – 90 km weighted OH column densities, derived from THOMAS observations, single limb scan MAHRSI results, and two photochemical model calculations. The single limb scan MAHRSI profiles cover the 50 km – 90 km altitude region and coincide in space and time with the respective THOMAS measurements. The denotation of the model results is analogous to Figure 2.

discussed in Section 3.3, the maximum systematic error for the OH column amounts observed by THOMAS is approximately 15%. The total systematic errors of the MAHRSI measurements are estimated to be +12.0/-15.5% for the 50 km – 90 km column and +19.0/-22.5% for the 40 km – 90 km column density (see also [*Conway et al., 1999*]).

## 5. CONCLUSIONS

The results of this first joint upper stratospheric and mesospheric OH measurement campaign, using two entirely different measurement techniques (different geometry of observation and different spectral regions), show that measurements of mesospheric OH are generally lower than expected from standard photochemical theory. This conclusion brings into question our understanding of the upper stratospheric $HO_x$ chemistry, since the same $HO_x$ chemical reactions dominate $OH/HO_2$ partitioning in both regions. In view of the latest balloon observations by *Jucks et al.* [1998], long term column measurements by *Burnett et al.* [1998] and the 1994/97 MAHRSI results, it seems unlikely that a rate change of a single reaction rate in $HO_x$ chemistry will lead to agreement between $HO_x$ models and measurements throughout the middle atmosphere.

*Acknowledgment.* The authors would like to thank R. Titz, M. Krocka, R. Nitsche, T. Weber and the CRISTA group (University of Wuppertal, Germany) for their support. C.R. Englert and B.A. Schimpf would like to thank Prof. K. Künzi for the supervision of their doctoral theses which largely contributed to this work. This work was partially supported by the Office of Naval Research, NASA OES Atmospheric Chemistry and Modeling and Analysis Program (ACMAP), NASA OSS Ionosphere, Thermosphere, and Mesosphere (ITM) Program, and DARA/DLR (Bonn).

## REFERENCES

Burnett C.R., and K. Minschwaner, Continuing development in the regime of decreased atmospheric column OH at Fritz Peak, *Geophys. Res. Let.*, *25*, 1313–1316, 1998.

Conway R.R. et al., Satellite measurements of hydroxyl in the mesosphere, *Geophys. Res. Lett.*, *23*, 2093–2096, 1996.

Conway R.R. et al., Middle Atmosphere High Resolution Spectrograph Investigation, *J. Geophys. Res.*, *104*, 16327–16348, 1999.

Conway R.R. et al., Satellite Observation of Upper Stratospheric and Mesospheric OH: The $HO_x$ Dilemma, *Geophys. Res. Lett.*, in press, 2000.

DeMore W.B. et al., Chemical kinetics and photochemical data for use in stratospheric modeling, *JPL Publication 97–4*, Jet Propulsion Laboratory, Pasadena, California, 1997.

Englert C.R., M. Birk, and H. Maurer, Antireflection coated, wedged, single–crystal silicon aircraft window for the far-infrared, *IEEE Trans. on Geoscience and Remote Sensing*, *37*, 1997–2003, 1999a.

Englert C.R., H. Maurer, and M. Birk, Photon induced far-infrared absorption in pure single crystal silicon, *Infrared Physics and Technology*, *40*, 447–451, 1999b.

Englert C.R., *Observation of OH in the middle atmosphere with an improved and characterized 2.5 THz heterodyne spectrometer*, Shaker Verlag, Aachen/Germany, 1999.

Hansen P.C., Numerical tools for analysis and solution of Fredholm integral equations of the first kind, *Inverse Problems*, *8*, 849–872, 1992.

Jucks K.W. et al., Observation of OH, $HO_2$, and $O_3$ in the upper stratosphere: Implications for $HO_x$ photochemistry, *Geophys. Res. Lett.*, *25*, 3935–3938, 1998.

Park K. et al., Pressure Broadening of the 83.869 $cm^{-1}$ Rotational Lines of OH by $N_2$, $O_2$, $H_2$, and He, *J. Quant. Spectrosc. Radiat. Transfer*, *55*, 285–287, 1996.

Rodgers C.D., Retrieval of atmospheric temperature and composition from remote measurements of thermal radiation, *Rev. Geophys. Space Phys.*, *14*, 609–624, 1976.

Röser H.–P., Heterodyne spectroscopy for submillimeter and far–infrared wavelengths from 100 $\mu$m to 500 $\mu$m, *Infrared Phys.*, *32*, 385–407, 1991.

Schimpf B., and F. Schreier, Robust and efficient inversion of vertical sounding atmospheric high–resolution spectra by means of regularization, *J. Geophys. Res.*, *102*, 16037–16055, 1997.

Schimpf B., Analyse von Fern–Infrarot Spektren zur Bestimmung der OH Konzentration in der mittleren Erdatmosphäre, Doc. thesis, Univ. of Bremen, Germany, 1999.

Summers M.E. et al., Implications of satellite OH observations for middle atmospheric $H_2O$ and ozone, *Science*, *277*, 1967–1970, 1997.

Titz R. et al., Observations of stratospheric OH at 2.5 THz with an airborne heterodyne system, *Infrared Phys. and Technol.*, *36*, 883–891, 1995a.

Titz R. et al., Stratospheric OH measurements with a 2.5 THz heterodyne spectrometer, *Proc. Third European Workshop on Polar Stratospheric Ozone*, Schliersee, Bavaria, Germany, 388–392, 1995b.

WMO – Global Ozone Research and Monitoring Project, *Report No.44, Scientific Assessment of Ozone Depletion: 1998*, World Meteorological Organization, Geneva, 1999.

C. R. Englert[1], B. A. Schimpf[2], M. Birk[2], F. G. Schreier[2], Institut für Optoelektronik, Deutsches Zentrum für Luft– und Raumfahrt e.V. (DLR), Oberpfaffenhofen, 82234 Weßling, Germany. (e-mail: englert@uap2.nrl.navy.mil)

R. R. Conway, M. H. Stevens, M. E. Summers, Upper Atmospheric Physics Branch, Space Science Division, Code 7640, E. O. Hulburt Center for Space Research, Naval Research Laboratory, Washington, D.C., 20375-5352.

# Numerical Simulations of Mountain Waves
# in the Middle Atmosphere over the Southern Andes

Kwok-Aun Tan

*School of Mathematics, University of New South Wales, Sydney, Australia*


Stephen D. Eckermann

*E. O. Hulburt Center for Space Research, Naval Research Laboratory, Washington, DC*

We use a two–dimensional nonlinear compressible mesoscale model to study mountain waves in the middle atmosphere over the southern Andes. Using realistic temperatures, winds and topography, the model generates large-amplitude long-wavelength breaking mountain waves in the middle atmosphere that compare favorably with satellite measurements. Modeled waves have preferred horizontal wavelengths. Spectral analysis reveals correspondences between these wavelengths and peaks in the spectrum of Andean topographic elevations. The shorter waves reach the stratosphere well before the longer ones, consistent with group velocity arguments, with longer wavelengths ultimately dominating. At later times we find evidence of downward propagating secondary waves produced by upper-level breaking of the primary waves.

## 1. INTRODUCTION

Theory and parameterization have indicated that gravity waves forced by flow over mountains are a major source of dynamical driving for the stratosphere and mesosphere [e.g., *Bacmeister*, 1993; *Boville*, 1995]. Detailed mesoscale model simulations have proved useful in modeling mountain wave dynamics and refining simpler parameterizations of these processes in the troposphere and lower stratosphere [e.g., *Laprise*, 1993; *Kim and Arakawa*, 1995; *Durran*, 1995; *Broad*, 1996; *Dörnbrack et al.*, 1998].

However, only a few such models have incorporated a middle atmosphere and investigated mountain wave dynamics up to and beyond the stratopause. One of the first relatively detailed studies was by *Schoeberl* [1985], who derived linear nonhydrostatic mountain wave solutions within realistic wind profiles from the ground up to ~70–80 km. More detailed numerical simulations were performed by *Bacmeister and Schoeberl* [1989] using a two-dimensional nonlinear time-dependent anelastic model of flow over an obstacle. They found zones of strong wave overturning in the middle atmosphere that generated secondary downward-propagating waves. Their model used idealized representations of the background atmosphere. *Satomura and Sato* [1999] used a fully nonlinear compressible two-dimensional model of flow over orography with realistic representations of the lower and middle atmospheres, based on climatology. They too simulated strong wave breaking in the middle stratosphere and generation of secondary waves from these unstable regions, although the wavelengths differed from those found by *Bacmeister and Schoeberl* [1989]. All of these studies considered flow over a bell-shaped obstacle and omitted Coriolis accelerations.

Here, we take further steps towards "real world" mesoscale model simulations of mountain waves in the middle atmosphere. Our simulations focus on a specific set of satellite

Atmospheric Science Across the Stratopause
Geophysical Monograph 123

measurements of mountain waves over the region shown in Plate 1a. Colored squares, labeled 1,2 and 3, show locations of three successive vertical temperature profiles acquired by the Cryogenic Infrared Spectrometers and Telescopes for the Atmosphere (CRISTA) on 6 November, 1994 at 6:24 UT. These profiles, reproduced in Plate 4 of *Grossman* [2000], reveal coherent stratospheric temperature oscillations. *Eckermann and Preusse* [1999] used theory and parameterization to argue that these oscillations were large-scale mountain waves generated by flow over the Andes.

In section 2, we show that two-dimensional mesoscale model simulations offer a potentially useful means of studying the generation and evolution of the mountain waves observed on this day. We describe our mesoscale model in section 3, and apply it in section 4 using "best" estimates of the large-scale winds, temperatures and orography. The results are analyzed in section 5 and the findings are summarized in section 6.

## 2. CRISTA OBSERVATIONS AND REGIONAL DOMAIN

Plate 1 shows that the large-scale topography here is quasi-two dimensional. The "ridge axis" in Plate 1a is a subjective fit to the orientation of the long axis of the Andes. The "perpendicular axis," aligned normal to this ridge axis, passes as close as possible to the three CRISTA profiles, and illustrates that the CRISTA data were acquired along an orbital segment that passed nearly orthogonally across the ridge axis of the Andes.

We take the blue dot in Plate 1 at the western end of the perpendicular axis as our "upstream point," since surface winds were eastward and the atmosphere here is far enough upstream to remain unaffected by the Andes [*Seluchi et al.*, 1998]. Figure 1a plots a hodograph of the horizontal wind profile at the upstream point on 6 November, 1994 at 6:00 UT, using data from NASA's Data Assimilation Office [*Coy and Swinbank*, 1997]. Upstream troposphere-stratosphere winds were roughly parallel to the perpendicular axis at this time, particularly near the surface and in the stratosphere, our primary regions of interest. Maximum deviations of ~45° occurred near the tropopause in Figure 1a.

Similar conditions occur when eastward surface winds flow in from the ocean across the Norwegian Mountains. In these situations, three-dimensional mesoscale models often predict long two-dimensional (plane) mountain waves in the lower stratosphere over Scandinavia, which compare well with aircraft and radiosonde data [*Leutbecher and Volkert*, 1996; *Dörnbrack et al.*, 1998, 1999]. Consequently, these waves can be modeled fairly accurately using a two-dimensional mesoscale model [*Volkert and Intes*, 1992]. We too use a two-dimensional mesoscale model as a first step in this study.

## 3. NUMERICAL MODEL AND DOMAIN

### 3.1. Model Description

The model used here was developed from an earlier version first described by *Tan and Leslie* [1998]. It solves the Euler equations for fully compressible nonlinear nonhydrostatic atmospheric flow over arbitrary terrain in a two-dimensional spatial domain. The velocity $\mathbf{u} = (u, v, w)$, potential temperature $\theta$ and Exner pressure $\pi$ are split into background and perturbation terms, as follows:

$$
\begin{aligned}
u(x, z, t) &= \bar{u}(z) + u'(x, z, t), \\
\theta(x, z, t) &= \bar{\theta}(z) + \theta'(x, z, t), \\
\pi(x, z, t) &= \bar{\pi}(z) + \pi'(x, z, t).
\end{aligned}
$$

The velocity components $v$ and $w$ contain perturbation terms only. A background temperature profile $\bar{T}(z)$ is used to initialize the background density $\bar{\rho}(z)$, potential temperature $\bar{\theta}(z)$ and Exner pressure $\bar{\pi}(z) = \bar{T}(z)/\bar{\theta}(z)$, assuming hydrostatic equilibrium. The cross-ridge background flow, $\bar{u}(z)$, is assumed to be in thermal wind balance. The momentum, thermodynamic and continuity equations become:

$$\frac{du'}{dt} + w'\frac{\partial \bar{u}}{\partial z} + c_p\theta\frac{\partial \pi'}{\partial x} - fv' = D(u'), \quad (1a)$$

$$\frac{dv'}{dt} + fu' = D(v'), \quad (1b)$$

$$\frac{dw'}{dt} + c_p\theta\frac{\partial \pi'}{\partial z} - g\frac{\theta'}{\bar{\theta}} = D(w'), \quad (1c)$$

$$\frac{d\theta'}{dt} + w'\frac{\partial \bar{\theta}}{\partial z} = D(\theta'), \quad (1d)$$

$$\frac{d\pi'}{dt} + w'\frac{\partial \bar{\pi}}{\partial z} + \frac{c_s^2}{c_p\bar{\rho}\bar{\theta}^2}\nabla \cdot (\bar{\rho}\bar{\theta}\mathbf{u}') = 0, \quad (1e)$$

where

$$
\begin{aligned}
\nabla \cdot (\bar{\rho}\bar{\theta}\mathbf{u}') &\equiv \left(\frac{\partial(\bar{\rho}\bar{\theta}u')}{\partial x} + \frac{\partial(\bar{\rho}\bar{\theta}w')}{\partial z}\right), \\
\frac{d}{dt} &\equiv \frac{\partial}{\partial t} + (\bar{u} + u')\frac{\partial}{\partial x} + w'\frac{\partial}{\partial z},
\end{aligned}
$$

$c_p$ and $c_v$ are the mass specific heats at constant pressure and volume, respectively, $R = c_p - c_v$ is the gas constant for dry air, $g$ is the gravitational acceleration, $c_s$ is the speed of sound and $f$ is the Coriolis parameter. The $D(X')$ terms in (1a)–(1d) are damping terms for the perturbation quantities $X'$, and represent the combined effects of a first-order turbulence closure scheme [*Lilly*, 1962], fourth-order computational damping (to suppress small-scale numerical modes),

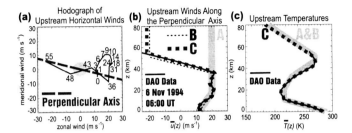

**Figure 1.** (a) Hodograph of horizontal winds over the upstream point in Figure 1, taken from DAO assimilated data on 6 November, 1994, 6:00 UT. The altitudes (in kilometers) of each wind value are labeled along the hodograph trace. Orientation of the perpendicular axis in Plate 1a is shown with a broken line; (b) black solid curve shows the upstream DAO winds along the perpendicular axis, $\bar{u}(z)$. Other labeled curves show model $\bar{u}(z)$ profiles used in Experiments A, B and C; (c) vertical profile of upstream DAO temperatures (solid black curve), and the model curves used in Experiments A, B and C.

and a sponge layer (Rayleigh damping) at upper levels [*Durran and Klemp*, 1983].

To incorporate topography at the bottom boundary, the model uses the terrain-following vertical coordinate

$$\zeta(x, z) = \frac{L\left[z - h(x)\right]}{L - h(x)}, \qquad (2)$$

where $L$ is the top model altitude and $h(x)$ is the topographic height function [*Gal-Chen and Sommerville*, 1975]. A free-slip lower boundary condition is used.

The model equations (1a)–(1e) are solved numerically using the Klemp-Wilhelmson time-splitting method on an Arakawa C-grid [*Klemp and Wilhelmson*, 1978; *Durran and Klemp*, 1983; *Wicker and Skamarock*, 1998]. The second-order Runge-Kutta method is used for the large time steps, $\Delta t_l$, and the forward-backward scheme is used for the small time steps, $\Delta t_s$. An implicit method is used in the vertical for the acoustic terms in the governing equations. A third-order upwind finite difference scheme is used for the advection terms and second-order centered finite differencing is used for the pressure gradient and divergence terms. The Miller-Thorpe radiation condition is used at the lateral boundaries to prevent side reflections [*Miller and Thorpe*, 1981]. The current model configuration was validated by reproducing mountain wave patterns from previous two-dimensional model experiments [e.g., *Doyle et al.*, 2000].

### 3.2. Model Domain and Initialization

The perpendicular axis $x$ in Plate 1 extended to 3000 km in our model with a grid resolution $\Delta x = 2$ km. We set the model lid $L = 80$ km and vertical resolution $\Delta z = 0.5$ km. Topographic elevations $h(x)$ in Plate 1b were smoothed to remove the two and four grid-point wavelengths that can

cause numerical aliasing. In all experiments, wind profiles were ramped up over 2 hours from an initial state of rest, to minimize the generation of transient oscillations [e.g., *Ikawa*, 1990]. The upper-level sponge was effective above ~55 km. Equations were forwarded using $\Delta t_l = 5$ s and $\Delta t_s = 1$ s.

## 4. MODEL SIMULATIONS

### 4.1. Experiments A, B and C

Three different model experiments, denoted A, B and C, were conducted. As shown in Figures 1b and 1c, each successive experiment more closely approximated the upstream DAO wind and temperature profiles at the time of the CRISTA measurements.

Experiment A used an idealized wind profile $\bar{u}(z)$ that omitted the westward shear layer in the DAO profile above 35 km (Figure 1b). This control experiment allowed us to study mountain wave generation and propagation without the complication of critical-level effects ($\bar{u}(z) = 0$).

In Experiment B, a linear westward shear layer was introduced above 35 km, yielding $\bar{u}(z) = 0$ at $z \approx 43$ km. Experiment B allowed us to introduce the critical layer, while maintaining an identical atmospheric situation to Experiment A below 35 km. To this end, both experiments used background temperatures $\bar{T}(z)$ from the 1976 U.S. Standard Atmosphere at mid-latitudes [*Minzner*, 1977], which fit the upstream DAO temperatures from ~0-55 km quite well (Figure 1c).

Experiment C used cubic spline fits to the DAO wind and temperature profiles from 0-55 km. This yielded a zero wind line at $z \approx 46$ km, slightly higher than in Experiment B. Lacking DAO data above 55 km, from 55-80 km we used constant winds and a linear temperature gradient equal to that in Experiments A and B.

### 4.2. Results

In all three experiments, mountain waves entered the middle atmosphere after a few hours and produced overturning isentropes after ~6 hours. We focus on temperatures since CRISTA measured temperature perturbations. Plate 2 plots potential temperatures $\theta(x, \zeta, t)$ and temperature perturbations $T'(x, \zeta, t)$ after $t = 18$ hours over the orographic region in each experiment. The critical levels in Experiments B and C efficiently absorb mountain waves, and yield rapid wave shortening and intense breaking at closely underlying altitudes. Despite the different wind profiles and the significant amounts of overturning and mixing by this time, the large-scale wave structures below 35 km are quite similar in each experiment.

The wave fields evolve significantly with time, as illustrated in Figure 2 using results from Experiment C. The middle atmosphere is dominated initially by short horizontal wavelengths, which trigger vigorous overturning and mixing by 8 hours. The turbulent zones persist and seem to move

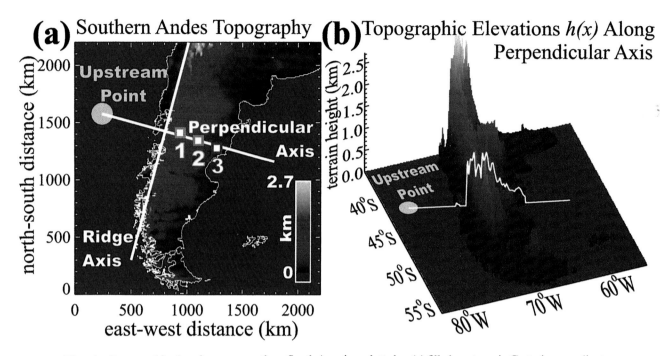

**Plate 1.** Topographic elevations over southern South America, plotted as (a) filled contours in Cartesian coordinates, (b) three-dimensional elevations in Mercator coordinates. Squares labeled 1,2 and 3 in (a) show locations of temperature profiles acquired by CRISTA. Raw elevations $h(x)$ along the perpendicular axis in (a) are plotted in (b). The blue dot in both figures is the upstream point. See text for further details.

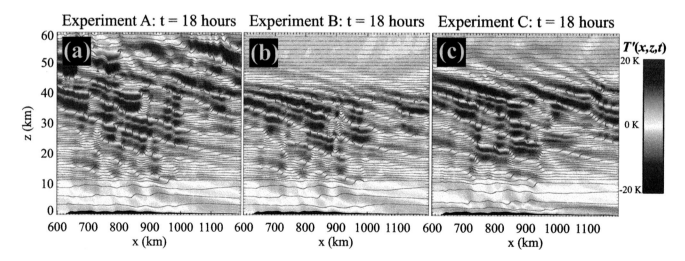

**Plate 2.** Temperature fields after 18 hours in Experiments A, B and C. Contours show isentropes $\theta(x, \zeta, t)$, with constant logarithmic separation between adjacent contours. Temperature perturbations $T'(x, \zeta, t)$ are overlayed using the blue-red color scheme shown to the right. Underlying topography is shown in black.

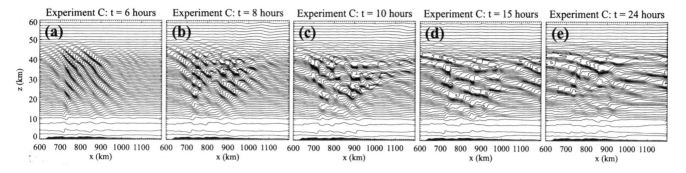

**Figure 2.** Isentropes $\theta(x, \zeta, t)$ from Experiment C after $t = 6, 8, 10, 15$ and 24 hours. Same constant logarithmic separation between contours is used in each panel. Underlying topography is shaded.

downstream after 10 hours. Thereafter, progessively longer horizontal wavelengths come to dominate the wave field, and less vigorous overturning is evident. After 24 hours, waves at lower stratospheric levels are suppressed compared to their earlier intensities.

## 5. ANALYSIS

### 5.1. Wavelength Selection

Preferred vertical wavelengths are evident in Plate 2 and Figure 2. The vertical wavelength of a stationary hydrostatic mountain wave is given theoretically by

$$\lambda_z(z) \approx \frac{2\pi \bar{u}(z)}{N(z)}, \qquad (3)$$

where $N$ is Brunt-Väisälä frequency and $\pi$ has its usual meaning here. At 15–30 km, $\bar{u}(z) = 20$ m s$^{-1}$ and $N = 0.021$ rad s$^{-1}$, which yields $\lambda_z = 6$ km, in agreement with the dominant vertical wavelengths found below 35 km in all three experiments. In Experiments B and C, where westward shear was introduced above 35 km, $\lambda_z$ decreases with height in accordance with (3).

Preferred horizontal wavelengths $\lambda_x$ also emerged in each experiment (Figure 2). We studied this spectrally by re-gridding $T'(x, \zeta, t)$ fields onto a regular height grid $z$, then computing two-dimensional power spectra $F_T(K, M)$ over the range $x = 250$–1250 km and $z = 12$–32 km. As expected from (3), for all $K$ a sharp spectral peak occurred at a vertical wavenumber $|\tilde{M}| \approx 2\pi(6 \text{ km})^{-1}$, enabling us to study overall spectral variability in these simulations using the one-dimensional spectral slice $F_T(K, M \approx \tilde{M})$.

Spectra from each experiment are plotted in Figure 3 after 6, 15 and 24 hours. The power spectrum of the topography $h(x)$ is plotted in Figure 3d, with evident peaks marked by light solid lines. These same lines are also plotted in the other panels of Figure 3, and show that peaks in stratospheric temperature spectra correspond quite well with peaks in the topographic spectrum. Thus, horizontal wavelength selection seems to be governed by spectral features in the Andean topography.

### 5.2. Time Evolution

Figures 2 and 3 show that the dominant horizontal wavelengths change with time. Short $\lambda_x$ values dominate the middle atmosphere initially, giving way to progressively longer horizontal wavelengths at later times. The vertical group velocity of any hydrostatic wave ($\lambda_z^2/\lambda_x^2 \ll 1$) is given by

$$c_{gz} = \frac{\partial \omega}{\partial m} = \frac{\lambda_z^2 N(z)}{2\pi \lambda_x}. \qquad (4)$$

From (3), all waves below 35 km have similar vertical wavelengths $\lambda_z$, and so the vertical group velocities of various

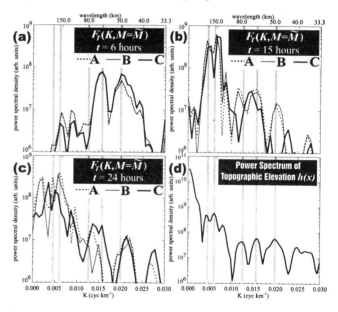

**Figure 3.** One-dimensional sections of the two–dimensional power spectra $F_T(K, M)$ at $M = \tilde{M} = 1/6$ cyc km$^{-1}$ from Experiments A, B and C after (a) 6 hours, (b) 15 hours, and (c) 24 hours. Spectra were computed from $T'(x, z, t)$ fields in the region $z = 12$–32 km and $x = 250$–1250 km. Since $\tilde{M}$ straddles the third and fourth harmonics of these spectra, we plot averages of the slices at these two harmonics. Panel (d) plots power spectral density of the topographic elevation $h(x)$. Local maxima in (d) are highlighted with gray vertical lines, and are also shown in the preceding panels.

mountain wave modes are inversely proportional to their horizontal wavelengths, $\lambda_x$.

Using (3) and (4), a wave of $\lambda_x = 50$ km has a group velocity $c_{gz} \approx 2.5$ m s$^{-1}$, and thus takes ~3.5 hours to propagate to $z = 30$ km. Spectra in Figure 3a confirm that the peak near $\lambda_x = 50$ km is fully developed after 6 hours. Conversely, a $\lambda_x = 150$ km wave takes ~10 hours to propagate to 30 km. Stratospheric mountain wave energy at this scale does not develop fully until ~10–15 hours into the simulation (Figure 3b). Note that $\bar{u}(z)$ was ramped up from zero during the first two hours, and was <20 m s$^{-1}$ below 17 km, making actual propagation times somewhat longer.

The mean width of the topography $h(x)$ is ~300-500 km (see Plate 1), which might be expected to force a $\lambda_x \sim 300$-500 km wave most strongly. Indeed, ridge widths here are similar to those in northern Scandinavia, where very long mountain waves often dominate the stratospheric wave field [*Dörnbrack et al.*, 1998, 1999]. Using (3) and (4), after 24 hours a 400 km wave would have propagated to a height of only 20 km or so, and thus would not be fully developed in the middle atmosphere. Spectra after 24 hours confirm this: while Experiments A and B show a peak at a wavelength ~400-500 km in Figure 3c, spectra from Experiment C peak most strongly nearer 200-250 km. Thus, longer simulations are needed to assess whether a $\lambda_x \sim$400 km wave eventually dominates the middle atmospheric wave fluctuations [see also *Dörnbrack et al.*, 1999].

Once each spectral peak has developed in Figure 3, its intensity tends to wane thereafter, due to wave breaking and increasing turbulent damping. As longer waves subsequently enter the region and break, the additional turbulence and interactions they engender probably act to damp these shorter horizontal wavelengths further.

### 5.3. Downward Feedback

Although Experiments A and B used identical background atmospheres below 35 km, the results in Plates 2a and 2b after 18 hours are identical only below ~20 km, and show significant differences above ~25 km. The time evolution of these differences can be gleaned from Figure 3. After 6 hours the spectra in Experiments A and B are identical, but show differences after 15 hours. By 24 hours (Figure 3c), the differences are as large in places as those from Experiment C.

These differences become evident below 35 km after waves start breaking. Implicit numerical diffusion in this model is very small, so these differences must originate from downward feedback of differing dynamics originating above 35 km, where different $\bar{u}(z)$ profiles yield different wave and turbulence fields. Turbulence itself cannot be advected this far downwards, since we include no mean vertical wind.

Two-dimensional spectra $F_T(K, M)$ separate upward and downward propagating waves [e.g., *Bacmeister and Schoeberl*, 1989]: only the upward components are plotted in Figure 3. After 12–24 hours, significant downward-propagating spectral density appeared in each experiment, most prominently at long horizontal wavelengths. Vertical velocity spectra $F_T(K, M)$ also showed downward propagating variance at shorter horizontal scales. The appearance of these features in Experiment A indicates that this is not a critical-level reflection effect, as expected since mean Richardson numbers are too large [*Jones*, 1968]. Rather, these findings are in qualitative agreement with previous simulations of secondary wave generation by mountain wave breaking [*Bacmeister and Schoeberl*, 1989; *Satomura and Sato*, 1999].

*Bacmeister and Schoeberl* [1989] found that their downward propagating secondary waves eventually led to reductions in overall wave intensity well below the regions of wave breaking. Attenuated wave activity at lower levels also arose in our simulations (Figure 2e). Whether the attenuation noted here results mostly from the effects of downward propagating secondary waves, or from breakdown of and additional turbulent damping by progressively longer upward-propagating primary waves, is unclear at present.

### 5.4. Comparison with CRISTA Profiles

The general form of the profiles over the Andes compares favorably with the CRISTA data. Figure 4 plots a sample temperature perturbation profile from Experiment C at $x = 845$ km, a location between points 1 and 3 in Plate 1a. The CRISTA profiles at these points are overlayed [see Plate 4 of *Grossmann*, 2000]. The wavelength, amplitude and general height variation of the model profile are all quite similar to the data. Oscillations in Experiment C seem to penetrate slightly higher prior to shortening and dissipating below the critical layer. Note too that the amplitudes of the CRISTA profiles are probably underestimated due to instrumental effects, and that CRISTA cannot resolve any waves of $\lambda_z \lesssim$ 3–5 km [*Eckermann and Preusse*, 1999].

Model profiles from locations directly over points 1 and 3 are not in phase with the CRISTA perturbations. This is due to somewhat different horizontal wavelengths in the model simulations. *Eckermann and Preusse* [1999] showed that the oscillations measured by CRISTA had a horizontal wavelength of either 400 km or 400/3 $\approx$ 130 km. Peaks somewhat near these values appear in Figure 3c, although other peaks also occur and there are variations among the experiments.

## 6. SUMMARY AND DISCUSSION

We have developed a high-resolution high-altitude nonlinear numerical model of compressible atmospheric flow over arbitrary topography, and have used it here to simulate mountain waves in the middle atmosphere over the southern Andes. The results have confirmed that stratospheric temperature perturbations measured by CRISTA on 6 November, 1994 were produced by intense breaking mountain waves forced by flow over the southern Andes, as argued by *Eckermann*

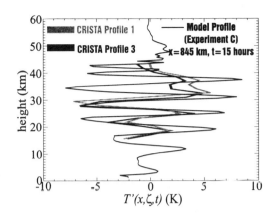

**Figure 4.** Black curve shows a vertical profile of temperature perturbations at $x = 845$ km, 15 hours into Experiment C. Two shaded solid curves reproduce the temperature perturbation profiles acquired by CRISTA at locations 1 and 3 in Plate 1a (after *Eckermann and Preusse* [1999]).

*and Preusse* [1999]. Amplitudes and vertical wavelengths from CRISTA compare fairly well with model profiles. The results showed that attenuated oscillations above 30 km were the result of vertical wavelength reductions and intense wave breaking, as conjectured by *Eckermann and Preusse* [1999]. Recent analysis of satellite data by *McLandress et al.* [2000] has shown that temperature fluctuations appear recurrently in the middle atmosphere over southern South America. Thus these breaking mountain waves may be relevant to the climatology of this region of the middle atmosphere [*Garcia and Boville*, 1994].

Presently we have a limited understanding of the amplitudes and wavelengths of mountain waves radiated into the middle atmosphere by complex topography. Existing parameterizations make use of topographic elevation spectra, focusing either on dominant peaks [e.g., *Bacmeister*, 1993] or their power-law shapes [e.g., *Shutts*, 1995] to specify radiated mountain wave fields. Our results, though limited to a specific two-dimensional section of the Andes, produced stratospheric wave fluctuations with distinct vertical and horizontal wavelengths. These horizontal wavelengths were closely related to peaks in the power spectrum of the topographic elevations $h(x)$, lending some support to spectral parameterization methods. However, the various horizontal wavelengths evolved differently with time in our model, due mostly to varying vertical group velocities. Wave fields were further complicated at later times by breaking and turbulence, leading not only to increased damping of primary waves but apparent generation of secondary waves. Even after 24 hours, when simulations were terminated due to computational constraints, mountain waves in the middle atmosphere were still evolving.

Actual topography and flow fields here are intrinsically three-dimensional [Plate 1b; Figure 1a; *Seluchi et al.*, 1998], which can influence mountain waves significantly [*Shutts*,

1998]. While two-dimensional models with turbulence parameterizations provide reasonable simulations of breaking gravity waves [*Liu et al.*, 1999; *Doyle et al.*, 2000], long-term fine-scale simulations of wave breakdown beneath a critical level seem to require three-dimensional models [*Winters and D'Asaro*, 1994; *Dörnbrack*, 1998; *Fritts and Werne*, 2000]. Additionally, the model here did not include mean flow evolution, which may be important in this problem. Thus, multiday interactive runs using three-dimensional models may be needed to describe these evolving mountain wave fields fully.

*Acknowledgments.* These simulations were made possible by a grant of computing time on the Australian National University Supercomputing Facility's Fujitsu VPP300 vector supercomputer. SDE acknowledges support for this research by the Office of Naval Research and by NASA through the Atmospheric Chemistry Modeling and Analysis Program and the UARS Guest Investigator Program. We thank D. Broutman, A. Dörnbrack, P. Preusse and two reviewers for helpful comments on earlier drafts.

## REFERENCES

Bacmeister, J. T., Mountain-wave drag in the stratosphere and mesosphere inferred from observed winds and a simple mountain-wave parameterization scheme, *J. Atmos. Sci., 50,* , 377-399, 1993.

Bacmeister, J. T., and M. R. Schoeberl, Breakdown of vertically propagating two-dimensional gravity waves forced by orography, *J. Atmos. Sci., 46,* 2109-2134, 1989.

Boville, B. A., Middle atmosphere version of the CCM2 (MACCM2): annual cycle and interannual variability, *J. Geophys. Res., 100,* 9017-9039, 1995.

Broad, A. S., High resolution numerical-model integrations to validate gravity-wave-drag parameterization schemes: a case study, *Q. J. R. Meteorol. Soc., 122,* 1625-1653, 1996.

Coy, L., and R. Swinbank, Characteristics of stratospheric winds and temperatures produced by data assimilation, *J. Geophys. Res. 102,* 25,763–25,781, 1997.

Dörnbrack, A., Turbulent mixing by breaking gravity waves, *J. Fluid Mech., 375,* 113-141, 1998.

Dörnbrack, A., M. Leutbecher, H. Volkert and M. Wirth, Mesoscale forecasts of stratospheric mountain waves, *Meteorol. Appl., 5,* 117-126, 1998.

Dörnbrack, A., M. Leutbecher, R. Kivi and E. Kyrö, Mountain-wave-induced record low stratospheric temperatures above northern Scandinavia, *Tellus, 51A,* 951-963, 1999.

Doyle, J. D., et al., An intercomparison of model-predicted wave breaking for the 11 January 1972 Boulder windstorm, *Mon. Wea. Rev., 128,* 901-914, 2000.

Durran, D. R., Do breaking mountain waves decelerate the local mean flow?, *J. Atmos. Sci., 52,* 4010-4032, 1995.

Durran, D. R.. and J. B. Klemp, A compressible model for the simulation of moist mountain waves, *Mon. Wea. Rev., 111,* 2341-2361, 1983.

Eckermann, S. D., and P. Preusse, Global measurements of stratospheric mountain waves from space, *Science, 286,* 1534-1537, 1999.

Fritts, D. C., and J. A. Werne, Turbulence dynamics and mixing due to gravity waves in the lower and middle atmosphere, *this volume.*

Gal-Chen, T., and R. C. T. Sommerville, On the use of a coordinate transformation for the solution of the Navier-Stokes equations, *J. Comp. Phys.*, *17*, 209-228, 1975.

Garcia, R. R., and B. A. Boville, "Downward control" of the mean meridional circulation and temperature distribution of the polar winter stratosphere, *J. Atmos. Sci.*, *51*, 2238-2245, 1994.

Grossmann, K. U., Recent improvements in middle atmosphere remote sounding techniques: the CRISTA-SPAS experiment, *this volume*.

Ikawa, M., High-drag states and foehns of a two-layered stratified fluid past a two-dimensional obstacle, *J. Meteorol. Soc. Japan.*, *68*, 163-182, 1990.

Jones, W. L., Reflexion and stability of waves in stably stratified fluids with shear flow: a numerical study, *J. Fluid Mech.*, *34*, 609-624, 1968.

Kim, Y.-J., and A. Arakawa, Improvement of orographic gravity wave parameterization using a mesoscale gravity wave model, *J. Atmos. Sci.*, *52*, 1875-1902, 1995.

Klemp, J. B., and R. B. Wilhelmson, The simulation of three-dimensional convective storm dynamics, *J. Atmos. Sci.*, *35*, 1070-1096, 1978.

Laprise, J. P. R., An assessment of the WKBJ approximation to the vertical structure of linear mountain waves: implications for gravity wave drag parameterization, *J. Atmos. Sci.*, *50*, 1469-1487, 1993.

Leutbecher, M., and H. Volkert, Stratospheric temperature anomalies and mountain waves: A three-dimensional simulation using a multi-scale weather prediction model, *Geophys. Res. Lett.*, *23*, 3329-3332, 1996.

Lilly, D. K., On the numerical simulation of buoyant convection, *Tellus*, *14*, 148-172, 1962.

Liu, H.-L., P. B. Hays, and R. G. Roble, A numerical study of gravity wave breaking and impacts on turbulence and mean state, *J. Atmos. Sci.*, *56*, 2152–2177, 1999.

McLandress, C., M. J. Alexander, and D. L. Wu, Microwave Limb Sounder observations of gravity waves in the stratosphere: a climatology and interpretation, *J. Geophys. Res.*, (in press), 2000.

Miller, M. J. and A. J. Thorpe, Radiation conditions for the lateral boundaries of limited-area numerical models, *Q. J. R. Meteorol. Soc.*, *107*, 615-628, 1981.

Minzner, R. A., The 1976 standard atmosphere and its relationship to earlier standards, *Rev. Geophys. Space Phys.*, *15*, 255-264, 1977.

Satomura, T., and K. Sato, Secondary generation of gravity waves associated with the breaking of mountain waves, *J. Atmos. Sci.*, *56*, 3847-3858, 1999.

Schoeberl, M. R., The penetration of mountain waves into the middle atmosphere, *J. Atmos. Sci.*, *42*, 2856-2864, 1985.

Seluchi, M., Y. V. Serafini and H. Le Treut, The impact of the Andes on transient atmospheric systems: a comparison between observations and GCM results, *Mon Wea. Rev.*, *126*, 895–912, 1998.

Shutts, G., Gravity-wave drag parameterization over complex terrain: the effect of critical level absorption in directional wind shear, *Q. J. R. Meteorol. Soc.*, *121*, 1005-1021, 1995.

Shutts, G. J., Stationary gravity-wave structure in flows with directional wind shear, *Q. J. R. Meteorol. Soc.*, *124*, 1421-1442, 1998.

Tan, K. A., and L. M. Leslie, Development of a non-hydrostatic model for atmospheric modelling, in *Computational Techniques and Applications: CTAC97*, edited by B.J. Noye, M.D. Teubner, and A.W. Gill, pp. 679-686, World Scientific Press, Singapore, 1998.

Volkert, H., and D. Intes, Orographically forced stratospheric waves over northern Scandinavia, *Geophys. Res. Lett.*, *19*, 1205-1208, 1992.

Wicker, L. J., and W. C. Skamarock, A time-splitting scheme for the elastic equations incorporating second-order Runge-Kutta time differencing, *Mon. Wea. Rev.*, *126*, 1992-1999, 1998.

Winters, K. B., and E. A. D'Asaro, Three-dimensional wave instability near a critical level, *J. Fluid Mech.*, *272*, 255-284, 1994.

K. Tan, School of Mathematics, University of New South Wales, Sydney, NSW 2052, Australia. (e-mail: K.Tan@unsw.edu.au)

S. D. Eckermann, E. O. Hulburt Center for Space Research, Code 7641.2, Naval Research Laboratory, Washington, DC 20375. (e-mail: eckerman@map.nrl.navy.mil)

# Planetary Wave Two Signatures in CRISTA 2 Ozone and Temperature Data

W.E. Ward

*CRESS/CRESTech, York University, North York, Ontario, Canada*

J. Oberheide, M. Riese, P. Preusse, D. Offermann

*Dept. of Physics, University of Wuppertal, D-42097 Wuppertal, Germany*

The Cryogenic Infrared Spectrometers and Telescopes for the Atmosphere (CRISTA) instrument has flown on two shuttle missions providing observations of constituents and temperature throughout the stratosphere and mesosphere at high vertical and horizontal resolution. During the second mission (CRISTA 2: August 7-17, 1997) a wave two signature was observed in the southern midlatitudes in the temperature data between 17 and 80 km and in the ozone data between 17 and 63 km (the limits of ozone data availability for the current data version). This signature has a period of ~12.5 days and a vertical wavelength of ~45 km. It is one of the few observational instances showing vertical and equatorward propagation of a planetary wave well into the mesosphere. The relative phase of the signature in the temperature and ozone varies with height with the two being in phase below 30 km and 180 degrees out of phase above 40 km. These signatures are consistent with those expected from a migrating planetary wave with the ozone signature being dynamically driven at lower altitudes and photochemically driven at higher altitudes. The CRISTA results above the middle stratosphere show a greater agreement with photochemical equilibrium calculations than previous analyses. These results clearly illustrate the penetration of a dynamical feature from the tropopause into the mesosphere and its effect on constituent distributions throughout this height range.

## 1. INTRODUCTION

The larger scale, long term systematic variation of ozone in the middle atmosphere has been a topic of major interest over the past 20 years [see Solomon, 1999; Jucks and Salawitch, this issue; Brasseur et al., this issue]. This activity has been fueled by evidence indicating that the total column amount of ozone is decreasing due to human use of chlorofluorocarbons thereby putting life on the planet at risk [Farman et al., 1985]. Much of this research has been concentrated on resolving questions associated with the chemistry of ozone, determining the morphology and time history of the large scale variations in its distribution and understanding the causes for these variations.

Ozone also varies over smaller spatial and shorter time scales. This variability is part of the natural response of ozone to dynamical and photochemical processes and is effectively a source of noise to those study-

Atmospheric Science Across the Stratopause
Geophysical Monograph 123

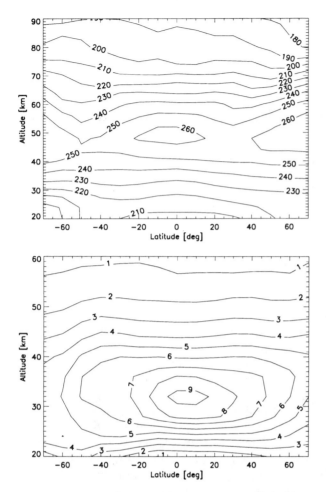

**Figure 1.** Zonal means of the temperature (K) (a) and ozone mixing ratio (ppmv) (b) calculated from data from August 14, 1997.

ing the long term systematic variations [Randel and Cobb, 1994]. An ability to model these smaller and shorter scale variations, however, provides an opportunity to confirm that the basic physics and chemistry associated with ozone is understood [Froidevaux et al., 1989; Randel et al., 1990; Smith, 1995; Allen et al., 1997].

In this paper, we present a simple analysis of temperature and ozone measurements from the 8 days during the CRISTA 2 mission (August 8-16, 1997) when CRISTA was actively taking measurements. At this time a large planetary wave 2 signature was present in the southern hemisphere. This wave extended from the tropopause to about 80 km in the temperature data and to at least 63 km in the ozone data (the upper boundary for ozone data in Version B01 data). Above 40 km the ozone and temperature perturbations are anti-correlated and below 30 km they are correlated. Similar correlations have been reported [Barnett et al.,

1975; Douglass et al., 1985; Froidevaux et al., 1989; Randel, 1993; Fishbein et al., 1993, Sabutis et al., 1997] and modelled [Hartmann and Garcia, 1979; Rood and Douglass, 1985; Rose and Brasseur, 1989; Smith 1995; Reddmann et al., 1999] and the CRISTA result conforms to the accepted understanding of this feature. Compared to previous results, the CRISTA data are of finer vertical resolution and greater vertical extent. The data presented here show that the planetary wave and its influence on the ozone concentration extend across the stratopause into the mesosphere.

The plan of this paper is as follows. First the CRISTA results are presented and the observed correlations described. These correlations are then analysed from a photochemical perspective [Froidevaux et al., 1989]. The paper concludes with some comments on the relevance of these results to the interpretation of high resolution satellite observations.

## 2. CRISTA OBSERVATIONS

The CRISTA instrument is designed to measure temperature and constituent distributions at high resolution by monitoring their thermal emissions (4-71 $\mu$m) using several IR grating spectometers [see Offermann et al., 1999; Riese et al., 1999; Grossmann (this volume) for details of the instrumentation and data analysis procedure]. The basic configuration for the CRISTA 2 flight was the same as the CRISTA 1 flight. CRISTA collected spectra at the limb behind the spacecraft using three telescopes oriented so that the two adjacent telescopes viewed at azimuths of 18° relative to the central telescope. The temperature and ozone data reported on in this paper were retrieved from the $CO_2$ 12.6 $\mu$m (for the stratosphere) and 15 $\mu$m (for the mesosphere) emissions and the ozone 12.7 $\mu$m emissions respectively. Temperature errors were of the order of 2 K and the ozone uncertainty was 10–15%. The data presented in this paper are a combination of observations from all three telescopes and ascending and descending nodes of the satellite orbit.

In Figure 1 zonal means of the temperature and ozone mixing ratio for this mission are presented. The mission date of mid-August corresponds to northern hemisphere (NH) late summer/southern hemisphere (SH) late winter. The temperature field is similar to summer solstice conditions with a warm stratopause and cold upper mesosphere in the summer (southern) hemisphere. The ozone field is roughly symmetric about the equator with a gradient of $\approx$ 1 ppm/10° of latitude at 50° S.

A planetary wave feature is seen in the CRISTA temperature and ozone fields. Figures 2 and 3 are SH polar views of these fields at heights of 20, 30, 40, 45, 50, and

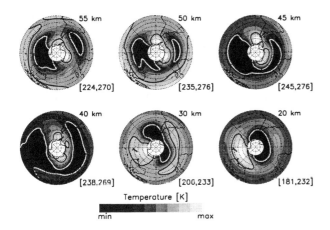

Figure 2. Polar views of the temperature field in the SH at 6 different altitudes. Dark contours are plotted at 70% and light contours are plotted at 30% of the full contour interval.

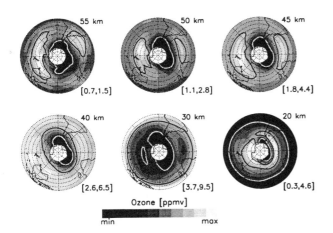

Figure 3. As in Figure 3 except for the ozone mixing ratio (ppmv).

55 km for August 15, 1997 (representative for all days of the mission). In each subplot, the range of ozone mixing ratio or temperature is listed in square brackets to the lower left. A wavenumber 2 signature is prominent in the temperature figures. A similar feature is seen clearly in the ozone fields at heights above 40 km and less distinctly at heights of 20 and 30 km. In each view, contours are used to isolate the regions of maximum and minimum temperature and ozone mixing ratio (dark contours at 70% of the full range and light contours at 30% of the full range). At 40 km and above, the regions of enhanced ozone correspond closely to regions of low temperature. Below 40 km the correlation is reversed.

Figure 4 shows the amplitude and phase of this wave 2 disturbance as a function of latitude and height for the temperature perturbation field (upper subplot) and the relative ozone variation (lower subplot) calculated by

fitting in a least-mean-squares (LMS) sense for the wave parameters using latitude/height bins of 5° by 1.5 km. The amplitude appears as the grey scale map and the phase as the superimposed black contouring. The phase corresponds to the longitude at which the maximum amplitude occurs on August 9.

The temperature perturbation field shows amplitude maxima near 50°S of ≈11, 13, and 9 K occuring at 25, 45 and 70 km respectively with equatorward propagation in the meosphere. Although expected from modelling studies (see Holton and Alexander, this issue, for a general description of planetary wave propagation characteristics) this is one of the few instances where this behaviour has been observed in atmospheric data. The minimum in the temperature at 35 km corresponds to the height of the maximum geopotential wave amplitude. The best fit for the wavenumber 2 signature was for an eastward propagating disturbance with a period of ~ 12.5 days. Using the 90° and -90° phase contours, the wavelength is seen to be ~45 km. LMS calculations also indicate the existence of a wavenumber 1 disturbance of approximately the same period and amplitude below 30 km poleward of the wave 2 disturbance. These features are consistent with previous descriptions of SH wave 2 planetary wave events in the late winter/early spring [Manney et al., 1991].

The ozone perturbation fields (Figure 4b) have significant amplitude above 25 km. The phase remains constant throughout this region at ~90° of longitude. Because the phase of the temperature perturbation varies with height, the ozone variations are roughly in phase below 30 km and out of phase above 40 km as expected from the plan views described above.

These data show that during the CRISTA 2 mission, the dynamics of the southern hemisphere mid to high latitudes are dominated by a wave 2 planetary wave extending from 20 km to 80 km with a maximum temperature amplitude of the order of 13 K at 45 km. Both the temperature and ozone fields show the planetary wave signature with the phase of the ozone signatures changing sign relative to the temperature at about 40 km.

## 3. DISCUSSION

The conceptual framework for temperature/ozone correlations such as those noted above was first developed by Hartmann and Garcia [1979]. They included both ozone chemistry, parameterized linearly in terms of its response to temperature perturbations and relaxation of odd oxygen perturbations, and transport effects. The behaviour of ozone was shown to vary with height as a result of the variation of its photochemical lifetime relative to dynamical timescale (taken to be ~

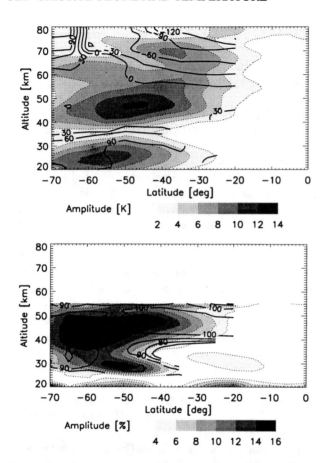

**Figure 4.** Height verses latitude cuts of the amplitude and phase of the wave 2 temperature (upper panel; in Kelvin) and ozone perturbations (lower panel; % change in ozone mixing ratio relative to the zonal mean value) in the Southern Hemisphere. The amplitudes are in grey scale and the phases are the labeled contours in degrees of longitude (see text for details).

1 day). Where its photochemical lifetime is short relative to dynamical time scales (the upper stratosphere and mesosphere, above ~ 35 km), the photochemistry would be expected to dominate and where its lifetime is slow relative to the dynamics (lower stratosphere, below ~ 35 km) transport effects would be expected to dominate. Because the ozone production rate increases with decreasing temperature, in regions dominated by photochemistry a negative correlation between temperature and ozone mixing ratio would occur. In regions dominated by transport, the effect was shown to vary. For large scale waves which penetrate to a sufficient height into the upper stratosphere (i.e. their maximum amplitude is above the transition region so that phase relations between the various velocity components associated with wave growth hold) they showed that the correlation would be positive. Work by Rood and Douglass [1985] demonstrated that care must be taken in

applying these concepts. In circumstances where there are strong dynamical processes (such as stratospheric warmings), the advective terms can dominate the correlations even in regions where photochemistry is fast.

An alternate treatment of the temperature dependence of the ozone photochemistry under photochemical equilibrium conditions excluding consideration of dynamical effects was introduced by Barnett et al. [1975] and developed further in a number of studies [Haigh and Pyle, 1982; Froidevaux et al., 1989; Smith, 1995]. They pointed out that the temperature dependence could be represented by

$$[O_3] = B \exp(\Theta_P/T) \qquad (1)$$

where $[O_3]$ is the ozone concentration, and $B$ and $\Theta_P$ are constants whose value depended on the concentrations of the reactants involved in the ozone chemistry. Values vary with $\Theta_P \approx 1405$ K for pure oxygen chemistry, $\approx 1200$ K for $NO_x$ chemistry, $\approx 400$ K for $HO_x$ chemistry and $\approx 200$ K for $Cl_x$ chemistry [Haigh and Pyle, 1982]. For the analysis of data at a given level, the derivative of the log of this expression provides the relationship

$$\frac{\Delta[O_3]}{[O_3]} = -\frac{\Theta_E}{T^2}\Delta T \qquad (2)$$

where $\Theta_E$ is an empirical constant to be determined. The advantage of this formulation is that it removes the dependence on the absolute value of the ozone mixing ratio and allows the temperature dependence of the chemistry to be investigated directly [Froidevaux et al., 1989]. Inclusion of additional feedback effects between temperature and ozone concentration arising from UV heating variations due to variations in ozone and radiative damping of temperature perturbations results in the relationship [Brasseur and Solomon, 1986]:

$$\frac{\Delta[O_3]}{[O_3]} = (\frac{\alpha}{P_H} - \frac{\Theta_E}{T^2})\Delta T \qquad (3)$$

where $\alpha$ is the radiative relaxation rate and $P_H$ is the gross UV heating rate. Approximate values for $P_H \approx 10$ K day$^{-1}$ from Brasseur et al. [1990] and $\alpha \approx .24$ day$^{-1}$ [Brasseur et al., 1987] are used in the calculations which follow.

Scatterplots of the ozone and temperature perturbations are provided in Figure 5 using SH data at 30 (upper panel) and 50 km (lower panel) from August 14, 1997. For this figure and the following analysis, temperature and ozone perturbations are calculated relative to the zonal mean values calculated for the corresponding latitude/height bin (see Figure 1 for the mean fields). At 30 km where the correlation is expected to be

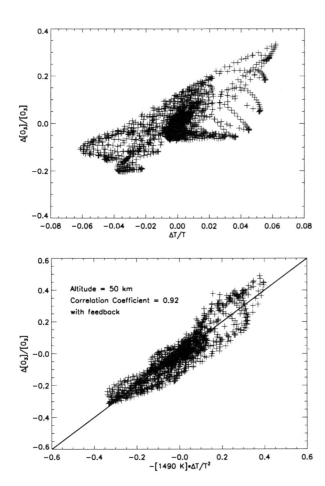

**Figure 5.** Scatter plots of normalized ozone perturbation vs normalized temperature perturbation at 30 km (upper panel) and -[1490 K] $\Delta T/T^2$ at 50 km (lower panel). See text for details.

per 10° of latitude at 50° as noted earlier), meridional transport is unlikely to contribute significantly to the observed correlations.

Optimal values of $\Theta_E$ are calculated as a function of height between 37 and 55 km for SH data for August 14 with and without feedback effects included (Figure 6, left subplot). The right hand subplot shows the correlation coefficient corresponding to this value. The correlation coefficient is greater than 0.8 above about 40 km but drops rapidly below this height. This is as expected since below about 40 km the photochemical time scale becomes of the same order as the dynamical time scale and the correlation would be expected to break down. The variation of $\Theta_E$ (feedback effects included) with height is similar to that calculated by Froidevaux et al. [1989] for photochemical equilibrium conditions (filled dots in Figure 4, left subplot) except that the peak at ≈40 km is higher (1900 K instead of 1500 K). The decrease with height of $\Theta$ seen here also appears in their calculations and is attributed to the increased influence of the $HO_x$ cycle with its lower temperature dependence on the observed $\Theta$. This is in contrast with the $\Theta_E$ calculated from the Limb Infrared Monitor of the Stratosphere (LIMS) observations which increases with height [Froidevaux et al., 1989]. The no feedback case also compares favourably with the corresponding Froidevaux et al. [1989] calculations. In terms of the quality of fit there appears to be little difference whether feedback is included or not. However, for data interpretation use of the $\Theta_E$ values calculated with feedback is physically more realistic.

dynamically driven $\Delta[O_3]/[O_3]$ verses $\Delta T/T$ is plotted (Figure 5a) and as expected, the correlation is positive. Further interpretation of this correlation is possible but requires a detailed dynamical analysis which we will not undertake here.

At 50 km the correlation is expected to be photochemically driven. In accordance with (3), values of $\Delta[O_3]/[O_3]$ and $-[\Theta_E/T^2]\Delta T$ are plotted (Figure 5b). The $\Theta_E$ value of 1490 K which is used is the value for which a maximum correlation is obtained (feedback effects are included). The resulting relationship is close to linear, suggesting that the photochemical description used here is appropriate and the dynamics play a second order role. Such an interpretation can also be supported on other grounds. As demonstrated by Froidevaux et al. [1989], since the dominant dynamical feature has a period of 12.5 days parcels are able to adjust to local photochemical equilibrium. In addition, since the zonal mean gradient in $O_3$ is gradual above 40 km (∼ 1 ppm

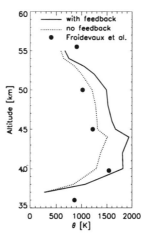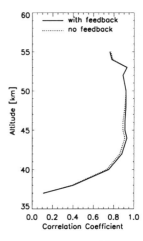

**Figure 6.** Plots of the value of $\Theta$ derived from SH data from August 14 with (solid line), without (dashed line) feedback effects and from the photochemical equilibrium calculations of Froidevaux et al., [1989] (solid circles, left panel) and the associated correlation coefficient (right panel).

Although CRISTA ozone data only extend to $\approx 60$ km at the present time, the observed correlation would be expected to be present at greater heights as the $HO_x$ remains important for ozone photochemistry thoughout the mesosphere [Allen et al., 1984; Summers and Conway, this issue]. At greater heights (above $\sim 65$ km) however day/night differences and tidal dynamics become important and need to be taken into account [see Zhu et al., this issue]. As a result, the data interpretation at these heights will become more complex .

## 4. CONCLUSIONS

Correlations between ozone and temperature have been reported in the literature for a number of years. Such correlations are reported in long term statistical studies [Randal and Cobb, 1994; Sabutis et al., 1997], studies of zonal means [Froidevaux et al., 1989], Fourier decomposition of large scale disturbances [Elson et al., 1994] and large scale wave breaking events [Fishbein et al., 1993]. In general ozone and temperature were observed to be positively correlated below $\approx 30$ km where the dynamical timescale is less than the photochemical timescale and negatively correlated above about 40 km where the photochemical timescale is less than the dynamical timescale. Modelling studies [Hartmann and Garcia, 1979; Rood and Douglass, 1985] confirmed these basic results but also indicated that a careful analysis of the dynamical contributions to the ozone transport was necessary before the observed correlations could be attributed to any specific process.

CRISTA observations provide another opportunity to examine this relationship between ozone and temperature. Variations in ozone and temperature data in the SH midlatitudes are shown to be associated with a wave 2 planetary wave. This wave extends throughout the data set from the lower stratosphere at 20 km to the mid-mesosphere at 80 km. The data are of sufficient accuracy that correlations may be calculated on a measurement-by-measurement basis. Correlations between these two parameters conform to previously published results and show ozone perturbations to be anticorrelated with temperature perturbations across the stratopause and correlated with temperature perturbations in the lower stratosphere. The CRISTA observations above 40 km agree closely with the photochemical equilibrium calculations of Froidevaux et al. [1989] in spite of the presence of strong planetary wave signatures.

Further work with this data set is certainly possible. The CRISTA observations are of higher vertical and horizontal resolution than previously published

work. Velocities may be derived from the temperature fields and the potential vorticity and Eliassen-Palm flux calculated and used to diagnose the dynamical conditions at this time [as in Manney et al., 1991]. A more extensive correlation analysis than that provided here could be undertaken and related more directly to the dynamical conditions. The development of a linear advective-photochemical model [Hartmann and Garcia, 1979; Douglass et al., 1985; Randel, 1990] would allow the contribution of the advective terms to be evaluated. Further evaluation of the contribution of the various photochemical cycles to the ozone variability could be undertaken by explicitly including other species measured by CRISTA into the photochemical equilibrium analysis undertaken here. It would be of interest to determine whether the close agreement with photochemical calculations noted here is a matter of superior data quality or the particular dynamical conditions present at the time of these measurements.

*Acknowledgments.* The Centre for Research in Earth and Space Technology is supported by the Technology Fund of the Province of Ontario. The CRISTA project was supported by grants 50 OE 8503 and 9501. of Deutsche Agentur für Weltraumangelegenheiten (DARA), Bonn, Germany. The CRISTA 2 instrument was flown as part of the Space Shuttle mission STS-85 of the National Aeronautics and Space Administration (NASA), USA.

## REFERENCES

Allen, D.R., et al., The 4-day wave as observed from the Upper Atmosphere Research Satellite Microwave Limb Sounder, *J. Atmos. Sci.*, *54*, 420–434, 1997.

Allen, M., et al., The vertical distibution of ozone in the mesosphere and lower thermosphere, *J. Geophys. Res.*, *89*, 4841-4872, 1984.

Barnett, J.J., J.T. Houghton, and J.A. Pyle, The temperature dependence of the ozone concentration near the stratopause, *Q. J. R. Meteorol. Soc.*, *101*, 245-257, 1975.

Brasseur, G. and S. Solomon, *Aeronomy of the Middle Atmosphere*, 441 pp., D. Reidel, Norwell, Mass., 1984.

Brasseur, G., et al., Response of middle atmosphere to short term solar ultra-violet variations: 2 Theory, *J. Geophys. Res.*, *92*, 903–914, 1987.

Brasseur, G.,et al., An interactive chemical dynamical radiative two-dimensional model of the middle atmosphere, *J. Geophys. Res.*, *95*, 5639–5655, 1990.

Brasseur et al., Natural and human-induced perturbations in the middle atmosphere, this volume, 2000.

Douglass, A.R., et al., Interpretation of ozone temperature correlations 2. Analysis of SBUV ozone data, *J. Geophys. Res.*, *90*, 10693–10708, 1985.

Elson, L.S., et al., Large-scale variations in ozone from the first two years of UARS MLS data, *J. Atmos. Sci.*, *51*, 2867–2876, 1994.

Farman, J.C., B.G. Gardner, and J.D. Shanklin, Large losses

of total ozone in Antarctica reveal seasonal ClOx/NOx interaction, *Nature*, *315*, 207–210, 1985.

Fishbein, E.F., et al., MLS observations of stratospheric waves in temperature and $O_3$ during the 1992 southern winter, *Geophys. Res. Lett.*, *20*, 1255–1258, 1993.

Froidevaux, L., et al., The mean ozone profile and its temperature sensitivity in the upper stratosphere and lower mesosphere: An analysis of LIMS observations, *J. Geophys. Res.*, *94*, 6389–6417, 1989.

Grossmann, K.U., Recent Improvements in middle atmosphere remote sounding techniques: The CRISTA-SPAS Experiment, this volume, 2000.

Haigh, J.D., and J.A. Pyle, Ozone perturbations experiments in a two-dimensional model, *Q. J. R. Meteorol. Soc.*, *108*, 551–574, 1982.

Hartmann, D.L., and R.R. Garcia, A mechanistic model of ozone transport by planetary waves in the stratosphere, *J. Atmos. Sci.*, *36*, 350–364, 1979.

Holton, J.R., and M.J. Alexander, The role of waves in the transport circulation of the middle atmosphere, this volume, 2000.

Jucks, K.W., and R.J. Salawitch, Future changes in upper stratospheric ozone, this volume, 2000.

Manney, G.L. et al., The behavior of wave 2 in the southern hemisphere stratosphere during late winter and early spring, *J. Atmos. Sci.*, *48*, 976–998, 1991.

Offermann, D., et al., The Cryogenic Infrared Spectrometers and Telescopes for the Atmosphere (CRISTA) experiment and middle atmosphere variability, *J. Geophys. Res.*, *104*, 16311–16325, 1999.

Randel, W.J., Kelvin wave-induced trace constituent oscillations in the equatorial stratosphere, *J. Geophys. Res.*, *95*, 18641–18652, 1990.

Randel, W.J., Global normal-mode Rossby waves observed in stratospheric ozone data, *J. Atmos. Sci*, *50*, 406–420, 1993.

Randel, W.J. and J.B. Cobb, Coherent variations of monthly mean total ozone and lower statospheric temperature, *J. Geophys. Res.*, *99*, 5433–5477, 1994.

Reddmann, T., R. Ruhnke, and W. Kouker, Use of coupled ozone fields in a 3-D circulation model of the middle atmosphere, *Ann. Geophysicae*, *17*, 415–429, 1999.

Riese, M., et al., Cryogenic Infrared Spectrometers and Telescopes for the Atmosphere (CRISTA) limb scan measurements, data processing, and atmospheric temperature and trace gas retrieval, *J. Geophys. Res.*, *104*, 16349–16367, 1999.

Rood, R.B. and A.R. Douglass, Interpretation of ozone temperature correlations 1. Theory, *J. Geophys. Res.*, *90*, 5733–5743, 1985.

Rose, K. and G. Brasseur, A three-dimensional model of chemically active trace species in the middle atmosphere during disturbed winter conditions, *J. Geophys. Res.*, *94*, 16387–16403, 1989.

Sabutis, J.L. et al., Wintertime planetary wave propagation in the lower stratosphere and its observed effect on northern hemisphere temperature-ozone correlations, *J. Geophys. Res.*, *102*, 21709–21717, 1997.

Smith, A.K., Numerical simulation of global variations of temperature, ozone, and trace species in the stratosphere, *J. Geophys. Res.*, *100*, 1253–1269, 1995.

Solomon, S., Stratospheric ozone depletion: A review of concepts and history, *Rev. Geophys.*, *37*, 275–316, 1999.

Summers, M.E., and R.R. Conway, Insights into middle atmospheric hydrogen chemistry from analysis of MAHRSI observations of hydroxyl (OH), this volume, 2000.

Zhu, X., J.-H. Yee, and D.F. Strobel, Coupled models of photochemistry and dynamics in the mesosphere and lower thermosphere, this volume, 2000.

W.E. Ward, CRESS/CRESTech, York University, North York, Ontario, Canada; email: william@stpl.cress.yorku.ca

J. Oberheide, M. Riese, P. Preusse, D. Offermann, Dept. of Physics, University of Wuppertal, D-42097 Wuppertal, Germany; email: jens@wpos2.physik.uni-wuppertal.de.

# Nighttime O$_2$ and O$_3$ Profiles Measured by MSX/UVISI Using Stellar Occultation Techniques

Jeng-Hwa Yee, Robert DeMajistre, Ronald J. Vervack, Jr., Frank Morgan,
James F. Carbary, Gerald J. Romick, Daniel Morrison, Steven A. Lloyd, Philip L. DeCola,
Larry J. Paxton, Donald E. Anderson, C. Krishna Kumar, and Ching-I Meng

*The Johns Hopkins University Applied Physics Laboratory, Laurel, Maryland*

Stellar occultation techniques are a powerful means of remotely sensing the Earth's atmosphere from space. In this paper, we show that a unique combination of extinctive and refractive stellar occultation methods can be used to retrieve atmospheric composition from the lower thermosphere down to the upper troposphere. The extinction measurements provide the composition information, while the refraction data are used to infer the total density, pressure, and temperature profiles in the lower atmosphere. Precise, simultaneous measurement of the total density profile in particular allows both Rayleigh scattering and refraction effects to be accurately accounted for in the interpretation of the lower atmosphere extinction measurements and also provides for direct calculation of mixing ratios. This combined technique has been successfully demonstrated using stellar occultation data obtained by the Ultraviolet and Visible Imagers and Spectrographic Imagers (UVISI) on the Midcourse Space Experiment (MSX) satellite. To illustrate the potential of the method, we present O$_2$ and O$_3$ profiles retrieved from two example occultations observed by MSX/UVISI in the polar vortex region during formation of the seasonal ozone hole in October 1996.

## 1. INTRODUCTION

Two molecules important in understanding the chemistry and physics of the Earth's atmosphere are O$_2$ and O$_3$. Knowledge of the O$_3$ profile in the stratosphere and upper troposphere is needed to monitor the Earth's ozone layer and to understand the mechanisms behind its loss. Measurement of the O$_3$ profile, especially in the lower stratosphere and upper troposphere, is thus a primary goal in remote sensing of the Earth's atmosphere. In the mesosphere and lower thermosphere (MLT), O$_2$

and O$_3$ play critical roles in photochemistry, airglow processes, and the MLT energy budget [*Mlynczak and Solomon*, 1993]. Accurate measurements of the O$_2$ and O$_3$ density profiles are therefore crucial to our understanding of the MLT region.

One method for measuring O$_2$ and O$_3$ density profiles in the Earth's atmosphere is stellar occultation. Occultation techniques have been used for many years to determine the composition and structure of the atmospheres of Earth, the other planets, and their satellites. The primary advantage of occultation techniques over other remote sensing methods is that they are relative measurements and therefore less sensitive to instrument degradation and changes in calibration.

The O$_2$ and O$_3$ density profiles in the lower thermosphere and mesosphere have been measured by several

Atmospheric Science Across the Stratopause
Geophysical Monograph 123
Copyright 2000 by the American Geophysical Union

authors using solar (e.g., *Johnson et al.* [1951]) or stellar (e.g., *Hays and Roble* [1973]) extinctive occultations. As discussed by *Roble and Hays* [1972], the retrieval of $O_2$ and $O_3$ density profiles in this region of the atmosphere is straightforward owing to the distinct nature of the absorption by these species in the Schumann-Runge and Hartley/Huggins spectral regions.

In the lower atmosphere, $O_3$ profiles have been determined from extinctive occultation measurements as well, primarily using the sun as a source because of the high signal-to-noise ratio it provides (e.g., *Cunnold et al.* [1989]; *Gunson et al.* [1990]; *Russell et al.* [1993]; *Rusch et al.* [1997]). However, solar occultations are limited both spatially and temporally to the terminator region. Stellar occultations, on the other hand, offer global coverage roughly every day at the expense of signal-to-noise. Despite the apparent complementarity of the two methods, however, stellar occultation methods have rarely been employed to measure lower atmospheric $O_3$ from space even though a theoretical study by *Hays and Roble* [1968] showed great promise.

A complication in the retrieval of the $O_3$ profile in the lower atmosphere is the presence of both Rayleigh scattering and atmospheric refraction. Refraction in particular leads to three effects that must be considered: bending of the light rays, attenuation caused by the flux divergence due to the bending, and scintillation related to small-scale density perturbations. Current extinctive occultation methods rely on climatological models of the total atmospheric density to correct for Rayleigh scattering and refraction effects. However, this results in an increase of the systematic uncertainty in the retrieved profiles, especially in the region below the ozone peak. This uncertainty can be minimized if the atmospheric density profile is simultaneously retrieved from measurements of stellar refraction.

In this paper, we present the analysis of combined extinctive and refractive stellar occultation data from the Ultraviolet and Visible Imagers and Spectrographic Imagers (UVISI) on the Midcourse Space Experiment (MSX). Spectrographic imagers are used to measure the atmospheric extinction of starlight, while a co-aligned imager is utilized to measure the stellar refraction angles in the lower atmosphere. The use of an imager to measure atmospheric refraction was originally proposed by *Jones et al.* [1962], but the suggestion has generally been overlooked. The simultaneous measurement of both the refraction and extinction of starlight along the same line of sight allows the constituent-dependent extinction measurements to probe the lower atmosphere more accurately than a purely extinctive method.

## 2. MSX/UVISI STELLAR OCCULTATIONS

Although not specifically designed for stellar occultation experiments, the MSX/UVISI complement of instruments combines five spectrographic imagers (SPIMs) covering the wavelength range of 120–900 nm and four imagers (two ultraviolet and two visible-light) with a spacecraft providing excellent pointing capabilities (1-$\sigma$ absolute pointing knowledge of $\sim$100 $\mu$rad and relative stability of $\sim$10 $\mu$rad over a time span of several minutes). Detailed descriptions of both MSX and UVISI may be found in *Mill et al.* [1994] and *Carbary et al.* [1994], respectively. We have taken advantage of the unique capabilities of MSX and UVISI to conduct proof-of-concept stellar occultation experiments demonstrating the viability of the combined extinctive and refractive method for the accurate retrieval of atmospheric composition. In these experiments, the SPIMs were used to measure the atmospheric extinction, while the narrow-field-of-view ($\sim$1°), visible-light imager (IVN) was used to measure the stellar refraction angles.

The geometry of a typical MSX/UVISI stellar occultation event is shown in Figure 1. During the occultation, the UVISI boresight remains fixed on the inertial position of the star, and the spectrograph entrance slits are held vertical with respect to the horizon. The star is acquired at high altitudes, and several hundred unattenuated $I_o$ spectra are obtained before the effects of the atmosphere are present. As the star sets through the atmosphere, the spatial position of its image in the spectrographs is unchanged until refraction sets in at lower altitudes (below 35 km), at which point the image moves slowly up the vertical slit until it disappears from the field of view.

As illustrated in Figure 1, the stellar signal is superimposed on the background airglow signal once the line of sight passes through the airglow layer. In contrast to the star, whose spatial position is fixed apart from refraction, airglow emissions appear in the bottom edge of the slit, move up across the forty spatial pixels, and disappear above the top edge as the boresight tangent altitude descends. Because each spatial pixel covers a field of view of $\sim$0.025° (about 1.5 km projected onto the limb), the UVISI SPIMs are essentially used as large aperture photometers ($\sim$100 cm$^2$) with an effective field of view of 0.1°×0.025°. With a star setting rate of roughly 0.07°/s ($\sim$3 km/s), the limb airglow emission at a given tangent altitude is sampled for approximately 10 seconds. On the ground, this multiple sampling of the airglow allows us to "shift" the SPIM pixels spatially and co-add them in a manner similar to

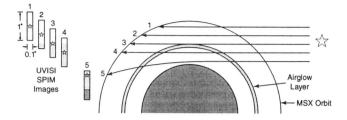

**Figure 1.** MSX/UVISI stellar occultation geometry. The UVISI instrument is oriented so that the slit is vertical with respect to the atmosphere horizontal. The center of the slit is fixed on the star's inertial position throughout the occultation; therefore, the star itself can refract up to 0.5° before disappearing from the field of view (FOV). The 1° FOV in the vertical direction corresponds to ~60 km at the limb. The superposition of the stellar and background airglow (shaded) signals once the line of sight has passed through the airglow layer is illustrated.

that used for a time-delay and integration (TDI) system, thereby improving the signal-to-noise of the observations and subsequent accuracy of the separation of the stellar and airglow signals.

Plate 1 shows a composite spectral image of data from all five SPIMs for a typical occultation along with the corresponding IVN image. In the SPIM images, scene brightness is represented by a color scale as a function of geometric tangent altitude (y-axis) and wavelength (x-axis). Because the star is a point source emitting a pseudo-continuum spectrum, it appears as a bright, narrow band running the length of the frame. The diffuse nature of airglow emission causes the airglow signal to cover an extended portion of the slit compared to the point source stellar signal. The SPIM images have been corrected for optical aberrations in the spectrographs and small, well-known deviations in the co-alignment of the five SPIMs so that the stellar signal at each wavelength is referenced to a common altitude. This results in the ragged horizontal edges of the composite image. The black regions of Plate 1 correspond to altitudes outside the SPIM fields of view, and the weak signal at the rightmost end of the spectral range is caused by a rapid decrease in the sensitivity of SPIM 5 at these wavelengths. In the IVN image, the star of interest is the bright dot near the image center.

The boresight tangent altitude of the images in Plate 1 is ~95 km. At this altitude, the effects of atmospheric extinction are most evident near 140 nm, where $O_2$ absorbs in the Schumann-Runge continuum. Although small in magnitude, extinction by $O_3$ in the Hartley band is also present near 250 nm. Thermospheric (e.g., $O(^1D)$ at 630 nm) and mesospheric (e.g., $O(^1S)$

"green line" at 557.7 nm, $O_2$ Atmospheric (0-0) band at 762 nm) emissions are seen, as are several geocoronal emissions (e.g., H Lyman $\alpha$ at 121.6 nm). The mesospheric emissions only cover the lower portion of the images because the upper parts of the spectrographic slits and IVN field of view still observe tangent heights above the mesospheric airglow emission layer.

At high altitudes where the SPIM images are generally free from airglow, the spatial intensity distribution of the star is seen to be strongly peaked with a spot size of approximately two pixels. When combined with the multiple sampling of the airglow emissions afforded by the design of the stellar occultation experiments, knowledge of this distribution provides the key to the separation of the stellar and airglow signals.

## 3. DATA ANALYSIS TECHNIQUE

Each MSX/UVISI occultation consists of SPIM and IVN images such as those presented in Plate 1. The analysis of these data consists of four general steps: 1) determination of the observed refraction angles from the IVN images; 2) separation of the stellar transmission spectra from the airglow spectra in the SPIM images; 3) retrieval of the total atmospheric density, pressure, and temperature profiles from the observed refraction angles; and 4) retrieval of the atmospheric composition profiles from the observed stellar transmission spectra. Although each step is executed separately, there are a number of critical interconnections among the various steps without which the analysis would be severely hampered. We briefly describe these steps and interconnections here. Steps 1 and 2 are discussed in further detail by *Yee et al.* [2000], while steps 3 and 4 are presented by *Vervack et al.* [2000] and *DeMajistre and Yee* [2000], respectively. The reader is referred to these papers for a complete description of the data analysis techniques, including detailed error analyses and a thorough discussion of the inherent assumptions.

*Step 1:* To determine the stellar refraction angles from the IVN images, the position of the star in each IVN image is first obtained through application of a centroid algorithm. This centroiding is highly accurate, and the resulting stellar position is known to better than 1/20th of an IVN pixel (0.005°/20 ⇒ 0.00025°). The stellar positions are then converted to absolute refraction angles in the occultation plane using knowledge of the imager field of view, satellite pointing information, and the inertial stellar position. The MSX satellite has excellent pointing capabilities, with a relative pointing stability of ~10 $\mu$rad or 0.0006° (1-$\sigma$) over a time span of several minutes. Accounting for the uncertainty in the centroiding process, pointing knowledge and stability, and coordinate transformations, the final net uncertainty in the observed refraction angles is 0.002° (1-$\sigma$), which corresponds to roughly 0.1 km on the limb. Thus, the altitude determination for the MSX/UVISI occultations is very precise.

*Step 2:* To separate the stellar and airglow signals in the measured extinction spectra, the SPIM images are first processed to remove a number of instrumental effects (e.g., changes in gain with altitude, dark subtraction). These corrected images are then input to an algorithm that uses the entire set of observations for a given wavelength to determine the stellar and airglow signals and their respective uncertainties simultaneously and self-consistently. Special care is taken to ensure that the uncertainties are estimated correctly because the signal registered by each SPIM pixel is a weighted sum of both stellar and airglow signals, which represent two different uncertainty probability distributions (point versus diffuse source). Spacecraft position and pointing information, knowledge of the instrument co-alignment matrices, and the absolute position of the star in the IVN images are used to establish the geometric relationships during the occultation event. The position of the star in the IVN images is particularly important because it provides the precise, spatial location of the star in the SPIM slits, knowledge of which is critical to the separation once the star begins to move vertically along the slit because of atmospheric refraction. Once separated, the stellar spectra are divided by an unattenuated stellar spectrum to obtain transmission spectra as a function of altitude.

Plate 2 shows the separated stellar transmission (left panel) and airglow volume emission (right panel) spectra for our typical occultation event. Absorption by $O_2$ and $O_3$ is clearly evident in the transmission spectra near 150 nm (thermospheric $O_2$, Schumann-Runge continuum), 250 nm (mesospheric $O_3$, Hartley band), and 600 nm (stratospheric/upper tropospheric $O_3$, Chappuis band). Atmospheric scintillation causes the alternating signal intensity in the atmospheric transmission below 40 km. Because of atmospheric refraction, the star is observed even after the geometric line of sight has passed below the horizon (i.e., at negative geometric tangent heights). The abrupt changes in the noise characteristics and visual appearance of both the transmission and airglow spectra (e.g., near 380 nm) occur at the spectral boundaries between the SPIMs and are caused by variations in instrument sensitivity, resolution, and wavelength sampling interval from one

SPIM to another. The peculiar structure near 760 nm is caused by difficulty in separating the stellar signal from the strong $O_2$ Atmospheric band emission.

The uncertainty in the measured transmission spectra is dependent on the magnitude and spectral type of the occulted star, the intensity of the airglow emissions, the characteristics of the SPIMs, and the tangent altitude of the measurements. Typically, the uncertainty for a given wavelength channel is 5–10% for a moderately bright star ($M_v \approx 2$–3); however, the uncertainty may be higher for channels with strong airglow contamination (e.g., the $O_2$ Atmospheric band), in which case those particular channels are eliminated from consideration. Although the measurement uncertainty for a given channel may seem relatively large, we note that each SPIM has 272 spectral channels and roughly 500–700 channels are used in each retrieval. The averaging effect of this multi-spectral sampling ultimately serves to reduce the uncertainty in the retrieved profiles.

*Step 3:* The refraction retrievals begin with the inversion of the total atmospheric density profile from the observed refraction angles using a constrained linear technique. For the MSX/UVISI total density retrievals, a second difference constraint is applied, and the altitude grid for the retrieval is fixed at 0–30 km with a 1 km spacing. Above 30 km, the initial guess for the inversion procedure — a MSISE-90 [*Hedin*, 1991] profile appropriate for the occultation time and tangent point location — is scaled so that the final retrieved profile is continuous across the boundary. Once the density profile has been retrieved, the pressure profile is obtained via integration of the hydrostatic equation. Finally, the ideal gas law is used to determine the temperature profile from the density and pressure profiles.

*Step 4:* The extinction retrievals are conducted using a two-stage process in which the spectral and spatial inversions are separated for computational efficiency. The wavelength range in the retrievals runs from 130 to 750–850 nm depending on the signal-to-noise ratio in the longer wavelength channels of SPIM 5. This ratio varies owing to strong airglow contamination by the $O_2$ Atmospheric band at 762 nm (see Plate 2) as well as the rapid decrease in the sensitivity of the long-wavelength end of SPIM 5 (see Plate 1). Because of the slightly different nature of the retrievals in the lower and upper atmosphere (see below), the retrievals in these two regions are conducted independently, and no effort is made to ensure the results are consistent.

In the first stage of the extinction retrievals, the transmission spectra at each altitude are independently fitted using a Marquardt-Levenberg, nonlinear, least-

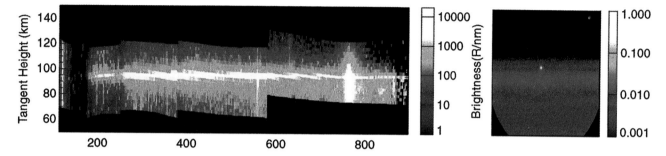

**Plate 1.** Typical example of processed MSX/UVISI stellar occultation data from the spectrographic imagers. The stellar spectrum is seen as a bright, narrow band running horizontally near the center of the images. These images correspond to a geometric tangent height of 95 km. Absorption by $O_2$ is clearly evident near 140 nm, while absorption by $O_3$ is just noticeable near 250 nm. Airglow and geocoronal emissions are visible throughout the spectrum. Also shown is the corresponding IVN image, in which the star of interest is the bright dot near the center.

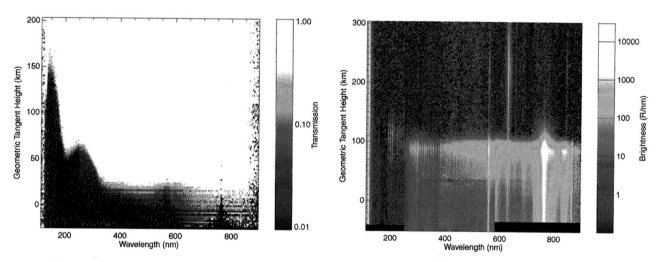

**Plate 2.** The left panel shows the stellar transmission spectra as a function of geometric tangent height as observed by MSX/UVISI during a typical occultation event. The right panel shows the nighttime airglow (or nightglow) spectra as a function of geometric tangent height for the same event. Note the different y-axis scales in the two panels.

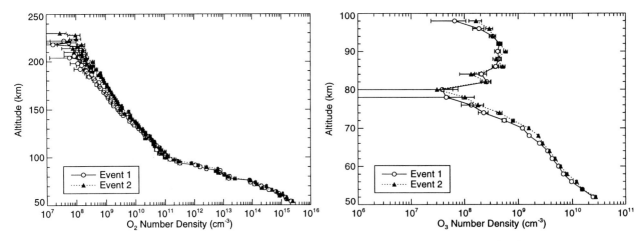

**Figure 2.** The left panel shows the retrieved $O_2$ number density profiles for two MSX/UVISI occultations. The right panel shows the $O_3$ profiles in the mesosphere retrieved from the same occultations.

squares technique. For a given spectrum, the various atmospheric parameters are fitted simultaneously. In the lower atmosphere, the output parameters of these fit spectra are the $O_3$ and $NO_2$ column densities, the total atmospheric column densities (from Rayleigh extinction), a combined refractive attenuation/scintillation profile (essentially a multiplicative factor to account for these approximately wavelength-independent processes at each altitude), and the column aerosol extinction profile. In performing the lower atmospheric spectral fits, the total density profile and associated covariance matrix retrieved from the refraction observations are considered as *a priori* information to constrain the retrieval of the total atmospheric column density profile, and the retrieved temperature profile is used to establish the correct temperature-dependent extinction cross section values, thereby reducing the systematic uncertainty that would be introduced by using a climatological profile. Finally, the aerosol extinction spectrum is assumed to have a wavelength dependence of $\lambda^{-1}$. Although the true aerosol extinction spectrum is likely more complex than this (e.g., see *Rusch et al.* [1998]), assuming a $\lambda^{-1}$ dependence simplifies the retrievals at the expense of a small increase in systematic uncertainty (see *DeMajistre and Yee* [2000] for a thorough discussion of aerosol extinction and other spectral retrieval issues). In the upper atmosphere, the spectral fitting is greatly simplified by the absence of refraction effects and aerosol extinction, and the output parameters are the $O_2$ and $O_3$ column densities only.

In the second stage, the various slant column profiles are spatially inverted to yield the corresponding vertical profiles using a constrained inversion technique similar to that in the refraction retrieval. As opposed to

methods that constrain the retrieved profile to an *a priori* profile (e.g., *Kyrölä et al.* [1993]), the MSX/UVISI inversions utilize a second difference *smoothness* constraint in the lower atmosphere retrievals, while the upper atmosphere retrievals are not constrained. The retrieval grids cover altitudes from 4–60 km and 50–250 km in the lower and upper atmosphere, respectively, with a spacing of 2 km in both cases. As part of the inversion process in the lower atmosphere, the total density profile retrieved from the refraction angle measurements is used to calculate the actual refracted path of the starlight through the atmosphere. This more accurately accounts for the increased extinction path length than use of a climatological density profile and correspondingly minimizes the systematic uncertainty in the extinction path calculation.

## 4. EXAMPLE RETRIEVALS

To demonstrate the potential of the stellar occultation technique for remotely sensing the Earth's atmosphere, we present the results from the analysis of two MSX/UVISI occultation events. Both occultations occurred under twilight conditions at 60°S latitude but at slightly different east longitudes (106° for Event 1; 117° for Event 2). The occultations took place on October 6 and October 12, 1996, respectively, and the same star ($\beta$ Tau, $M_v$=1.65, B-type) was used in each case.

The results of the lower thermosphere/mesosphere retrievals for these two occultations are presented in Figure 2. The left panel shows the retrieved $O_2$ number density profiles, while the right panel shows the $O_3$ profiles. The minimum in the $O_3$ profiles in the mesosphere is easily seen, and the $O_2$ and $O_3$ profiles are consistent

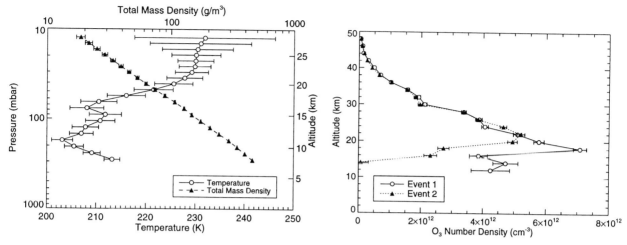

**Figure 3.** The left panel shows the total density and temperature profiles retrieved from the refraction angle measurements for Event 2 as a function of both pressure and altitude. The right panel shows the retrieved $O_3$ number density profiles in the stratosphere and upper troposphere for both occultation events. The rapid decrease below the ozone peak in the profile for Event 2 results from the line of sight passing through the seasonal Antarctic ozone hole.

between the two occultations. Simultaneous measurements of both the $O_2$ and $O_3$ profiles in this region of the atmosphere are of great importance to studies of mesospheric photochemistry and airglow processes. In combination with the coincident airglow measurements, these profiles may be used to investigate the physics and chemistry of the MLT region self-consistently.

The results of the lower atmospheric retrievals are presented in Figure 3. The left panel shows total density and temperature profiles retrieved from the refraction angle measurements for Event 2 only; the right panel shows the retrieved $O_3$ profiles for both occultations. As opposed to the mesospheric $O_3$ profiles, the stratospheric $O_3$ profiles show remarkable differences despite the occurrence of the two occultations at roughly the same geographic location and time. Further investigation revealed that the two occultations probed the region near the seasonal Antarctic ozone hole and that Event 2 probed more deeply into the hole itself. These profiles demonstrate that the stellar occultation technique is capable of sensing the dramatic changes in the shape of the ozone profile associated with the seasonal hole, especially during polar night.

The primary advantage of the combined occultation technique is the improvement in accuracy of the lower atmospheric $O_3$ retrievals owing to a better accounting of Rayleigh scattering and refraction effects (through use of a simultaneously measured total density profile) and temperature-dependence in the extinction cross sections (through use of the simultaneously measured temperature profile). However, two other benefits also war-

rant mention. First, retrieval of the total atmospheric density allows direct determination of the $O_3$ mixing ratio profile. In the absence of such measurements, mixing ratios must be inferred using a climatological density profile. Second, the pressure-altitude relationship is well established because the atmospheric pressure is retrieved and the altitude determined from the occultation geometry. This relationship is often difficult to infer from ground-based and other space-based methods but is often important in atmospheric modeling.

Finally, we note that although the lower and upper atmosphere retrievals are currently independent, the $O_3$ profiles shown here merge smoothly across the boundary near 50 km between the two retrievals. When converted to an $O_2$ profile assuming standard atmospheric mixing ratios, the retrieved total density profile also merges consistently with the $O_2$ profile in the upper atmosphere. The consistency of these independently-retrieved profiles lends great confidence to our retrieval techniques and suggests that the accuracy of future retrievals, particularly in the lower atmosphere, may be improved through joint consideration of both regions.

## 5. CONCLUDING REMARKS

Although MSX/UVISI was not optimized for occultation experiments, it provided a valuable opportunity to test retrieval methods on a proof-of-concept basis owing to its unique combination of instrumentation and pointing capabilities. Using MSX/UVISI observations, we have demonstrated that the stellar occultation technique is a viable method for probing the Earth's atmosphere from the lower thermosphere down to the upper troposphere. To carry out the retrievals in the lower atmosphere, where the effects of Rayleigh scattering and atmospheric refraction are significant, we have developed a combined extinctive and refractive occultation technique. This technique is only briefly discussed above but is described in great detail by *Yee et al.* [2000], with the associated refraction and extinction retrieval algorithms presented in *Vervack et al.* [2000] and *DeMajistre and Yee* [2000], respectively.

The results presented here show that stellar occultation measurements can yield the $O_3$ density profile from roughly 100 km down to 10 km altitude, while the $O_2$ profile can be determined from 220 km down to ~5 km through combination with the total density profile retrieved from the refraction data. Profiles such as these are valuable in numerous atmospheric studies. Accurate measurements of the $O_3$ profile in the lower atmosphere are critical in long-term monitoring and understanding of the Earth's ozone layer. In the upper atmosphere, the retrieved $O_2$ and $O_3$ profiles can be combined with the simultaneous measurements of the airglow emission profiles to study upper atmospheric physics and chemistry self-consistently.

To date, roughly 200 stellar occultation events have been conducted by MSX/UVISI under various conditions (e.g., high-latitude ozone measurements). More than 40 of these occultations were carried out over ground-based facilities or at near coincidence with other space-based instruments (e.g., SAGE II, POAM III, UARS) for the purposes of validation. In addition, approximately 30 events were performed from June through early October 1999 during the Antarctic polar night to investigate the key parameters that contribute to the formation of the seasonal ozone hole. Finally, over 50 occultations were performed during January–March 2000 in support of the SOLVE campaign. Future papers will discuss our analysis of these other events, validation of the retrieved results through comparison to ground-based and satellite data, and investigations into the photochemistry of the upper atmosphere.

*Acknowledgments.* We thank the MSX teams for operation of both MSX and UVISI. Funding for the satellite operation was from the Department of Defense's Ballistic Missile Defense Organization. The stellar occultation analysis was supported by internal JHU/APL basic research funding and NASA Grant NAG5-7552 to JHU/APL.

## REFERENCES

Carbary, J. F., E. H. Darlington, T. J. Harris, P. J. McEvaddy, M. J. Mayr, K. Peacock, and C. I. Meng, Ultraviolet and visible imaging and spectrographic imaging instrument, *Appl. Opt.*, *33*, 4201–4213, 1994.

Cunnold, D. M., W. P. Chu, R. A. Barnes, M. P. McCormick, and R. E. Veiga, Validation of SAGE II ozone measurements, *J. Geophys. Res.*, *94*, 8447–8460, 1989.

DeMajistre, R., and J.-H. Yee, Atmospheric remote sensing using a combined extinctive and refractive stellar occultation technique III. Inversion method and error analysis for the retrieval of atmospheric composition from spectroscopic measurements, *J. Geophys. Res.*, *submitted*, 2000.

Gunson, M. R., C. B. Farmer, R. H. Norton, R. Zander, C. P. Rinsland, J. H. Shaw, and B.-C. Gao, Measurements of $CH_4$, $N_2O$, CO, $H_2O$ and $O_3$ in the middle atmosphere by the Atmospheric Trace Molecule Spectroscopy Experiment on Spacelab 3, *J. Geophys. Res.*, *95*, 13,867–13,882, 1990.

Hays, P. B., and R. G. Roble, Stellar spectra and atmospheric composition, *J. Atmos. Sci.*, *25*, 1141–1153, 1968.

Hays, P. B., and R. G. Roble, Stellar occultation measurements of molecular oxygen in the lower thermosphere, *Planet. Space Sci.*, *21*, 339–348, 1973.

Hedin, A. E., Extension of the MSIS thermosphere model into the middle and lower atmosphere, *J. Geophys. Res.*, *96*, 1159–1172, 1991.

Johnson, F. S., J. D. Purcell, and R. Tousey, Measurements of the vertical distribution of atmospheric ozone from rockets, *J. Geophys. Res.*, *56*, 583–594, 1951.

Jones, L. M., F. F. Fischbach, and J. W. Peterson, Satellite measurements of atmospheric structure by refraction, *Planet. Space Sci.*, *9*, 351–352, 1962.

Kyrölä, E., E. Sihvola, Y. Kotivuori, M. Tikka, and T. Tuomi, Inverse theory for occultation measurements 1. Spectral inversion, *J. Geophys. Res.*, *98*, 7367–7381, 1993.

Mill, J. D., R. R. O'Neil, S. Price, G. J. Romick, O. M. Uy, E. M. Gaposchkin, G. C. Light, W. W. Moore, Jr., T. L. Murdock, and A. T. Stair, Jr., Midcourse Space Experiment: Introduction to the spacecraft, instruments, and scientific objectives, *J. Spacecr. Roc.*, *31*, 900–907, 1994.

Mlynczak, M. G., and S. Solomon, A detailed evaluation of the heating efficiency in the middle atmosphere, *J. Geophys. Res.*, *98*, 10,517–10,541, 1993.

Roble, R. G., and P. B. Hays, A technique for recovering the vertical number density profile of atmospheric gases from planetary occultation data, *Planet. Space Sci.*, *20*, 1727–1744, 1972.

Rusch, D. W., R. M. Bevilacqua, C. E. Randall, J. D. Lumpe, K. W. Hoppel, M. D. Fromm, D. J. Debrestian, J. J. Olivero, J. H. Hornstein, F. Guo, and E. P. Shettle, Validation of POAM ozone measurements with coincident MLS, HALOE, and SAGE II observations, *J. Geophys. Res.*, *102*, 23,615–23,627, 1997.

Rusch, D. W., C. E. Randall, M. T. Callan, M. Horanyi, R. T. Clancy, S. C. Solomon, S. J. Oltmans, B. J. Johnson, U. Koehler, H. Claude, and D. D. Muer, A new inversion for stratospheric aerosol and gas experiment II data, *J. Geophys. Res.*, *103*, 8465–8475, 1998.

Russell, J. M., III, L. L. Gordley, J. H. Park, S. R. Drayson, W. D. Hesketh, R. J. Cicerone, A. F. Tuck, J. E. Frederick, J. E. Harries, and P. J. Crutzen, The Halogen Occultation Experiment, *J. Geophys. Res.*, *98*, 10,777–10,797, 1993.

Vervack, R. J., Jr., J.-H. Yee, J. F. Carbary, and F. Morgan, Atmospheric remote sensing using a combined extinctive and refractive stellar occultation technique II. Inversion method and error analysis for the retrieval of bulk atmospheric properties from visible-light refraction angle measurements, *J. Geophys. Res., submitted*, 2000.

Yee, J.-H., R. DeMajistre, R. J. Vervack, Jr., F. Morgan, J. F. Carbary, G. J. Romick, D. Morrison, S. A. Lloyd, P. L. DeCola, D. G. Kupperman, L. J. Paxton, D. E. Anderson, C. K. Kumar, and C.-I. Meng, Atmospheric remote sensing using a combined absorptive and refractive stellar occultation technique I. Overview and results from MSX/UVISI observations, *J. Geophys. Res., submitted*, 2000.

# Coupled Models of Photochemistry and Dynamics in the Mesosphere and Lower Thermosphere

Xun Zhu and Jeng-Hwa Yee

Applied Physics Laboratory, The Johns Hopkins University, Laurel, MD 20723-6099

Darrell F. Strobel

Department of Earth and Planetary Sciences, The Johns Hopkins University, Baltimore, MD 21218

Techniques to model coupled photochemistry and dynamics with both fast and slowly varying timescales of transport are described in the context of a chemical solver used in the JHU/APL two-dimensional model and tidal model. A one-dimensional photochemical-diffusive-advective model is formulated to explicitly include the coupling between the diurnally forced photochemistry and the transport by tidal waves in the middle atmosphere. The coupling between the photochemistry and the slowly varying transport is examined systematically and illustrated by several examples of modeling techniques.

## 1. INTRODUCTION

Numerical models that couple photochemistry with dynamics are major tools for understanding temporal and global behavior of the atmospheric states. The coupling between the chemistry and zonal mean dynamics through the non-localized chemical heating becomes especially important in the mesosphere and lower thermosphere (MLT). Around the mesopause, the dynamically controlled O distribution generates a latitudinal chemical heating rate that counters the radiative heating rate gradient [$Zhu$ $et$ $al.$ 1999b]. Furthermore, $O_3$ concentrations in the MLT are expected to be determined collectively by downward transport of O atoms from the lower thermosphere, upward transport of $H_2O$ from the stratosphere, and the local $O_x$-$HO_x$ photochemistry. One-dimensional chemical-transport models and multi-dimensional chemical-dynamical models have been widely used to simulate middle atmosphere photochemistry and its coupling with dynamical transport [e.g., $Allen$ $et$ $al.$, 1984; $Garcia$ $and$ $Solomon$, 1983; $Ko$ $et$ $al.$ 1984; $Summers$ $et$ $al.$, 1997;

$Smith$, 1995]. Most coupled models can be divided into two categories. In general, one-dimensional models couple the diurnally forced photochemistry with the transport by vertical diffusion under fixed lower boundary conditions of long-lived tracers. The integration time step for the diurnally forced photochemistry is relatively small, especially near twilight. Two- or three- dimensional models couple the slowly varying photochemistry with the slowly-varying transport processes on a climatological timescale from months to years with a relatively large time step on integration.

In the last few years, the Johns Hopkins University Applied Physics Laboratory (JHU/APL) has developed three major modeling tools for studying dynamical and physical processes in the MLT region: (1) a two-dimensional dynamical model [$Zhu$ $et$ $al.$, 1997]; (2) a linear spectral tidal model [$Zhu$ $et$ $al.$, 1999a]; and (3) a photochemical solver [$Zhu$ $et$ $al.$ 1999b]. In this article, we mainly review and describe how the photochemical solver is coupled with the transport induced by tidal waves and by the slowly varying zonal mean meridional circulation. In section 2, we formulate a one-dimensional photochemical-diffusive-advective model that explicitly couples the diurnally forced photochemistry with the transport by tidal waves. Section 3 presents several modeling techniques that couple the photochemistry with the slowly varying transport. Section 4 briefly describes the dynamical module and the chemical solver used for JHU/APL coupled 2-D model. Two examples of the modeled fields are shown in section 5. Our brief conclusions are given in section 6.

Atmospheric Science Across the Stratopause
Geophysical Monograph 123

337

## 2. COUPLING BETWEEN THE PHOTOCHEMISTRY AND TIDAL WAVES

If we denote $\chi_i$ as the mixing ratio of an atmospheric tracer, then its temporal variation is often represented by the following one-dimensional continuity equation,

$$\frac{\partial \chi_i}{\partial t} = (P_i - L\chi_i) + \frac{1}{\rho_0}\frac{\partial}{\partial z}\left(\rho_0 K_{zz}\frac{\partial \chi_i}{\partial z}\right) \quad (1)$$

where t is time, z is altitude, $K_{zz}$ is the vertical diffusion coefficient, $\rho_0$ is the air density, $P_i$ and $L\chi_i$ are the photochemical production and loss terms for $\chi_i$, respectively. One important application of (1) is to calculate the diurnal variations of chemical species at a given geographic location. In this case, $K_{zz}$ includes the effect of eddy diffusion by small and meso-scale gravity waves whose periods are much less than 1 day. We also need an initial condition for $\chi_i$ in order to solve (1). The initial condition can generally be derived from a multi-dimensional model (e.g. JHU/APL coupled 2-D model) that describes the slowly varying photochemical and transport processes. The timescale of atmospheric tidal waves is ~1 day. This makes it difficult to effectively incorporate transport by tidal waves either in (1) by a parameterized diffusion or in the multi-dimensional model where the diurnal variability of the species are not explicitly simulated. Furthermore, in the MLT, the large amplitudes of temperature and wind perturbations of tidal waves contribute significantly to diurnal variations of the observed species such as $O_3$ and $O$ derived from the airglow emissions.

When the advective transport by tidal waves are included (1) can be revised into

$$\frac{\partial \chi_i}{\partial t} + \frac{u}{a\cos\phi}\frac{\partial \chi_i}{\partial \lambda} + \frac{v}{a}\frac{\partial \chi_i}{\partial \phi} + w\frac{\partial \chi_i}{\partial z} = C_i + D_i \quad (2)$$

where $a$ is the Earth's radius, $\lambda$ and $\phi$ are longitude and latitude, respectively. The two terms on the right-hand-side are same as in (1) that represent the net photochemical production rate and the total vertical diffusion, respectively. The added three terms on the left-hand-side are the advective transport of tracer $\chi_i$ by a tidal wind (u, v, w) in the horizontal ($\lambda$, $\phi$) and vertical (z) directions, respectively. For motions with short timescales, the transport of tracers is more important in vertical than in horizontal because of the strong vertical gradient in the photochemical production. Furthermore, a one-dimensional chemical-transport model similar to (1) is much easier to solve. However, the atmosphere is nearly incompressible to the planetary scale motions. The horizontal divergence and vertical convergence of the total air mass cancel each other. Therefore, one can not introduce the vertical advection without introducing at least one term of the horizontal advection to balance the continuity equation. One

way to introduce the advective transport by vertical velocity in a one-dimensional frame is to neglect the meridional ($\phi$) advection in (2) and introduce the following moving coordinate that transforms the universal time (t) into the local time (t*)

$$t^* = t + \frac{\lambda}{\Omega}, \ \chi_i(t,\lambda,z) = \chi_i(t+\Omega^{-1}\lambda, z) = \chi_i(t^*,z) \quad (3)$$

$$\frac{\partial}{\partial t} = \frac{\partial}{\partial t^*}, \ \frac{\partial}{\partial \lambda} = \frac{1}{\Omega}\frac{\partial}{\partial t^*} \quad (4)$$

where $\Omega$ (= $7.292\times10^{-5}$ s$^{-1}$) is the Earth's rotation frequency. Substituting (3) and (4) into (2) and neglecting the meridional advection term, we obtain

$$\left(1+\frac{u}{a\Omega\cos\phi}\right)\frac{\partial \chi_i}{\partial t} + w\frac{\partial \chi_i}{\partial z} = C_i + D_i. \quad (5)$$

For simplicity, we have omitted the asterisk that denotes the local time. The required horizontal velocity (u) is derived from w by applying the moving coordinate transform (3)-(4) to a two-dimensional continuity equation

$$\frac{\rho_0}{a\Omega\cos\phi}\frac{\partial u}{\partial t} + \frac{\partial}{\partial z}(\rho_0 w) = 0. \quad (6)$$

For the migrating tidal waves, the vertical velocity can be written as the following function of the local time t [Zhu et al. 1999a]:

$$w(z,t) = \hat{w}_c(z)\cos(s\Omega t) + \hat{w}_s(z)\sin(s\Omega t), \quad (7)$$

where s=1 and s=2 correspond to diurnal and semi-diurnal tides, respectively. Furthermore, $\hat{w}_c(z)$ and $\hat{w}_s(z)$ are related to the tidal wave amplitude, $\hat{w}(z)$, and its phase, $\delta(z)$, through

$$\hat{w}_c(z) = \hat{w}(z)\cos[\delta(z)], \ \hat{w}_s(z) = \hat{w}(z)\sin[\delta(z)]. \quad (8)$$

Substitution of (7) into (6) yields the horizontal velocity used in (5)

$$u(z,t) = a\cos\phi[\hat{u}_c(z)\cos(s\Omega t) + \hat{u}_s(z)\sin(s\Omega t)], \quad (9)$$

where

$$\hat{u}_c(z) = \frac{1}{s}\left[\frac{\partial \hat{w}_s(z)}{\partial z} - \frac{\hat{w}_s(z)}{H}\right], \quad (10)$$

$$\hat{u}_s(z) = -\frac{1}{s}\left[\frac{\partial \hat{w}_c(z)}{\partial z} - \frac{\hat{w}_c(z)}{H}\right], \quad (11)$$

where H is the scale height of the air density $\rho_0$. Expressions similar to (10) and (11) can be easily derived when the vertical velocity (7) is a linear superposition of several tidal wave components, such as diurnal and semidiurnal tides. Equation (5) is the one-dimensional model that couples the diurnally forced photochemistry with the vertical transport of short timescales. It can be rewritten as

$$A(w)\frac{\partial \chi_i}{\partial t} + w\frac{\partial \chi_i}{\partial z} = C_i + D_i, \quad (12)$$

where the amplification factor A(w) due to the advection is

$$A(w) = 1 + \frac{u(w)}{a\Omega\cos\phi}. \quad (13)$$

Note that A(0)=1 when w≡0 and (1) is recovered from (12). When w≠0, (12) indicates an amplification of time dependent term when the horizontal advection is to be included implicitly in the one-dimensional photochemical-transport model. In addition, the effect of tidal temperature perturbation is incorporated in $C_i$ for calculating the rate coefficients. In general, (12) is a one-dimensional photochemical-diffusive-advective model with the vertical velocity satisfying the continuity equation.

Finally, we also note that the diffusive transport defined in (1) does not include the diffusive separation when the molecular diffusion becomes dominant. The current version of JHU/APL chemical-dynamical model extends to 130 km in which the vertical fluxes from the molecular diffusion and the eddy diffusion are treated separately. In general, the diffusive transport term for the mixing ratio of a minor tracer, $\chi_i$, can be approximated as

$$D_i \approx \frac{1}{\rho_0}\frac{\partial}{\partial z}\left(\rho_0 K_{zz}\frac{\partial \chi_i}{\partial z}\right) - (\beta_i K_m)\frac{\partial \chi_i}{\partial z} \quad (14)$$

where $K_{zz}$ (= $K_e$ + $K_m$) is the sum of the eddy ($K_e$) and molecular ($K_m$) diffusion coefficients and

$$\beta_i = H_0^{-1} - H_i^{-1} = \frac{g}{kT}(M_0 - M_i). \quad (15)$$

In the above, H and M are the scale height and mass of molecules, respectively. The subscripts "0" and "i" denote air and species i. Equation (14) indicates that diffusive transport term $D_i$ for the mixing ratio has both an advective component with $\beta_i K_m$ as equivalent vertical velocity and a diffusive component with $K_e+K_m$ as vertical diffusion coefficient.

## 3. COUPLING BETWEEN THE PHOTOCHEMISTRY AND SLOWLY VARYING TRANSPORT

In order to integrate a photochemical-dynamical system for climatological timescales we have to use a time step much greater than the one used for integrating the diurnally forced model described in the last section. For a general velocity field (u,v,w), (2) is a primitive continuity equation that couples the photochemistry with the transport by atmospheric motions at all timescales. An accurate and stable numerical scheme requires integration time

steps to be less than the smallest timescale unless the system is in a chemical equilibrium with respect to the smallest timescale. Therefore, a large time step in integrating a photochemical-dynamical system will be accomplished if one of the following two conditions holds: (i) the system only consists of slowly varying states; (ii) the system is confined near an equilibrium state. These two conditions are a result of restrictions on the following two different types of timescales in a chemical-dynamical system. One is the internal timescales that are the inverse of the linear damping rates to a prescribed perturbation from an equilibrium state. The other is the timescales of the external forcing. The photochemical timescales discussed in the literature usually refer to the internal ones that describe how quickly the system will be adjusted toward a chemical equilibrium state under a fixed external forcing [e.g., *Brasseur and Solomon*, 1984]. The external forcing that continuously destroys the approaching chemical equilibrium is due to the transport and the diurnally forced photolysis. Both the internal timescales and external timescales of the photolysis rates for different species may differ by many orders of magnitude in the middle atmosphere. Many radical species in the middle atmosphere also show almost discontinuous variations near twilight [e.g., *Brasseur and Solomon*, 1984], indicating very short timescales. Therefore, one has to introduce slowly varying states in the continuity equation to couple the photochemistry with the slowly varying transport.

Replacing the mixing ratio of an individual species $\chi_i$ in (2) with a slowly varying state $\hat{\chi}_i$ we can write the continuity equation for the $\hat{\chi}_i$

$$\frac{\partial \hat{\chi}_i}{\partial t} = C_i(\hat{\chi}_j, J_{ij}) + T_i(\nabla\hat{\chi}_i), \quad (16)$$

where $C_i(\hat{\chi}_j, J_{ij})$ indicates that in general the net photochemical production of state $\hat{\chi}_i$ depends on the photolysis rates ($J_{ij}$) and all the states ($\hat{\chi}_j$). Furthermore, $T_i(\nabla\hat{\chi}_i)$ includes both advective and diffusive transports that are generally dependent on $\nabla\hat{\chi}_i$.

Various assumptions and approximations have been made to couple the fast varying photochemistry with slowly varying transport. Different approaches of coupling rest on particular choices of slowly varying states. Here, we briefly discuss the five cases as listed follows: A: {$\hat{\chi}_i$} are family species that slowly vary with time [e.g., *Brasseur and Solomon*, 1984], for example, $\hat{\chi}[O_x]$ = $\chi[O_3]$ + $\chi[O(^3P)]$ + $\chi[O(^1D)]$); B: {$\hat{\chi}_i$} are the solutions of (2) with general velocity field but under the diurnally averaged photolysis rate [e.g., *Zhu et al.*, 1999b]; C: {$\hat{\chi}_i$} are associated with the diurnal means of solutions from a photochemical model [e.g., *Turco and Whitten*, 1978;

*Summers et al.*, 1997]; D: $\{\hat{\chi}_i\}$ are associated with the diurnal means of solutions from a one-dimensional photochemical-transport model such as (1); E: $\{\hat{\chi}_i\}$ are the solutions of zonal mean of the primitive continuity equation (16). The JHU/APL 2-D model couples its photochemical solver with the slowly varying dynamics by use of either case B (optional) or case D (standard) for the slowly varying state.

The important differences between the cases A and B have been extensively discussed in *Zhu et al.* [1999b]. Case A is based on the fact that the internal timescales for family species are usually much greater than those of individual species. An extreme example for the family species is the sum of all the species, the photochemical timescale of the family is infinity in this case and the air density is solely determined by the dynamics. Because the external forcing of the diurnally averaged photolysis in case B is slowly varying which always confines the system near an equilibrium state, we can still take large time step in integration though its internal timescales of solution are extremely short.

In case C and case D, the nonlinear correlations in the chemical production and loss rates at short timescales have been included. They can be parameterized by readjusting the photolysis rates ($J_{ij}$) and the rate coefficients ($k_{ij}$). The correction factors $\alpha_{ij}$ and $\beta_{ij}$ are calculated by [*Turco and Whitten*, 1978]

$$\alpha_{ij} = \frac{\overline{J_{ij}\chi_j}}{\overline{J_{ij}}\hat{\chi}_j} \quad , \quad \beta_{ij} = \frac{\overline{k_{ij}\chi_i\chi_j}}{\overline{k_{ij}}\hat{\chi}_i\hat{\chi}_j}, \qquad (17)$$

where $\overline{(\,)}$ denotes the diurnal average, $\chi_i$ is derived from the diurnally forced model, and $\hat{\chi}_i$ is the solution of (16). The major gain of case D over case C is that the coupling between the photochemistry and the diffusive transport has been much more accurately calculated at short timescales. Such a coupling could be important in the MLT region for its large $K_{zz}$ values. The case E corresponds to an ideal situation where the solution of (16) for the slowly varying state is the actual mean state of the modeled species and is directly comparable with measurements. The major difference between case D and case E is the neglect of several nonlinear correlation terms in the zonally averaged equation (e.g., $\overline{u\chi_i}$ and $\overline{v\chi_i}$). In addition, introduction of $\alpha_{ij}$ and $\beta_{ij}$ in (17) also assumes a clear separation between the short and long timescales in cases C and D. However case E can not be integrated directly since it does not form a closed system. One encounters a similar closure problem as in turbulence studies if we attempt to derive the measurable quantity of $\overline{\chi}_i$ directly from (16). Finally, we like to point out that though the slowly varying states derived by different approaches may differ significantly they could all lead to similar solutions for the diurnally forced model of (1) or (5).

## 4. JHU/APL TWO-DIMENSIONAL DYNAMICAL MODEL AND PHOTOCHEMICAL SOLVER

A splitting-up technique is used to solve (2) and (16) for transport and chemical operators. The nominal time steps that couple chemistry and dynamics are 15 minutes for (2) and 12 hours for (16), respectively. The correction factors $\alpha_{ij}$ and $\beta_{ij}$ are updated every 15 days. The dynamical module of the JHU/APL 2-D model was developed in *Zhu et al.* [1997]. The central idea of the globally balanced dynamical model is to solve the two elliptic equations for the stream function of the meridional circulation and the geopotential tendency. At each time step, the mass continuity equation and thermal wind balance are satisfied automatically.

Our photochemical-dynamical coupled model (2) or (16) is formulated directly in terms of species mixing ratios. The photochemical solver derives the solution for the stiff set of ordinary differential equations (ODEs) under a given net photochemical production in (2) or (16)

$$\frac{d\chi_i}{dt}\bigg|_{chem} = C_i. \qquad (18)$$

For every photodissociation or chemical reaction the added net photochemical production in (2) or (16) can be expressed as [*Turco and Whitten*, 1978]

$$C_i \leftarrow C_i \pm \alpha_{ij}J_{ij}\chi_j \quad \text{or} \quad C_i \leftarrow C_i \pm \beta_{jk}k_{jk}\chi_j\chi_k \qquad (19)$$

where $\alpha_{ij}$ and $\beta_{jk}$ equal unity for the diurnally forced model and for the case B in the slowly varying model. Otherwise the correction factors are updated and stored as functions of altitude and latitude in JHU/APL 2-D model. The plus or minus sign of the added term in (19) depends on whether the species $\chi_i$ is produced or destroyed. *Zhu et al.* [1999b] describe the calculation of the photolysis rates and the solution of (18). Here, we like to emphasize that the needed Jacobian matrix is calculated analytically:

$$\frac{\partial C_i}{\partial \chi_j} \leftarrow \frac{\partial C_i}{\partial \chi_j} \pm \alpha_{ij}J_{ij} \quad \text{or} \quad \frac{\partial C_i}{\partial \chi_j} \leftarrow \frac{\partial C_i}{\partial \chi_j} \pm \beta_{jk}k_{jk}\chi_k. \qquad (20)$$

To efficiently solve a stiff set of ODEs usually requires variable step sizes and an implicit scheme. This makes the calculation of the Jacobian matrix unavoidable [*Press et al.* 1992]. Our approach of analytically assigning the Jacobian matrix based on (20) makes the chemical solver both computationally efficient and programmatically easy to adjust when adding new photochemical reactions.

## 5. PRELIMINARY MODEL RESULTS

We show in panels (a) and (b) of Fig. 1 the temperature and vertical velocity perturbations at 95 km on March 15

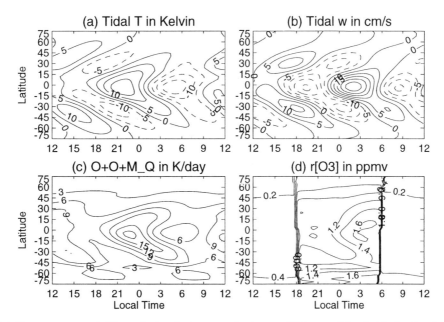

**Figure 1.** (a) Temperature and (b) vertical velocity perturbations at 95 km on March 15 simulated by JHU/APL linear spectral tidal model; corresponding (c) chemical heating rate by the exothermic reaction O+O+M and (d) ozone mixing ration (ppmv) as a function of local time and latitude simulated by JHU/APL one-dimensional photochemical-diffusive-advective model.

simulated by JHU/APL linear spectral tidal model. The perturbations are superposition of diurnal and semidiurnal tides whose amplitudes are comparable around 95 km [*Zhu et al.* 1999a]. The temperature and vertical velocity perturbations are coupled to the one-dimensional photo-chemical-diffusive-advective model (12) for the rate coefficient calculations and advective transport, respectively. Figure 1 also shows (c) the corresponding chemical heating rate by the exothermic reaction O+O+M and (d) ozone mixing ratio (ppmv) derived from the model. At 95 km, the atomic oxygen has the chemical timescale longer than 1 day. Furthermore, the zonal mean of atomic oxygen mixing ratio exhibits a large vertical gradient near 90 km [*Zhu et al.* 1999b]. Therefore, its local time variation is mostly determined by the vertical advective transport. On the other hand, ozone variations at 95 km show distinctively two timescales. Near twilight, the extremely short chemical timescale produces almost a discontinuous variation in its mixing ratio. Away from twilight and near the equator, the local time variation of ozone is strongly affected by the destruction reaction $O+O_3 \rightarrow O_2+O_2$ with its rate coefficient $k_{13}$ increasing with the temperature. Thus, the local time ozone is strongly anti-correlated to the temperature perturbations. From the airglow model simulations, it is found that the chemical heating rate at 95 km shown in Fig. 1c is proportional to the limb brightness of $O_2$ atmospheric band at 90-km tangent height. Our model simulation of Fig. 1c reproduces several measured features of limb brightness of $O_2$ atmospheric band at the 90-km tangent height [*Burrage et al.*, 1994]. These include: the ratio of the maximum emission to the minimum emission is about a factor 4, the correct local times corresponding to maximum and minimum emission, and the asymmetric feature with respect to equator of the maximum emission that spread into the mid-latitudes.

Figure 2 shows the responses of temperature and several species on December 15 to the long-term solar flux variations. The radiative heating rate and photolysis of Lyman-$\alpha$ and Schumann-Runge continuum (SRC) are parameterized according to the F10.7 cm solar flux in the JHU/APL coupled 2-D model. The model is run for two sets of solar fluxes corresponding to F10.7 equal to 60 and 260, respectively. The differences in temperature, $H_2O$, and O shown in Fig. 2 mainly result from the changes in solar uv energy input, Lyman-$\alpha$ photodissociation of $H_2O$ and SRC photodissociation of $O_2$. The $H_2O$ sensitivity is consistent with a similar simulation by *Fleming et al.* [1995]. On the other hand, the difference in $O_3$ results collectively from the increase in both O and $HO_x$, and the temperature dependence in rate coefficients. Above 95 km, the decrease in $O_3$ is mainly due to the increase of rate coefficient $k_{13}$ due to increase of temperature. Below 95 km, the increases in O and $HO_x$ contribute almost equally to ozone production and loss. However, the strong meridional gradient in O variation around 90 km due to the transport produces a different $O_3$ response to the solar flux variations in summer and winter hemispheres.

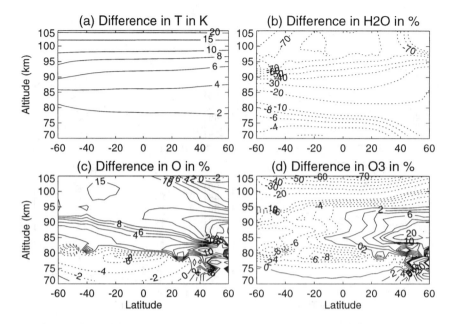

**Figure 2.**    (a) Difference in zonal mean temperatures on December 15, T(F_max)-T(F_min), between the solar flux maximum (F_max) and solar flux minimum (F_min) conditions simulated by JHU/APL coupled 2-D model. The panels (b)-(d) are the corresponding fractional differences in percent with respect to global mean of (b) water vapor, (c) atomic oxygen, and (d) ozone respectively.

## 6. CONCLUSIONS

We have developed a one-dimensional photochemical-diffusive-advective model to explicitly couple diurnally forced photochemistry with the advective transport by tidal waves. Furthermore, the coupling between the photochemistry and the slowly varying transport is examined systematically and illustrated by several examples of modeling techniques. The critical numerical procedure that makes the JHU/APL chemical solver accurate and efficient is explicitly presented. Using two model examples of photochemical-transport coupling corresponding to diurnally forced timescales and slowly varying transport timescales we show how the coupling processes may significantly alter tracer distributions in the MLT region.

*Acknowledgments.* This research was supported by the TIMED project sponsored by NASA under contract NAS5-97179 to the Johns Hopkins University Applied Physics Laboratory and in part by NSF Grant ATM-9419683.

## REFERENCES:

Allen, M. et al., The vertical distribution of ozone in the mesosphere and lower thermosphere. *J. Geophys. Res.*, *89*, 4841-4872, 1984.

Brasseur, G. and S. Solomon, *Aeronomy of the Middle Atmosphere: Chemistry and Physics of the Stratosphere and Mesosphere.* D. Reidel Publishing Company, Boston, 441 pp, 1984.

Burrage, M. D et al., 1994: Observations of the $O_2$ atmospheric band nightglow by the High Resolution Doppler Imager. *J. Geophys. Res.*, *99,* 15,017-15,023.

Fleming, E. L. et al., The middle atmospheric response to short and long term solar uv variations: analysis of observations and 2D model results. *J. Atmos. Terr. Phys.*, *57,* 333-365, 1995.

Garcia, R. R. and S. Solomon, A numerical model of the zonally averaged dynamical and chemical structure of the middle atmosphere. *J. Geophys. Res.*, *88*, 1379-1400, 1983.

Ko, M. K. et al., The seasonal and latitudinal behavior of trace gases and $O_3$ as simulated by a two-dimensional model of the atmosphere. *J. Atmos. Sci.*, *41*, 2381-2408, 1984.

Press, W. H. et al., *Numerical Recipes in Fortran. The Arts of Scientific Computing. 2nd Edition.* Cambridge University Press, 963 pp, 1992.

Smith, A. K., Numerical simulation of global variations of temperature, ozone, and trace species in the stratosphere. *J. Geophys. Res.*, *100*, 1253-1269, 1995.

Summers, M. E. et al., Seasonal variation of middle atmospheric $CH_4$ and $H_2O$ with a new chemical-dynamical model. *J. Geophys. Res.*, *102*, 3503-3526, 1997.

Turco, R. P. and R. C. Whitten, A note on the diurnal averaging of aeronomical models. *J. Atmos. Terr. Phys.*, *40*, 13-20, 1978.

Zhu, X. et al., A globally balanced two-dimensional middle atmosphere model. Dynamical studies of mesopause meridional circulation and stratosphere-mesosphere exchange. *J. Geophys. Res.*, *102*, 13,095-13,112, 1997.

Zhu, X. et al., On the numerical modeling of middle atmosphere tides. *Quart. J. Roy. Meteorol. Soc.*, *125,* 1825-1857, 1999a.

Zhu, X. et al., Numerical modeling of chemical-dynamical coupling in the upper stratosphere and mesosphere. *J. Geophys. Res., 104,* 23,995-24,011, 1999b.

D. F. Strobel, Department of Earth and Planetary Sciences, Johns Hopkins University, Baltimore, MD 21218.

J.-H Yee, and X. Zhu, Applied Physics Laboratory, Johns Hopkins University, 11100 Johns Hopkins Road, Laurel, MD 20723-6099. (xun.zhu@jhuapl.edu)